Scattering by Obstacles

Mathematics and Its Applications

A. G. Ramm
Mathematics Department, Kansas State University, Manhattan, U.S.A.

Scattering by Obstacles

D. Reidel Publishing Company
A MEMBER OF THE KLUWER ACADEMIC PUBLISHERS GROUP
Dordrecht / Boston / Lancaster / Tokyo

Library of Congress Cataloging in Publication Data

CIP

Ramm, A. G. (Alexander G.)
Scattering by obstacles.

(Mathematics and its applications)
Bibliography: p.
Includes index.
1. Wave-motion, Theory of. 2. Scattering (Mathematics)
I. Title. II. Series: (Mathematics and its applications (D. Reidel Publishing Company))
QA927.R36 1986 531'.1133 86–3281
ISBN 90–277–2103–3

Published by D. Reidel Publishing Company,
P.O. Box 17, 3300 AA Dordrecht, Holland

Sold and distributed in the U.S.A. and Canada
by Kluwer Academic Publishers,
190 Old Derby Street, Hingham, MA 02043, U.S.A.

In all other countries, sold and distributed
by Kluwer Academic Publishers Group,
P.O. Box 322, 3300 AH Dordrecht, Holland

Printed in The Netherlands.

To my parents
with love

SERIES EDITOR'S PREFACE

Approach your problems from the right end and begin with the answers. Then one day, perhaps you will find the final question.

'The Hermit Clad in Crane Feathers' in R. van Gulik's *The Chinese Maze Murders*.

It isn't that they can't see the solution. It is that they can't see the problem.

G.K. Chesterton. *The Scandal of Father Brown* 'The point of a Pin'.

Growing specialization and diversification have brought a host of monographs and textbooks on increasingly specialized topics. However, the "tree" of knowledge of mathematics and related fields does not grow only by putting forth new branches. It also happens, quite often in fact, that branches which were thought to be completely disparate are suddenly seen to be related.

Further, the kind and level of sophistication of mathematics applied in various sciences has changed drastically in recent years: measure theory is used (non-trivially) in regional and theoretical economics; algebraic geometry interacts with physics; the Minkowsky lemma, coding theory and the structure of water meet one another in packing and covering theory; quantum fields, crystal defects and mathematical programming profit from homotopy theory; Lie algebras are relevant to filtering; and prediction and electrical engineering can use Stein spaces. And in addition to this there are such new emerging subdisciplines as "experimental mathematics", "CFD", "completely integrable systems", "chaos, synergetics and large-scale order", which are almost impossible to fit into the existing classification schemes. They draw upon widely different sections of mathematics. This programme, Mathematics and Its Applications, is devoted to new emerging (sub)disciplines and to such (new) interrelations as exempla gratia:

- a central concept which plays an important role in several different mathematical and/or scientific specialized areas;
- new applications of the results and ideas from one area of scientific endeavour into another;
- influences which the results, problems and concepts of one field of enquiry have and have had on the development of another.

The Mathematics and Its Applications programme tries to make available a careful selection of books which fit the philosophy outlined above. With such books, which are stimulating rather than definitive, intriguing rather than encyclopaedic, we hope to contribute something towards better communication among the practitioners in diversified fields.

Scattering theory, the area of science which investigates what happens when a wave encounters an obstacle is old, well established and flourishing. It is basic to many subspecialisms in physics and electrical engineering and has many-fold, and sometimes surprising links with various other parts of mathematics.

Inverse scattering, the problem of reconstructing an object from knowledge about its scattering properties is equally important. Indeed in some sciences (think of geophysical exploration) inverse scattering type data are just about the most important kind of data available.

And of course both scattering and inverse scattering are most important to soliton-instanton-monopole theory, though this book does definitely not go into this direction. In fact it has practically zero intersection with all books I know about direct and inverse scattering.

Most of what it contains can only be found in the research journal literature and is exposed here for the first time within the larger framework of a coherent book. Yet the book is basically introductory and selfcontained and also aimed at users of the theory such as electrical engineers, physicists and mathematically inclined geologists.

Especially noteworthy is the inclusion of scattering by infinite obstacles.

The unreasonable effectiveness of mathematics in science ...

Eugene Wigner

Well, if you know of a better 'ole, go to it.

Bruce Bairnsfather

What is now proved was once only imagined.

William Blake

As long as algebra and geometry proceeded along separate paths, their advance was slow and their applications limited.

But when these sciences joined company they drew from each other fresh vitality and thenceforward marched on at a rapid pace towards perfection.

Joseph Louis Lagrange.

Bussum, March 1986 Michiel Hazewinkel

Table of Contents

Preface

This book deals with the mathematical theory of wave scattering by obstacles. The obstacles we consider are bodies with finite or infinite boundaries. Much was written about potential scattering, but less about scattering by obstacles. The subject is vast and this book is not aimed at a complete presentation of all of the known results. We will discuss, mostly for scalar wave scattering, the following basic topics: (1) existence and uniqueness of solutions to the scattering problems, (2) eigenfunction expansions and analytic properties of the S-matrix, (3) time-dependent problems, singularity and eigenmode expansions, complex poles and their properties, limiting amplitude principle, (4) scattering by obstacles with infinite boundaries, (5) scattering by small bodies including the case of many ($\sim 10^{23}$) small bodies, that is the case of a medium consisting of many small bodies, (6) inverse scattering problems. Much of the material has not been presented previously in book form. The material in Sections I.7; II.2.3; III.2—5; IV; V, VI, VII.4—6, Appendices 3—5, 7, was previously published by the author in a number of papers (see Bibliographical notes). The author tried to present all the material in the book as simply as possible. There are many unsolved problems in this area. A number of open problems are formulated in the book.

The book can be used by mathematicians, physicists and electrical engineers. It can be understood by graduate students. Part of the material was used in lectures given by the author at various universities and research laboratories.

Acknowledgement

It is a pleasure to thank Dr. H. D. Alber for reading the manuscript and for many helpful remarks. The author is grateful to many colleagues with whom he discussed some of the problems studied in this monograph. Among these colleagues are H. D. Alber, C. L. Dolph, M. Fiddy, G. Kristensson, R. Leis, P. Martin, F. Miller, R. Picard, S. Ström, P. Werner, F. Ursell, A. Weglein and others. The author is grateful to the Universities of Bonn, Göteborg, Stuttgart, Manchester and London, at which some of the results were presented.

The author thanks Academic Press, American and British Institutes of Physics, and IEEE for permission to use his papers in this book. In particular, Appendices 4, 5, 7 and Chapter 6 are the papers Ramm [11d], [6n], [6m], [13], and the tables in Chapter IV are taken from the paper Kristensson, G., Ramm, A. G., Ström, S. [1].

The author thanks Dr. R. M. Fitzgerald and Office of Naval Research for support.

Introduction

In the preface the topic of this book was described. Consider the problem

$$(\nabla^2 + k^2)\, u = 0 \text{ in } \Omega, \quad k > 0, \quad \Omega \subset R^3 \tag{1}$$

$$u = 0 \text{ on } \Gamma = \partial\Omega, \tag{2}$$

$$u = u_0 + v, \tag{3}$$

where

$$u_0 = \exp\{-ikn \cdot x\} \tag{4}$$

$$r\left(\frac{\partial v}{\partial r} - ikv\right) \to 0, \quad r = |x| \to \infty. \tag{5}$$

The problem (1)—(5) is the scattering problem which corresponds to the scattering of the scalar plane wave (4), propagating in the direction of the unit vector $-n$, by the obstacle $D = R^3 \backslash \Omega$ with boundary $\Gamma = \partial\Omega$. We consider usually the three-dimensional problem. Some attention will be paid to the two-dimensional problem mostly in connection with the analytic properties of the corresponding scattering matrix (there is a logarithmic branch point at $k = 0$). Unless otherwise stated the single obstacle D is assumed to be a finite body with a smooth connected boundary $\Gamma = \partial D = \partial\Omega$, and Ω is the exterior domain. Scattering by infinite obstacles and by many finite obstacles will also be studied. The Neumann boundary condition (N points into Ω)

$$\frac{\partial u}{\partial N} = 0 \text{ on } \Gamma, \tag{6}$$

and the Robin boundary condition,

$$\frac{\partial u}{\partial N} + \eta u = 0 \text{ on } \Gamma, \quad \eta = \eta(s) \in C(\Gamma), \operatorname{Im} \eta(s) \geqslant 0 \tag{7}$$

will also be considered instead of (2).

Problem (1)—(5) is of interest in many applications. Most of the results for this problem can be obtained in the same way for electromagnetic wave scattering.

The basic questions about this problem include: (1) existence and uniqueness of the solution to (1)—(5), (2) analytic properties of the solution as a function of k, (3) properties of the scattering amplitude defined by the formula

$$v \sim \frac{\exp(ikr)}{r} f(-n, \nu, k) \quad \text{as } |x| = r \to \infty, xr^{-1} = \nu, \tag{8}$$

(4) inverse problem of finding the surface of the obstacle from the knowledge of the scattering amplitude (the data will be specified later), (5) eigenfunction expansions for the Laplacians in exterior domains, construction of the wave operators, properties of the resolvent kernel of the Dirichlet Laplacian in exterior domains, (6) principles of limiting amplitude and limiting absorption, (7) numerical solution of exterior boundary value problems, (8) scattering by small bodies $ka \ll 1$ where a is the diameter of the obstacle, (9) many-body problems, (10) scattering by obstacles with infinite boundaries. The last problem requires some special attention. Even the statement of the scattering problem for obstacles with infinite boundaries is not obvious for a number of reasons. Firstly, one can not look for a solution of the form (3), (5) because the scattered field may not satisfy the radiation condition (5). Example: take D to be the half-space $x_3 < 0$. In this case the solution is well known and is of the form $u = \exp\{-ikn \cdot x\} - \exp\{-ik\bar{n} \cdot x\}$, where $n = (n_1, n_2, n_3)\bar{n} = (n_1, n_2, -n_3)$. The scattered field does not satisfy the condition (5).

Secondly, the uniqueness of the solution to the problem

$$(\nabla^2 + k^2)\, v = 0 \text{ in } \Omega, \quad v = 0 \text{ on } \Gamma, \quad v \text{ satisfies (5)}, \tag{9}$$

does not hold for some domains with infinite boundaries, for example, for the domain $0 \leqslant x_3 \leqslant 1$, or for "bottlenecks", that is for domains such that $\text{meas}_2 \{x: x_3 = R, x \in \Omega\} \to 0$ as $R \to \infty$. Here meas_2 is the area of the part of the plane $x_3 = R$ which lies in Ω. In such domains the spectrum of the Dirichlet Laplacian is discrete.

The questions (1)—(10) will be discussed in this book. Not all of these questions can be discussed in detail because one needs several books to do this. There is an extensive literature on the subject. So, the author thinks it is worthwhile to explain what are the contents of this book and to give a very brief review of the literature.

Mostly integral equation methods were used in the study of existence and uniqueness problems, see Kupradze [1, 2], Leis [1], Werner [1],

Müller [1], Weyl [1], Colton, Kress [1], but other methods were also of importance. In particular the limiting absorption method is a powerful tool (see Eidus [3], Vainberg [1], Wilcox [1]). In Lax, Phillips [1] a theory based on the conservation of energy for the time dependent wave equation is given. The author thinks that the easiest way to obtain the basic results of the scattering theory for the Helmholtz equation (1) is by way of integral equations, and this way is chosen for our presentation.

The eigenfunction expansions, and the spectral and scattering theories for the Schrödinger equation in the whole space were given in Ikebe [1] and in domains with infinite boundaries in Ramm [1], where a different method was used. This latter method is used in our presentation. The results of Ikebe [1] were generalized by Shenk [1] Shenk and Thoe [1], Wilcox [1] and others. Analytic properties of solutions to problem (1)—(5) and, more generally, to the problem

$$(\ell - k^2)\, v = f \text{ in } \Omega,\quad v = 0 \text{ on } \Gamma,\quad v \text{ satisfies (5)},\ \ell = -\nabla^2 + q(x), \tag{10}$$

were studied in Ramm [2] where these properties were linked with the asymptotic behavior for $t \to \infty$ of the solution to the time dependent problem

$$u_{tt} + \ell u = 0,\ x \in \Omega,\ t > 0,\ u = 0 \text{ on } \Gamma,\ u(x, 0) = 0,\ u_t(x, 0) = f(x). \tag{11}$$

Unless otherwise stated it is assumed in this book that $q(x) = 0$, $\ell = -\nabla^2$, and the function $f(x) \in L^2(\Omega)$ in (10) and (11) is compactly supported in Ω. Under these assumptions one can prove in a simple way, as in Ramm [2], that the solution to (10) (with $q(x) = 0$ or $q(x)$ compactly supported) is a meromorphic function of k with poles k_j in the half-plane $\text{Im } k < 0$. These poles are called resonances, or natural frequencies, or complex poles of the resolvent kernel G:

$$[\nabla^2 + k^2 - q(x)]\, G(x, y, k) = -\delta(x - y) \text{ in } \Omega,\ G = 0 \text{ on } \Gamma \tag{12}$$

$$G \in L^2(\Omega) \text{ if } \text{Im } k > 0. \tag{13}$$

The location (especially the values of $\text{Im } k_j$) and properties of these poles (their existence, asymptotic behavior as $j \to \infty$, etc.) are of great interest theoretically and practically. The location of the poles can be used for identification of the scatterer.

There is a close connection between the resonances (complex poles) and the asymptotic behavior of solutions to problem (11) as $t \to \infty$.

This connection leads to the so called singularity expansion method (SEM). To describe SEM let us look for a solution to (11) of the form $u(x, t) = (2\pi)^{-1} \int_{-\infty}^{\infty} v(x, k) \exp(-ikt)\, dk$. Then v solves problem (10) and can be written as

$$v(x, k) = \int_{\Omega} G(x, y, k) f(y)\, dy. \tag{14}$$

This v is a meromorphic function of k. Suppose that its poles k_j satisfy the condition

$$|\mathrm{Im}\, k_1| \leqslant |\mathrm{Im}\, k_2| \ldots \leqslant |\mathrm{Im}\, k_j| \to \infty \text{ as } j \to \infty. \tag{15}$$

Then there are only a finite number (N, say) of poles in any strip $-b < \mathrm{Im}\, k \leqslant 0$, $b = \text{const} > 0$. Assume also that

$$|v| < c(b)\,(1 + |k|)^{-a}, \quad a > \tfrac{1}{2}, |\mathrm{Re}\, k| \to \infty, 0 > \mathrm{Im}\, k > -b. \tag{16}$$

Then one can move the contour in the formula

$$u(x, t) = (2\pi)^{-1} \int_{-\infty}^{\infty} \exp(-ikt)\, v\,(x, k)\, dk \tag{17}$$

down in the complex k plane and obtain the asymptotic SEM expansion

$$u(x, t) = \sum_{j=1}^{N} c_j(x, t) \exp(-ik_j t) + 0\,(\exp\{-|\mathrm{Im}\, k_{N+1}|t\}), \quad t \to \infty \tag{18}$$

where

$$c_j(x, t) \exp(-ik_j t) = \operatorname*{Res}_{k=k_j} \{v(x, k) \exp(-ikt)\}, c_j(x, t) = 0(t^{m_j - 1})$$

as $t \to +\infty$, and m_j is the multiplicity of the pole k_j.

One can prove (15) and (16) for some shapes of the obstacle (e.g., for convex smooth obstacles or for star-like obstacles; a star-like obstacle is such an obstacle that there exists a point in D from which every point of Γ can be seen). For trapping obstacles, that is for obstacles with boundary that can support a standing wave (a ray bouncing between two points of the boundary in geometrical optics approximation) the statement (15) is not true. For example, for an obstacle consisting of two strictly convex smooth reflecting bodies (i.e., (2) holds) Ikawa claims that there exist infinitely many poles k_j on the line $\operatorname{Im} k_j = -b < 0$, where $b > 0$ is a constant (Ikawa [1], [2]). The SEM asymptotic formula (18) is not valid for such obstacles. It is an interesting open problem whether one can write (under certain assumptions including (15)) $u = \sum_{j=1}^{\infty} c_j(x, t) \exp(-ik_j t)$, that is whether one can take $N = \infty$ in (18). Another interesting problem is to prove existence of infinitely many complex not purely imaginary resonances (a proof of existence of infinitely many purely imaginary resonances is given in this book) and the inverse problem of recovering the scatterer from the knowledge of its complex poles and some information about the corresponding principal parts of $G(x, y, k)$. In Dolph–Scott [1] one can find a review of the use of complex poles. For some scatterers (e.g. star-like) the solution to the problem

$$\begin{aligned} &u_{tt} + \ell u = f(x) \exp(ikt), \; x \in \Omega, \; t > 0, \\ &u(x, 0) = u_t(x, 0) = 0, \; u = 0 \text{ on } \Gamma \end{aligned} \tag{19}$$

behaves asymptotically as $t \to +\infty$ like $v(x) \exp(ikt)$, where $v(x)$ is called the limiting amplitude, and solves the stationary problem (10). There is a close relationship between the limiting amplitude principle and the limiting absorption principle which says that the limit $\lim_{\varepsilon \downarrow 0} (\ell - k^2 - i\varepsilon)^{-1} f$ exists in a suitable topology and is a solution to (10). It turns out that the asymptotics of $T^{-1} \int_0^T u(x, t) \exp(-ikt) \, \mathrm{d}t$ as $T \to \infty$ is more naturally related to the limiting absorption principle than the asymptotics of $u(x, t) \exp(-ikt)$. There is an extensive literature on the limiting absorption and limiting amplitude principles (see Eidus [1], Vainberg [1], Mishnaevskii [1], [2], Ramm [1a], [2a–l] and references in these papers). Most of the existing works on the limiting absorption and limiting amplitude principles deal with partial differential operators. But a few results in abstract operator-theoretical settings are also available: Eidus [1] gave a theory for the case when ℓ is an abstract self-adjoint operator on a Hilbert space and the resolution of identity for ℓ satisfies some smoothness conditions; Ramm [2k] gave necessary and sufficient conditions for the validity of the limiting amplitude principle for some linear nonselfadjoint operators. Mishnaevskii [1] gave sufficient conditions for the limiting absorption principle to hold for an operator Sturm–Liouville equation. His results generalize the earlier results of Kato [1], Jäger [1], Saito [1] and others. Vainberg [1, 2] justified the limiting absorption and limiting amplitude principles for a wide class of partial differential equations.

A nonclassical approach to the exterior boundary value problem is as follows. Consider the problem

$$(-\nabla^2 - k^2)\, v = 0 \text{ in } \Omega, \quad v = f \text{ on } \Gamma, v \text{ satisfies (5).} \tag{20}$$

Let

$$v = \int_\Gamma g\sigma \,\mathrm{d}s, \quad g = \frac{\exp(ik|x-y|)}{4\pi|x-y|}. \tag{21}$$

Then v solves (20) iff

$$Q\sigma = \int_\Gamma g\sigma \,\mathrm{d}s = f. \tag{22}$$

The operator Q is compact in $H = L^2(\Gamma)$ but nonselfadjoint. If Q has a complete (in H) set of eigen- and root vectors and this set forms a basis for H then one can solve (22) using an expansion of σ in root vectors of Q, that is vectors satisfying the equation $(Q - \lambda_j)^m \psi = 0$, where m is an integer, λ_j is an eigenvalue of Q, and m may be different for various ψ. If $m = 1$ then ψ is an eigenvector. This method is analogous to solving the equation $A\sigma = f$ with a self-adjoint compact operator A by the formula $\sigma = \sum_{j=1}^{\infty} \lambda_j^{-1}(f, \psi_j)\psi_j$ where $A\psi_j = \lambda_j\psi_j$ and the set $\{\psi_j\}$ forms an orthonormal basis of H. A non-selfadjoint compact operator may have no eigenvalues (e.g. Volterra's operator, quasinilpotent operators). If the set of root vectors of Q forms a basis of H then we say that Q has the basisness property. It is an interesting problem to give sufficient conditions for a compact nonselfadjoint operator to have the basisness property. The eigenvalues of Q are functions of k. One can prove (Ramm [3], [6]) that the resonances are precisely the zeros of the eigenvalues of Q. Other characterizations of the resonances and methods for their computation are given in Ramm [3].

The inverse diffraction problem consists in finding the obstacle from the scattering data. The scattering data are the values of the scattering amplitude on some subsets of the five-dimensional space k, ν, n, $0 < k < \infty$, $\nu, n \in S^2$, where S^2 is the unit sphere. A well known theorem of M. Schiffer (see e.g. Colton, Kress [1]) says that Γ is uniquely defined if the scattering amplitude $f(n_0, \nu, k)$ is known for $a \leqslant k \leqslant b$, $b > a$ and $\nu \in S^2$. We prove the uniqueness theorem assuming that the scattering data are $f(n_j, \nu, k_0)$, $k_0 > 0$, $\nu \in \tilde{S}$, where $\tilde{S}$ a solid angle, $\{n_j\}$ is a countable set of unit vectors such that $n_j \neq n_i$ if $j \neq i$. It is easy to see that if f is known for $\nu \in \tilde{S}$ then f is uniquely defined for all $\nu \in S^2$. The main point in our result is, of course, the fact that k is fixed, $k = k_0$, so that the surface is recovered from the above data. If k_0 is taken sufficiently large and Γ is convex then an effective algorithm for recovering the surface is given. Our presentation is based on the papers Ramm [8b, g, k]. The method consists of explicit recovery of the support function of the convex scatterer from the high frequency asymptotics of the scattering amplitude. The support function $a(\ell)$ of a convex body D is defined to be $a(\ell) = \sup_{s \in \Gamma}(\ell, s)$ if the origin is inside D. Here ℓ is a unit vector. The usual and nonconstructive approach in earlier works was to recover the Gaussian curvature of the convex scatterer from the high-frequency asymptotics of the scattering amplitude, and then Minkowski's theorem (which says that a convex smooth closed surface Γ is uniquely defined by its Gaussian curvature known as a function of the unit normal on Γ) yields the uniqueness theorem for the inverse scattering problem. This approach is not constructive because Minkowski's theorem is a uniqueness theorem, it does not provide the algorithm to recover the surface. The inverse scattering problem for nonconvex bodies is of great interest but there are no results concerning the constructive solution of this problem presently (except Alber, Ramm [1]). In the literature (e.g. Colton, Kress [1]), there is a discussion of the inverse scattering problem in the case when the scattering data are known for small k. But for small $k(ka \ll 1$ where a is the diameter of the scatterer) the scattering amplitude is determined basically by the dipole radiation (corresponding to the first term of the expansion of the scattering amplitude in powers of ka). Therefore for $ka \ll 1$ a body of arbitrary shape scatters waves as an ellipsoid, and practically only the 3 scalars (semiaxes of the ellipsoid) can be recovered from the scattering amplitude in the case $ka \ll 1$ (Ramm [10]).

Having discussed the basic qualitative questions as uniqueness and existence of the solution to the scattering problem, properties of the solution, of the Green function, eigenfunction expansions in solutions to the scattering problem (distorted plane waves, eigenfunctions of continuous spectrum of the Dirichlet Laplacian in the exterior domain), the time dependent problem and the limiting amplitude and limiting absorption principles, the singularity and eigenmode expansions, and the inverse scattering problem, we turn to the numerical schemes for solving exterior boundary value problems. The emphasis is on the *T*-matrix scheme. This scheme can be described as follows. Let $\{\psi_n\}$ be a system of functions satisfying equation (1) and the radiation condition (5). Such functions are called outgoing waves. Applying Green's formula to ψ_n and v, the solution of (20), yields

$$\int_\Gamma \psi_n \, h \, \mathrm{d}s = f_n, \ \forall n, \ f_n = \int_\Gamma f \frac{\partial \psi_n}{\partial N} \, \mathrm{d}s, \quad h = \frac{\partial v}{\partial N}, \tag{23}$$

and N is the outer unit normal on Γ. Let $\{\phi_j\}$ be a linearly independent and complete system of functions in $L^2(\Gamma) = H$. Substitute $h_m = \sum_{j=1}^m c_j^{(m)} \phi_j$, where $c_j^{(m)}$ are constants, in place of h in (23) to obtain

$$\sum_{j=1}^m a_{nj} c_j^{(m)} = f_n, \quad 1 \leqslant n \leqslant m, \quad a_{nj} = \int_\Gamma \psi_n \phi_j \, \mathrm{d}s. \tag{24}$$

The system (24) is to be solved for $c_j^{(m)}$ and it is to be proved that $h_m \to h$ in H as $m \to \infty$, where h solves (23). If h is known then the solution of (20) can be found by the Green formula. Practically the last step involving integration over the surface can be avoided. This is shown in Chapter IV where other questions of interest are discussed: does $c_j^{(m)} \to c_j$ as $m \to \infty$? is the convergence uniform in j? does the equality $h = \sum_{j=1}^\infty c_j \phi_j$ hold?; how does the rate of convergene depend on the choice of the systems $\{\psi_j\}$ and $\{\phi_j\}$?

There is an extensive literature on the *T*-matrix approach (see Varadan [1]) but for about two decades convergence of this numerical method was not proved. A proof of its convergence and a theoretical analysis of the method including answers to the above question are given in Ramm [7] (see also Kristensson, Ramm, Ström [1]) and presented in Chapter IV. We briefly discuss in Chapter IV the other major numerical methods: integral equation methods and finite elements methods. In particular, integral equations, which are solvable at the exceptional values of k corresponding to the spectrum of an interior Laplacian, are derived. Several methods for deriving such equations (Brakhage-Werner [1], Leis [1], Ursell [1], Ramm [1b]) are given in Chapter 1.

Wave scattering by small bodies is a classical topic. The basic work was done by Rayleigh [1871] who continued this work until his death. The physics of the problem is well understood: the main term in the scattered field is the dipole radiation. The problem is to compute this dipole radiation for a body of arbitrary shape and to study the dependence of the scattering amplitude on the boundary conditions. This was done by the author and summarized in Ramm [4], where references can also be found. If the scatterers are small compared with the wavelength $ka \ll 1$, and $d \gg a$, where d is the average distance between the scatterers, then the many body problem can be greatly simplified. The scattering amplitude in this case is determined by a finite number of functionals of the field, that is by some numbers rather than functions as in the case of high frequency scattering. For these numbers one can derive an algebraic linear system rather than a system of integral equations. This system can be easily solved by iterations as shown in Ramm [4]. As in application of the formula for the scattering matrix for wave scattering by a small body of arbitrary shape one can solve an inverse radiation problem. This problem is to find the initial field at the point x_0 from the measurements of the field scattered by a small probe placed at the point x_0 in the initial field. This material is presented in Chapter V and the presentation is based on Ramm [4]. In Werner [2], [3], Colton, Kress [1], MacCamy [1], Favorin [1], and references in these papers, the behavior as $k \to 0$ of solutions to exterior boundary value problems was studied. In the two-dimensional case the Green's function has a logarithmic branch point at $k = 0$. In Ramm [2f] it was shown that Green's function of the Schrödinger Dirichlet operator with $q(x) \geqslant 0$ in the exterior domain has a limit as $k \to 0$. If $q = 0$ then Green's function of the exterior Neumann Laplacian does not have a limit as $k \to 0$. These questions are discussed in Section V.4, as well as the case of the Robin Laplacian.

We do not discuss some ideas of homogenization theory (Marchenko, Khruslov [1], Sanchez-Palencia, E. [1]). This theory deals, for example with the situation when the medium consists of many small bodies and the problem is to describe the medium as a continuous medium (homogenized medium) with some "averaged" effective parameters. We do not discuss the quantum-mechanical scattering by delta-function potentials (Albeverio, S. *et al.* [1], [2] and references therein).

In chapter 6 we study inverse scattering problems for partial differential equations. These problems are of special interest in geophysics and nondestructive control. A typical example is the following problem. Let $[\nabla^2 + \omega^2 n(x)]u(x, y, k) = -\delta(x - y)$ in R^3 $n(x) = n_0 + v(x)$, $n_0 = \text{const} > 0$, $v(x) = 0$ if $|x| > R$, where $R > 0$ is an arbitrary large number, $v(x) \in L^2$. The function $v(x)$ describes an inhomogeneity. The problem is to find $v(x)$ from a knowledge of the field $u(x, y, k)$ on the plane $x_3 = 0$, i.e., for all positions of $x^1 = (x_1, x_2, 0)$ and $y^1 = (y_1, y_2, 0)$ and for $0 < k < k_0$, where k_0 is an arbitrary small number. The data $u(x, y, k)$ can be also given on a sphere or on a cylinder. The problem can be complicated by the assumption that $n_0 = n_1$ if $x_3 > 0$, $n_0 = n_2$ if $x_3 < 0$, where $n_1 \neq n_2$, $n_j = \text{const}$, $j = 1, 2$. We discuss also the inverse problem for two unknown functions, the density of the medium and the velocity profile. In this case the differential equation is

$$(\nabla^2 + \omega^2 - \omega^2 a_1(x) - \nabla \cdot a_2(x)\nabla)G = -\delta(x - y),$$

and the functions $a_j(x)$, $j = 1, 2$, are to be found from the scattering data. A model problem of the induction logging is also studied. The inverse scattering problem by a compactly supported potential $q(x)$, $x \in R^3$, in the Born approximation can be reduced to the integral equation for q:

$$\int_{|y| \leqslant R} \exp\{-ik(s' - s) \cdot y\} q(y)\,\mathrm{d}y = F(k(s' - s)),$$

where $s', s \in S^2$, $k > 0$ is fixed, and the right side is known. This problem occurs in all the inverse problems in Chapter VI. We give an analytic solution to this problem with a prescribed accuracy and the error estimates.

In Chapter VII the scattering problem in domains with infinite boundaries is discussed. This is a difficult subject. Even the formulation of the scattering problem presents some difficulties as was mentioned in the beginning of this introduction. Some results on the nature of the spectrum of the Laplace and Schrödinger operators in domains with infinite boundaries are given in Glazman [1] and Eastham, Kalf [1]. In the earlier works (Rellich [1, 2], Jones [1]) the emphasis was on the nature of the spectrum of the Laplacians. An important result is a proof of the absence of positive eigenvalues. This result was obtained for the Dirichlet Laplacian in domains with infinite boundaries which are illuminated everywhere near infinity from a point 0 placed in D. This assumption can be written as $N \cdot x < 0$ on Γ. Here N is the normal to Γ directed into D and x is the radius vector in the coordinate system with the origin at the point 0. This assumption is a global assumption about the surface Γ. Rellich's method does not work for the Schrödinger operator. For the first time the absence of positive eigenvalues for Schrödinger operators with $|q(x)| = o(|x|^{-1})$ as $|x| \to \infty$ in domains with infinite boundaries which are illuminated from some point 0 only near infinity (that is $N \cdot x \leqslant 0$ for $|x| > R$, where $R > 0$ is an arbitrarily large number) was proved in Ramm [1a] (see also [1d]). The method used in this paper was given in Kato [1] where for the first time similar results were obtained for the Schrödinger operator in the whole space. It is interesting that for the class of infinite boundaries introduced in Ramm [1a] the problem of absence of positive eigenvalues can be treated with only local assumptions near infinity. For example, this is not possible if the boundary is a half-tube: a perturbation of a finite part of the tube can put a positive eigenvalue on the continuous spectrium of the Dirichlet Laplacian in such domain (Rellich [2]). In Tayoshi [1] the absence of positive eigenvalues was proved for the Dirichlet Laplacian in domains with smooth boundaries such that $\Omega(R) \subset \{x: (|x|^2 - x_1^2) < cx_1^{2\ell}\}$, $N(x) \cdot a(x) \leqslant 0$ for $x \in \Gamma \cap \partial\Omega(R)$, where $R > 0$ is an arbitrarily large fixed constant, $c > 0$, $\ell < 1$, $\Omega(R) = \Omega \cap \{x: x_1 > R\}$, $x = (x_1, \ldots, x_n)$, $N(x)$ is the normal directed into D at the point $x \in \Gamma$, and $a(x) = (x_1, \ell x_2, \ldots, \ell x_n)$. Such domains, for example, may be paraboloids. In this case the assumption $N \cdot x \leqslant 0$ for $|x| \geqslant R$ does not hold (the interior of the paraboloid is Ω). All of these results are very special cases of the theorems proved in Mishnaevskii [1—2]. He developed a powerful technique for proving the absence of positive eigenvalues for abstract Sturm—Liouville equations, gave a deep spectral analysis of such equations, proved eigenfunction expansion theorems, a limiting absorption principle, and showed how these results can be obtained for general second-order elliptic operators in a wide class of domains with infinite boundaries. He considered also various boundary conditions. Before his work only in Ramm [1f] for the Schrödinger equation in some domains with infinite boundaries was the Robin boundary condition treated. Proofs of the remarkable results due to Mishnaevskii are difficult and will not be presented in this book, where we try to keep the technical level of the presentation simple. We give a general investigation of the spectral properties of the Laplacians in domains with infinite boundaries. Necessary and sufficient conditions for the discreteness of the spectrum of the Dirichlet Laplacian L are given. The spectrum of L in semi-infinite tubes is studied. Sufficient conditions for the continuous spectrum of L to fill the ray $(0, \infty)$ and to have no eigenvalues are given. The notion of a canonical domain is introduced. A canonical domain is a domain in which Green's function of L exists and admits certain estimates. For example, if Green's function of L can be constructed by the separation of variables method and satisfies certain estimates, then the domain is canonical. A cone is a canonical domain. For such domains we develop a perturbation theory. This theory allows one to study the spectral properties of L in domains whose boundaries are compact perturbations of the boundary of a canonical domain. We prove the limiting absorption principle and the eigenfunction expansions theorem for L in such domains. In appendices some auxiliary material is given. Part of this material is not available in book form.

It is now in order to explain what audience the author had in mind and what is original in this book. This book is intended to serve as an introduction to the subject for an audience with various interests: mathematicians, physicists, and engineers interested in the theory. The presentation is essentially self-contained. The author tried to present material in a simple way and avoid technically difficult proofs. Although the book is an introduction to the subject, most of the material is not available in book form and the author's research interests influence the choice of the material considerably. The material in Sections I.7., II.2, 3., III.2—5., IV, V, VI, VII, 4—6, Appendices 3—5, 7, was published by the author in research papers. We discuss scalar wave scattering, but nearly all of our methods and results can be rewritten for electromagnetic wave scattering.

We do not discuss in this book several important topics: (1) exponential decay of the solutions to the wave equation in exterior domains, (2) high frequency asymptotics of the solutions to scattering problems and rigorous justification of the geometrical optics, (3) distribution of the complex poles of the scattering amplitude, (4) scattering by periodic and rough surfaces, and (5) scattering by moving boundaries. See references in Bibliographical notes. The bibliography is not complete, but if the reader consults the references he or she will have reasonably complete information about the subject.

There are practically no intersections between the material and presentation in this book and in the books by Newton [1], Lax, Phillips [1], Wilcox [1], Reed, Simon [1], etc.

In references to the formulas of the same section we give only the number of the formula, to the formulas of the different section of the same chapter we give the number of the section and the number of the formula, to the formulas of the different chapter we give the number of the chapter, the section and the number of the formula. Formulas of the introduction are referred to as (0.x) where x is the number of the formula in the introduction. The bibliography, arranged in alphabetical order, serves as the author index. The table of contents serves as the subject index.

Chapter I

Scattering by an Obstacle

1. STATEMENT OF THE PROBLEM. BASIC INTEGRAL EQUATIONS

1.1 In this section we introduce some notations, formulate some properties of the potentials of the single and double layers, and discuss the integral equations for solving the exterior boundary value problem (0.20) resulting from Green's formula. This discussion is a preliminary step in a study of the solvability of the problem (0.20). The classical approach leads to integral equations which are not always solvable for some k. Therefore one wishes to derive some unconditionally solvable integral equations for solving (0.20), equations which are equivalent to the boundary value problem (0.20).

The scattering problem (0.1—5) is a particular case of the boundary value problem (0.20) with $f = -u_0 = -\exp(-ikn \cdot s)$ on Γ. The classical approach to this problem is to look for a solution of the form

$$v(x) = D\mu \equiv \int_\Gamma \mu(t) \frac{\partial g(x, t)}{\partial N_t} \, dt, \quad g = \frac{\exp(ik|x-y|)}{4\pi|x-y|}. \tag{1}$$

Then Equation (0.20) and the radiation condition (5) are satisfied and from the boundary condition (0.20) one obtains an integral equation for μ (N_t is the exterior normal pointing into Ω):

$$\mu = -A'\mu + 2f, \quad A'\mu = 2\int_\Gamma \mu(t) \frac{\partial g(s, t)}{\partial N_t} \, dt, \quad s \in \Gamma \tag{2}$$

Here and below, the following well known formulas from potential theory are used (see e.g. Günter [1]):

$$\text{(a)} \quad v_\pm = (D\mu)_\pm = \frac{A'\mu \mp \mu}{2}, \quad \text{(b)} \quad \frac{\partial w}{\partial N_\pm} = \frac{\partial (S\sigma)}{\partial N_\pm} = \frac{A\sigma \pm \sigma}{2}, \tag{3}$$

$$\frac{\partial (D\mu)}{\partial N_-} = \frac{\partial (D\mu)}{\partial N_+}, \tag{3'}$$

where

$$w(\sigma) \equiv S\sigma \equiv \int_\Gamma g(x, t)\, \sigma(t)\, \mathrm{d}t, \quad A\sigma = 2\int_\Gamma \sigma(t) \frac{\partial g(s, t)}{\partial N_s}\, \mathrm{d}t, \quad s \in \Gamma \tag{4}$$

the sign + (−) corresponds to the limit value on Γ from the interior domain D (exterior domain Ω), and to the upper (lower) sign in the left side of (3) there corresponds the upper (lower) sign in the right-hand side of (3). The operator $A' = \bar{A}^*$, where the bar denotes complex conjugation of the kernel of the operator and the operator A^* is adjoint to A in $L^2(\Gamma) \equiv H^0$. Formulas (3) are obtained in Günter [1] under the assumption that μ and σ are continuous and the surface Γ belongs to $C^{1,\lambda}$ that is in a neighborhood of any point of Γ one can introduce a local cartesian coordinate system such that the x_3 — axis is directed along the normal to Γ at this point and the function ϕ in the equation of the surface $x_3 = \phi(x_1, x_2)$ has first derivatives satisfying the Hölder condition with exponent λ. A function $\psi(x)$ satisfies the Hölder condition with exponent λ if $|\psi(x) - \psi(y)| \leqslant c|x - y|^\lambda$, $c = \text{const}$, $0 < \lambda \leqslant 1$. Formula (3′) means that

$$\lim_{\varepsilon \to +0} \left. \frac{\partial v(x)}{\partial N_s} \right|_{x=s+\varepsilon N_s} - \left. \frac{\partial v(x)}{\partial N_s} \right|_{x=s-\varepsilon N_s} = 0, \text{ where } v(x) = D\mu, \quad s \in \Gamma$$

If μ is only continuous, then

$$\lim_{\varepsilon \to +0} \left. \frac{\partial v(x)}{\partial N_s} \right|_{x=s\pm\varepsilon N}$$

may not exist. These limits exist and are Hölder-continuous if $\mu \in C^{1,\lambda}(\Gamma)$, $\lambda > 0$. In Appendix 1 more information on potential theory is given. In particular, the operators A and A^* are compact in the spaces $L^2(\Gamma) = H^0(\Gamma)$ and $C^\alpha(\Gamma)$ if $\Gamma \in C^{1,\lambda}$, where $C^0(\Gamma) = C(\Gamma)$ and $C^\alpha(\Gamma)$ is the space of Hölder-continuous functions with the norm $|f| = \max_{x\in\Gamma} |f(x)| + \sup_{x,y\in\Gamma} |f(x) - f(y)|\,|x - y|^{-\alpha}$, $0 < \alpha \leqslant 1$, $\mu \in C^{1,\lambda}(\Gamma)$ means that the first derivatives of μ belong to $C^\lambda(\Gamma)$.

Therefore, Fredholm's alternative holds for Equation (2). In Burago-Maz'ya [1], Fabes *et al.* [1], and Kral [1], the potential theory for irregular domains, i.e. domains with nonsmooth boundaries is given.

1.2. If the solution to the problem (0.20) exists, then it can be written by Green's formula

$$v(x) = \int_\Gamma \left(v_- \frac{\partial g}{\partial N} - g \frac{\partial v}{\partial N_-} \right) \mathrm{d}s, \quad x \in \Omega. \tag{5}$$

We will show later that the problem (0.20) is solvable for any $f \in C(\Gamma)$ or $L^2(\Gamma)$. Therefore one can always find the solution to (0.20) in the form

$$v(x) = Df - Sh \tag{6}$$

where

$$v_- = f, \tag{7a}$$

$$\frac{\partial v}{\partial N_-} = h. \tag{7b}$$

Let us derive integral equations for f and h. From (6), (7) and (3) it follows that

$$f = \frac{A'f + f}{2} - Qh, \quad \text{i.e., } Qh = \frac{A'f - f}{2}, \quad Qh = \int_\Gamma g(s, t) h(t)\, \mathrm{d}t \tag{8}$$

and

$$h = \frac{\partial(Df)}{\partial N_-} - \frac{Ah - h}{2}, \quad \text{i.e., } h + Ah = 2 \frac{\partial(Df)}{\partial N_-}. \tag{9}$$

Remark 1. Condition (7a) is equivalent to Equation (8). Indeed, we have derived (8) from (7a). Conversely, if (8) holds then one writes the right hand side of the first Equation (8) as $(Df)_- - (Sh)_- = v_-$, and obtain (7a). Condition (7b) is equivalent to Equation (9). Indeed, we have derived (9) from (7b). Conversely, if (9) holds then the right hand side of the first equation (9) is $\partial v / \partial N_-$, where v is defined by (6).

LEMMA 1. *If $k^2 \notin \sigma(-\Delta_D)$, $[k^2 \notin \sigma(-\Delta_N)]$, i.e. k^2 is not an eigenvalue of the Dirichlet* [*Neumann*] *Laplacian in D, and h is a solution to* (8) [(9)] *then the function* (6) *satisfies equations* (7) *and solves problem* (0.20).

Proof. (1) Suppose h solves (8). Then by Remark 1, Equation (7a) holds for v defined by (6). Therefore the function (6) (with h solving (8)) solves problem (0.20). We wish to show that Equation (7b) holds. Let us write Green's formula (5) for the solution to problem (0.20). We will prove later (Lemma I.2.2) that the problem (0.20) has at most one solution. Therefore the functions (5) and (6) are identical in Ω. Since (7a) has already been established, we conclude that

$$S\sigma = 0 \text{ in } \Omega, \quad \sigma \equiv \frac{\partial v}{\partial N_-} - h. \tag{10}$$

If $k^2 \notin \sigma(-\Delta_D)$ then $S\sigma = 0$ in D. Therefore by the jump relation (3b), $\sigma = 0$. So, Equation (7b) holds.

(2) Suppose h solves (9). Then by Remark 1 Equation (7b) holds. Define $v(x)$ by formula (6) and write Green's formula (5) for this v. Since the function (5) and (6) are identical in Ω, and (7b) holds, one concludes that

$$D\mu = 0 \text{ in } \Omega, \quad \mu = f - v_-. \tag{11}$$

By formula (3′), $\partial u/\partial N_+ = 0$ where $u(x) = D\mu$. Therefore $(\nabla^2 + k^2)u = 0$ in D, $\partial u/\partial N_+ = 0$. Since $k^2 \notin \sigma(-\Delta_N)$, one has $u = 0$ in D. This, (11) and the jump relation (3a) imply that $\mu = 0$. So, Equation (7a) holds. □

Remarks 2. If h solves both Equations (8) and (9), then both Equations (7a) and (7b) hold with $v(x)$ defined by (6), and this v solves problem (0.20).

LEMMA 2. *There is at most one function which solves both Equations* (8) *and* (9).

Proof. It is sufficient to prove that $Qh = 0$ and $h + Ah = 0$ imply $h = 0$. These equations mean that $w = Sh$ satisfies the equations $w_+ = 0$, $\partial w/\partial N_+ = 0$. Therefore $w(x) = 0$ in D. Since $w_+ = w_- = 0$, by Lemma I.2.2 $w = 0$ in Ω. Thus

$$h = \frac{\partial w}{\partial N_+} - \frac{\partial w}{\partial N_-} = 0. \qquad \square$$

Remark 3. If $k^2 \in \sigma(-\Delta_D)$ and h solves (8) then Equation (7b) may not hold. Indeed, if $k^2 \in \sigma(-\Delta_D)$, then $\dim N(Q) = r$, where r is the dimension of the eigenspace of $-\Delta_D$ corresponding to the eigenvalue k^2. Therefore Equation (8) has many solutions. On the other hand, f

defines uniquely the solution $v(x)$ to problem (0.20) (by Lemma I.2.2), so that $\partial v/\partial N_-$ is uniquely defined by f. Therefore, if $k^2 \in \sigma(-\Delta_D)$ one can find a (unique) solution to Equation (8) for which (7b) holds. This solution is the one which solves both Equations (8) and (9) (see Lemma 2).

If $k^2 \in \sigma(-\Delta_N)$ then Equation (9) has many solutions. Let h_j, $j = 1, 2$ solve Equation (9), $h_1 \not\equiv h_2$, $v_j = Df - Sh_j$, $v = v_1 - v_2$, $h = h_1 - h_2 \not\equiv 0$. Then $v = -Sh$, $\partial v/\partial N_- = h$, $h + Ah = 0$, i.e. $\partial v/\partial N_+ = 0$. Therefore, $v_+ = (Sh)_+ \not\equiv 0$, $v_+ = v_- \not\equiv 0$ (because if $v_+ = 0$ then $v = 0$ in D and in Ω, and therefore $h = 0$, which is a contradiction) $v_{1-} \not\equiv v_{2-}$ if $h \not\equiv 0$. As above, the only solution to (9) for which function (6) satisfies Equation (7a) is the solution which satisfies both Equations (8) and (9).

We give several proofs of uniqueness and existence of the solution to exterior boundary value problem (in particular, problem (0.20)) in Section I.1.2. In Section I.7 we study the possibility of the representations of the double (single) layer potentials as a single (double) layer potential. We can give a simple proof of existence of the solution to the problem (0.20), based on the Equations (8) and (9).

LEMMA 3. *Problem* (0.20) *is solvable for any* $k > 0$.

Proof. It is sufficient to prove that Equation (8) has a solution. If $k^2 \notin \sigma(-\Delta_D)$ then Equation (8) is uniquely solvable (Lemma (I.7.2) and function (6) solves problem (0.20) (Lemma 1) and Equation (9) holds. If $k^2 \in \sigma(-\Delta_D)$ then Equation (8) is solvable iff

$$\int_\Gamma (A'f - f)\sigma_j \, dt = 0, 1 \leqslant j \leqslant r, \quad \sigma_j = \frac{\partial u_j}{\partial N_+},$$

where u_j, $1 \leqslant j \leqslant r$, form a basis in the eigenspace of $-\Delta_D$ corresponding to the eigenvalue k^2 (Lemma I.7.3). We have $A'f - f = 2\, v_+(f)$, where $v(f) = Df$ is given by (3). Thus

$$\int_\Gamma (A'f - f)\sigma_j \, dt = 2 \int_\Gamma v_+ \frac{\partial u_j}{\partial N_+} \, dt = 2 \int_\Gamma \frac{\partial v}{\partial N_+} u_j \, dt = 0.$$

Here we used Green's formula and the boundary condition $u_j = 0$. Therefore Equation (8) is always solvable, although nonuniquely if $k^2 \in \sigma(-\Delta_D)$. Let h be its solution. Define v by formula (6). Then v

satisfies Equation (7a) (by Remark 1) and, therefore, solves problem (0.20).

This concludes the proof of existence of the solution to problem (0.20) for any f, $f \in L^2(\Gamma)$ or $C(\Gamma)$. □

Remark 4. Equation (9) is solvable iff $\int_\Gamma \partial(Df)/\partial N_- \, \mu_j \, dt = 0$ (*) where μ_j, $1 \leqslant j \leqslant r'$, form a basis of $N(I + A')$. If $\mu_j + A'\mu_j = 0$, and $u_j = D\mu_j$, where $D\mu$ is defined in (1), then $u_{j-} = 0$, and $\mu_j = u_{j-} - u_{j+} = u_{j+}$. Let $v = Df$. Then, by (3′), $\partial v/\partial N_- = \partial v/\partial N_+$. Thus $\int_\Gamma \partial v/\partial N_+ \, u_{j+} \, dt = \int_\Gamma v \, \partial u_j/\partial N_+ \, dt = 0$, since $\partial u_j/\partial N_+ = \partial u_j/\partial N_- = 0$. The last equality follows from the fact that $u_j = 0$ in Ω. Therefore (*) holds for any f (for which $\partial(Df)/\partial N_-$ is defined, i.e. for $f \in C^{1,\lambda}(\Gamma)$), and Equation (9) is always solvable.

LEMMA 4. *The solution to the system of Equations* (8) *and* (9) *exists and is unique.*

Proof. Uniqueness has been established in Lemma 2. Let us prove existence. We have proved (Lemma 3) that probem (0.20) has a solution $v(x)$. Let us write formula (5) for this solution, take into account that $v_- = f$, and denote $\partial v/\partial N_-$ by h. Then (6) holds and, as after formula (7b), one derives Equations (8) and (9) which are satisfied by $h = \partial v/\partial N_-$. Thus, existence of the solution to the system of Equations (8) and (9) is established. □

1.2. The classical potential theory integral equation for solving (0.20) can be obtained by looking for v in the form (1), applying the boundary condition (0.20) and using (3). This yields

$$\mu = -A'\mu + 2f. \tag{12}$$

This equation is studied in Section I.2. The operator A' in (12) is compact in $L^2(\Gamma)$ and in $C^\alpha(\Gamma)$, $\alpha \geqslant 0$ if the surface Γ is smooth. By Fredholm's alternative, (12) is solvable for any f iff $N(I + \bar{A}) = \{0\}$, where $N(A)$ is the null space of a linear operator A. Suppose $\dim N(I + \bar{A}) = r'$, $r' > 0$, and let $\{\nu_j\}$ form a basis of $N(I + \bar{A})$, $\nu_j + \bar{A}\nu_j = 0$, $1 \leqslant j \leqslant r'$, then the necessary and sufficient conditions for solvability of (12) are

$$(f, \nu_j) = 0, \; 1 \leqslant j \leqslant r'. \tag{13}$$

It is natural to look for an integral equation for solving (0.20) that is unconditionally solvable. There are several ways to derive such equations.

1.3. One way is to use a Green's function $g_\varepsilon(x, y, k)$ in place of g in (12). Here $g_\varepsilon(x, y, k)$ is the solution to the following problem

$$(-\nabla^2 - k^2)g_\varepsilon = \delta(x - y) \text{ in } \Omega_\varepsilon,\ g_\varepsilon = 0 \text{ in } \partial\Omega_\varepsilon,\ g_\varepsilon \text{ satisfies (0.5)}, \tag{14}$$

$\Omega_\varepsilon = \{x: |x| \geqslant \varepsilon\}$, $\partial\Omega_\varepsilon$ is the boundary of Ω_ε, and $\varepsilon > 0$ is so chosen that $N(I + \bar{A}_\varepsilon) = \{0\}$, where the operators A_ε and A'_ε are defined as in (3), (4) but with g_ε in place of g. The ball $B_\varepsilon = \{x: |x| \leqslant \varepsilon\}$ is chosen inside D. If $N(I + \bar{A}_\varepsilon) = \{0\}$ then Equation (12) with A'_ε in place of A_ε is uniquely solvable for any f. In Section I.2 it will be shown that it is possible to choose $\varepsilon > 0$ such that $N(I + A_\varepsilon) = \{0\}$.

1.4. The second way to derive an unconditionally solvable integral equation for solving (0.20) is to look for a solution of the form

$$v(x) = D\mu + i\eta S\mu \tag{15}$$

where $\eta > 0$ is a parameter. Applying the boundary condition (0.20) and formula (3) yields:

$$\frac{A'\mu + \mu}{2} + i\eta Q\mu = f. \tag{16}$$

This is an equation of Fredholm's type. It is unconditionally solvable if the homogeneous equation

$$\frac{\bar{A}\sigma + \sigma}{2} - i\eta\bar{Q}\sigma = 0 \tag{17}$$

has only the trivial solution. Equation (17) can be written as

$$\frac{\partial}{\partial N_+}\bar{S}\sigma - i\eta\bar{Q}\sigma = 0. \tag{18}$$

Thus the function $u(x) = \bar{S}\sigma$ solves the problem

$$(\nabla^2 + k^2)u = 0 \text{ in } D,\ \frac{\partial u}{\partial N_+} - i\eta u = 0 \text{ on } \Gamma. \tag{19}$$

Since $k^2 > 0$ and $\eta > 0$ one obtains from (19) that $u \equiv 0$ in D, $u = 0$ on Γ and, therefore, $u(x)$ solves the homogeneous exterior problem (0.20). Thus, $u(x) \equiv 0$ in Ω, and $\sigma = \partial u/\partial N_+ - \partial u/\partial N_- = 0$. Therefore Equation (16) is unconditionally solvable (for any $\eta > 0$) and the function $v(x)$ defined by (15) solves (0.20).

1.5. The third way to derive an unconditionally solvable integral equation for solving (0.20) is to look for the solution in the form

$$v(x) = D\mu + T\rho,\ T\rho \equiv \int_D g(x, y, k)\rho(y)\,dy. \tag{20}$$

Applying boundary condition (0.20) yields

$$\frac{A'\mu + \mu}{2} + T\rho = f. \tag{21}$$

Let us impose a second equation relating μ and ρ:

$$\rho = iT\rho + iD\mu, \text{ in } D. \tag{22}$$

The second equation can be chosen in many other ways but the following two requirements should be met: (1) the system of Equations (21), (22) is to be of Fredholm's type; (2) the homogeneous system should have only the trivial solution. These requirements are met for Equations (21)–(22). Indeed, set $\phi = \binom{\mu}{\rho}$. Then (21) and (22) can be written as

$$\phi + B\phi = F, \quad F = \begin{pmatrix} 2f \\ 0 \end{pmatrix}, \quad B = \begin{pmatrix} A' & 2T \\ -iD & -iT \end{pmatrix} \tag{23}$$

where D is the notation for the operator (1), B is a compact linear operator on $H = L^2(\Gamma) \oplus L^2(D)$. Thus, the homogeneous equation (23) has only the trivial solution. Indeed, the homogeneous equation (21) shows that $(D\mu)_- + T\rho = 0$. Therefore the function $v(x)$ defined in (20) solves the homogeneous problem (0.20), and this implies that $v(x) \equiv 0$ in Ω. Equation (22) can be written as

$$(-\nabla^2 - k^2)v = iv \text{ in } D. \tag{24}$$

Multiply (24) by $\bar{v}$ and integrate over D to obtain

$$\int_D |\nabla v|^2\,dx - \int_\Gamma \frac{\partial v}{\partial N_+}\bar{v}\,ds - k^2\int_D |v|^2\,dx = i\int_D |v|^2\,dx. \tag{25}$$

Since $v \equiv 0$ in Ω, one has $\partial v/\partial N_- = 0$. This, Equation (3′), and the fact that $T\rho \in C^1(R^3)$ imply that $\partial v/\partial N_+ = 0$. Therefore, the surface integral in (25) vanishes and, since the left hand side of (25) is real-valued, one concludes that $v \equiv 0$ in D. This implies that $\rho = (-\nabla^2 - k^2)v = 0$ in D and $\mu = (D\mu)_- - (D\mu)_+ = v_- - v_+ = 0$.

1.6. The fourth way to derive an unconditionally solvable integral equation for solving (0.20) is to consider the function g_M in place of g in the basic equation (12) of potential theory, where

$$g_M = g + ik \sum_{n=0}^{\infty} \sum_{m=-n}^{n} c_{nm} \, \psi_{nm}(x) \, \overline{\psi}_{nm}(y) \tag{26}$$

where c_{nm} are constants to be chosen soon, and $\psi_{nm}(x)$ are outgoing wave functions

$$\psi_{nm}(x) = h_n^{(1)}(k|x|) \, Y_{nm}\left(\frac{x}{|x|}\right), \tag{27}$$

$h_n^{(1)}(kr)$ are the spherical Hankel functions, $Y_{nm}(\theta, \phi)$ are the normalized spherical harmonics, and (θ, ϕ) are the spherical coordinates on the unit sphere S^2. It is known that

$$g = \frac{\exp(ik|x-y|)}{4\pi|x-y|} = ik \sum_{n=0}^{\infty} \sum_{m=-n}^{n} j_n(k|x|) h_n^{(1)}(k|y|) \, Y_{nm}\left(\frac{x}{|x|}\right) \overline{Y}_{nm}\left(\frac{y}{|y|}\right),$$

$$|x| < |y| \tag{28}$$

where $j_n(r)$ are the spherical Bessel functions.

If one looks for a solution to problem (0.20) in the form (1) but with g_M in place of g, then for μ one obtains Equation (12) with $A' = A'_M$ defined by (2) with g_M in place of g.

Let us formulate the assumption about c_{nm} which ensure that the homogeneous equation

$$\mu = -A'_M \mu \tag{29}$$

has only the trivial solution. It follows from (29), (3) and the uniqueness of the solution to the homogeneous problem (0.20) that

$$v \equiv v_M \equiv \int_\Gamma \mu(t) \frac{\partial g_M(x, t)}{\partial N_t} \, dt \equiv 0 \text{ in } \Omega. \tag{30}$$

From this and (3′) it follows that $\partial v/\partial N_+ = 0$. Let B_R be a ball of radius $R > 0$ with center at the origin, $B_R \subset D$, and let $D_R = D \setminus B_R$. One has

$$0 = \int_{D_R} \left(v \frac{\partial \bar{v}}{\partial N} - \bar{v} \frac{\partial v}{\partial N} \right) \mathrm{d}s$$

$$= \int_{\Gamma} \left(v \frac{\partial \bar{v}}{\partial N_+} - \bar{v} \frac{\partial v}{\partial N_+} \right) \mathrm{d}s - \int_{\partial B_R} \left(v \frac{\partial \bar{v}}{\partial r} - \bar{v} \frac{\partial v}{\partial r} \right) \mathrm{d}s.$$

Since $\partial v/\partial N_+ = 0$, this implies that

$$0 = \int_{\partial B_R} \left(v \frac{\partial \bar{v}}{\partial r} - \bar{v} \frac{\partial v}{\partial r} \right) \mathrm{d}s. \tag{31}$$

From (30) it follows that

$$v = \left[ik \sum_{n=0}^{\infty} \sum_{m=-n}^{n} j_n(k|x|)\, Y_{nm}\left(\frac{x}{|x|} \right) + \right.$$
$$\left. + \, ik \sum_{n=0}^{\infty} \sum_{m=-n}^{n} c_{nm} h_n^{(1)}(k|x|)\, Y_{nm}\left(\frac{x}{|x|} \right) \right] \mu_{nm} \tag{32}$$

where

$$\mu_{nm} \equiv \int_{\Gamma} \mu(t) \frac{\partial \overline{\psi}_{nm}(t)}{\partial N_+} \mathrm{d}t. \tag{33}$$

From (31), (32), orthogonality properties of Y_{nm} and the formula for the Wronskian $W[j_n(k|x|) + c_{nm}h_n^{(1)}(k|x|),\, j_n(k|x|) + \bar{c}_{nm}\bar{h}_n^{(1)}(k|x|)] = (|\frac{1}{2} + c_{nm}|^2 - \frac{1}{4})\, W[h_n^{(1)}(k|x|),\, \bar{h}_n^{(1)}(k|x|)]$, $k > 0$, it follows that

$$0 = \sum_{n=0}^{\infty} \sum_{m=-n}^{n} b_{nm} |\mu_{nm}|^2, \qquad b_{nm} = |c_{nm} + \tfrac{1}{2}|^2 - \tfrac{1}{4}. \tag{34}$$

If $b_{nm} > 0$ then $\mu_{nm} = 0$. If $\mu_{nm} = 0$ for all $n \geqslant 0$, $-n \leqslant m \leqslant n$, then $\mu = 0$. Indeed, multiply (33) by $ik\, j_n(k|x|)\, Y_{nm}(x/|x|)$, use (28) and sum over n, m to obtain

$$u(x) \equiv \int_\Gamma \mu \frac{\partial g(x,t)}{\partial N_t} \, dt = 0, \quad x \in B_R. \tag{35}$$

Since $u(x)$ is a solution to Helmholtz's equation in D, (35) implies that $u(x) \equiv 0$ in D. Therefore $\partial u/\partial N_+ = 0$ and (3′) shows that $\partial u/\partial N_- = 0$. Thus, $u(x)$ solves the homogeneous problem

$$(\nabla^2 + k^2)u = 0 \text{ in } \Omega, \quad \frac{\partial u}{\partial N_-} = 0, \; u \text{ satisfies (0.5)}. \tag{36}$$

The solution to this problem is unique. Therefore $u(x) = 0$ in Ω, and $\mu = u_- - u_+ = 0$. Thus, if $b_{nm} > 0$ (or $b_{nm} < 0$), then Equation (29) has only the trivial solution.

1.7. The fifth way to derive an unconditionally solvable integral equation for solving (0.20) is to look for a solution of the form

$$v(x) = D\mu + S\sigma \tag{37}$$

where D and S are defined in (1) and (4). Applying the boundary condition (0.20) one obtains an indeterminate equation, i.e. one equation for two unknown functions:

$$\frac{A'\mu + \mu}{2} + Q\sigma = f \tag{38}$$

where Q is defined in (8). This is an equation in $H = L^2(\Gamma)$ for the two unknown functions μ and σ. One can prove that this equation is solvable. This is done in Section I.2.

1.8. In Section I.2 we also give a proof of existence of the solution to (0.20) using an a priori estimate (rather than integral equations), although the proof via integral equation methods is much simpler. Integral equation methods have the great advantage of being applicable for the numerical solution of the problem.

1.9. All the methods that we have discussed are applicable without any serious changes to the exterior boundary value problem with the Neumann boundary condition or the Robin boundary condition (impedance boundary condition):

$$\frac{\partial u}{\partial N_-} + \eta u_- = f, \tag{39}$$

where η is a given constant (possibly complex valued).

1.10. Consider the transmission boundary value problem:

$$(\nabla^2 + k_0^2)u = 0 \text{ in } \Omega, \quad k_0 > 0 \tag{40}$$

$$(\nabla^2 + k_1^2)u = h(x) \text{ in D}, \quad k_1 > 0 \tag{41}$$

$$u_+ = u_- \text{ on } \Gamma \tag{42}$$

$$\rho \frac{\partial u_+}{\partial N} = \frac{\partial u_-}{\partial N} + f, \text{ on } \Gamma, \quad \rho = \text{const} > 0, \tag{43}$$

$$\int_{|s|=R} \left| \frac{\partial u}{\partial r} - ik_0 u \right|^2 \mathrm{d}s \to 0 \text{ as } R \to \infty. \tag{44}$$

This is of interest in the problem of wave scattering by a homogeneous penetrable body D. The scattering problem can be formulated as follows: find u which satisfies Equations (40)—(43) with $f = h = 0$, which is of the form $u = u_0 + v$, where v satisfies the radiation condition (44) and u_0 is a given incident field which is often of the form $u_0 = \exp\{-ikn \cdot x\}$, the plane wave propagating in the direction of the unit vector $-n$.

2. EXISTENCE AND UNIQUENESS OF THE SOLUTION TO THE SCATTERING PROBLEMS

In this section we prove that the problems (0.20) and (1.40)—(1.44) are uniquely solvable.

2.1. First let us prove the following well-known lemma.

LEMMA 1. *If* $u \in L^2(\Omega_R)$, *or* $\int_{S_r} |u|^2 \, \mathrm{d}s \to 0$, *where* $r \to \infty$, $S_r = \{x: |x| = r\}$, *and*

$$(\nabla^2 + k^2)u = 0 \text{ in } \Omega_R = \{x: |x| \geqslant R\}, \quad k > 0, \tag{1}$$

then

$$u \equiv 0 \text{ in } \Omega_R \subset R^3.$$

First proof. Solutions to Helmholtz's equation are locally smooth. Multiply Equation (1) by $\overline{Y}_{nm}$ and integrate over the unit sphere S^2 to obtain ($x = (r, \theta, \mu)$):

$$u''_{nm} + \frac{2}{r}\, u'_{nm} + \left(k^2 - \frac{n(n+1)}{r^2}\right) u_{nm} = 0, \quad u_{nm}(r) \equiv \int_{S^2} u\, \overline{Y}_{nm}\, \mathrm{d}s, \tag{2}$$

where $u'_{nm} = \mathrm{d}u_{nm}/\mathrm{d}r$. Thus, $u_{nm} = c_{1n} h_n^{(1)}(kr) + c_{2n} h_n^{(2)}(kr)$, where c_{1n}, c_{2n} are constants and $h_n^{(j)}(kr)$, $j = 1, 2$ are the spherical Hankel functions, $h_n^{(j)}(z) = [\pi/(2z)]^{1/2} H_{n+\frac{1}{2}}^{(j)}(z)$, and $H_p^{(j)}(z)$ are the usual Hankel functions. The Parseval equality

$$\int_{S^2} |u(r, \theta, \phi)|^2\, \mathrm{d}s = \sum_{n=0}^{\infty} \sum_{m=-n}^{n} r^2 |u_{nm}(r)|^2 < \infty \tag{4}$$

and the asymptotics (as $r \to +\infty$)

$$u_{nm} = \frac{(2\pi^{-1})^{1/2}\left\{c_{1n} \exp\left[i\left(kr - \frac{\pi}{2}n - \frac{\pi}{2}\right)\right] + c_{2n} \exp\left[-i\left(kr - \frac{\pi n}{2} - \frac{\pi}{2}\right)\right]\right\}}{kr} + o\left(\frac{1}{kr}\right) \tag{5}$$

imply that $c_{1n} = c_{2n} = 0$, $n \geqslant 0$. Thus $u \equiv 0$ in Ω_R.

This proof is valid in R^r, $r \geqslant 2$. With obvious modification the proof is valid if $u \in L^2(\Omega_R)$. Notice that the condition $\liminf_{r\to\infty} \int_{S_r} |u|^2\, \mathrm{d}s = 0$ is not sufficient for the conclusion of Lemma 1 to hold. Example: $u = r^{-1} \sin kr$, $r_n = n\pi k^{-1} \to \infty$, $u(r_n, k) = 0$. However, if $\liminf_{r\to\infty} \int_{S_r} \{|u|^2 + |\partial u/\partial r|^2\}\, \mathrm{d}s = 0$ and (1) holds, then $u \equiv 0$.

Second proof. From (1) it follows that $\nabla u \in L^2(\Omega_R)$. Indeed, define $F(r) = \int_R^r \mathrm{d}r\, r^2 \int_{S^2} |u(r, \omega)|^2\, \mathrm{d}s$, $\omega = (\theta, \phi)$. Then $F(r)$ increases monotonically and $F(\infty) < \infty$. Furthermore,

$$F'(r) = r^2 \int_{S^2} |u|^2 \, ds > 0;$$

$$F''(r) = 2r \int_{S_2} |u|^2 \, ds + 2r^2 \operatorname{Re} \int_{S_2} \frac{\partial u}{\partial r} \bar{u} \, ds. \tag{6}$$

Since $F(\infty) < \infty$, and $F' > 0$, clearly either $F''(r) \to 0$ as $r \to \infty$ or there exists a sequence $r_n \to \infty$ such that $F''(r_n) \to 0$. If $F''(r_n) \to 0$ for some sequence $r_n \to \infty$, then (6) implies that

$$\operatorname{Re} \int_{S_{r_n}} \frac{\partial u}{\partial r} \bar{u} \, ds \to 0, \quad r_n \to \infty, \quad S_{r_n} = \{x: |x| = r_n\}. \tag{7}$$

From Equation (1) it follows that

$$\int_{\Omega_{R,r_n}} |\nabla u|^2 \, dx = k^2 \int_{\Omega_{R,r_n}} |u|^2 \, dx + \int_{S_{r_n}} \frac{\partial u}{\partial r} \bar{u} \, ds - \int_{S_R} \frac{\partial u}{\partial r} \bar{u} \, ds, \tag{8}$$

$\Omega_{R,r} = \{x: R \leqslant |x| \leqslant r\}$.

From (7), (8) and the assumption $u \in L^2(\Omega_R)$, it follows that

$$\int_{\Omega} (|u|^2 + |\nabla u|^2) \, dx < \infty. \tag{9}$$

Therefore one can apply Green's formula to obtain

$$u(x) = \int_{S_R} \left(g^+ \frac{\partial u}{\partial N} - u \frac{\partial g^+}{\partial N} \right) ds = \int_{S_R} \left(g^- \frac{\partial u}{\partial N} - u \frac{\partial g^-}{\partial N} \right) ds \tag{10}$$

where $g^{\pm} = \exp(\pm ik|x-y|)/(4\pi|x-y|)$. Let $x = (r, \omega)$. Consider analytic continuation of $u(r, \omega)$ on the complex plane $z = r \exp(i\phi)$. Since S_R is a bounded set, $|x - s| = (r^2 - 2r|s| \cos x\hat{}s + |s|^{1/2})$ is analytic in r, $|r| > R$, $u(z, \omega)$ is analytic near infinity and

$$u(z, \omega) = \frac{\exp(ikz)}{z} f^{+}(z, \omega) = \frac{\exp(-ikz)}{z} f^{-}(z, \omega) \quad (11)$$

where $f^{\pm}(z, \omega)$ are analytic and bounded near infinity on the complex plane z. One can write explicit expressions for $f^{\pm}$ but we do not need them. From (11) it follows that $f^{+} = f^{-} = 0$. Indeed, take $z = iy$, $y \to +\infty$. The left hand side of (11) decays exponentially while the right hand side grows exponentially unless $f^{+} = f^{-} = 0$. Thus $u(z, \omega) = 0$ and $u(r, \omega) = 0$. □

Remark 1. No boundary conditions are imposed on u. From Lemma 1 and Equation (14) the conclusion of Lemma 2 follows.

LEMMA 2. *Problem* (0.20) *has at most one solution.*

Proof. Let us show that the homogeneous problem (0.20) has only the trivial solution. Let w solve the homogeneous problem (0.20). From the radiation condition (0.5) or (1.44) one obtains

$$0 = \lim_{R \to \infty} \left\{ \int_{S_R} \left(\left| \frac{\partial w}{\partial r} \right|^2 + k^2 |w|^2 \right) ds - \right.$$
$$\left. -ik \int_{S_R} \left(w \frac{\partial \bar{w}}{\partial r} - \bar{w} \frac{\partial w}{\partial r} \right) ds \right\}. \quad (12)$$

From Equation (0.20) and Green's formula it follows that

$$\int_{S_R} \left(w \frac{\partial \bar{w}}{\partial r} - \bar{w} \frac{\partial w}{\partial r} \right) ds = \int_{\Gamma} \left(w \frac{\partial \bar{w}}{\partial N} - \bar{w} \frac{\partial w}{\partial N} \right) ds = 0. \quad (13)$$

Therefore

$$\lim_{R \to \infty} \int_{S_R} \left(\left| \frac{\partial w}{\partial r} \right|^2 + k^2 |w|^2 \right) ds = 0. \quad (14)$$

From Lemma 1 and Equation (14) the conclusion of Lemma 2 follows. □

2.2. Existence of the solution to the boundary value problem (0.20) can be established most easily by means of integral equation methods.

THEOREM 1. *Problem* (0.20) *is uniquely solvable for any* $f \in L^2(\Gamma)$.

Proof. Uniqueness is proved in Lemma 2. Existence is proved in Lemma 1.3. We give another proof here. To prove existence assume first that $k^2 \notin \sigma(-\Delta_N)$, that is k^2 is not an eigenvalue of the Neumann Laplacian in D. Then, looking for a solution to (0.20) of the form (1.1) one obtains Equation (1.2). By Fredholm's alternative, this equation is uniquely solvable for μ iff the equation $\sigma = -\bar{A}\sigma$ has only the trivial solution $\sigma = 0$. If $\sigma + \bar{A}\sigma = 0$ and $\sigma \not\equiv 0$, then (1.3) shows that $w = S\bar{\sigma}$ solves the problem

$$(\nabla^2 + k^2)w = 0 \text{ in } D, \ \frac{\partial w}{\partial N_+} = 0. \tag{15}$$

From our assumption $k^2 \notin \sigma(-\Delta_N)$ and (15) it follows that $w = 0$ in D, $w = 0$ on Γ. Since $w = S\bar{\sigma}$ is continuous across Γ one sees that w solves the problem

$$(\nabla^2 + k^2)w = 0 \text{ in } \Omega, \quad w_- = 0, \quad w \text{ satisfies (1.44) with } k_0 = k. \tag{16}$$

Thus $w = 0$ in Ω by Lemma 2. Therefore $\bar{\sigma} = \partial w/\partial N_+ - \partial w/\partial N_- = 0$, and the proof of the existence is complete for the case $k^2 \notin \sigma(-\Delta_N)$.

Suppose $k^2 \in \sigma(-\Delta_N)$. Then our argument is exactly as above but instead of g in (1.1) we use g_ε defined in (1.14). In Lemma 3 below we prove that there exists an $\varepsilon > 0$ such that the problem (15) in $D_\varepsilon = D \backslash B_\varepsilon$ with $w = 0$ on ∂B_ε has only the trivial solution. Therefore the solution to problem (0.20) exists, is unique, and can be found in the form (1.1) with g_ε in place of g. □

LEMMA 3. *For any* $k > 0$, *there exists an* $\varepsilon > 0$ *such that the problem*

$$(\nabla^2 + k^2)u = 0 \ in \ D \backslash B_\varepsilon, \ u = 0 \ on \ \partial B_\varepsilon, \ \frac{\partial u}{\partial N_+} = 0 \ on \ \Gamma \tag{17}$$

has only the trivial solution.

Remark 2. There are infinitely many such ε.

Proof. If problem (17) has a non-trivial solution for $\varepsilon = \varepsilon_0$ then for $\varepsilon \neq \varepsilon_0$ and $|\varepsilon - \varepsilon_0|$ sufficiently small, problem (17) has only the trivial solution. This follows from Lemma 4. □

LEMMA 4. *Let* $k_1^2(\varepsilon) \leqslant k_2^2(\varepsilon) \leqslant \ldots \leqslant k_n^2(\varepsilon) \leqslant \ldots$ *be the eigenvalues of problem* (17), *counted according to their multiplicities. If* $\varepsilon_1 < \varepsilon_2$ *then* $k_n^2(\varepsilon_1) < k_n^2(\varepsilon_2)$.

Proof. From the variational principle for eigenvalues it follows that $k_n^2(a) \leqslant k_n^2(b)$ if $a \leqslant b$. Suppose that $a < b$ and $k_n^2(a) = k_n^2(b)$. Since $k_n^2(a) \leqslant k_n^2(\varepsilon) \leqslant k_n^2(b)$ for $a \leqslant \varepsilon \leqslant b$ one obtains that $k_n^2 \equiv k_n^2(\varepsilon) = k_n^2(a) = k_n^2(b)$, $a \leqslant \varepsilon \leqslant b$. Pick m such that $k_m^2(a) > k_n^2(a)$ and consider a sequence $a = \varepsilon_1 < \varepsilon_2 < \varepsilon_3 < \ldots < \varepsilon_m < b$. Let u_j be the eigenfunction of the problem (17) in $D_j = D \backslash B_{\varepsilon_j}$ corresponding to the eigenvalue k_n^2. Let $u_j = 0$ in the domain $\varepsilon_1 \leqslant |x| \leqslant \varepsilon_0$, then $u_j \in H^1(D_1)$. The function $u_1, \ldots u_m$ are linearly independent in $L^2(D_1)$. Indeed, if $\sum_{j=1}^m c_j' u_j(x) = 0$, $c_j' = \text{const}$, $x \in D_1$, then $u_1(x) = 0$ in $\varepsilon_1 < |x| < \varepsilon_2$, unless $c_1' = 0$. Repeating this argument, one derives that $c_j' = 0$, $1 \leqslant j \leqslant m$. Define $u(x) = \sum_{j=1}^m c_j u_j(x)$ and choose the coefficients c_j so that $(u, u_j)_{L^2(D_1)} = 0$, $1 \leqslant j \leqslant m-1$, $\|u\|_{L^2(D_1)} = 1$. Since u is an admissible function, it follows from the variational principle that $\int_{D_1} |\nabla u|^2 \, dx \geqslant k_m^2(a)$. On the other hand,

$$\int_{D1} |\nabla u|^2 \, dx = \sum_{i,j=1}^{m} c_i \bar{c}_j \int_{D_1} \nabla u_i \nabla \bar{u}_j \, dx$$

$$= \sum_{i,j=1}^{m} c_i \bar{c}_j \, k_n^2 \int_{D_1} u_i \bar{u}_j \, dx = k_n^2 \int_{D_1} |u|^2 \, dx = k_n^2.$$

Thus $k_n^2 \leqslant k_n^2(a)$. This is a contradiction. □

2.3. Existence of solutions to exterior boundary value problems with Neumann's and impedance boundary conditions can be proved as in Theorem 1.

2.4. Let us give another proof of the existence of the solution to problem (0.20) using Equation (1.38).

LEMMA 5. *Let A and B be linear compact operator on a Hilbert space H, and*

$$N(I + A^*) \cap N(B^*) = \{0\}. \tag{18}$$

Then the equation $(I + A)\mu + B\sigma = f$ is solvable for any $f \in H$. Condition (18) is also necessary for the solvability of the equation $(I + A)\mu + B\sigma = f$.

Proof. Let $T = I + A$. Since A is linear and compact $R(T)$ is closed, $\dim N(T^*) = \operatorname{codim} R(T) < \infty$. It is clear that equation $T\mu + B\sigma = f$ is densely solvable, i.e. solvable for any f in a dense subset of H, if (18) holds. Indeed, if $h \in N(T^*) \cap N(B^*)$, $h \neq 0$, then $(f, h) = 0$ is a necessary condition for solvability of the equation $T\mu + B\sigma = f$. Thus, this equation is not densely solvable. On the other hand, if the set $\{T\mu + B\sigma\}$ is not dense in H, then there exists $h \in H$, $h \neq 0$, such that $(T\mu + B\sigma, h) = 0$ for all μ and σ. Taking $\sigma = 0$ one sees that $h \in N(T^*)$ and similarly $h \in N(B^*)$. Thus (18) does not hold. Since condition (18) is necessary for dense solvability it is necessary for everywhere solvability, i.e. solvability for any $f \in H$. Let us prove the sufficiency. Let P be the orthogonal projection onto $R(T)$. Then

$$T\mu + P B\sigma = Pf \tag{19}$$

$$(I - P)B\sigma = (I - P)f. \tag{20}$$

If (20) is solvable then (19) is solvable and

$$T\mu + B\sigma = f. \tag{21}$$

Indeed, let σ solve (20) and let $T\mu = P(f - B\sigma)$. Since the right hand side in the last equation belongs to $R(T)$, this equation is solvable and

$$T\mu + B\sigma = Pf - P B\sigma + B\sigma = Pf + (I - P)f = f. \tag{22}$$

To prove that Equation (21) is solvable for any f if (18) holds let us use the already established dense solvability of (21) and, therefore, (20). But (20) is a linear equation in a finite dimensional space, since $\operatorname{codim} R(T) < \infty$. If such an equation is densely solvable then it is everywhere solvable. But then, as was shown above, Equations (19) and (21) are solvable for any f. □

Let us apply Lemma 5 to Equation (1.38). We have to prove that $N(I + \bar{A}) \cap N(Q^*) = \{0\}$. Let $\sigma \in N(I + \bar{A}) \cap N(Q^*)$ and $w = S\bar{\sigma}$. Then $w = 0$ on Γ and $\partial w/\partial N_+ = 0$. By the uniqueness of the solution to Cauchy's problem for elliptic equations, $w = 0$ in D. Since w solves problem (13) Lemma 2 says that $w = 0$ in Ω. Thus

$$\sigma = \frac{\partial w}{\partial N_+} - \frac{\partial w}{\partial N_-} = 0.$$

We have proved

PROPOSITION 1. *The solution to problem* (0.20) *can be found in the form* (1.37).

2.5. Another proof of existence of the solution to problem (0.20) is given in Section I.1.5.

2.6. Consider the transmission boundary value problem (1.40)–(1.44).

LEMMA 6. *Problem* (1.40)—(1.44) *has at most one solution.*

Proof. It is sufficient to prove that the corresponding homogeneous problem has only the trivial solution. From Green's formula it follows that

$$0 = \lim_{R\to\infty} \int_{|r|=R} \left(u \frac{\partial \bar{u}}{\partial r} - \bar{u} \frac{\partial u}{\partial r} \right) ds - \int_{\Gamma} \left(u \frac{\partial \bar{u}}{\partial N_-} - \bar{u} \frac{\partial u}{\partial N_-} \right) ds. \tag{23}$$

Furthermore

$$\int_{\Gamma} \left(u \frac{\partial \bar{u}}{\partial N_-} - \bar{u} \frac{\partial u}{\partial N_-} \right) ds = \rho \int_{\Gamma} \left(u \frac{\partial \bar{u}}{\partial N_+} - \bar{u} \frac{\partial u}{\partial N_+} \right) ds = 0. \tag{24}$$

Thus

$$\lim_{R\to\infty} \int_{|r|=R} \left(u \frac{\partial \bar{u}}{\partial r} - \bar{u} \frac{\partial u}{\partial r} \right) ds = 0. \tag{25}$$

From (25), (1.44), and (14) one derives (16) with $k = k_0$.

Now, Lemma 1 says that $u = 0$ in Ω. Therefore $u = 0$ on Γ, $\partial u/\partial N_+ = 0$, and $u = 0$ in D. □

Let

$$\mathscr{H}^q = H^q \oplus \tilde{H}^{q-1/2}, \quad H^q = W^{2,q}(D), \quad \tilde{H}^q = W^{2,q}(\Gamma).$$

The space $\mathscr{H}^q$ consists of ordered pairs $w = \binom{u}{\sigma}$ where $u \in H^q$, $\sigma \in \tilde{H}^{q-1/2}$, $(w_1, w_2)_{\mathscr{H}^q} = (u_1, u_2)_{H^q} + (\sigma_1, \sigma_2)_{\tilde{H}^{q-1/2}}$.

Assume that $h \in H^q, f \in \tilde{H}^{q-1/2}$

THEOREM 2. *Problem* (1.40)—(1.44) *is uniquely solvable.*

Proof. Uniqueness follows from Lemma 6. In order to prove existence it is sufficient to find a Fredholm-type integral equation for the solution to problem (1.40) (1.44) which is equivalent to this problem. Since problem (1.40)–(1.44) has no more than one solution, the equivalent integral equation has no more than one solution. Since this equation is of Fredholm's type it is solvable. To derive a suitable integral equation let us use Green's formula in D:

$$\int_{R^3} \{g(\nabla^2 + k_0^2)u - u(\nabla^2 + k_0^2)g\} \, dy = u(x) + \kappa \int_D gu \, dy + \int_D gh \, dy, \quad \kappa = k_0^2 - k_1^2 \tag{26}$$

where $g = g(x, y, k_0)$. The left hand side of this equation can be written as

$$\lim_{R \to \infty} \int_{S_R} \left(g \frac{\partial u}{\partial N} - u \frac{\partial g}{\partial N_-} \right) ds - \int_\Gamma \left(g \frac{\partial u}{\partial N_-} - u \frac{\partial g}{\partial N_-} \right) ds + \int_\Gamma \left(g \frac{\partial u}{\partial N_+} - u \frac{\partial g}{\partial N_+} \right) ds = \int_\Gamma g \left(\frac{\partial u}{\partial N_+} - \frac{\partial u}{\partial N_-} \right) ds \tag{27}$$

$$= (1 - \rho) \int_\Gamma g \frac{\partial u}{\partial N_+} ds + \int_\Gamma gf \, ds.$$

Combine (26) and (27) to obtain

$$u(x) - F(x) = \kappa Tu + (1 - \rho) S\sigma, \tag{28}$$

where

$$\sigma = \frac{\partial u}{\partial N_+}, \tag{29}$$

and

$$F(x) \equiv -Th + Sf. \tag{30}$$

For any σ the function u which solves (28) satisfies (1.40)–(1.42) and (1.44). It satisfies (1.43) iff

$$(\rho-1)\frac{\partial F}{\partial N}-\kappa(\rho-1)\frac{\partial Tu}{\partial N}+ \\ +(1-\rho)\left[\rho\frac{A\sigma+\sigma}{2}-\frac{A\sigma-\sigma}{2}\right]=f, \tag{31}$$

or

$$\sigma=\frac{1-\rho}{1+\rho}A\sigma-\frac{2\kappa}{1+\rho}\frac{\partial Tu}{\partial N}+\frac{2}{1+\rho}\left(\frac{\partial F}{\partial N}+\frac{f}{1-\rho}\right) \tag{32}$$

where formula (3) was used. Equations (32) and (29) are equivalent if u satisfies (28). Let

$$w=\binom{u}{\sigma}. \tag{33}$$

Then (28) and (32) can be written as

$$w=Bw+h_1 \tag{34}$$

where

$$B=\begin{pmatrix}-\kappa T & (1-\rho)S\\ -\dfrac{2\kappa}{1+\rho}\dfrac{\partial}{\partial N}T & \dfrac{1-\rho}{1+\rho}A\end{pmatrix},\quad h_1=\begin{pmatrix}F\\ \dfrac{2}{1+\rho}\left(\dfrac{\partial F}{\partial N}+\dfrac{f}{1-\rho}\right)\end{pmatrix}, \tag{35}$$

$h_1\in\mathscr{H}^q$.

It remains to be proved that Equation (34) is a Fredholm-type equation in $\mathscr{H}^q$ and that this equation is equivalent to problem (1.40)–(1.44). This is done in Lemmas 7 and 8 below. □

LEMMA 7. *Let* $\mathscr{H}^q=H^q\oplus\tilde{H}^{q-1/2}$. *Then* $B:\mathscr{H}^q\to\mathscr{H}^q$ *is compact.*

Proof. It is well known (see e.g., Eskin [1]) that the embeddings $H^q\to H^p$, $\tilde{H}^q\to H^p$ and $H^\ell\to\tilde{H}^s$ are compact if $q>p$ and $\ell>s+\frac{1}{2}$. It is also well known (see, e.g., Eskin [1]) that T: $H^q\to H^{q+2}$ is continuous and therefore $\partial/\partial N\;T$: $H^q\to H^{q+1}$ is continuous,

$A: \tilde{H}^q \to \tilde{H}^{q+2}$ is continuous, $S: \tilde{H}^q \to H^{q+3/2}$ is continuous. From these facts the conclusion of Lemma 7 follows. □

LEMMA 8. *Equation* (34) *is equivalent to problem* (1.40)–(1.44).

Proof. It was already proved that (34) follows from (1.40)–(1.44). Suppose $\binom{u}{\sigma}$ solves (34), then Equations (28) and (32) hold. Equation (28) implies Equations (1.40)–(1.42) and (1.44). Equation (32) implies (1.43). □

2.7. Consider the transmission scattering problem

$$(\nabla^2 + k_0^2)u = 0 \text{ in } \Omega, \quad k_0 > 0 \tag{36}$$

$$(\nabla^2 + k_1^2)u = 0 \text{ in } D, \quad k_1 > 0 \tag{37}$$

$$u_+ = u_- \text{ on } \Gamma \tag{38}$$

$$\rho \frac{\partial u}{\partial N_+} = \frac{\partial u}{\partial N_-} \quad \text{on } \Gamma \tag{39}$$

$$u = u_0 + v \tag{40}$$

$$\lim_{R \to \infty} \int_{S_R} \left| \frac{\partial v}{\partial r} - ikv \right|^2 \mathrm{d}s = 0 \tag{41}$$

where $u_0(x)$ is an incident field which satisfies Equation (36) in the whole space, for example,

$$u_0 = \exp(-ik_0 n \cdot x). \tag{42}$$

The scattered field v satisfies problem (1.40)–(1.44) with

$$h = (k_0^2 - k_1^2)u_0, \qquad f = (1 - \rho)\frac{\partial u_0}{\partial N}. \tag{43}$$

Thus, Theorem 2 states that problem (36)–(41) has a solution and this solution is unique.

2.8. Let us finally sketch yet another approach to the proof of the existence of solutions to exterior boundary value problems.

Let

$$(\nabla^2 + k^2)u = f \text{ in } \Omega \subset R^3, \quad u|_\Gamma = 0, \quad u \text{ satisfies (41)}. \tag{44}$$

Let $f = 0$ if $|x| \geqslant R$, $\Gamma \subset B_R$, $v = \int_{R^3} gh \, dy$, $h = 0$ if $|x| \geqslant R$, and w be defined as the solution to the problem.

$$\nabla^2 w + \lambda^2 w = 0 \text{ in } \Omega \cap B_R, \; w = v \text{ on } \Gamma, \; w = 0 \text{ on } S_R. \quad (45)$$

We take R such that problem (45) is uniquely solvable, i.e., λ^2 is not an eigenvalue of the homogeneous problem (45). Let $\phi \in C_0^\infty(\Omega \cap B_R)$, $\phi = 1$ on Γ and $\phi = 0$ in a neighborhood of S_R. Then the function $u = v - \phi w$ solves (44) iff

$$\nabla^2 u + k^2 u = h - 2\nabla\phi\nabla w - w\nabla^2\phi - (k^2 - \lambda^2)\phi w = f.$$

This can be written as

$$h - Bh = f, \quad (46)$$

where

$$Bh = 2\nabla\phi\nabla w + w\nabla^2\phi + (k^2 - \lambda^2)\phi w. \quad (47)$$

If we show that B is compact in $L^2(\Omega \cap B_R)$ and the homogeneous Equation (46) has only the trivial solution, we conclude that Equation (46) is solvable for any $f \in L^2(\Omega \cap B_R)$, and therefore problem (44) is solvable. The mapping $h \to Bh$ is a composition of the mappings $(w \to (k^2 - \lambda^2)\phi w + 2\nabla w \nabla\phi + w\nabla^2\phi) \circ (v \to w) \circ (h \to v)$ which acts as follows: $(H^{q+2} \to H^{q+1}) \circ (H^{q+2} \to H^{q+2}) \circ (H^q \to H^{q+2})$, where $H^q = H^q(\Omega \cap B_R)$. Therefore $h \to Bh$ acts from H^q into H^{q+1}. Since the domain $\Omega \cap B_R$ is compact, the imbedding i: $H^{q+1} \to H^q$ is compact. Therefore B: $H^q \to H^q$ is compact. If h solves the homogeneous equation (46) then $v - \phi w$ solves the homogeneous problem (44). Thus $u = 0$ in Ω, $v = \phi w$ in Ω. Therefore $v = 0$ in Ω_R, $v = 0$ on S_R, and if one defines $\alpha(x) = v$ in D, $\alpha(x) = w$ in $\Omega \cap B_R$, then

$$(\nabla^2 + k^2)\alpha = 0 \text{ in } D, \; (\nabla^2 + \lambda^2)\alpha = 0 \text{ in } \Omega \cap B_R. \quad (48)$$

Multiplying (48) by $\bar{\alpha}$ and integrating over B_R yields

$$\int_{B_R} (|\nabla\alpha|^2 - p|\alpha|^2) \, dx = 0, \quad p = \begin{cases} k^2 & \text{in } D \\ \lambda^2 & \text{in } \Omega \cap B^{\cdot} \end{cases} \quad (49)$$

If $\operatorname{Im} \lambda^2 \neq 0$ then (49) implies that $\alpha = 0$ in $\Omega \cap B_R$. Therefore $w = 0$ in $\Omega \cap B_R$ and $v = 0$ in $\Omega \cap B_R$. But then $h = 0$.

2.9. Let us prove uniqueness of the solution to the impedance boundary value problem

$$(\nabla^2 + k^2)u = 0 \text{ in } \Omega, \quad \operatorname{Im} k \geqslant 0, \quad \operatorname{Re} k \geqslant 0, \quad k \neq 0 \tag{50}$$

$$u_N + \eta(s)u = f \text{ on } \Gamma, \quad u_N = \frac{\partial u}{\partial N} \tag{51}$$

$$\operatorname{Im}(\bar{k}\eta(s)) \geqslant 0 \text{ on } \Gamma \tag{52}$$

$$\int_{S_r} |u_r - iku|^2 \, ds \to 0, \quad r \to \infty, \quad u_r = \frac{\partial u}{\partial r}. \tag{53}$$

LEMMA 9. *Problem* (50)—(53) *has no more than one solution.*

Proof. Consider the homogeneous problem, i.e. $f = 0$. Condition (53) yields

$$\int_{S_r} (|u_r|^2 + |k|^2|u|^2) \, ds - 2 \operatorname{Im} \int_{S_r} \bar{k}\bar{u}u_r \, ds \equiv \varepsilon(r) \to 0, \quad r \to \infty. \tag{54}$$

Apply Green's formula to u and $\bar{u}$ to obtain:

$$0 = \int_{B_r \cap \Omega} \bar{u}(\nabla^2 + k^2)u \, dx = \int_{B_r \cap \Omega} (k^2|u|^2 - |\nabla u|^2) \, dx + \\ + \int_{S_r} \bar{u}u_r \, ds - \int_\Gamma \bar{u}u_N \, ds. \tag{55}$$

From (55) it follows that

$$-\bar{k} \int_{S_r} \bar{u}u_r \, ds = \int_{B_r \cap \Omega} (\bar{k}k^2|u|^2 - \bar{k}|\nabla u|^2) \, dx - \int_\Gamma \bar{k}\bar{u}u_N \, ds. \tag{56}$$

If $\operatorname{Im} k \geqslant 0$ then $\operatorname{Im} \bar{k} \leqslant 0$. It follows from (56) that

$$-\operatorname{Im} \bar{k} \int_{S_r} \bar{u}u_r \, ds = \operatorname{Im} \int_{B_r \cap \Omega} (k|k|^2|u|^2 - \bar{k}|\nabla u|^2) \, dx - \\ - \operatorname{Im} \int_\Gamma \bar{k}\bar{u}u_N \, ds. \tag{57}$$

If

$$-\mathrm{Im}\,\bar{k}\int_{\Gamma}\bar{u}u_N\,\mathrm{d}s \geqslant 0 \tag{58}$$

then the right-hand side of (57) is nonnegative and

$$-\mathrm{Im}\,\bar{k}\int_{S_r}\bar{u}u_r\,\mathrm{d}s \geqslant 0. \tag{59}$$

It follows from (54) and (59) that

$$\int_{S_r}(|u_r|^2+|k|^2|u|^2)\,\mathrm{d}s \to 0, \quad r\to\infty, \quad k\neq 0. \tag{60}$$

This implies that $u = 0$ in Ω. It remains to be noticed that the homogeneous condition (51) and inequality (52) imply that

$$-\mathrm{Im}\,\bar{k}\int_{\Gamma}\bar{u}u_N\,\mathrm{d}s = \mathrm{Im}\,\bar{k}\int_{\Gamma}\eta(s)\,|u|^2\,\mathrm{d}s \geqslant 0.$$

This completes the proof. □

2.10. In this section we prove that (50) and (53) imply that

$$\int_{S_r}|u|^2\,\mathrm{d}s = 0(1), \quad r\to\infty. \tag{61}$$

This is of interest because often in the literature the radiation condition is stated as two conditions

$$u_r - iku = o(r^{-1}), \quad r\to\infty \tag{62}$$

$$u = 0(r^{-1}), \quad r\to\infty. \tag{63}$$

These two conditions imply that Green's formula is applicable in the infinite domain Ω because

$$\int_{S_r}(uv_r - u_rv)\,\mathrm{d}s \to 0, \quad r\to\infty \tag{64}$$

if u and v satisfy conditions (62) and (63). In this case there is no need to assume that u and v solve Equation (50). One has

$$\int_{S_r} (uv_r - u_r v)\,\mathrm{d}s = \int_{S_r} \{u(v_r - ikv) + v(iku - u_r)\}\,\mathrm{d}s.$$

Applying Schwarz's inequality and conditions (62) and (63) yields (64). However, if (62) holds and u solves Equation (50) then (63) holds automatically. This is the reuslt we want to prove. Notice that (62) implies (53).

LEMMA 10. *Equations* (50) *and* (53) *imply* (61) *and* (63) *if* $\operatorname{Im} k \geqslant 0$.

Proof. It follows from (54) and (57) that

$$\tfrac{1}{2}\,\varepsilon(r) + \operatorname{Im}\int_\Gamma \bar{k}\bar{u}u_N\,\mathrm{d}s = \operatorname{Im}\int_{B_r\cap\Omega} (k|k|^2|u|^2 - \bar{k}|\nabla u|^2)\,\mathrm{d}s + \\ + \tfrac{1}{2}\int_{S_r} (|u_r|^2 + |k|^2|u|^2)\,\mathrm{d}s, \qquad \varepsilon(r) \to 0, \qquad r \to \infty. \tag{65}$$

If $\operatorname{Im} k \geqslant 0$ then both terms in the right-hand side of (65) are nonnegative and the left-hand side of (65) is bounded as $r \to \infty$.

Therefore

$$\int_{S_r} (|u_r|^2 + |k|^2|u|^2)\,\mathrm{d}s = 0(1), \quad r \to \infty. \tag{66}$$

If $\operatorname{Im} k > 0$ then it follows from (65) that

$$\int_\Omega (|u|^2 + |\nabla u|^2)\,\mathrm{d}x < \infty. \tag{67}$$

Equation (66) implies (61) and the representation formula for u, which follows immediately from Green's formula:

$$u(x) = \int_\Gamma \left\{ u(s)\,\frac{\partial g(x, s)}{\partial N} - g(x, s)\,\frac{\partial u(r)}{\partial N} \right\} \mathrm{d}s. \tag{68}$$

From (68) equation (63) follows immediately. □

Another proof of Lemma 10 for the most interesting case $k > 0$ is similar to the proof of Lemma 1 in Section 2.1. If u satisfies Equation (50) then $u_{nm}(r) \equiv \int_{S_2} u\overline{Y}_{nm}\, \mathrm{d}r$, where Y_{nm} are the spherical harmonics, satisfies Bessel's equation

$$u''_{nm} + \frac{2}{r} u'_{nm} + \left(k^2 - \frac{n(n+1)}{r^2} \right) u_{nm} = 0, \quad u'_{nm} = \frac{\mathrm{d}u_{nm}}{\mathrm{d}r}.$$

Therefore $u_{nm} = a_{nm} h_n^{(1)}(kr) + b_{nm} h_n^{(2)}(kr)$, where a_{nm} and b_{nm} are constants, and $h_n^{(j)}(z)$ are the spherical Hankel functions.

Condition (53) implies that

$$\sum_{n=0}^{\infty} \sum_{m=-n}^{n} |u'_{nm} - iku_{nm}|^2 r^2 \to 0, \quad r \to \infty.$$

Therefore $r(u'_{nm} - iku_{nm}) \to 0, r \to \infty$ for $n = 1, 2 \ldots$.

This implies that $b_{nm} = 0$, $n = 1, 2, \ldots, -n \leqslant m \leqslant n$, and

$$u = \sum_{n=0}^{\infty} \sum_{m=-n}^{n} a_{nm} h_n^{(1)}(kr)\, Y_{nm}(\theta, \phi).$$

One has

$$\int_{S_r} |u|^2 \,\mathrm{d}s = r^2 \sum_{n=0}^{\infty} \sum_{m=-n}^{n} |h_n^{(1)}(kr)|^2 |a_{nm}|^2. \tag{69}$$

If $k > 0$ then (Gradshteyn-Ryzhik [1] 8.478)

$$r_1^2 |h_n^{(1)}(kr_1)|^2 \leqslant r_2^2 |h_n^{(1)}(kr_2)|^2, \quad r_1 \geqslant r_2, \tag{70}$$

and one has

$$\int_{S_{r_1}} |u|^2 \,\mathrm{d}s \leqslant \int_{S_{r_2}} |u|^2 \,\mathrm{d}s, \quad r_1 \geqslant r_2. \tag{71}$$

This implies (61). Inequality (71) is of independent interest: it gives a maximum principle for solutions to Equation (50) satisfying condition (53). □

LEMMA 11. *If u solves equation* (50) *in* Ω_R, $R > 0$, *and satisfies condition* (53) *then*

$$\max_{r \leqslant R} \|u\|_{L^2(S_r)} = \|u\|_{L^2(S_R)}, \quad k > 0. \tag{72}$$

The following simple lemma is often useful.

LEMMA 12. *If*

$$(\nabla^2 + k^2)u = 0 \text{ in } \Omega, \quad k > 0; (53) \textit{ holds} \tag{73}$$

and

$$\lim_{r \to \infty} \operatorname{Im} \int_{S_r} \bar{u} u_r \, \mathrm{d}s = 0 \tag{74}$$

then $u = 0$ in Ω.

Proof. If $k > 0$ then (74) and (54) imply (60). From this and Lemma 1 it follows that $u \equiv 0$ in Ω. □

Remark 3. If

$$\nabla^2 u = 0 \text{ in } \Omega, \ u_N + \eta u = 0 \text{ on } \Gamma, \ \operatorname{Re} \eta \leqslant 0, \ u(\infty) = 0,$$

then $u \equiv 0$ in Ω.

Proof. One has

$$0 = \int_{\Omega'_R} \bar{u} \nabla^2 u \, \mathrm{d}x = -\int_{\Omega'_R} |\nabla u|^2 \, \mathrm{d}x + \int_\Gamma \eta(r) |u|^2 \, \mathrm{d}s +$$

$$+ \int_{S_R} u \frac{\partial u}{\partial r} \, \mathrm{d}s, \quad \Omega'_R = \Omega \cap B_R.$$

If $u(\infty) = 0$ then $u = 0(|x|^{-1})$ and $|\nabla u| = 0(|x|^{-2})$ as $|x| \to \infty$.

Therefore, if $R \to \infty$, one obtains

$$0 = -\int_\Omega |\nabla u|^2 \, \mathrm{d}x + \int_\Gamma \eta(s) |u|^2 \, \mathrm{d}s.$$

If $\operatorname{Re} \eta \leqslant 0$ this implies that $u = 0$ in Ω. □

2.11. In this section we present yet another approach to the existence of the solution to exterior boundary value problems. This approach is based on the limiting absorption principle. In order to prove the existence of the solution to problem (0.20) consider the problem

$$(-\nabla^2 - k^2 - i\varepsilon)v_\varepsilon = 0 \text{ in } \Omega, \quad v_\varepsilon \in L^2(\Omega), \quad \varepsilon > 0, \quad v_\varepsilon = f \text{ on } \Gamma. \tag{75}$$

For $\varepsilon > 0$ the existence and uniqueness of the solution to problem (75) is easy to prove. The limiting absorption principle says that if the limit $\lim v_\varepsilon = v$, $\varepsilon \to 0$, exists in some norm, then this limit is the solution to problem (0.20). The justification of this principle consists of proving the existence of the $\lim v_\varepsilon = v$ and in checking that v solves (0.20).

LEMMA 13. *Problem* (75) *is uniquely solvable.*

Proof. Look for a solution of the form (I.1.1) with $g(x, y, \sqrt{k^2 + i\varepsilon})$ in place of $g(x, y, k)$. The argument in Section I.1.2 shows that the solution of this form does exist and is unique because the equation $v + \bar{A}_\varepsilon v = 0$ has only the trivial solution $v = 0$. Here A_ε is defined by formula (1.4) with $g(x, y, \sqrt{k^2 + i\varepsilon})$ in place of $g(x, y, k)$. To prove that $v = 0$ consider $w = S_\varepsilon \bar{v}$, where $S_\varepsilon v$ is defined by formula (1.4) with $g(x, y, \sqrt{k^2 + i\varepsilon})$ in place of $g(x, y, k)$.

Then

$$(\nabla^2 + k^2 + i\varepsilon)w = 0 \text{ in } D, \quad \frac{\partial w}{\partial N_+} = 0 \text{ on } \Gamma. \tag{76}$$

Since the spectrum of the Neumann Laplacian in D is real, Equation (76) implies that $w = 0$ in D. Thus $w_+ = 0$, $w_- = 0$ and $(\nabla^2 + k^2 + i\varepsilon)w = 0$ in Ω. Therefore $w = 0$ in Ω and $\bar{v} = \partial w/\partial N_+ - \partial w/\partial N_- = 0$. □

Remark 4. One can prove that

$$(\nabla^2 + k^2 + i\varepsilon)w = 0 \text{ in } \Omega, \quad w_- = 0, \quad \varepsilon > 0 \tag{77}$$

implies that $w = 0$ in Ω as follows. From the representation of $w = S_\varepsilon \bar{v}$ it is seen that $w \in L^2(\Omega)$ if $\varepsilon > 0$, and $|w| \leqslant c \exp(-a|x|)$, $c = \text{const} > 0$, $a = a(\varepsilon, k) > 0$. Multiply (77) by $\bar{w}$ and integrate over $B_R \setminus D$ to obtain

$$\int_{B_R \setminus D} (k^2|w|^2 + i\varepsilon|w|^2 - |\nabla w|^2)\, \mathrm{d}x + \int_{S_R} \frac{\partial w}{\partial N} \bar{w}\, \mathrm{d}s = 0. \tag{78}$$

Take $R \to \infty$. The surface integral vanishes in the limit and (78) yields

$$\int_\Omega (k^2|w|^2 + i\varepsilon|w|^2 - |\nabla w|^2)\,\mathrm{d}x = 0. \tag{79}$$

Take the imaginary part to obtain $\int_\Omega |w|^2\,\mathrm{d}x = 0$. Thus $w = 0$ in Ω.

THEOREM 3. (*Limiting absorption principle*). *There exists the limit in* $L_{2,1+\delta}(\Omega) = L_{2,1+\delta}$ *where* $\|v\|^2_{L_{2,1+\delta}} = \int_\Omega |v|^2(1+|x|)^{-(1+\delta)}\,\mathrm{d}x \equiv \|v\|^2$, ($\delta > 0$ *is arbitrary*),

$$\lim_{\varepsilon \to 0} v_\varepsilon = v. \tag{80}$$

The function v *solves problem* (0.20). *In any compact strictly inner subdomain* $\tilde{\Omega} \subset \Omega$ *convergence in* (80) *holds in* $C^m(\tilde{\Omega})$, $m = 1, 2, \ldots$.

Proof. (i) Suppose first that $\sup\|v_\varepsilon\| \leqslant c$, $0 < \varepsilon \leqslant 1$. Then $\|\Delta v_\varepsilon\| \leqslant c_1$. If $f \in H^{\ell+2}(\Gamma)$, then

$$\|v_\varepsilon\|_{H^{\ell+5/2}_{(B_R\setminus D)}} \leqslant c(R, k)\,(c + \|f\|_{H^{\ell+2}(\Gamma)}). \tag{81}$$

This inequality is a particular case of the apriori estimates for elliptic operators (see e.g. Berezanskii [1] or Eskin [1]). In particular, if one takes $\ell = -\frac{1}{2}$ in (81) then it follows from (81) that $\|v_\varepsilon\|_{H^2(B_R\setminus D)} \leqslant c_1$, where c_1 does not depend on ε. From the embedding theorem (see e.g. Berezanskii [1]) it follows that there is a subsequence, which is denoted again v_ε such that

$$v_\varepsilon \to v \text{ in } H^\ell(B_R\setminus D), \quad \ell < 2, \quad \varepsilon \to 0, \quad v_\varepsilon \to v \text{ in } H^\ell(\Gamma), \quad \ell < 3/2.$$

Therefore one can pass to the limit $\varepsilon \to 0$ in the Green formula for v_ε and obtain

$$v(x) = \int_\Gamma \left(f\frac{\partial g}{\partial N} - g\frac{\partial v}{\partial N} \right) \mathrm{d}s. \tag{82}$$

This formula shows that v satisfies Helmholtz's equation in Ω and the radiation condition. Passing to the limit in the boundary condition shows that $v = f$ on Γ. Thus, the limit function does exist and solves problem (0.20). Since the solution to problem (0.20) is unique, every subsequence v_ε converges to v and (80) holds in $C^m(\bar{\Omega})$ $m = 1, 2, 3, \ldots$, and in $H^{\ell'+5/2}(B_R \setminus D)$ for any $\ell' < \ell$, where $f \in H^\ell(\Gamma)$. From the Green formula for v_ε it follows that

$$|v_\varepsilon| \leqslant \frac{c}{1+|x|} \tag{83}$$

where $c > 0$ does not depend on ε. Estimate (83) and the local convergence of v_ε to v (convergence on compact domains) show that (80) holds in $L_{2,1+\delta}$ with arbitrary $\delta > 0$.

(ii) Suppose now that

$$\|v_{\varepsilon_n}\| \to \infty, \quad \varepsilon_n \to 0. \tag{84}$$

Consider $w_n = v_n \|v_n\|^{-1}$, $v_n \equiv v_{\varepsilon_n}$. Clearly $\|w_n\| = 1$, $w_n = f\|v_n\|^{-1}$ on Γ, so that $\|w_n\|_{H^\ell(\Gamma)} \to 0$ as $n \to \infty$. The argument in part (i) of the proof shows that $W_n \underset{L_{2,1+\delta}}{\rightarrow} w$, $n \to \infty$, where w solves the homogeneous problem (0.20) since $w = 0$ in Ω. On the other hand, passing to the limit in $L_{2,1+\delta}$ in the equation $\|w_n\| = 1$ yields $\|w\| = 1$. This contradiction proves that assumption (84) cannot hold. The limiting absorption principle is proved. □.

Remark 5. Theorem 3 says that the problem

$$\begin{aligned} &(\nabla^2 + k^2)v = 0 \text{ in } \Omega,\ k > 0; \quad v = f \text{ on } \Gamma; \\ &v(x, k + i\varepsilon) \in L^2(\Omega), \quad \varepsilon > 0 \end{aligned} \tag{85}$$

has a solution $v = \lim v(x, k + i\varepsilon)$, $\varepsilon \to +0$, and this solution is unique.

3. EIGENFUNCTION EXPANSION THEOREM

3.1. Consider the solutions to the scattering problem

$$(\nabla^2 + k^2)u = 0 \text{ in } \Omega,\ \ k > 0,\ \ u = 0 \text{ on } \Gamma,\ \ u = \exp(-ikn \cdot x) + v, \tag{1}$$

$$\lim_{R \to \infty} \int_{S_R} \left| \frac{\partial v}{\partial r} - ikv \right|^2 \mathrm{d}s = 0. \tag{2}$$

The solution to problem (1)—(2) does exist and is unique. This was proved in Section 1.2. The aim of this section is to study analytic properties of this solution, to derive this solution from the resolvent kernel of the Dirchlet Laplacian in Ω, which is a self adjoint non-negative operator in $L^2(\Omega)$, and to prove the eigenfunction expansion theorem. The resolvent kernel is defined as the solution to the problem.

$$(\nabla^2 + k^2) G(x, y, k) = -\delta(x-y) \text{ in } \Omega, \quad G = 0 \text{ on } \Gamma, \quad G \text{ satisfies (2)}. \tag{3}$$

From the results in Section I.2 it follows that problem (3) is uniquely solvable and G can be represented as

$$G(x, y, k) = g - \int_\Gamma g(x, s, k)\mu(s, y, k)\, \mathrm{d}s, \quad \mu = \frac{\partial G(s, y, k)}{\partial N_s}, s \in \Gamma \tag{4}$$

where $g = \exp(ik|x-y|)/(4\pi|x-y|)$. Taking the normal derivative on Γ and using formula (1.3) one obtains from (4) the equation for μ:

$$\mu = -A\mu + 2\frac{\partial g}{\partial N_s}, y \in \Omega, \tag{5}$$

where A is defined in (1.4). It was shown in Section 1.1 that the solution to Equation (5), which satisfies the equation

$$g(s, y, k) = \int_\Gamma g(s, s', k)\mu(s', y, k)\, \mathrm{d}s', \tag{6}$$

is unique. Equation (6) is equivalent to the boundary condition (3) on Γ.

3.2. Note that

$$g(x, y, k) = \frac{\exp(ik|y|)}{4\pi|y|} \exp(-ikn \cdot x)\,(1 + o(1)) \text{ as } |y| \to \infty, \quad \frac{y}{|y|} = n, \tag{7}$$

and the factor $\exp(-ikn \cdot x)$, the plane wave, is the solution to the scattering problem in the free space (in this case there is no scattering, no outgoing wave v). Actually, $o(1)$ in (7) is $O(|y|^{-1})$. For the basic theorem below it is sufficient to have $o(1)$ in (7). We prove that the solution to the scattering problem (1) can be constructed in the same way.

LEMMA 1. *The following formula holds*

$$G(x, y, k) = \frac{\exp(ik|y|)}{4\pi|y|}\, u(x, n, k)\,(1 + o(1)), \quad |y| \to \infty, \quad \frac{y}{|y|} = n \tag{8}$$

where $u(x, n, k)$ is the solution to problem (1), *and* $o(1) = 0(|y|^{-1})$.

Proof. It follows from (4) that

$$G(x, y, k) = \frac{e^{ik|x|}}{4\pi|x|}\left\{\exp(-ikn \cdot y) - \right.$$
$$\left. - \int_\Gamma \exp(-ikn \cdot s)\mu(s, y, k)\, \mathrm{d}s\right\}(1 + o(1)) \tag{9}$$

as $|x| \to \infty$, $x/|x| = n$. Let us denote by $\tilde{u}(y, n, k)$ the expression in the braces and show that $\tilde{u}$ solves problem (1). If $u(x, n, k)$ solves (1) then Green's formula for v yields

$$v(x, n, k) = -\int_\Gamma \left(G\frac{\partial v}{\partial N} - v\frac{\partial G}{\partial N}\right)\mathrm{d}s + \lim_{R\to\infty}\int_{S_R}\left(G\frac{\partial v}{\partial r} - v\frac{\partial G}{\partial r}\right)\mathrm{d}s$$
$$= \int_\Gamma v\,\frac{\partial G}{\partial N}\,\mathrm{d}s = -\int_\Gamma u_0(s, n, k)\mu(s, x, k)\,\mathrm{d}s, \tag{10}$$

where

$$u_0(x, n, k) = \exp(-ikn \cdot x). \tag{11}$$

Thus

$$u(x, n, k) = u_0(x, n, k) - \int_\Gamma u_0(s, n, k)\mu(s, x, k)\,\mathrm{d}s = \tilde{u}(x, n, k). \tag{12}$$

where μ is defined in (4). □

Let us use the symmetry $G(x, y, k) = G(y, x, k)$. It follows from (4) that

$$G(x, y, k) = G(y, x, k) = g(x, y, k) - \int_\Gamma g(x, s, k)\mu(s, y, k)\,\mathrm{d}s. \tag{4'}$$

Therefore, as $|y| \to \infty$, $y|y|^{-1} = n$, one obtains

$$G(x, y, k) = \frac{\exp(ik|y|)}{4\pi|y|} \left\{ \exp(-ikn \cdot x) - \int_\Gamma g(x, s, k) \frac{\partial u(s, n, k)}{\partial N_s} \, ds \right\} (1 + o(1)). \tag{8'}$$

Comparing (8), (8′), and (12) yields

$$\begin{aligned} u(x, n, k) &= u_0(x, n, k) - \int_\Gamma u_0(s, n, k) \mu(s, x, k) \, ds \\ &= u_0(x, n, k) - \int_\Gamma g(x, s, k) \frac{\partial u(s, n, k)}{\partial N_s} \, ds \end{aligned} \tag{12'}$$

where u_0 is given by (11).

In particular

$$\int_\Gamma \exp(-ikn \cdot s) \frac{\partial G(s, x, k)}{\partial N_s} \, ds = \int_\Gamma g(x, s, k) \frac{\partial u(s, n, k)}{\partial N_s} \, ds. \tag{12''}$$

In the following theorem an analogue of the Fourier transform is constructed which gives an eigenfunction expansion in solutions to the scattering problem for the Dirichlet Laplacian in Ω.

THEOREM 1. *Let* $f(x) \in L^2(\Omega)$. *Define*

$$Ff = \tilde{f}(\kappa) = \frac{1}{(2\pi)^{3/2}} \int_\Omega f(x) \, \overline{u(x, \kappa)} \, dx \tag{13}$$

where $u(x, \kappa) = u(x, n, k)$ *is the solution to the scattering problem given in* (12).

Then

$$F^{-1} \tilde{f} = f(x) = \frac{1}{(2\pi)^{3/2}} \int_{R^3} \tilde{f}(\kappa) \, u(x, \kappa) \, d\kappa = F^* \tilde{f}. \tag{14}$$

Here $\kappa = kn$, $n \in S^2$, $k \geqslant 0$, *the integrals in* (13) *and* (14) *converge in* L^2, *and the mapping* $F: f \to \tilde{f}$ *is an isometric isomorphism of* $L^2(\Omega)$ *onto* $L^2(R^3)$,

$$f = F^{-1}\tilde{f}, \quad F^{-1} = F^*, \text{ and } F(-\nabla^2)F^{-1} = k^2, \tag{15}$$

that is the Dirichlet Laplacian in Ω *is unitarily equivalent to the operator of multiplication by* k^2.

Proof. Step 1. The spectral theorem for selfadjoint operators says that (see Appendix 2, p. 378)

$$\frac{1}{\pi}\int_0^\infty \operatorname{Im} G(x, y, \sqrt{\lambda + io})\,d\lambda = \frac{2}{\pi}\int_0^\infty \operatorname{Im} G(x, y, k)k\,dk = \delta(x-y) \tag{16}$$

where $G(x, y, k)$ is the resolvent kernel of the Dirichlet Laplacian in Ω, $\lambda = k^2$. We compute $\operatorname{Im} G(x, y, k)$ in terms of the solutions to the scattering problem and then Equation (16) will lead to the eigenfunction expansion theorem. Applying Green's formula to $G(x, y, k)$ and $\overline{G(x, y, k)}$ yields

$$\frac{1}{\pi}\operatorname{Im} G(x, y, k) = \frac{1}{2\pi i}\int_{|r|=R} ds \left\{ G(x, s, k)\,\overline{\frac{\partial G(s, y, k)}{\partial |s|}} - \right.$$
$$\left. - \overline{G(x, s, k)}\,\frac{\partial G(s, y, k)}{\partial |s|} \right\}. \tag{17}$$

If $R \to \infty$ it follows from (17) and (8) that

$$\frac{1}{\pi}\operatorname{Im} G(x, y, k) = \frac{1}{2\pi i}\int_{S^2} d\omega\, \frac{2ik}{16\pi^2}\, u(x, \omega, k)\,\overline{u(y, \omega, k)}$$
$$= \frac{k}{16\pi^3}\int_{S^2} d\omega\, u(x, \omega, k)\,\overline{u(y, \omega, k)}, \quad \omega = \frac{s}{|s|}. \tag{18}$$

From (16) and (18) one obtains the completeness relation for the eigenfunctions $u(x, k) = u(x, n, k)$:

$$\frac{1}{(2\pi)^3}\int_{R^3} u(x, \kappa)\,\overline{u(y, \kappa)}\,d\kappa = \delta(x-y) \tag{19}$$

where $d\kappa = k^2\, dk\, d\omega$, and the integral is understood as the kernel of the identity operator in $L^2(R^3)$, that is in the same sense as (16)

Step 2. Define the operator

$$Ff \equiv \tilde{f}(k) = \frac{1}{(2\pi)^{3/2}} \int_\Omega f(x)\, \overline{u(x, k)}\, dx \tag{20}$$

on the set $C_0^\infty(\Omega)$. Let us show that (14) holds if $f \in C_0^\infty$, and F is an isometry. We have

$$\lim_{N\to\infty} \frac{1}{(2\pi)^{3/2}} \int_{|k|\le N} dk\, \tilde{f}(k) u(x, k) = \text{l.i.m.} \frac{1}{(2\pi)^3} \int_\Omega dy\, f(y) \times$$
$$\times \int_{|k|\le N} u(x, k)\overline{u(y, k)}\, dk = f(x) \tag{21}$$

where (19) was used. Thus the eigenfunction expansion theorem is proved for $f \in C_0^\infty(\Omega)$. The operator F is isometric on $C_0^\infty(\Omega)$:

$$\|Ff\|^2 = \frac{1}{(2\pi)^3} \int_{R^3} |\tilde{f}(k)|^2\, dk$$
$$= \underset{N\to\infty}{\text{l.i.m.}} \int_{|k|\le N} dk \int_\Omega \int_\Omega dx\, dy\, f(x)\overline{f(y)} \cdot \overline{u(x, k)}\, u(y, k) \tag{22}$$
$$= \int_\Omega \int_\Omega dx\, dy\, f(x)\overline{f(y)}\, \delta(x-y) = \int_\Omega |f|^2\, dx.$$

Let us extend $F\colon L^2(\Omega) \to L^2(R^3)$ by continuity. Since $C_0^\infty(\Omega)$ is dense in $L^2(\Omega)$ the extended operator, which is denoted by F also, is closed, defined on all of $L^2(\Omega)$, and isometric. Therefore F is invertible, its range $R(F)$ is closed, and to complete the proof one should demonstrate that F is sujective, that is $R(F) = L^2(R^3)$.

Step 3. It is clear that the adjoint operator $F^*\colon L^2(R^3) \to L^2(\Omega)$ is of the form

$$F^*h = \int_{R^3} d\kappa\, h(\kappa) u(x, \kappa), \quad h \in C_0^\infty(R^3). \tag{23}$$

Since F is an isometry $R(F)$ is closed. Therefore $R(F) = L^2(R^3)$ if $N(F^*) = \{0\}$, where $N(A)$ is the null space of the operator A. In order to prove that $F^*h = 0$, $h \in L^2(R^3)$, implies that $h = 0$ we need some lemmas. □

LEMMA 2. *If* $h \in L^2(R^3)$, $0 < a < b < \infty$, *and*

$$f(x) = \int_a^b dk\, k \int_{S^2} d\omega\, h(\kappa)u(x, \kappa), \quad \kappa = k\omega \tag{24}$$

then

$$(E_b - E_a)f = f \tag{25}$$

where E_λ *is the spectral family of the Dirichlet Laplacian in* $L^2(\Omega)$.

Proof. Let $\zeta^2 = \kappa^2 + i\varepsilon$, $\varepsilon > 0$. Green's formula yields

$$u(x, k) = (k^2 - \zeta^2) \int_\Omega G(x, y, \zeta)u(y, k)\, dy. \tag{26}$$

From (24) and (26) it follows that

$$\int_\Omega G(x, y, \zeta)f(y)\, dy = \int_a^b dk\, k \int_{S^2} d\omega\, h(\kappa)u(x, \kappa)\,(k^2 - \zeta^2)^{-1}. \tag{27}$$

Since

$$\begin{aligned}(E_b - E_a)f &= \frac{1}{\pi}\int_a^b \operatorname{Im} G(\sqrt{\lambda + i0})f\, d\lambda \\ &= \frac{1}{\pi}\int_a^b \frac{G(\sqrt{\lambda + i0})f - G(\sqrt{\lambda - i0})f}{2i}\, d\lambda\end{aligned} \tag{28}$$

where $G(z)$ is the operator in $L^2(\Omega)$ with the kernel $G(x, y, \sqrt{z})$ and $\operatorname{Im}\sqrt{z} > 0$,

$$\lim_{\varepsilon \downarrow 0} \frac{1}{\pi} \operatorname{Im} \frac{1}{k^2 - \kappa^2 - i\varepsilon} = \delta(k^2 - \kappa^2) \tag{29}$$

and $k > 0$, Equation (25) follows from (24), (27), and (29). □

LEMMA 3. $N(F^*) = \{0\}$.

Proof. If

$$f(x) \equiv F^*h = \int_{R^3} d\kappa\, h(\kappa)u(x, \kappa) = 0 \tag{30}$$

then for any $0 < a < b < \infty$ one has

$$0 = (E_b - E_a)f = \int_a^b dk\, k \int_{S^2} d\omega\, h(\kappa)u(x, \omega, k). \tag{31}$$

Therefore

$$0 = \gamma(x, k) \equiv \int_{S^2} d\omega\, h(k, \omega)u(x, \omega, k), \quad x \in \Omega \tag{32}$$

for almost all $k > 0$. Consider the identity

$$\begin{aligned} 0 &= \int_{S_r} \left(\gamma(y, k) \frac{\partial g(x, y, k)}{\partial |y|} - g(x, y, k) \frac{\partial \gamma}{\partial |y|} \right) dS_r \\ &= \int_{S^2} d\omega\, h(k, \omega) \int_{S_r} \left(u(y, \omega, k) \frac{\partial g}{\partial |y|} - g(x, y, k) \frac{\partial u(y, \omega, k)}{\partial |y|} \right) dS_r \\ &= \int_{S^2} d\omega\, h(k, \omega) \left\{ \int_{S_r} \left(u_0 \frac{\partial g}{\partial |y|} - g \frac{\partial u_0}{\partial |y|} \right) dS_r + \right. \\ &\quad \left. + \int_{S_r} \left(v \frac{\partial g}{\partial |y|} - g \frac{\partial v}{\partial |y|} \right) dS_r \right\}. \end{aligned} \tag{33}$$

Taking $r \to \infty$ one obtains from (33) the identity

$$0 = \int_{S^2} d\omega\, h(k, \omega)u_0(x, \omega, k), \quad x \in \Omega \tag{34}$$

for almost all $k > 0$, where $u_0 = \exp(-ik\omega \cdot x)$ and the radiation condition for v was used: the first integral in the braces on the right hand side of (34) is equal to u_0 while the second goes to zero as $r \to \infty$. Multiplying (34) by $k^2\,dk$ and integrating in k over $(0, \infty)$ one obtains that the Fourier transform of $h(k, \omega) = h(\kappa)$ is zero. Therefore $h = 0$.

Let us finish the proof of Theorem 1. It remains to be proved that (15) holds.

Step 4. Let $\tilde{f} \in C_0^\infty(R^3)$. Then $-\nabla^2 F^{-1}\tilde{f} = F^{-1}(k^2\tilde{f})$. Applying F to this equation and using the density of $C_0^\infty(R^3)$ in $L^2(R^3)$ one obtains (15).

3.3. In the same fashion the eigenfunction expansion theorem can be proved for the Neumann Laplacian in Ω and for the Laplacian corresponding to the Robin boundary condition $\partial u/\partial N + \eta u = 0$ on Γ, $\operatorname{Im} \eta = 0$. The condition $\operatorname{Im} \eta = 0$ rather that $\operatorname{Im} \eta \geqslant 0$ is imposed for the corresponding operator to be selfadjoint.

4. PROPERTIES OF THE SCATTERING AMPLITUDE

4.1. The scattering amplitude is defined by the formula

$$f(-n, \nu, k) = \lim \beta^{-1}(r) v(x, n, k), \quad r = |x| \to \infty, xr^{-1} = \nu \quad (1)$$

where v is defined in (3.1) and $\gamma(r) = r^{-1} \exp(ikr)$.

It follows from (3.9) and (3.10) that

$$f(-n, \nu, k) = -(4\pi)^{-1} \int_\Gamma \exp(-ikn \cdot s) \frac{\partial u(s, \nu, k)}{\partial N}\, ds \quad (2)$$

while (1), (3.1), (3.12′), and (3.12″) imply

$$f(-n, \nu, k) = -(4\pi)^{-1} \int_\Gamma \exp(-ik\nu \cdot s) \frac{\partial u(s, n, k)}{\partial N}\, ds. \quad (2')$$

It follows from (2) and (2′) that $f(-n, \nu, k) = f(-\nu, n, k)$. An alternative proof of this reciprocity relation is given below as a part of the proof of the following theorem in which the bar denotes complex conjugate.

THEOREM 1. *The following formulas hold*

$$f(n, \nu, -k) = \overline{f(n, \nu, k)}, \quad k > 0 \tag{3}$$

$$f(n, \nu, k) = f(-\nu, -n, k), \tag{4}$$

$$\frac{f(n, \nu, k) - \overline{f(\nu, n, k)}}{2i} = \frac{k}{4\pi} \int_{S^2} f(n, \omega, k) \overline{f(\nu, \omega, k)} \, \mathrm{d}\omega. \tag{5}$$

In particular

$$\operatorname{Im} f(n, n, k) = \frac{k}{4\pi} \int_{S^2} |f(n, \omega, k)|^2 \, \mathrm{d}\omega, \tag{6}$$

this is called the optical theorem.

Note that $f(n, \nu, k)$ is meromorphic in k on all of the complex plane (see Theorem 2 below). This and the condition (3) *imply*

$$f(n, \nu, -k_1 + ik_2) = \overline{f(n, \nu, k_1 + ik_2)} \tag{3'}$$

where k_1 and k_2 are real numbers. Equation (3′) *implies that the zeros and poles of the scattering amplitude occur in pairs symmetric with respect to the imaginary axis on the complex k-plane.*

Proof. (1) Equation (3) follows from the real-valuedness of the Helmholtz equation and the boundary condition (3.1).

Indeed, if u solves (3.1) then

$$u(x, n, k) = u_0 + v, \ u_0 = \exp(-ikn \cdot x), \tag{7}$$

$$v \sim f(-n, n', k)\gamma(r), \quad r = |x| \to \infty, \quad n' = xr^{-1}. \tag{8}$$

Furthermore, $u(x, n, -k)$ solves the same problem as $\overline{u(x, n, k)}$, namely $u(x, n, -k)$ solves (3.1) and

$$v(x, n, -k) \sim f(-n, n', -k)\bar{\gamma}(r), \ r \to \infty, \quad xr^{-1} = n'. \tag{9}$$

Since $\overline{u_0(x, n, k)} = u_0(x, n, -k)$ and since the solution to this exterior boundary value problem is unique, one concludes that

$$u(x, n, -k) = \overline{u(x, n, k)} \text{ in } \Omega. \tag{10}$$

Equation (3) follows from (10) immediately.

(2) Let us prove (4). Let $u_j = u(x, n_j, k)$, $j = 1, 2$, solve problem (3.1) with $n = n_j$,

$$u_j = u_{0j} + v_j = u_{0j} + f_j\gamma + 0(r^{-2}), \quad r \to \infty, \quad \omega = xr^{-1} \tag{11}$$

where $u_{0j} = \exp(-ikn_j x)$ and $f_j = f(-n_j, \omega, k)$.

Green's formula yields

$$\begin{aligned} 0 &= \lim_{r\to\infty} \int_{S_r} \left(u_1 \frac{\partial u_2}{\partial r} - u_2 \frac{\partial u_1}{\partial r} \right) ds \\ &= \lim_{r\to\infty} \left\{ \int_{S_r} \left(u_{01} \frac{\partial u_{02}}{\partial r} - u_{02} \frac{\partial u_{01}}{\partial r} \right) ds + \right. \\ &\quad + \int_{S_r} \left(v_1 \frac{\partial v_2}{\partial r} - v_2 \frac{\partial v_1}{\partial r} \right) ds + \\ &\quad \left. + \int_{S_r} \left(u_{01} \frac{\partial v_2}{\partial r} - v_2 \frac{\partial u_{01}}{\partial r} \right) ds + \int_{S_r} \left(v_1 \frac{\partial u_{02}}{\partial r} - u_{02} \frac{\partial v_1}{\partial r} \right) ds \right\}. \end{aligned} \tag{12}$$

The first integral in the right side of (12) is equal to zero since u_{0j} solve the Helmholtz equation everywhere in the ball $B_r = \{x : |x| \leqslant r\}$. The limit of the second integral in the right side of (12) is zero because v_j satisfies the radiation condition. Therefore

$$\begin{aligned} 0 &= \lim_{r\to\infty} \int_{S_r} (u_{01}\, ik\gamma f_2 + f_2\gamma ikn_1 \cdot \omega u_{01} + f_1\gamma u_{02}(-ikn_2 \cdot \omega) - \\ &\quad - u_{02} ik\gamma f_1)\, ds = \lim_{r\to\infty} ikr^2\gamma \int_{S^2} [f_2 u_{01}(1 + n_1\omega) - \\ &\quad - f_1 u_{02}(1 + n_2 \cdot \omega)]\, d\omega. \end{aligned} \tag{13}$$

Let us use the following formula which is proved later as Lemma 1:

$$\int_{S^2} \exp(ikn \cdot \omega r) f(\omega)\, d\omega = \frac{2\pi i}{k} [\bar{\gamma} f(-n) - \gamma f(n)] + o\left(\frac{1}{r}\right), \tag{14}$$

$r \to \infty$

where $f \in C^1(S^2)$. We will prove later that $f(n, \omega, k)$ is analytic in ω in the sense which will be specified at the end of this section. From (7), (13), and (14) one obtains

$$0 = \lim_{r\to\infty} ikr^2\gamma \frac{2\pi i}{k} [\bar{\gamma}f(-n_2, n_1, k)(1 + n_1 \cdot n_1) - \\ - \gamma f(-n_2, -n_1, k)(1 - n_1 \cdot n_1) - \bar{\gamma}f(-n_1, n_2, k)(1 + n_2 \cdot n_2) + \\ + \gamma f(-n_1, n_2, k)(1 - n_2 \cdot n_2) + o(r^{-1})]. \tag{15}$$

Since $n_1 \cdot n_1 = n_2 \cdot n_2 = 1$, one concludes from (15) that $f(-n_2, n_1, k) = \phi(-n_1, n_2, k)$, which is Equation (4). □

Let us prove (14).

LEMMA 1. *Formula* (14) *holds.*

Proof. Let $\omega = (\theta, \phi)$, $n \cdot \omega = \xi = \cos\theta$. Then

$$\int_{S^2} \exp(ikrn \cdot \omega) f(\omega)\, d\omega$$

$$= \int_0^{2\pi} d\phi \int_{-1}^{1} d\xi \exp(ikr\xi) f(\arccos \xi, \phi)$$

$$= \int_0^{2\pi} d\phi \left[\frac{\exp(ikr)f(0, \phi) - \exp(-ikr)f(\pi, \phi)}{ikr} + o\left(\frac{1}{r}\right)\right]$$

$$= \frac{2\pi}{ik}[\gamma f(n) - \bar{\gamma}f(-n)] + o\left(\frac{1}{r}\right), \quad r \to \infty.$$

One can write (14) as

$$\lim_{r\to\infty} \exp(ikrn \cdot \omega) = 2\pi ik^{-1}[\bar{\gamma}\delta(n + \omega) - \gamma\delta(n - \omega)]. \tag{16}$$

□

(3) Let us prove formulas (5) and (6). Formula (6) is a particular case of (5). In order to prove (5) start with the identity:

$$0 = \int_{S_r} \left(u_1 \frac{\partial \bar{u}_2}{\partial r} - \bar{u}_2 \frac{\partial u_1}{\partial r} \right) ds = \int_{S_r} \left[\left(u_{01} \frac{\partial \bar{u}_{02}}{\partial r} - \bar{u}_{02} \frac{\partial u_{01}}{\partial r} \right) + \right.$$

$$+ \left(v_1 \frac{\partial \bar{u}_{02}}{\partial r} - \bar{u}_{02} \frac{\partial v_1}{\partial r} \right) + \left(u_{01} \frac{\partial \bar{v}_2}{\partial r} - \bar{v}_2 \frac{\partial u_{01}}{\partial r} \right) + \tag{17}$$

$$\left. + \left(v_1 \frac{\partial \bar{v}_2}{\partial r} - \bar{v}_2 \frac{\partial v_1}{\partial r} \right) \right] ds \equiv T_1 + T_2 + T_3 + T_4.$$

Green's formula shows that $T_1 = 0$. The remaining integrals can be written as in formula (13)

$$0 = \lim_{r \to \infty} \int_{S_R} (\gamma f_1 ikn_2 \cdot \omega \bar{u}_{02} - ik\gamma f_1 \bar{u}_{02} -$$

$$- ik\bar{\gamma}\bar{f}_2 u_{01} + \bar{\gamma}\bar{f}_2 ikn_1 \cdot \omega u_{01} - 2ik|\gamma|^2 f_1 \bar{f}_2)\, ds. \tag{18}$$

Applying formula (14) to this equation, one obtains

$$0 = \lim_{r \to \infty} \int_{S_R} [ik\gamma(n_2 \cdot \omega - 1) f_1 \bar{u}_{02} - ik\gamma(n_1 \cdot \omega - 1)\bar{f}_2 u_{01} - 2ik|\gamma|^2 f_1 \bar{f}_2]\, ds$$

$$= \lim_{r \to \infty} \left[4\pi|\gamma|^2 r^2 f(-n_1, -n_2, k) - 4\pi|\gamma|^2 r^2 \overline{f(-n_2, -n_1, k)} - \right.$$

$$\left. - 2ik|\gamma|^2 r^2 \int_{S^2} f_1 \bar{f}_2\, d\omega \right].$$

Thus

$$4\pi f(-n_1, -n_2, k) - 4\pi \overline{f(-n_2, -n_1, k)}$$

$$= 2ik \int_{S^2} f(-n_1, \omega, k) \overline{f(-n_2, \omega, k)}\, d\omega. \tag{19}$$

This formula is equivalent to (5). □

We use the most widely accepted definition of the scattering amplitude $f(n, \nu, k)$. This scattering amplitude does not satisfy the symmetry relation $f(n, \nu, k) = f(\nu, n, k)$ in general. In Lax—Phillips [1], another definition of the scattering amplitude is used. Their scattering amplitude satisfies the symmetry relation but does not satisfy the reciprocity relation (4), it is analytic in the lower half-plane $\text{Im } k \leqslant 0$ of the complex k-plane, while our scattering amplitude is analytic in the upper half-plane $\text{Im } k \geqslant 0$.

4.2. The scattering amplitude is often called the radiation pattern by engineers. It is a function of five variables $n, \nu, k \in S^2 \times S^2 \times R_+$, where $R_+ = (0, \infty)$. Let us study its properties as a function of k. We assume as usual in this chapter that the scatterer is a bounded body with a smooth boundary.

THEOREM 2. *The scattering amplitude is a restriction on* R_+ *of a function* $f(n, \nu, k)$ *which is analytic in the half-plane* $\text{Im } k > 0$, *and meromorphic on the whole complex k-plane.*

Proof. Because of (2) it is sufficient to prove that $\partial u(s, \nu, k)/\partial N$ has these properties as a function of k, or that $G(x, y, k)$ has these properties as a function of k (see formula (3.8), for example). From formula (3.4) it follows that G has these properties iff $\mu = \partial G/\partial N_s$ does. From formula (3.5) and Lemma 2 below it follows that μ does have these properties. □

LEMMA 2. *Let* $B(k)$ *be a bounded linear mapping of a Banach space X into a Banach space Y which is analytic in k in a connected domain* Δ *of the complex k-plane.*

Assume that $B(k) \in$ *Fred* (X, Y), *that is* $R(B(k)) = \overline{R(B(k))}$, $r =$ *dim* $N(B(k)) =$ *dim* $N(B^*(k)) < \infty$. *Then either* $B^{-1}(k)$ *does not exist at any point* $k \in \Delta$, *or* $B^{-1}(k)$ *does exist at all points of* Δ *except for a discrete set* k_j. *In a neighborhood of any point* $k \neq k_j$, $k \in \Delta$ *the operator* $B^{-1}(k)$ *is analytic. The points* k_j *are poles of* $B^{-1}(k)$ *and the Laurent coefficients* B_p, $-m_j \leqslant p \leqslant -1$ *of* $B^{-1}(k) = \sum_{p=-m_j}^{\infty} B_p(k - k_j)^p$ *around* k_j *are finite rank operators. This conclusion remains valid if* $B(k)$ *is meromorphic in k provided that the coefficient* b_0 *in the expansion* $B(k) = (k - k_0)^{-m_0} \sum_{p=0}^{\infty} b_p(k - k_0)^p$ *around any pole* k_0 *is a Fredholm operator,* $b_0 \in$ *Fred* (X, Y).

Remark 1. The operator $B(k) = I + A(k)$ in Equation (3.5) is analytic in k on the whole complex plane (which is Δ in this case). Since $A(k)$ is compact in $C(\Gamma)$ the operator $B(k) \in$ Fred $(C(\Gamma))$, where Fred $(X) \equiv$ Fred (X, X). Finally, $B^{-1}(k)$ does exist for Im $k > 0$. Thus, Lemma 2 is applicable to Equation (3.5).

To prove Lemma 2 we need Lemma 3.

LEMMA 3. *Let $B \in$ Fred (X, Y). Then there exists a finite rank operator K such that $B + K$ is an isomorphism between X and Y.*

Proof of Lemma 3. Let $\{\phi_j\}$, $1 \leqslant j \leqslant r$ be a basis of $N(B)$, $\{\tilde{\phi}_j\} \in X^*$, $1 \leqslant j \leqslant r$, be a system of functionals which is biorthogonal to $\{\phi_j\}$, $(\phi_i, \tilde{\phi}_j) = \delta_{ij} = \left\{\begin{smallmatrix}1 & i=j\\ 0 & i\neq j\end{smallmatrix}\right.$, $\{\psi_j\}$, $1 \leqslant j \leqslant r$, be a basis of $N(B^*)$, and $\{\tilde{\psi}_j\}$, $1 \leqslant j \leqslant r$, be a system of elements of Y which is biorthogonal to $\{\psi_j\}$, B: $X \to Y$, B^*: $Y^* \to X^*$, X^* is the adjoint space for X, that is the space of linear bounded functionals on X, and (ϕ, ψ) is the value of the functional ψ at the element ϕ. Define

$$Kf = \sum_{j=1}^{r} (f, \tilde{\phi}_j)\tilde{\psi}_j. \tag{20}$$

Let us prove that the operator $B + K$ is an isomorphism of X onto Y. The operator $B + K$ is injective. Indeed, if $(B + K)f = 0$ then $Bf = -Kf = -\sum_{j=1}^{r} (f, \tilde{\phi}_j)\tilde{\psi}_j$. Necessary conditions for the solvability of this equation are

$$(Kf, \psi_j) = 0, 1 \leqslant j \leqslant r. \tag{21}$$

Since $\{\phi_j\}$ and $\{\tilde{\phi}_j\}$ are biorthogonal systems (21) implies that

$$(f, \tilde{\phi}_j) = 0, \quad 1 \leqslant j \leqslant r \quad \text{and} \quad Bf = 0. \tag{22}$$

If $Bf = 0$ then $f = \sum_{j=1}^{r} c_j\phi_j$ and $(f, \tilde{\phi}_j) = c_j = 0$, where the biorthogonality of $\{\phi_j\}$ and $\{\tilde{\phi}_j\}$ is used. Thus $f = 0$ and $B + K$ is injective. Since $R(B)$ is closed and K is a finite rank operator one concludes that $R(B + K)$ is closed. Indeed, assume $Bu_n + Ku_n \to f$. Then, since $(Bu_n + Ku_n, \psi_j) = (u_n, \tilde{\phi}_j) \to (f, \psi_j)$ as $n \to \infty$, one sees that Ku_n converges. Therefore Bu_n converges. Since $R(B)$ is closed, $\lim Bu_n = Bu$. One can choose u to be the limit of u_n because $\dim N(B) < \infty$. To do this, decompose X into a direct sum $X = N(B) \oplus X_r$ (here we use the basic assumption $\dim N(B) < \infty$), and take $u_n \in X_r$. The operator B_r which is the restriction of B onto X_r is injective, $B_r u_n = Bu_n$. Consider B_r: $X_r \to R(B) = R(B_r)$. Since $R(B)$ is closed, the Banach theorem about inverse operators says that B^{-1}: $R(B) \to X_r$

is continuous. Therefore $Bu_n \to Bu$ implies $u_n \to u$. Furthermore, $Ku_n \to Ku$ and $Bu = f - Ku$. Thus $R(B + K)$ is closed. To prove that $B + K$ is surjective take any $y \in Y$ and consider the equation

$$Bu = y - Ku = y - \sum_{j=1}^{r} (u, \tilde{\phi}_j)\tilde{\psi}_j. \tag{23}$$

Since $R(B)$ is closed this equation is solvable iff $(y - Ku, \psi_m) = 0$, $1 \leqslant m \leqslant r$. Since $(\tilde{\psi}_j, \psi_m) = \delta_{jm}$ one obtains

$$0 = (y, \psi_m) - (u, \tilde{\phi}_m), 1 \leqslant m \leqslant r. \tag{24}$$

Let

$$u = u_r + \sum_{j=1}^{r} c_j\phi_j, \tag{25}$$

where $u_r \in X_r$, c_j are constants and $\{\phi_j\}$, $1 \leqslant j \leqslant r$ is the basis of $N(B)$. Notice that $Bu = Bu_r$. Conditions (24) can be written as

$$(y, \psi_m) - (u_r, \tilde{\phi}_m) = \sum_{j=1}^{r} c_j(\phi_j, \tilde{\phi}_m) = c_m, \quad 1 \leqslant m \leqslant r. \tag{26}$$

Therefore Equation (23) can be written as a solvable equation for u_r:

$$\begin{aligned} Bu_r &= y - \sum_{j=1}^{r} (u_r, \tilde{\phi}_j)\tilde{\psi}_j - \sum_{j=1}^{r} \left(\sum_{m=1}^{r} [(y, \psi_m) - (u_r, \tilde{\phi}_m)]\phi_m, \tilde{\phi}_j \right) \tilde{\psi}_j \\ &= y - \sum_{j=1}^{r} (u_r, \tilde{\phi}_j)\tilde{\psi}_j - \sum_{j=1}^{r} [(y, \psi_j) - (u_r, \tilde{\phi}_j)]\tilde{\psi}_j = y - \sum_{j=1}^{r} (y, \psi_j)\tilde{\psi}_j. \end{aligned} \tag{27}$$

If u_r is the solution to (27) (there is only one solution to (27) in X_r) then (25) solves (23) provided that $c_j = (y, \psi_j) - (u_r, \tilde{\phi}_j)$. Note that our proof of surjectivity does not use the closedness of $R(B + K)$ and gives an independent proof of this closedness. □

Proof of Lemma 2. Suppose that $B^{-1}(k_0)$ exists, $k_0 \in \Delta$. Then $B^{-1}(k)$ exists and is analytic in k in some neighborbood $|k - k_0| < \delta$. Indeed $B(k) = B(k_0)\,[I + B^{-1}(k_0)\,(B(k) - B(k_0))]$. Thus

$$\begin{aligned} &B^{-1}(k) = \sum_{j=0}^{\infty} \{B^{-1}(k_0)\,[B(k) - B(k_0)]\}^j\, B^{-1}(k_0) \\ &\text{if} \quad \|B^{-1}(k_0)\|\,\|B(k) - B(k_0)\| < 1. \end{aligned} \tag{27'}$$

If $B(k)$ is analytic then formula (27) shows that $B^{-1}(k)$ is analytic in a neighborhood $|k - k_0| < \delta$, where $\delta > 0$ is sufficiently small so that the inequality in (27′) holds. Suppose now that $B^{-1}(k_0)$ does not exist. Let us show that in this case either k_0 is a pole of $B^{-1}(k)$ and the Laurent coefficients of $B^{-1}(k)$ around k_0 are finite rank operators, or $B(k)$ is not invertible in a neighborhood of k_0. Since $B(k_0) \in$ Fred (X, Y) the operator $B(k_0) + K$ is invertible by Lemma 2, where K is a finite rank operator which does not depend on k (it depends on k_0). Thus

$$B^{-1}(k) = (B(k_0) + K + B(k) - B(k_0) - K)^{-1} = (I - \Gamma K)^{-1}\Gamma \tag{28}$$

where

$$\Gamma = \Gamma(k) = (B(k_0) + K + B(k) - B(k_0))^{-1}. \tag{29}$$

Since $B(k_0) + K$ is an isomorphism and $\|B(k) - B(k_0)\| \to 0$ as $|k - k_0| \to 0$, the operator $\Gamma(k)$ is analytic in a neighborhood $|k - k_0| < \delta$ if $\delta > 0$ is sufficiently small. We will prove that the operator $(I - \Gamma K)^{-1}$ is either analytic in $0 < |k - k_0| < \delta$ and has a pole at $k = k_0$, or does not exist in the neighborhood $|k - k_0| < \delta$. Consider the equation

$$(I - \Gamma K)u = f, \quad u = f + \sum_{j=1}^{r} (u, \tilde{\phi}_j)\Gamma\tilde{\psi}_j \tag{30}$$

where $r = \dim N(B(k_0))$ and we use the notations introduced in the proof of Lemma 3, $B = B(k_0)$. From (30) it follows that

$$u = f + \sum_{j=1}^{r} c_j\eta_j, \quad \eta_j = \Gamma(k)\tilde{\psi}_j \tag{31}$$

$$c_m = f_m + \sum_{j=1}^{r} c_j\, a_{mj}, \quad a_{mj} \equiv (\eta_j(k), \tilde{\phi}_m), \; f_m = (f, \tilde{\phi}_m), \; 1 \leqslant m \leqslant r. \tag{32}$$

Equations (32) and (30) are equivalent. The solution to Equation (30) is given by formula (31) iff c_m, $1 \leqslant m \leqslant r$, solve (32). The solution to (32) can be computed by the formula

$$c_m = \frac{\Delta_m}{\Delta_0}, \quad 1 \leqslant m \leqslant r, \quad \Delta_0 = \det(\delta_{mj} - a_{mk}(k)) \tag{33}$$

where Δ_m is the determinant of the matrix $\delta_{mj} - a_{mj}(k)$ in which the m-th column is substituted by the column vector $\{f_m\}$, $1 \leqslant m \leqslant r$. Since $\Gamma(k)$ is analytic in the neighborhood $|k - k_0| < \delta$, the functions Δ_m, $0 \leqslant m \leqslant r$, are analytic in this neighborhood. Therefore either $\Delta_0(k) \equiv 0$ in the disk C_δ: $|k - k_0| < \delta$, or it has only isolated zeros of finite order.

In the first case the system (32) is not solvable at any $k \in C_\delta$ and some $\{f_m\}$. In the second case the system (32) is solvable for all k except $k = k_j$, where the set $\{k_j\}$ is the discrete set of zeros of $\Delta_0(k)$. Taking $\delta > 0$ sufficiently small one can assume that in the second case all the points k_j lie outside of the region $0 < |k - k_0| \leqslant \delta$. Since equation $(I - \Gamma(k_0)K)u = 0$ has nontrivial solutions, the function $\Delta_0(k)$ has a zero at $k = k_0$. Let m_0 be the order of this zero. Then it follows from (33) that the coefficients $c_m = c_m(k)$ have a pole at $k = k_0$. The order of this pole is not more than m_0 and is equal to m_0 for some choices of f_m. From formulas (32) and (33) it is seen that the Laurent coefficients of the operator $(I - \Gamma(k)K)^{-1}$ are finite rank operators. The last statement of Lemma 2 follows from the formula $B^{-1}(k) = (k - k_0)^{m_0} \{\sum_{p=0}^{\infty} b_p(k - k_0)^p\}^{-1}$ and the already proven part of Lemma 2. □

Remark 3. In Reed, Simon [1b], p. 108 and p. 369, problem 34, it is stated that the assumption, that the coefficients A_p, $-m_0 \leqslant p \leqslant -1$, of the operator $A(k) = \sum_{p=-m_0}^{\infty} A_p(k - k_0)^p$ are finite-rank, is crucial for the validity of Theorem XIII.13 in their book. Lemma 2 shows that the Fredholm property of A_{-m_0}, rather than the assumption that A_{-m_0} is finite-rank, is important.

Remark 4. The set of poles of the scattering amplitude is the same as the set of poles of the resolvent kernel G. These poles are called resonances in the physics literature and natural frequencies (sometimes scattering frequencies) in the engineering literature. The location and properties of these poles (resonances) are of interest theoretically and practically (see Appendix 7).

The fact that the resonances lie in the half-plane $\text{Im } k < 0$ does not depend on the geometry of the surface Γ. On the other hand most of what is known about resonances is proved for convex or star-like scatterers. The scatterer D is called star-like (star shaped) if there exists a point $0 \in D$ such that any point of Γ can be seen from 0. The set of convex scatterers is a proper subset of the set of star-like scatterers.

For convex and star-like obstacles (and, more generally, for non-trapping obstacles) it is proved that:

there is a strip $-a < \operatorname{Im} k \leqslant 0$ which is free from complex poles (34)

and

$$|Im\, k_j| \to \infty \quad \text{as} \quad j \to \infty. \tag{35}$$

Because of (3) the poles are symmetrical with respect to the imaginary axis: if $k_j = a_j - ib_j$, $b_j > 0$ is a pole of multiplicity m_j then $-a_j - ib_j$ is a pole of the same multiplicity. A non-trapping obstacle is such an obstacle that singularities of the solution to the nonstationary problem

$$E_{tt} = \nabla^2 E \text{ in } \Omega, \quad t > 0, \tag{36}$$

$$E(t, x, y) = 0 \text{ on } \Gamma, \quad t > 0, \quad y \in \Omega \tag{37}$$

$$E(0, x, y) = 0, \quad E_t(0, x, y) = \delta(x - y) \tag{38}$$

propagate to infinity as $t \to +\infty$. This condition is difficult to check. There are some alternative conditions which are formulated in terms of the stationary problem, but these are also difficult to check (see, e.g., Melrose [1] and Morawetz, Ralston, Strauss [1]).

Condition (35) implies that in any half plane $\operatorname{Im} k > -b$ there is a finite number of complex poles. Some properties of these poles are given in Appendix 7.

4.3. Let us study the properties of the scattering amplitude as a function of ν. The basic result is the following.

THEOREM 3. *If $f(n, \nu, k)$ is known for a fixed $n = n_0$, fixed $k = k_0$ and all ν in a solid angle $\tilde{S}^2$ then $f(n_0, \nu, k_0)$ is uniquely defined on all of the unit sphere S^2.*

Proof. From (2′) and the compactness of Γ it follows that $f(n, \nu, k_0)$ is an entire function $F(\nu_1, \nu_2, \nu_3)$ of the three complex variables ν_1, ν_2, ν_3 where ν_j, $1 \leqslant j \leqslant 3$, are the Cartesian components of vector ν. This function is known on the manifold which can be described in an appropriate coordinate system as $-a \leqslant \nu_1 \leqslant a$, $-a \leqslant \nu_2 \leqslant a$, $\nu_3 = \sqrt{1 - \nu_1^2 - \nu_2^2}$, where $a > 0$ is a small number. The function $F(\nu_1, \nu_2, \sqrt{1 - \nu^2 - \nu_2^2}) \equiv \phi(\nu_1, \nu_2)$ is analytic in the domain $|\nu_1|^2 + |\nu_2|^2 < 1$ and is known in a neighborhood of the origin. By the uniqueness of analytic continuation, this function is uniquely defined in the region $|\nu_1|^2 + |\nu_2|^2 \leqslant 1$. Therefore $F(\nu_1, \nu_2, \nu_3)$ is uniquely defined on $S^2 = \{\nu: \nu_1^2 + \nu_2^2 + \nu_3^2 = 1\}$. □

5. THE S-MATRIX AND WAVE OPERATORS

5.1. Let F_0 be the usual Fourier transform, F_0: $L^2(R^3_x) \to L^2(R^3_k)$, $F_0 f = (2\pi)^{-3/2} \int_{R^3} f(x) \exp(ikn \cdot x)\, dx$, F_+ be the Fourier transform corresponding to formulas (3.13), (3.14), F_+: $L^2(\Omega) \to L^2(R^3_k)$, and F_- be the Fourier transform corresponding to the solutions of the problem

$$(\nabla^2 + k^2)u_-(x, n, k) = 0 \text{ in } \Omega, \quad k > 0 \tag{1}$$

$$u_- = 0 \text{ on } \Gamma \tag{2}$$

$$u_- = \exp(-ikn \cdot x) + v_- \tag{3}$$

$$\int_{S_r} \left| \frac{\partial v}{\partial r} + ikv_- \right|^2 ds \to 0, \quad r \to \infty. \tag{4}$$

The functions u_- are "incoming" solutions to the scattering problem, while $u_+ = u$, solutions to (3.1)—(3.2), are "outgoing" solutions. One can prove formulas (3.13)—(3.14) with $u_-(x, k)$ in place of $u(x, k)$ in the same way as in the proof of Theorem 1 in Section I.3.

Define the wave operators by the formula

$$\Omega_\pm = F_0^* F_\pm \tag{5}$$

and the S operator by the formula

$$S = \Omega_+^* \Omega_- = F_+^* F_-. \tag{6}$$

Since F_0, F_+ and F_- are unitary, the operators $\Omega_\pm$ and S are unitary, $\Omega_\pm$: $L^2(\Omega) \to L^2(R^3_x)$, S: $L^2(\Omega) \to L^2(\Omega)$.

It follows from (3.15) that

$$F_\pm H F_\pm^* = k^2 = F_0 H_0 F_0^* \tag{7}$$

where H is the Dirichlet Laplacian in $L^2(\Omega)$, and H_0 is the Laplacian in $L^2(R^3)$. Therefore

$$F_0^* F_\pm H = H_0 F_0^* F_\pm, \tag{8}$$

or

$$\Omega_\pm H = H_0 \Omega_\pm. \tag{9}$$

This is the intertwining relation for the wave operators. In the same way one can prove that

$$\Omega_\pm f(H) = f(H_0) \Omega_\pm \tag{10}$$

where $f(H)$ is a function of the selfadjoint operator H.

The physical meaning of the S operator is clear from the formula

$$u_+ = Su_-. \tag{11}$$

Since $u_\pm$ does not belong to $L^2(\Omega)$ one should explain the mathematical meaning of (11), but from the physical point of view the S operator transforms incoming solutions to the scattering problem into outgoing solutions. The wave operators relate solutions in the free space, $u_0 = \exp(-ikn \cdot x)$, and the solutions $u_\pm$ to the scattering problem

$$\Omega_+ u_+ = u_0, \quad \Omega_- u_- = u_0. \tag{12}$$

Formula (11) follows from (12) and (6): apply Ω_+^* to the first formula (12) and use the unitarity of Ω_+, $\Omega_+^*\Omega_+ = I$. Therefore it is sufficient to explain the mathematical meaning of (12). Let $h(k) \in C_0^\infty(R_k^3)$, F_+ be defined by formula (3.13). The first equality (12) means that

$$\Omega_+ F_+^* h = F_0^* h, \quad \forall h \in C_0^\infty(R_k^3). \tag{13}$$

This formula is equivalent to (12) and has a clear mathematical meaning since $F_+^* h \in L^2(\Omega)$, $F_0^* h \in L^2(R_x^3)$, Ω_+: $L^2(\Omega) \to L^2(R^3)$. A similar argument holds for the second equation (12).

5.2. We now turn to the definition of the S-matrix. Physicists often define the S-matrix $S(k)$ as follows. For a fixed $k > 0$ the functions $u_\pm = u_\pm(x, -n, k)$ have asymptotics

$$u_\pm \sim \exp(ikn \cdot x) + f_\pm(n, \nu, k)\gamma^\pm, \quad \gamma^\pm = \frac{\exp(\pm ikr)}{r}, \quad r = |x| \to \infty, \quad xr^{-1} = \nu. \tag{14}$$

Formula (4.16) shows that (14) can be written for $r \to \infty$ as

$$u_+ \sim \frac{2\pi i}{k}\delta(n+\nu)\gamma^- - \left[\frac{2\pi i}{k}\delta(n-\nu) - f_+(n, \nu, k)\right]\gamma^\pm \equiv \psi_-\gamma^- - \psi_+\gamma^+. \tag{15}$$

Define the kernel of the scattering matrix by the formula

$$\psi_+(n) = \int_{S^2} S(n, -n', k)\, \psi_-(n')\, \mathrm{d}n'. \tag{16}$$

It follows from (15) and (16) that

$$\frac{2\pi i}{k}\,\delta(n-\nu) - f_+(n,\nu,k) = \frac{2\pi i}{k}\,S(n,\nu,k). \tag{17}$$

Thus

$$S(n,\nu,k) = \delta(n-\nu) - \frac{k}{2\pi i}\,f_+(n,\nu,k). \tag{18}$$

The S operator (6) is an operator in $L^2(\Omega)$. The S-matrix $S(k)$ is an operator in $L^2(S^2)$ with the kernel (18), i.e.

$$S(k) = I - \frac{k}{2\pi i}\,F \tag{19}$$

where I is the identity, and F is an operator in $L^2(S^2)$ with the kernel $f_+(n, \nu, k)$. Since $f_+(n, \nu, k)$ is a smooth function of n and ν, the operator F belong to the trace-class operators. Note that $f_+(n, \nu, k) = f(n, \nu, k)$ where $f(n, \nu, k)$ is defined by (4.1).

6. INEQUALITIES FOR SOLUTIONS TO HELMHOLTZ'S EQUATION FOR LARGE FREQUENCIES

These inequalities were proved in Alber [1] for $D \subset R^2$. His proof holds for $D \subset R^3$ and we follow his arguments closely.

Let $D \subset R^3$ be a bounded domain with a smooth boundary Γ. Assume that

$$x \cdot N \geqslant \beta > 0 \text{ on } \Gamma \tag{1}$$

where N is the unit normal to Γ pointing into Ω, $\Omega = R^3 \setminus D$, β is a constant, and the origin is inside D. Let

$$\Omega_R = \{x : |x| \geqslant R\}, \quad \|u\| = \|u\|_{L^2(\Omega)}, \quad \|u\|_R = \|u\|_{L^2(\Omega_R)},$$
$$\|u\|_{\partial\Omega} = \|u\|_{L^2(\partial\Omega)}.$$

THEOREM 1. *If u solves the problem*

$$(\nabla^2 + k^2)u = f \text{ in } \Omega, \quad k > 0, \tag{2}$$

$$u = \mu \text{ on } \Gamma, \tag{3}$$

$$r\left(\frac{\partial u}{\partial r} - iku\right) \to 0, \quad r \to \infty, \tag{4}$$

and

$$\|rf\| < \infty \tag{5}$$

Then

$$\|\nabla u\|_R^2 + k^2 \|u\|_R^2 \leqslant CR^{1+\alpha}\{\|rf\|^2 + \|\partial_t \mu\|_{\partial\Omega}^2 + (1+k^2)\|\mu\|_{\partial\Omega}^2\} \tag{6}$$

and

$$\|r^{-1/2-\varepsilon}\nabla u\|^2 + k^2\|r^{-1/2-\varepsilon}u\|^2 \leqslant C(1+\varepsilon^{-1})\{\|rf\|^2 + \|\partial_t\mu\|_{\partial\Omega}^2 + \\ + (1+k^2)\|\mu\|_{\partial\Omega}^2\} \tag{7}$$

where $\partial_t\mu$ *is the tangential derivative on* $\partial\Omega$, *c denotes various constants which do not depend on k, R and* ε.

In order to prove inequalities (6)—(7) one needs two lemmas.

LEMMA 1. *If* (2) *holds then*

$$\int_{\partial\Omega_R} [2\,\mathrm{Re}\, ru_r\bar{u}_N - (|\nabla u|^2 - k^2|u|^2)N \cdot x + 2u\bar{u}_N]\,\mathrm{d}s \\ = \int_{\Omega_R} [|\nabla u|^2 + k^2|u|^2 + 2u\bar{f} + 2\,\mathrm{Re}(u_r r\bar{f})]\,\mathrm{d}x. \tag{7}$$

Here u_r *and* u_N *are the radial and normal derivatives of u, and the bars denote complex conjugation.*

Proof. Notice that

$$\nabla(u_r) = \nabla(\nabla u \cdot xr^{-1}) = (\nabla u)_r + r^{-1}\nabla u - xr^{-2}u_r. \tag{8}$$

The Green formula and (8) imply that

$$\int_{\partial\Omega_R} r(u_r\bar{u}_N + \bar{u}_r u_N)\,\mathrm{d}s = \int_{\Omega_R} r(u_r\Delta\bar{u} + \bar{u}_r\Delta u)\,\mathrm{d}x + \\ + \int_{\Omega_R} \{r[(\nabla u)_r \cdot \nabla\bar{u} + (\nabla\bar{u})_r \cdot \nabla u] + 2|\nabla u|^2\}\,\mathrm{d}x. \tag{9}$$

This formula, the identity

$$\int_{\partial\Omega_R} N \cdot x(|\nabla u|^2 - k^2|u|^2)\, \mathrm{d}s = \int_{\Omega_R} \{r[(|\nabla u|^2)_r - k^2(|u|^2)_r] + 3(|\nabla u|^2 - k^2|u|^2)\}\, \mathrm{d}x, \tag{10}$$

and Equation (2) imply:

$$\int_{\partial\Omega_R} [r(u_r\bar{u}_N + \bar{u}_r u_N) - N \cdot x(|\nabla u|^2 - k^2|u|^2)]\, \mathrm{d}s$$

$$= \int_{\Omega_R} [2\,\mathrm{Re}(u_r r\bar{f}) + 3\,k^2|u|^2 - |\nabla u|^2]\, \mathrm{d}x = \int_{\Omega_R} [2\,\mathrm{Re}(ru_r\bar{f}) + |\nabla u|^2 + k^2|u|^2]\, \mathrm{d}x + 2\int_{\Omega_R} [k^2|u|^2 - |\nabla u|^2]\, \mathrm{d}x = \int_{\Omega_R} [|\nabla u|^2 + k^2|u|^2 + 2u\bar{f} + 2\,\mathrm{Re}(r\bar{f}u_r)]\, \mathrm{d}x - 2\int_{\partial\Omega_R} u\bar{u}_N\, \mathrm{d}s. \tag{11}$$

This formula is equivalent to (7). □

LEMMA 2 (*Morawetz-Ludwig* [1]). *If* (1)–(5) *holds then*

$$\frac{\beta}{4}\int_\Gamma |u_N|^2\, \mathrm{d}s + \int_\Omega [|\nabla u|^2 - |u_r|^2 + \tfrac{1}{2}|u_r - iku + r^{-1}u|^2]\, \mathrm{d}x \leqslant c\|rf\|^2 + c\beta^{-1}\int_\Gamma \{(s\cdot t)^2 + |s\cdot N|\,|u_t|^2 + k^2|u|^2\}\, \mathrm{d}s. \tag{12}$$

where $k \geqslant \delta > 0$, $c = \text{const} > 0$, β *is the constant from* (1), *the origin is inside D and a ball* $|x| \leqslant a$ *belongs to D for some positive* a, *the constant c does not depend on k and* β *but depends on* δ *and* a.

Proof of Theorem 1. Let

$$y(R) = \int_{\Omega_R} (|u_r|^2 + k^2|u|^2)\,\mathrm{d}x, \quad y' = \int_{S_R \cap \Omega} (|u_r|^2 + k^2|u|^2)\,\mathrm{d}s. \tag{13}$$

Let us write formula (7) as

$$\begin{aligned} F = \int_{S_R \cap \Omega} [r(|u_r|^2 + k^2|u|^2) + 2\,\mathrm{Re}\, u\bar{u}_r]\,\mathrm{d}s = \int_{\Omega_R} (|u_r|^2 + k^2|u|^2)\,\mathrm{d}x + \\ + \int_{\Omega_R} [|\nabla u|^2 - |u_r|^2 + 2\,\mathrm{Re}(u\bar{f}) + 2\,\mathrm{Re}(ru_r\bar{f})]\,\mathrm{d}x + \\ + \int_{S_R \cap \Omega} r(|\nabla u|^2 - |u_r|^2)\,\mathrm{d}s - \int_{\Gamma \cap \partial\Omega_R} [2\,\mathrm{Re}(ru_r\bar{u}_N) - \\ - (|\nabla u|^2 - k^2|u|^2)N \cdot x + 2\,\mathrm{Re}\, u\bar{u}_N]\,\mathrm{d}s. \end{aligned} \tag{14}$$

From (14) and the elementary inequality $ab \leqslant \varepsilon a^2 + (4\varepsilon)^{-1}b^2$, $\varepsilon > 0$, one obtains

$$\begin{aligned} F \leqslant y(R) + \int_{\Omega_R} [|\nabla u|^2 - |u_r|^2 + \varepsilon(|u|^2 + |u_r|^2) + \\ +(4\varepsilon)^{-1}(|f|^2 + |rf|^2)]\,\mathrm{d}x + F_1 \end{aligned} \tag{15}$$

where F_1 is the sum of the surface integrals in (14). Therefore

$$F \leqslant (1+\varepsilon)y(R) + J_1(R) + J_2(r), \tag{16}$$

where

$$J_2 = \int_{\Omega_R} [|\nabla u|^2 - |u_r|^2 + (4\varepsilon)^{-1}(1 + r^2)|f|^2]\,\mathrm{d}x. \tag{17}$$

Let

$$\psi(r) = \max\left(\frac{r}{2}, r - 1\right), \quad \eta(r) = \begin{cases} 2r^{-3-2\varepsilon} & 0 < r \leqslant 2 \\ (r-1)^{-2-\varepsilon}\, 2^{-2-2\varepsilon} & 2 \leqslant r < \infty \end{cases}. \tag{18}$$

Then

$$(\psi\eta)' = -(1+\varepsilon)\eta. \tag{19}$$

Notice that

$$\psi(R)y'(R) \leqslant (R-1)y' = Ry' - y' \leqslant R\int_{S_R\cap\Omega}(|u_r|^2 + k^2|u|^2)\,ds - \\ - \int_{S_R\cap\Omega}(|u_r|^2 + k^2|u|^2)\,ds \leqslant \int_{S_R\cap\Omega}[r(|u_r|^2 + k^2|u|^2) + 2\,\mathrm{Re}(u\bar{u}_r)]\,ds \tag{20}$$

provided that

$$R \geqslant 2, \quad k \geqslant 1. \tag{21}$$

If

$$R \leqslant 2, \quad k \geqslant 1, \tag{22}$$

then $\psi(R) = R/2$,

$$\psi(R)y'(R) = Ry' - \frac{R}{2}y' \leqslant \int_{S_R\cap\Omega}[r(|u_r|^2 + k^2|u|^2) + \\ + 2\,\mathrm{Re}(u\bar{u}_r)]\,ds. \tag{23}$$

From (20), (23), and (16) it follows that

$$\psi(R)y'(R) \leqslant F \leqslant (1+\varepsilon)y(R) + J_1(R) + J_2(R). \tag{24}$$

Multiply (24) by η and use (19) to obtain

$$(\eta\psi y)' \leqslant \eta(J_1 + J_2). \tag{25}$$

Therefore

$$y(R) \leqslant \eta^{-1}(R)\psi^{-1}(R)\int_a^R \eta(r)\,(J_1(r) + J_2(r))\,ds. \tag{26}$$

Here $a > 0$ is chosen so that $y(a) = 0$. This choice is possible because $S \cap \Omega = \emptyset$ for sufficiently small $a > 0$. One has

$$\int_a^R \eta J_2 \, dr \leqslant \sup_{a \leqslant r \leqslant R} |J_2| \int_a^R \eta \, dr \leqslant c \sup_{r \geqslant a} |J_2(r)| \leqslant$$

$$\leqslant c\varepsilon^{-1} \int_\Omega (1 + r^2)|f|^2 \, dx + c \int_\Gamma |u_t|^2 \, ds + ck^2 \int_\Gamma |u|^2, \tag{27}$$

where inequality (12) was used.

Furthermore, using inequality (12) again, one obtains

$$\left| \int_a^R \eta J_1 \, dr \right| \leqslant \int_0^R dr \, \eta(r) \int_{S_r \cap \Omega} r(|\nabla u|^2 - |u_r|^2) \, ds +$$

$$+ \int_a^R \eta \, dr \sup_{a \leqslant r \leqslant R} \int_{\Gamma \cap S_r} [2r|u_r u_N| + |\nabla u|^2 + k^2 |u|^2 +$$

$$+ 2|uu_N|] \, ds \leqslant c\|rf\|^2 + c \int_\Gamma |u_t|^2 \, ds + ck^2 \int_\Gamma |u|^2 \, ds. \tag{28}$$

The constants c in (27) and (28) depend on β but not on $k \geqslant \delta > 0$ and R. Notice that

$$\eta^{-1}(R)\, \psi^{-1}(R) \leqslant cR^{1+\varepsilon}, \quad R \geqslant 2. \tag{29}$$

Inequalities (26)—(29) imply (6). Inequality (7) follows from (26) and (29). □

Remark 1. If $D = B_a$ is a ball of radius a, then the solution to problem (2)—(4) with $f = 0$ satisfies the following estimate (a maximum principle for solutions to Helmholtz's equation).

$$\max_{r \geqslant a} \int_{S_r} |u(r, \omega)|^2 \, ds_r = \int_{S_a} |u(a, \omega)|^2 \, ds_a, \tag{30}$$

where $ds_r = r^2\, d\omega$, and $d\omega$ is the element of the area of the unit sphere S^2.

Proof. If $f = 0$ then

$$u = \sum_{n=0}^{\infty} \sum_{m=-n}^{n} c_{nm} h_n^{(1)}(kr) Y_{nm}(\omega),$$
$$\int_{S_2} |u(r, \omega)|^2\, d\omega = \sum_{n=0}^{\infty} \sum_{m=-n}^{n} |c_{nm}|^2 |h_n^{(1)}(kr)|^2 \tag{31}$$

where Y_{nm} are the spherical harmonics normalized in $L^2(S^2)$, and $h_n^{(1)}(kr)$ are the spherical Hankel functions. The following properties of the spherical Hankel functions (see Gradshteyn-Ryzhik [1], 8.478)

$$r^2 |h_n^{(1)}(kr)|^2 \leqslant a^2 |h_n^{(1)}(ka)|^2 \quad \text{if} \quad r \geqslant a \quad \text{and } n \geqslant 0 \tag{32}$$

and (31) yields (30). □

7. REPRESENTATION OF SOLUTIONS TO HELMHOLTZ'S EQUATION

There is an extensive literature on the equivalence of the integral equations of potential theory and the corresponding boundary value problems. The results of this section show, for example, that the solution to the interior Dirichlet problem can be obtained in the form of a single layer potential as easily as in the standard form of a double layer potential.

7.1. The aim of this section is to give conditions necessary and sufficient for a potential of double layer v to be equal to a potential of single layer w, where v and w are defined by the formulas (1.1) and (1.4) respectively. The basic results can be formulated as the following theorems.

THEOREM 1. *For any $w(\sigma)$ $[v(\mu)]$ there exists a $v(\mu)$ $[w(\sigma)]$ such that $w(\sigma) \equiv v(\mu)$ in the interior domain D. The $v(\mu)$ $[w(\sigma)]$ is uniquely determined.*

THEOREM 2. *Suppose that $w(\sigma)$ is given. Then a necessary and sufficient condition for this $w(\sigma)$ to be identically equal to a $v(\mu)$ in Ω is*

$$(w, \bar{h}_j) = \int_\Gamma w h_j \, dt = 0, \quad 1 \leqslant j \leqslant r', \quad (I + A)h_j = 0, \tag{1}$$

where the system $\{h_j\}$ forms a basis of $N(I + A)$, and A is defined by the formula (1.4).

Suppose that $v(\mu)$ is given. Then a necessary and sufficient condition for this $v(\mu)$ to be identically equal to a $w(\sigma)$ in Ω is

$$(\mu, \bar{\sigma}_j) = 0, \quad 1 \leqslant j \leqslant r, \quad A\sigma_j - \sigma_j = 0, \tag{2}$$

where the system $\{\sigma_j\}$ forms a basis of $N(A - I)$, and A is defined by formula (1.4).

Condition (1) *can be written as a condition on σ:*

$$\int_\Gamma \sigma Q h_j \, dt = 0, \quad 1 \leqslant j \leqslant r', \quad (I + A)h_j = 0, \tag{1'}$$

where Q is defined in (1.8), *or*

$$\int_\Gamma \sigma \mu_j \, dt = 0, \quad 1 \leqslant j \leqslant r', \tag{1''}$$

where μ_j are defined in Lemma 1 *below.*

COROLLARY 1. *If the problem*

$$(\nabla^2 + k^2)u = 0 \text{ in } D, \quad k > 0, \quad u_+ = f \tag{3}$$

is solvable then the solution can be represented as $u = w(\sigma)$ and also as $u = v(\mu)$. The same is true for the problem

$$(\nabla^2 + k^2)u = 0 \text{ in } D, \quad k > 0, \quad \frac{\partial u}{\partial N_+} = f. \tag{4}$$

7.2. In this section we give some lemmas which are used in the proof of the Theorems 1, 2.

LEMMA 1. *Let problem* (4_0), *i.e. problem* (4) *with* $f = 0$, *have* r' *linearly independent solutions* u_j, $1 \leqslant j \leqslant r'$. *Then the equation*

$$A'\mu + \mu = 0 \tag{5}$$

has precisely r' *linearly independent solutions* $\mu_j = \overset{+}{u_j}$, *and* $u_j = v(\mu_j)$.

Let problem (3_0) *i.e. problem* (3) *with* $f = 0$, *have* r *linearly independent solutions* U_j, $1 \leqslant j \leqslant r$. *Then the equation*

$$A\sigma - \sigma = 0 \tag{6}$$

has precisely r *linearly independent solutions* $\sigma_j = \partial U_j/\partial N_+$, *and* $U_j = w(\sigma_j)$.

LEMMA 2. *Assume that* $\Gamma \in C^\infty$. *If* (3_0) *has only the trivial solution then the operator* Q *defined by* (1.8), $Q\colon \tilde{H}^q \to \tilde{H}^{q+1}$ *is an isomorphism for any real* q.

LEMMA 3. *If* (3_0) *has* r *linearly independent solutions then the equation*

$$Q\sigma = f \tag{7}$$

is solvable iff

$$\int_\Gamma f\sigma_j \,\mathrm{d}t = 0, \quad 1 \leqslant j \leqslant r$$

where σ_j *are defined in Lemma 1.*

Proof of Lemma 1. Both parts of the Lemma can be proved similarly. Let us prove the first part. We prove that to each solution to problem (4_0) there corresponds a solution to (5) and vice versa, and this correspondence preserves linear independence.

Let u solve (4_0). Green's formula yields

$$u(x) = -\int_\Gamma u \frac{\partial g(x, t)}{\partial N_+} \,\mathrm{d}t = -v(u_+). \tag{8}$$

Let $\mu = u_+$. Then $v(\mu)$ solves problem (4_0). Thus, by (1.3′), one obtains

$$0 = \frac{\partial v}{\partial N_+} = \frac{\partial v}{\partial N_-}. \tag{9}$$

Since the homogeneous exterior Neumann problem has only the trivial solution satisfying the radiation condition, one concludes that

$$v(\mu) = 0 \text{ in } \Omega. \tag{10}$$

Therefore, by (1.3), μ solves Equation (5). Thus, every nontrivial solution to (4_0) generates a nontrivial solution to (5) by the formula $\mu = u_+$.

Let $\mu \not\equiv 0$ solve (5). Then the potential $v(\mu)$ solves the problem

$$(\nabla^2 + k^2)v = 0 \text{ in } \Omega, \quad \frac{\partial v}{\partial N_-} = 0, \quad k > 0, \tag{11}$$

and v satisfies the radiation condition. Therefore $v = 0$ in Ω, $0 = \partial v/\partial N_- = \partial v/\partial N_+$, and the function $u(x) = v(\mu) \not\equiv 0$ solves problem (4_0) in D. Thus, every nontrivial solution to (5) generates a nontrivial solution to (4_0). The number of linearly independent solutions to (4_0) and (5) is the same. This follows from the fact that the correspondence between solutions to (4_0) and (5) preserves linear independence. Indeed, suppose that the system $\{u_j\}$, $1 \leqslant j \leqslant r'$, of solutions to ($4_0$) is linearly independent, that is

$$\left\{ \sum_{j=1}^{r'} c_j u_j(x) = 0 \text{ in } D, \quad c_j = \text{const} \right\} \Rightarrow \forall c_j = 0. \tag{12}$$

Then the system $\{\mu_j\}$, $1 \leqslant j \leqslant r'$, $\mu_j = u_j^+$, of solutions to (5) is linearly independent. To prove this, assume that

$$\sum_{j=1}^{r'} c_j \mu_j = 0. \tag{13}$$

Then

$$0 = v\left(\sum_{j=1}^{r'} c_j \mu_j \right) = \sum_{j=1}^{r'} c_j v(\mu_j) = \sum_{j=1}^{r'} c_j u_j. \tag{14}$$

From (12) and (14) it follows that $\forall c_j = 0$. The first part of Lemma 1 is proved.

To prove the second part, let U solve (3_0). Green's formula yields

$$U(x) = \int_\Gamma g \frac{\partial U}{\partial N_+} \, \mathrm{d}t. \tag{15}$$

Let $\sigma = \partial U/\partial N_+$. Then $w(\sigma)$ solves problem (3_0). Since $0 = w_+ = w_-$, and the exterior homogeneous Dirichlet problem

$$(\nabla^2 + k^2)U = 0 \text{ in } \Omega, \quad k > 0, \quad U_- = 0 \tag{16}$$

has only the trivial solution satisfying the radiation condition, one concludes that $U(x) = 0$ in Ω. Therefore

$$0 = \frac{\partial U}{\partial N_-} = A\sigma - \sigma$$

so that σ solves (6). Thus, every nontrivial solution to (3_0) generates a nontrivial solution to (6). If $\sigma \not\equiv 0$ solves (6) then $\partial w(\sigma)/\partial N_- = 0$. This implies that $w = 0$ in Ω. Therefore $0 = w_- = w_+$. So, $U(x) = w(\sigma)$ solves problem (3_0) and $U \not\equiv 0$ in D. Thus, every nontrivial solution to (6) generates a nontrivial solution to (3_0). This correspondence between the solutions to (3_0) and (6) preserves linear independence. A proof is left to the reader. Lemma 1 is proved. □

Proof of Lemma 2. If the problem (3_0) has only the trivial solution, then Q is injective. Indeed, if $Q\sigma = 0$ and $\sigma \not\equiv 0$ then $w(\sigma)$ solves (3_0) and $w(\sigma) \not\equiv 0$ in D. Indeed, if $w \equiv 0$ in D then $\sigma = \partial w/\partial N_+ - \partial w/\partial N_- = -\partial w/\partial N_-$. But $w(\sigma) = 0$ in Ω since it solves (16) and satisfies the radiation condition. Thus $\partial w/\partial N_- = 0$, and $\sigma = 0$. It remains to be proved that $Q: \tilde{H}^q \to \tilde{H}^{q+1}$, and that Q is surjective.

That $Q: \tilde{H}^q \to \tilde{H}^{q+1}$ can be checked as follows. We have $Q = Q_\varepsilon + Q'_\varepsilon$. The kernels of Q_ε and Q'_ε are $(4\pi r_{st})^{-1}\ \eta_\varepsilon$ and $(1 - \eta_\varepsilon)(4\pi r_{st})^{-1}$ respectively, where $\eta \in C^\infty$, $\eta(r) = 1$ if $0 \leqslant r \leqslant 1$, $\eta(r) = 0$ if $r \geqslant 2$, $0 \leqslant \eta \leqslant 1$, $\eta_\varepsilon = \eta(r\varepsilon^{-1})$, $\varepsilon > 0$. The operator $Q'_\varepsilon: \tilde{H}^q \to \tilde{H}^\infty$. The smoothing property of the operator Q_ε is determined by the singularity of the kernel. In the local coordinates, in which the x_3 axis is the outer normal to Γ at the point s, the operator Q_ε is of the form

$$Q_\varepsilon \sigma = \int |s - t|^{-1}\ q_\varepsilon(s - t, t)\sigma(t)\ \mathrm{d}t, \qquad \int = \int_{R^2} \tag{17}$$

where $q_\varepsilon(s - t, t) = 1/4\pi + 0(|s - t|)$, $|s - t| < \varepsilon$.

Therefore the smoothing properties of Q_ε are the same as for the operator

$$q_\varepsilon \sigma = \frac{1}{4\pi} \int |s - t|^{-1}\eta_\varepsilon(|s - t|)\sigma(t)\ \mathrm{d}t. \tag{18}$$

Take the Fourier transform of (18) to obtain

$$\widetilde{q_\varepsilon \sigma} = 2^{-1}(|\lambda|^{-1} * \tilde{\eta}_\varepsilon)\tilde{\sigma}, \tag{19}$$

where $*$ denotes convolution, and

$$\tilde{\sigma} = \int \sigma(t) \exp(i\lambda \cdot t)\,\mathrm{d}t, \quad \int |t|^{-1} \exp(i\lambda \cdot t)\,\mathrm{d}t = 2\pi|\lambda|^{-1}. \tag{20}$$

It follows from (19) that $|\widetilde{q_\varepsilon \sigma}| \leqslant c\,|\lambda|^{-1}|\tilde{\sigma}|$ for $|\lambda| \to \infty$, $c = \text{const} > 0$. Therefore q_ε is a pseudodifferential operator of order -1 (see, e.g., Taylor [1]), that is

$$\|q_\varepsilon \sigma\|_{\tilde{H}^{q+1}} \leqslant c\|\sigma\|_{\tilde{H}^q}. \tag{21}$$

Therefore

$$\|Q\sigma\|_{\tilde{H}^{q+1}} \leqslant c\|\sigma\|_{\tilde{H}^q}. \tag{22}$$

Let us prove that $R(Q) = \tilde{H}^{q+1}$.

Take any $f \in \tilde{H}^q$. Solve the problem

$$(\nabla^2 + k^2)u = 0 \text{ in } \Omega, \quad \frac{\partial u}{\partial N_-} = f,\ r\left(\frac{\partial u}{\partial r} - iku\right) \to 0, \quad r \to \infty \tag{23}$$

This problem is uniquely solvable and the solution can be found in the form $u = w(\sigma)$, where

$$A\sigma - \sigma = 2f. \tag{24}$$

Equation (24) is uniquely solvable since k^2 is not an eigenvalue of the problem (3_0). Since A is a smoothing operator (if Γ is smooth), it follows from (24) that $\sigma \in \tilde{H}^q$. Therefore $u_- = Q\sigma \in \tilde{H}^{q+1}$. To show that $R(Q) = \tilde{H}^{q+1}$ take any $h \in \tilde{H}^{q+1}$ and solve the problem

$$(\nabla^2 + k^2)u = 0 \text{ in } \Omega,\ u_- = h, \quad r\left(\frac{\partial u}{\partial r} - iku\right) \to 0, \quad r \to \infty. \tag{25}$$

Calculate $\partial u/\partial N_- = f \in \tilde{H}^q$. Find the unique $\sigma \in \tilde{H}^q$ from (24). Then $w(\sigma) = u$ in Ω, $Q\sigma = w_- = u_- = h \in \tilde{H}^{q+1}$. Thus $Q\sigma = h$ for any $h \in \tilde{H}^{q+1}$ and a uniquely defined $\sigma \in \tilde{H}^q$. Therefore $Q\colon \tilde{H}^q \to \tilde{H}^{q+1}$ is an isomorphism. □

Proof of Lemma 3. If $Q\sigma = f$, $Q\sigma_j = 0$, then $(f, \bar{\sigma}_j) = (Q\sigma, \bar{\sigma}_j) = (\sigma, \overline{Q\sigma_j}) = 0$, where we used the symmetry of the kernel of Q. This proves the necessity. Let us prove the sufficiency. The operator Q: $\tilde{H}^q \to \tilde{H}^{q+1}$ is a Fredholm operator with index $Q \equiv$ ind $Q \equiv \dim N(Q) - \operatorname{codim} R(Q) = \dim N(Q) - \dim N(Q^*) = 0$. Therefore there are $r = \dim N(Q)$ orthogonality conditions

$$(f, q_j)_{\tilde{H}^{q+1}} = 0, \quad 1 \leqslant j \leqslant r \tag{26}$$

(where q_j are some functions), which are necessary and sufficient for $f \in R(Q)$. Consider Q as an operator from $\tilde{H}^{-1}$ onto $\tilde{H}^0$. Then conditions (26) can be written as

$$(f, \tilde{q}_j)_{\tilde{H}^0} = 0, \quad 1 \leqslant j \leqslant r, \tag{27}$$

where $\{\tilde{q}_j\}$, $1 \leqslant j \leqslant r$ is a linearly independent system of functions in $\tilde{H}^0$. Since the system $\{\bar{\sigma}_j\}$, $1 \leqslant j \leqslant r$, is linearly independent in $\tilde{H}^0$ and conditions

$$(f, \bar{\sigma}_j)_{\tilde{H}^0} = 0, \quad 1 \leqslant j \leqslant r, \tag{28}$$

are necessary for $f \in R(Q)$ one concludes that

$$\operatorname{span}\{\tilde{q}_1, \ldots, \tilde{q}_r\} = \operatorname{span}\{\bar{\sigma}_1, \ldots, \bar{\sigma}_r\}. \tag{29}$$

Therefore conditions (28) are also sufficient for $f \in R(Q)$. □

7.3. In this section we prove Theorems 1 and 2.

Proof of Theorem 1. Assume first that

$$r = 0, \tag{30}$$

i.e. problem (3_0) has only the trivial solution. If

$$w(\sigma) = v(\mu) \text{ in } D, \tag{31}$$

then

$$A'\mu - \mu = 2w(\sigma). \tag{32}$$

If $v(\mu)$ is given then (32) is an equation for σ. By Lemma 2 this equation is uniquely solvable for σ if (30) holds.

If σ is its solution then

$$w_+ = v_+. \tag{33}$$

From (33) it follows that (31) holds, and σ is determined uniquely.

If $w(\sigma)$ is given then (32) is an equation for μ. This equation is uniquely solvable if (30) holds. Indeed, by Lemma 1 if $r = 0$ then Equation (6) has only the trivial solution. Therefore the homogeneous equation (32) has only the trivial solution. Thus, Equation (32) is uniquely solvable by Fredholm's alternative, μ is uniquely determined, and, as before, (31) holds.

(2) Assume now that $r > 0$. If $w(\sigma)$ is given then (32) is an equation for μ. This equation is solvable in $\tilde{H}^0$ iff

$$(w(\sigma), \bar{\sigma}_j)_{\tilde{H}^0} = 0, \quad \text{where} \quad \bar{A}\bar{\sigma}_j - \bar{\sigma}_j = 0, \quad 1 \leqslant j \leqslant r. \tag{34}$$

Here we used Fredholm's alternative, the equation $A'^* = \bar{A}$, where $\bar{A}$ is the operator whose kernel is the complex conjugate of the kernel of A, and the fact that A and $\bar{A}$ have null spaces of the same dimension r. In fact, if $\sigma \in N(\bar{A})$ then $\bar{\sigma} \in N(A)$. By Lemma 1, $\bar{\sigma}_j = \partial U_j/\partial N_+$. Therefore

$$(w(\sigma), \sigma_j)_{\tilde{H}^0} = \int_\Gamma w(\sigma) \frac{\partial U_j}{\partial N_+} \, ds = \int_\Gamma \frac{\partial w}{\partial N_+} U_j \, ds = 0 \tag{35}$$

since $U_j = 0$ on Γ. Thus, Equation (32) is solvable for μ. The potential $v(\mu)$ satisfies the condition

$$v_+(\mu) = w(\sigma) \text{ on } \Gamma. \tag{36}$$

The requirement (31) determines μ uniquely.

Indeed, if

$$v(\mu_1) = v(\mu_2) = w(\sigma) \text{ in } D \tag{37}$$

then

$$v(\mu) = 0 \text{ in } D, \quad \mu = \mu_1 - \mu_2. \tag{38}$$

Thus $\partial v/\partial N_+ = 0$. Since $\partial v/\partial N_- = \partial v/\partial N_+$ one has $\partial v/\partial N_- = 0$.

Therefore $v(\mu) = 0$ in Ω since it solves problem (11). Thus $\mu = v_- - v_+ = 0$, and $\mu_1 = \mu_2$.

If $v(\mu)$ is given then (32) is an equation for σ. By Lemma 3, this equation is solvable iff

$$\int_\Gamma (A'\mu - \mu)\sigma_j \, ds = 0, \quad 1 \leqslant j \leqslant r, \tag{39}$$

where σ_j solves (6). One has

$$\int_\Gamma (A'\mu - \mu)\sigma_j \, ds = \int_\Gamma (A\sigma_j - \sigma_j)\mu \, ds = 0, \tag{40}$$

so that (39) holds. Therefore Equation (32) is solvable for σ. The requirement (31) determines σ uniquely. Indeed, if

$$w(\sigma_1) = w(\sigma_2) = v(\mu) \text{ in } D \tag{41}$$

then

$$w(\sigma) = 0 \text{ in } D, \quad \sigma = \sigma_1 - \sigma_2. \tag{42}$$

Therefore $w_-(\sigma) = w_+(\sigma) = 0$, $w(\sigma) = 0$ in Ω, $\sigma = \partial w/\partial N_+ - \partial w/\partial N_- = 0$, and $\sigma_1 = \sigma_2$. Theorem 1 is proved. □

Proof of Theorem 2. (1) Suppose that $v(\mu)$ is given and

$$w(\sigma) = v(\mu) \text{ in } \Omega. \tag{43}$$

Then

$$Q\sigma = w_-(\sigma) = v_-(\mu) = \frac{A'\mu + \mu}{2} \tag{44}$$

is an equation for σ. By Lemma 3, this equation is solvable for σ iff

$$\int_\Gamma (A'\mu + \mu)\sigma_j \, ds = 0, \quad 1 \leqslant j \leqslant r, \quad \sigma_j = A\sigma_j. \tag{45}$$

Condition (45) can be written as

$$\int_\Gamma \mu\sigma_j \, ds = 0, \quad 1 \leqslant j \leqslant r, \tag{47}$$

which is condition (2). If (44) holds then (43) holds since the exterior Dirichlet problem has at most one solution satisfying the radiation condition. The second statement in Theorem 2 is proved.

(2) Suppose that $w(\sigma)$ is given and (43) holds. Then (44) is an equation for μ. By Fredholm's alternative, this equation is solvable iff

condition (1) holds. If (44) has a solution then (43) holds. This proves the first statement in Theorem 2. To prove (1″) one writes

$$0 = \int_\Gamma w(\sigma)h \, \mathrm{d}t = \int_\Gamma \sigma w(h_j) \, \mathrm{d}t = \int_\Gamma \sigma v_+(\mu_j) \, \mathrm{d}s$$

$$= \int_\Gamma \sigma \frac{A'\mu_j - \mu_j}{2} \, \mathrm{d}s = \int_\Gamma \sigma\mu_j \, \mathrm{d}s, \text{ where we used Lemma 1.} \qquad \square$$

Remark 1. Our argument shows that if $v(\mu)$ is given, $r > 0$, and (47) holds, then Equation (44) is solvable for σ and the general solution to (44) is of the form

$$\sigma = \sigma_0 + \sum_{j=1}^{r} c_j \sigma_j \tag{48}$$

where σ_0 is a particular solution to (44), c_j are arbitrary constants, and σ_j, $1 \leqslant j \leqslant r$, is a basis of $N(Q)$. The set of r linearly independent solutions to Equation (6) forms a basis of $N(Q)$.

Indeed, if $\sigma \in N(Q)$ then $w(\sigma) = 0$ in Ω. Thus, $0 = \partial w/\partial N_- = A\sigma - \sigma/2$, so that σ solves (6). Conversely, if σ solves (6), then $\partial w(\sigma)/\partial N_- = 0$. Thus $w(\sigma) = 0$ in Ω so that $Q\sigma = w_-(\sigma) = 0$, and $\sigma \in N(Q)$.

Therefore Equation (44) has many solutions. Nevertheless, $w(\sigma)$ for any solution to (44) is equal to $v(\mu)$ in Ω as we proved earlier. This can also be seen from the fact that $w_-(\sigma_j) = 0$ implies $w(\sigma_j) = 0$ in Ω.

If $w(\sigma)$ is given, $r' > 0$, and (1) holds, then Equation (44) is solvable for μ, and its general solution is

$$\mu = \mu_0 + \sum_{j=1}^{r'} c_j \mu_j \tag{49}$$

where c_j are constants. For any μ the function $v(\mu)$ is the same in Ω, $v(\mu) = w(\sigma)$ in Ω. This was proved earlier and can also be seen from the fact that

$$v(\mu_j) = 0 \text{ in } \Omega. \tag{50}$$

To prove (50) notice that μ_j solves (5) and therefore $v_-(\mu_j) = 0$. Thus $v(\mu_j)$ solves (16), and so (50) holds.

Proof of Corollary 1. If (3) is solvable then its solution can be written (by Green's formula) as

$$u(x) = \int_\Gamma \left(g \frac{\partial u}{\partial N_+} - u \frac{\partial g}{\partial N_+} \right) \mathrm{d}t = w\left(\frac{\partial u}{\partial N_+} \right) - v(u_+). \tag{51}$$

By Theorem 1 the right-hand side of (51) can be represented as $v(\mu)$ and as $w(\sigma)$. The same argument holds for problem (4). □

7.4. In this section we assume that $D \subset R^2$, $\Gamma = \partial D$, $g_0 = 1/2\pi \ln(1/|x-y|)$; $S\sigma = \int_\Gamma g_0(x, t)\sigma(t)\, \mathrm{d}t$; $Q\sigma = \int_\Gamma g_0(s, t)\sigma(t)\, \mathrm{d}t$, $s \in \Gamma$; v, A_0 and A_0' are defined as in Section I.1 with $g = g_0$. In R^3 the equation

$$\int_\Gamma \frac{\sigma(t)\, \mathrm{d}t}{4\pi|s-t|} = 0, \quad s \in \Gamma$$

implies that $\sigma = 0$. In R^2 the analogous equation $Q\sigma = 1/2\pi \int_\Gamma \ln(1/|s-t|)\, \sigma(t)\, \mathrm{d}t = 0$, $s \in \Gamma$, may have nontrivial solutions for some Γ. Example: $\sigma = 1$, $\Gamma = \{t: |t| = R\}$. For a suitable $R > 0$ one has

$$\int_{|t|=R} \ln \frac{1}{|s-t|}\, \mathrm{d}t = 0. \tag{52}$$

One can prove that $\dim N(Q) = 1$. This holds even in the case when Γ consists of several components (see Section V.4). We wish to prove the following proposition, which will be useful in Section V.4.

PROPOSITION. *The operator $Q: \tilde{H}^q \to \tilde{H}^{q+1}$ is an isomorphism iff $N(Q) = \{0\}$.*

Proof. The proof consists of three parts. First, one shows that Q acts from $\tilde{H}^q$ into $\tilde{H}^{q+1}$. This can be done as in the proof of Lemma 7.2.2, p. 75. Secondly, one notices that if Q_0 is an isomorphism then $N(Q) = \{0\}$. Finally, one proves that if $N(Q) = \{0\}$ then Q is surjective. We now turn to the proof of this last claim.

Let us prove that if $N(Q) = \{0\}$ then Q is an isomorphism between $\tilde{H}^q$ and $\tilde{H}^{q+1}$. Since Q acts from $\tilde{H}^q$ into $\tilde{H}^{q+1}$, we need to prove that

$R(Q) = \tilde{H}^{q+1}$. Take any $f \in \tilde{H}^{q+1}$ and look for the solution of the problem

$$\Delta u = 0 \text{ in } \Omega, \quad u_- = f \text{ on } \Gamma \tag{53}$$

of the form

$$u = c \ln \frac{1}{|x - x_0|} + v, \tag{54}$$

where $x_0 \in D$ and $v = D\mu$ is the double layer potential.

Then

$$f = c \ln \frac{1}{|s - x_0|} + \frac{A'\mu + \mu}{2}, \quad s \in \Gamma. \tag{55}$$

The equation $A\sigma + \sigma = 0$ has precisely one linearly independent solution $\sigma = \text{const} \neq 0$. Therefore (55) is solvable iff

$$\int_\Gamma f \, \mathrm{d}s = c \int_\Gamma \ln \frac{1}{|s - x_0|} \, \mathrm{d}s.$$

This equation determines c uniquely iff $\int_\Gamma \ln (1/|s - x_0|) \, \mathrm{d}s \neq 0$. The last inequality holds for some $x_0 \in D$ iff $N(Q) = \{0\}$. Indeed, suppose that

$$\int_\Gamma \ln \frac{1}{|s - x|} \, \mathrm{d}s = 0 \text{ for all } x \in D. \text{ Then } u(x) = S\,1 \equiv 0 \text{ in } D,$$

and, therefore, $Q\,1 = 0$. So, $N(Q) \neq \{0\}$ contrary to the assumption. We have proved that if $N(Q) = \{0\}$ then for any given $f \in \tilde{H}^{q+1}$ the problem (53) is solvable and its solution can be found in the form (54) where x_0 is a point in D, and $v = D\mu$, so that $\Delta v = 0$ in Ω, $v(\infty) = 0$.

The idea of the proof that $R(Q) = \tilde{H}^{q+1}$ is as follows. Compute $h = \partial u / \partial N_-$ where u is given by (54). Thus u solves the problem

$$\Delta u = 0 \text{ in } \Omega, \quad \frac{\partial u}{\partial N_-} = h, \quad u(\infty) = 0. \tag{56}$$

Look for the solution of (56) of the form

$$u(x) = S\sigma. \tag{57}$$

Then

$$\frac{A\sigma - \sigma}{2} = h. \tag{58}$$

This equation is uniquely solvable since $N(A - I) = \{0\}$. The problem (56) has at most one solution, so the function (57) (with σ which solves (58)) is the same as the function (54). Thus

$$Q\sigma = f. \tag{59}$$

Since $\sigma = \partial u/\partial N_+ - \partial u/\partial N_-$ and $\partial u/\partial N_\pm \in \tilde{H}^q$ provided that $f \in \tilde{H}^{q+1}$, once concludes that $\sigma \in \tilde{H}^q$. The proof is complete. □

Chapter II

The Inverse Scattering Problem

1. STATEMENT OF THE PROBLEM AND UNIQUENESS THEOREMS

1.1. The problem is to find the shape of the obstacle and the boundary condition on its surface Γ from the scattering data. The surface Γ may consist of several connected closed components. The data are the values of the scattering amplitude on a subset in the space $S^2 \times S^2 \times R_+$ of the variables n, ν, k; $n, \nu \in S^2$, $k > 0$. This subset can be chosen in a number of ways:

$$n = n_0 \text{ is fixed}, \nu \in \tilde{S}^2, 0 \leqslant a \leqslant k \leqslant b, b > a, \tag{1}$$

where $\tilde{S}^2$ is a solid angle, i.e. an open set in S^2,

$$n \text{ and } \nu \text{ run through a subset of } S^2, k_0 > 0 \text{ is fixed}, \tag{2}$$

$$n_0 \text{ is fixed}, k_0 > 0 \text{ is fixed}, \nu \in \tilde{S}^2. \tag{3}$$

The boundary condition on the surface Γ of the obstacle is assumed to be one of the three types

$$u = 0 \text{ on } \Gamma, \tag{4}$$

$$\frac{\partial u}{\partial N} = 0 \text{ on } \Gamma, \tag{5}$$

$$\frac{\partial u}{\partial N} + \eta(s)u = 0 \text{ on } \Gamma, \tag{6}$$

where

$$\operatorname{Im} \eta \geqslant 0, \quad \eta \in C(\Gamma). \tag{7}$$

Condition (7) is sufficient for the exterior Robin problem

$$(\nabla^2 + k^2)u = 0 \text{ in } \Omega, \quad \operatorname{Im} k \geqslant 0, \quad k \neq 0 \tag{8}$$

$$u \text{ satisfies (6), (7) and the radiation condition}, \tag{9}$$

to have only the trivial solution $u = 0$ in Ω. If $k = 0$, Im $\eta = 0$, $\eta \leqslant 0$ and $u = 0(r^{-1})$ as $r \to \infty$, then the problem

$$\nabla^2 u = 0 \text{ in } \Omega, \quad \frac{\partial u}{\partial N} + \eta u = 0 \text{ on } \Gamma, \quad \eta \leqslant 0, \quad u = 0(r^{-1}) \tag{10}$$

has only the trivial solution as well. If $\eta > 0$, this is no longer true. Example: $\Omega = \{x: |x| \geqslant 1\}$, $u = r^{-1}$, $\eta = 1$. Notice that from Theorem (I.4.3) it follows that if $f(n_0, \nu, k_0)$ is known for $\nu \in \tilde{S}^2$, then it is uniquely determined for $\nu \in S^2$.

1.2. Let us prove.

LEMMA 1. *The scattering data* (1) *determine* Γ *and the boundary condition* (4), (5) *or* (6) *uniquely.*

Proof. Suppose there are two obstacles D_1 and D_2 with the same scattering amplitude $f(n_0, \nu, k)$, $\nu \in S^2$, $0 \leqslant a \leqslant k \leqslant b$, $b > a$ and the boundary condition is any one of (4)—(6), the same for D_1 and D_2.

Assume first that $D_1 \cap D_2 = \emptyset$. Let $u_j(x, n_0, k)$ solve problem (0.1)—(0.5) with $D = D_j$, $j = 1, 2$. Consider the function $u = u_1 - u_2$ in Ω_R, where $R > 0$ is sufficiently large so that $(D_1 \cup D_2) \subset B_R$. Since $u_j = u_{0j} + f\gamma + o(r^{-1})$ as $r \to \infty$, $\gamma = r^{-1}\exp(ikr)$, one sees that $u = o(r^{-1})$ as $r \to \infty$. Since u solves Equation (I.2.1) and $ru \to 0$ as $r \to \infty$, one obtains from Lemma 1, Section I.2, that $u \equiv 0$ in Ω_R.

By the unique continuation property, $u \equiv 0$ in $\Omega_{12} \equiv R^3 \setminus (D_1 \cup D_2)$. In particular $u = 0$ on $\Gamma_1 \cup \Gamma_2$. Therefore

$$u_1 \equiv u_2 \text{ in } \Omega_{12}, \quad u_1 = 0 \text{ on } \Gamma_2.$$

Thus u_1 solves problem

$$(\nabla^2 + k^2)u_1 = 0 \text{ in } D_2, \quad u_1 = 0 \text{ on } \Gamma_2, \quad a \leqslant k \leqslant b, \quad b > a \tag{11}$$

and $u_1 \not\equiv 0$ in D_2 because otherwise $u_1 = 0$ in $\Omega_1 = R^3 \setminus D_1$, which is impossible since

$$|u_1| = |u_0 + v_1| \geqslant |u_0| - |v_1| = 1 - 0\left(\frac{1}{r}\right) > 0 \quad \text{as} \quad r \to \infty. \tag{12}$$

Equation (11) shows that the Dirichlet Laplacian in D_2 has a continuous spectrum, since any number k^2, $a \leqslant k \leqslant b$ is an eigenvalue of this Dirichlet Laplacian. This is a contradiction since the spectrum of the Dirichlet Laplacian in a compact domain is known to be discrete. This contradiction proves that the assumption $D_1 \neq D_2$, $D_1 \cap D_2 = \emptyset$ does not hold. If $D_1 \neq D_2$ but $D_1 \cap D_2 \neq \emptyset$ then consider a connected component Δ of the set $D_2 \setminus (D_1 \cap D_2)$. The boundary of Δ consists of $\tilde{\Gamma}_2 \cup \tilde{\Gamma}_1$, where $\tilde{\Gamma}_j$ is a subset of Γ_j, j = 1, 2. One derives as above that the spectrum of the Dirichlet Laplacian in Δ is continuous. This contradiction proves that $D_1 = D_2$. In the above argument we assume that the Dirichlet condition on Γ_j holds. The argument is the same for the other boundary conditions (5) and (6), because the Laplacian defined by either of the boundary conditions (5)–(6) in a compact domain has a discrete spectrum. Finally, if the scattering amplitude $f(n_0, \nu, k)$ defines the obstacle uniquely and, as was proved above, defines $u(x, n_0, k)$ uniquely in Ω, then the function η in the boundary condition (6) is also uniquely defined:

$$\eta = -\frac{1}{u}\frac{\partial u}{\partial N} \text{ on } \Gamma. \tag{13}$$

Notice that if u satisfies boundary condition (6) and η is a bounded piecewise continuous function then u cannot vanish on an open set $\tilde{\Gamma}$ because if $u = 0$ on $\tilde{\Gamma}$, then it follows from (6) that $\partial u/\partial N = 0$ on $\tilde{\Gamma}$. This and the uniqueness of the solution to the Cauchy problem for elliptic equations imply that $u \equiv 0$ in Ω which is impossible. The solution u cannot vanish on a closed subset Γ_1 of Γ which is dense in an open subset of Γ. Indeed, if Γ_1 is dense in some open subset $\tilde{\Gamma}$ of Γ, then, by continuity of u, one concludes that $u = 0$ on $\tilde{\Gamma}$, and this leads to a contradiction as above. If Γ_1 is nowhere dense in Γ, then for every point $s \in \Gamma_1$ one finds a sequence of points $s_n \to s$, $u(s_n) \neq 0$. At the points s_n the function η is uniquely defined by (13). At the point s the function η is uniquely defined by continuity: $\eta(s) = \lim \eta(s_n)$. Therefore, both the scatterer and the function $\eta(s)$, $s \in \Gamma$ are uniquely determined by the data (1). □

Remark 1. The uniqueness theorem that we used at the end of the proof, and which says that a solution of the Helmholtz equation with the zero Cauchy data on an open set of a smooth surface Γ is identically zero in Ω, is a particular case of the Holmgren uniqueness theorem (see e.g. John [1], p. 83).

Let us specify the data (2). Let $\tilde{S}_0^2$ denote a sequence of distinct unit vectors n_j (i.e. $n_j \neq n_i$ if $j \neq i$) with a limit point n_0. Assume that the scattering data $f(n, \nu, k)$ are given for

$$n \in \tilde{S}_0^2, \quad \nu \in \tilde{S}^2, \quad \text{and} \quad k = k_0 > 0 \text{ fixed.} \tag{2'}$$

THEOREM 1. *The scattering data* (2′) *determine* Γ *and the boundary condition* (4), (5) *or* (6) *uniquely.*

Proof. The proof goes as in Lemma 1 and we conclude that $u_j = u_j(x, n_j, k)$ solves the problem

$$(\nabla^2 + k_0^2)u_j = 0 \text{ in } \Delta, \quad j = 1, 2, \ldots \tag{14}$$

$$u_j \text{ satisfies a homogeneous boundary condition on } \partial\Delta. \tag{15}$$

The system of functions $\{u_j\}$ is linearly independent in $L^2(\Delta)$. This is proved below. Therefore k_0^2 is an eigenvalue of infinite multiplicity of the Laplacian defined by the boundary conditions (15) in a bounded domain Δ. Since this Laplacian has eigenvalues of finite multiplicity only, we have a contradiction which shows that $D_1 = D_2$.

Let us prove that the system $\{u_j\}$ is linearly independent in $L^2(\Delta)$. Assume that

$$\sum_{j=1}^{n} c_j u_j = 0 \text{ in } L^2(\Delta) \tag{16}$$

where $c_j = \text{const}$ and n is an arbitrary integer.

The unique continuation property of the solutions to Helmholtz's equation and (16) imply that

$$\sum_{j=1}^{n} c_j u_j = 0 \text{ in } \Omega_R. \tag{17}$$

Thus

$$\sum_{j=1}^{n} c_j\{\exp(-ikn_j \cdot s) + 0(|x|^{-1})\} = 0 \text{ in } \Omega_R, \tag{18}$$

and

$$c_m\{1 + 0(|x|^{-1})\} + \sum_{j \neq m} c_j\{\exp[ik(n_m - n_j) \cdot x] + 0(|x|^{-1})\} = 0 \text{ in } \Omega_R. \tag{19}$$

Since $n_j \neq n_i$ if $i \neq j$, one has

$$\min |n_m - n_j| > 0, \quad j \neq m. \tag{20}$$

Let

$$n_m - n_j = a_{mj} \cdot n_{mj}, \quad a_{mj} = |n_m - n_j|, \quad n_{mj} = (n_m - n_j) a_{mj}^{-1}. \tag{21}$$

Notice that

$$\int_{|x - x_0| \leqslant a} \exp(ika_{mj} n_{mj} x) \, dx$$

$$= \exp(ika_{mj} n_{mj} \cdot x_0) \int_{|y| \leqslant a} \exp(ika_{mj} |y| \cos\theta) \, dy$$

$$= \exp(ika_{mj} n_{mj} \cdot x_0) \int_0^a dr\, r^2 \left(\frac{\pi}{2rka_{mj}}\right)^{1/2} 4\pi \left(\frac{2}{\pi k a_{mj} r}\right)^{1/2} \sin(ka_{mj} r) \tag{22}$$

$$= 4\pi \exp(ika_{mj} n_{mj} \cdot x_0) \frac{1}{ka_{mj}} \int_0^a r \sin(ka_{mj} r) \, dr = 0(a).$$

Integrate (19) over the ball $|x - x_0| \leqslant a$, choose x_0 large, so that $0(|x|^{-1}) < 1$, and take a large. Then the first term in (19) after integration is $0(|c_m| a^3)$ while the sum after integration is $0(a)$ according to (22). Therefore $c_m = 0$. Since m is arbitrary we proved that $c_j = 0, 1 \leqslant 1 \leqslant n$. □

Remark 2. The argument in Lemma 1 shows that there are no two (or more) disjoint obstacles with the same scattering data $f(n_0, \nu, k_0)$, $\nu \in S^2$. Notice that k is fixed here, $k = k_0$.

1.3. In this section we show that the knowledge of $f(n_0, \nu, k_0)$ determines $\partial^m f(n_0, \nu, k)/\partial k^m$ at $k = k_0$ for all $m = 1, 2, 3 \ldots$ uniquely. Since $f(n_0, \nu, k)$ is analytic in k in a neighborhood of the real axis on the complex k-plane, the data $f(n_0, \nu, k)$ determine $f(n_0, \nu, k_0)$ uniquely. However, the mapping $f(n_0, \nu, k_0) \to f(n_0, \nu, k)$ depends on Γ. This, unfortunately, does not allow one to derive the uniqueness theorem with the data $f(n_0, \nu, k_0)$.

As in Section 1.2 we conclude that the data $f(n_0, \nu, k_0)$ determine uniquely the scattering solution $u = u(x, k_0) = u(x, -n_0, k)$ in Ω. The function $u(x, k)$ is analytic in k in a neighborhood of any point k, $\operatorname{Im} k = 0$. Let $\dot{u} = \partial u(x, k)/\partial k|_{k=k_0}$. Then differentiating the Helmholtz equation in k yields

$$\nabla^2 \dot{u} + k_0^2 \dot{u} = -2k_0 u(x, k_0) \text{ in } \Omega, \tag{23}$$

$$\dot{u} = 0 \text{ on } \Gamma, \tag{24}$$

$$\dot{u} = \dot{u}_0 + \dot{v}, \qquad u_0 = \exp(-ik_0 n_0 \cdot x), \tag{25}$$

where

$$\dot{v}(x, k_0 + i\varepsilon) \in L^2(\Omega) \quad \text{if} \quad \varepsilon > 0. \tag{26}$$

Thus, the solution to problem (23)–(26) is defined by the limiting absorption principle: one solves the problem

$$L_\varepsilon w_\varepsilon \equiv [\nabla^2 + (k_0 + i\varepsilon)^2] w = -2k_0 v(x, k_0) \equiv F, \text{ in } \Omega, \tag{27}$$

$$w_\varepsilon = -\int_\Omega GF \, dy - \int_\Gamma \dot{u}_0 \frac{\partial G}{\partial N} \, ds, \ \varepsilon > 0, \text{(see (39))}, \tag{28}$$

$$w_\varepsilon = -\dot{u}_0 \text{ on } \Gamma, \tag{29}$$

which has a unique solution for any $\varepsilon > 0$, and then defines the solution of (23)–(26) as the limit

$$\lim_{\varepsilon \to +0} w_\varepsilon = \dot{v}. \tag{30}$$

The problem is to prove that this limit exists in $H^2_{\text{loc}}(\Omega)$. If it exists it is unique and v, defined by (30), solves (23)–(26).

We know that

$$L_\varepsilon \dot{v}_\varepsilon = -2(k_0 + i\varepsilon) v(x, k_0 + i\varepsilon) \equiv F_\varepsilon \text{ in } \Omega, \tag{31}$$

$$\dot{v}_\varepsilon \in L^2(\Omega), \ \varepsilon > 0, \tag{32}$$

$$\dot{v}_\varepsilon = -\dot{u}_{0\varepsilon} \text{ on } \Gamma, \quad u_{0\varepsilon} = u_0(x, k_0 + i\varepsilon), \tag{33}$$

$$\lim_{\varepsilon \to +0} \dot{v}_\varepsilon = \dot{v}, \tag{34}$$

where the limit exists in $H^2_{\text{loc}}(\Omega)$. Therefore (30) is established if one proves that

$$\lim_{\varepsilon \to +0} (w_\varepsilon - \dot{v}_\varepsilon) = 0. \tag{35}$$

We cannot use problem (31)—(33) for finding $\dot{v}$ because our data are F and $\dot{u}_0$ and not F_ε and $\dot{u}_{0\varepsilon}$. The existence of the $\lim_{\varepsilon\to 0} w_\varepsilon$ does not follow from the known results on the limiting absorption principle because these results are established for F which decay at infinity as $0(|x|^{-\alpha})$, $\alpha > 1$, while in our case $F = 0(|x|^{m-1})$ as $|x| \to \infty$. Our proof of the existence of the lim w_ε makes use of the special structure of F which will be clear from the argument below. If we prove that $\dot{v} = \lim w_\varepsilon$ then $\dot{v}$ is determined uniquely in Ω by the data F and $\dot{u}_0$. In the same way $\partial^m v(x, k_0)/\partial k^m$, $m > 1$ are determined in Ω. We know that v is analytic in a neighborhood of any point $k_0 > 0$ and the knowledge of $\partial^m v(x, k_0)/\partial k^m$ in Ω determines $v(x, k)$ in Ω for all $k > 0$. Thus $f(n_0, \nu, k)$ is uniquely determined. In order to complete the proof we should establish (35).

From (27) and (31) it follows that

$$\begin{aligned} &L_\varepsilon(w_\varepsilon - \dot{v}_\varepsilon) = F - F_\varepsilon, \\ &\quad w_\varepsilon - \dot{v}_\varepsilon = \dot{u}_{0\varepsilon} - \dot{u}_0 \text{ on } \Gamma. \end{aligned} \tag{36}$$

Let $G(x, y, k_0 + i\varepsilon)$ be the resolvent kernel of the Dirichlet Laplacian in Ω:

$$[\nabla_x^2 + (k_0 + i\varepsilon)^2]G(x, y, k_0 + i\varepsilon) = -\delta(x - y) \text{ in } \Omega, \tag{37}$$

$$G = 0 \text{ on } \Gamma, \quad G \in L^2(\Omega_x) \quad \text{if} \quad \varepsilon > 0. \tag{38}$$

The kernel $G(x, y, k)$ is meromorphic in k on the whole complex plane k and analytic in a neighborhood of the ray $k > 0$. It follows from (36) that

$$\begin{aligned} w_\varepsilon = \dot{v}_\varepsilon - \int_\Gamma \frac{\partial G(x, s, k_0 + i\varepsilon)}{\partial N_s} (\dot{u}_0 - \dot{u}_{0\varepsilon})\, ds - \\ - \int_\Omega G(x, y, k_0 + i\varepsilon)\,(F - F_\varepsilon)\, dy. \end{aligned} \tag{39}$$

Since

$$\dot{u}_{0\varepsilon} \to \dot{u}_0 \text{ in } H^2_{\text{loc}} \quad \text{as} \quad \varepsilon \to 0, \tag{40}$$

one concludes from (34), (39), and (40) that (35) holds provided that

$$J \equiv \int_\Omega G(x, y, k_0 + i\varepsilon)(F - F_\varepsilon)\,dy \to 0 \text{ in } H^2_{\text{loc}} \quad \text{as} \quad \varepsilon \to 0. \tag{41}$$

Let us prove (41). Using (27), (31) and the formula

$$v = -\int_\Gamma u_0 \frac{\partial G}{\partial N_s}\,ds \tag{42}$$

one can write J as

$$J = \int_\Gamma ds \left[2k_0 u_0 G_\varepsilon * \frac{\partial G}{\partial N_s} - 2(k_0 + i\varepsilon) u_{0\varepsilon} G_\varepsilon * \frac{\partial G_\varepsilon}{\partial N_s} \right] \tag{43}$$

where

$$G_\varepsilon * \frac{\partial G}{\partial N_s} = \int_\Omega G(x, s, k_0 + i\varepsilon) \frac{\partial G(s, y, k_0)}{\partial N_s}\,ds \tag{44}$$

and $G_\varepsilon * \partial G_\varepsilon/\partial N_s$ is defined similarly. It is at this point that the structure of F is used. The idea of the rest of the proof is to transform integrals of the type (44), using the resolvent identity for a linear operator A:

$$(A - \lambda I)^{-1} - (A - \mu I)^{-1} = (\lambda - \mu)(A - \lambda I)^{-1}(A - \mu I)^{-1}, \tag{45}$$

into integrals over Γ in order to avoid integration over the noncompact set Ω and justify the passage to the limit $\varepsilon \to 0$. It follows from (45) that

$$\begin{aligned} G_\varepsilon * G &= [(k_0 + i\varepsilon)^2 - k_0^2]^{-1}(G_\varepsilon - G), \\ G_\varepsilon * G_\varepsilon &= \left(\frac{1}{2k} \frac{\partial G}{\partial k} \right)_{k = k_0 + i\varepsilon} \end{aligned} \tag{46}$$

Therefore

$$G_\varepsilon * \frac{\partial G}{\partial N_s} = [(k_0 + i\varepsilon)^2 - k_0^2]^{-1} \left[\frac{\partial G(x, s, k_0 + i\varepsilon)}{\partial N_s} - \frac{\partial G(x, s, k_0)}{\partial N_s} \right] \quad (47)$$

$$G_\varepsilon * \frac{\partial G_\varepsilon}{\partial N_s} = \frac{1}{2(k_0 + i\varepsilon)} \frac{\partial}{\partial N_s} \frac{\partial G(x, s, k_0 + i\varepsilon)}{\partial k}. \quad (48)$$

Note that $G(x, s, k) = 0$, $s \in \Gamma$, and the symbol $(\partial G(x, s, k))/\partial N_s$ means that one computes the derivative $(\partial G(x, z, k))/\partial N_z$ at the point $z \notin \Gamma$ in the direction N_z such that $N_z \to N_s$ when $z \to s$, and then passes to the limit $z \to s$. The symbol $\partial/\partial N_s\ (\partial G(x, s, k))/\partial k$ has a similar meaning. In particular, $\partial/\partial N_s\ (\partial G(x, s, k)/\partial k) = \partial/\partial k\ (\partial G(x, s, k)/\partial N_s)$, $\operatorname{Im} k \geqslant 0$. From (43), (47), (48), and (40) it follows that (41) holds because the integral in (43) is taken over the compact set Γ and the integrand converges to zero uniformly in $x \in \Omega$, where $\Omega \subset \Omega$ is any compact domain, and $0 \leqslant a \leqslant k_0 \leqslant b$, where a and b are arbitrary positive numbers.

This completes the proof of (35).

Remark 3. For the transmission problem the uniqueness theorem with the data $f(n_0, \nu, k_0)$ does not hold: one can have the same scattering amplitude $f(n_0, \nu, k_0)$ and various transmission constants and k_1^2. For example, let $D = \{x: |x| \leqslant 1\}$,

$$(\nabla^2 + 1)u = 0 \text{ in } \Omega, \quad (\nabla^2 + k_1^2)u = 0 \text{ in } D,$$
$$u_+ = u_-, \quad \alpha \frac{\partial u}{\partial N_+} = \frac{\partial u}{\partial N} \quad (49)$$

$$u = f \exp(ir) r^{-1}, \quad r \geqslant 1, \quad f = \text{const},$$
$$u = cr^{-1} \sin(k_1 r), \quad r \leqslant 1, c = \text{const}. \quad (50)$$

Equation (49) is satisfied if

$$f \exp i = c \sin k_1, \quad f(i-1) \exp i = \alpha c(k_1 \cos k_1 - \sin k_1). \quad (51)$$

Equations (51) are satisfied if

$$k_1 \cot k_1 = 1 + \alpha^{-1}(i-1). \quad (52)$$

For a fixed α Equation (52) can have several roots k_{1m}, $m = 1, 2, \ldots$. Thus various triples (D, α, k_{1m}), $m = 1, 2, \ldots$ produce the same field in Ω and the same scattering amplitude f.

Remark 4. The uniqueness theorem does not hold for the inverse source problem. This problem consists in finding the source $h(x)$ from the scattering amplitude $f(\nu, k_0)$. Here

$$(\nabla^2 + k_0^2)v = h(x), \quad h = 0 \text{ in } \Omega_R \tag{53}$$

$$v \sim \gamma(r) f(\nu, k_0), \quad r \to \infty, \quad \nu = xr^{-1}. \tag{54}$$

To prove that $f(\nu, k_0)$ does not determine h uniquely, consider $h(x) \not\equiv 0$ which produces $v \equiv 0$ in Ω_R and therefore $f = 0$. Such h are called nonradiating sources.

Take any $v \in C_0^2(B_R)$ and set $h = (\nabla^2 + k_0^2)v$. Then $v = 0$ in Ω_R and $f = 0$.

Remark 5. Consider the inverse problem of potential scattering. Let $q \in C_0^\infty(R^3)$

$$(\nabla^2 + k_0^2 - q(x))u = 0 \text{ in } R^3, \quad k_0 > 0, \tag{55}$$

$$u = u_0 + v, \quad v \sim r^{-1} \exp(ik_0 r) f(n_0, \nu, k_0). \tag{56}$$

Do the data $f(n_0, \nu, k_0)$ determine $q(x)$ uniquely? The answer is no. To prove this consider $v = r^{-1} \exp(ik_0 r)$ in Ω_R, where $R > 0$ is a fixed number, continue v in B_R so that $v \in C^2(R^3)$ and $v + u_0 \neq 0$ in B_R. There are infinitely many continuations v which satisfy these requirements. Define $q = ((\nabla^2 + k_0^2)v)/(v + u_0)$. Then $q \in C_0^\infty(B_R)$, $q \not\equiv 0$, and the corresponding scattering amplitude is $f = 1$ for any of the potentials q.

One can easily prove, however, that the data $f(n_0, \nu, k)$, $0 < a \leqslant k \leqslant b$, $b > a$, determines the potential $q(x) \in C_0^\infty(R^3)$ uniquely. Actually this can be proved for $q \in C_0(R^3)$.

Remark 6. If one knows $f(n, \nu, k)$ for small k, $0 < k < k_0$, $n, \nu \in S^2$, then, by analyticity, $f(n, \nu, k)$ is uniquely determined for all k, and the obstacle is determined uniquely. But practically only the first one or two coefficients in the expansion of $f(n, \nu, k)$ in the power series about $k = 0$ can be measured. Practically, in general, any body scatters as an ellipsoid if $ka \ll 1$ (see Chapter 5), so that the scattering data allows one to determine the axes of this ellipsoid, i.e. three scalar parameters. This is the reason why most of the applied work on inverse scattering problems is done in the high frequency region.

2. CONSTRUCTION OF OBSTACLES FROM THE SCATTERING DATA AT HIGH FREQUENCIES

2.1. Let D be a convex bounded domain with a smooth connected, boundary Γ. Let ψ: $\Gamma \to S^2$ be such a map (spherical map of the surface) that $\psi(s)$ is the unit exterior (positive) normal to Γ at the point s. This unit vector $\psi(s)$ is identified with a point on the unit sphere S^2, namely with the end point of the vector $\psi(s)$ the initial point of which is placed at the center of the unit ball with the surface S^2. Let $K_1(s)$, $K_2(s)$ and $K(s) = K_1K_2$ be the principal curvatures and the Gaussian curvature of Γ at the point s, $H = (K_1 + K_2)/2$ be the mean curvature. Let 0 be the origin of the coordinate system, Γ_+ be the part of Γ which is seen from 0, the illuminated part of Γ, and Γ_- is the shadowed part of Γ. Let ℓ be a unit vector. If $0 \in D$ then $\Gamma_+ = \Gamma$. Define the support function of the convex surface Γ with respect to the origin 0 by the formulas

$$a(\ell) = \max_{s \in \Gamma} (\ell, s), \quad 0 \in D \tag{1}$$

$$a(\ell) = -\max_{s \in \Gamma_+} (\ell, s), \quad 0 \notin D \tag{2}$$

Note that if v runs through a subset of negative numbers then $-\max v = \min |v|$. The support function is the distance from the origin to the tangent plane to Γ perpendicular to the vector ℓ. If $0 \notin D$, $s \in \Gamma_+$, and $\ell = \psi(s)$ is the unit positive normal to Γ at the point s, then $(\ell, s) < 0$. The support function defined by formula (2) is the distance from the origin 0 to the tangent plane to Γ_+ perpendicular to ℓ.

Let us derive the equation of the surface Γ from the support function. It is clear from the definition that $a(\ell)$ can be extended from S^2 to R^3 as a homogeneous function of order 1 so that

$$a(t\ell) = ta(\ell), \quad t > 0. \tag{3}$$

Since the surface is smooth $a(\ell)$ is smooth. It follows from (3) that

$$\sum_{j=1}^{3} \frac{\partial a}{\partial \alpha_j} \alpha_j = a(\ell), \quad \ell = (\alpha_1, \alpha_2, \alpha_3). \tag{4}$$

If $0 \in D$ so that formula (1) holds, then the equation of the surface Γ is of the form

$$x_j = \partial a/\partial \alpha_j, \quad 1 \leqslant j \leqslant 3. \tag{5}$$

This is a parametric equation of the surface, and the role of the parameters is played by the unit vector $\ell = (\alpha_1, \alpha_2, \alpha_3)$. In order to derive (5) let us write the equation of the tangent plane to Γ perpendicular to ℓ at the point s at which $(\ell, s) = a(\ell)$:

$$\sum_{j=1}^{3} x_j \alpha_j = a(\ell). \tag{6}$$

The surface Γ is the envelope of the family of the tangent planes (6). Therefore its equation can be obtained by the formula

$$\frac{\partial}{\partial \alpha_j}\left(\sum_{j=1}^{3} x_j \alpha_j - a(\alpha_1, \alpha_2, \alpha_3)\right) = 0. \tag{7}$$

Equation (7) is equivalent to (5). If $0 \notin D$ and formula (2) holds, then the same argument shows that the equation of the tangent plane is

$$\sum_{j=1}^{3} x_j \alpha_j = -a(\ell), \tag{8}$$

and the equation of the surface is

$$x_j = -\partial a/\partial \alpha_j, \quad 1 \leqslant j \leqslant 3. \tag{9}$$

EXAMPLE. Consider a sphere $\sum_{j=1}^{3} (x_j - b_j)^2 = R^2$. The tangent planes to the sphere orthogonal to the unit vector $\sum_{j=1}^{3} \alpha_j e_j = \ell$, where e_j are the orthogonal unit vectors of the coordinate system, are $\sum_{j=1}^{3} (x_j - b_j \pm R\alpha_j)\alpha_j = 0$. Assuming that the origin is outside of the sphere, i.e. $|b|^2 = \sum_{j=1}^{3} |b_j|^2 > R^2$, one has $a(\ell) = -(b, \ell) - R$ on Γ_+. Therefore Equation (9) is the parametric equation of Γ_+:

$$x_j = b_j + R\alpha_j, \quad 1 \leqslant j \leqslant 3. \tag{10}$$

One should have in mind that the constant $R = R(\alpha_1^2 + \alpha_2^2 + \alpha_3^2)^{1/2}$ should be considered as a restriction on S^2 of the homogeneous function of order 1 defined on R^3 and equal to constant R on S^2 so that $\partial R/\partial \alpha_j = R\alpha_j$.

If the support function can be obtained from the scattering data then the surface can be easily recovered by formula (9). Usually it is convenient to place the origin at some known point and in this case $0 \notin D$, where D is the scatterer. This is the reason why formula (9) rather than (5) is used.

2.2. Let us derive the high frequency asymptotics of the scattering amplitude assuming that the scatterer is convex. The idea is to use the Kirchhoff approximation which is known to give the correct high-frequency asymptotics.

The starting point is the formula (I.4.2) for the scattering amplitude. The function $h = \partial u/\partial N$ in this formula satisfies the equation analogous to (I.3.5).

$$h + Ah = 2\,\frac{\partial \exp(-ik\nu\,.\,s)}{\partial N}, \tag{11}$$

where A is defined in (I.1.4).

In what follows, it is convenient to use $-\nu$ in place of ν

The Kirchhoff approximation is

$$h = 0 \text{ on } \Gamma_-, \quad h = 2ik(N_s, \nu)\exp(ik\nu\,.\,s) \text{ on } \Gamma_+, \tag{12}$$

where $\Gamma_+(\Gamma_-)$ is the illuminated (shadowed) part of Γ, i.e. $(N_s, \nu) < 0$ on Γ_+, $(N_s, \nu) > 0$ on Γ_- (see Fig. 1). Substitute (12) into (I.4.2) to obtain

$$f = -\frac{ik}{2\pi}\int_{\Gamma_+} \exp\{ik(\nu - n)\cdot s\}(N_s, \nu)\,\mathrm{d}s. \tag{13}$$

Here $f = f(-n, -\nu, k) = f(\nu, n, k)$.

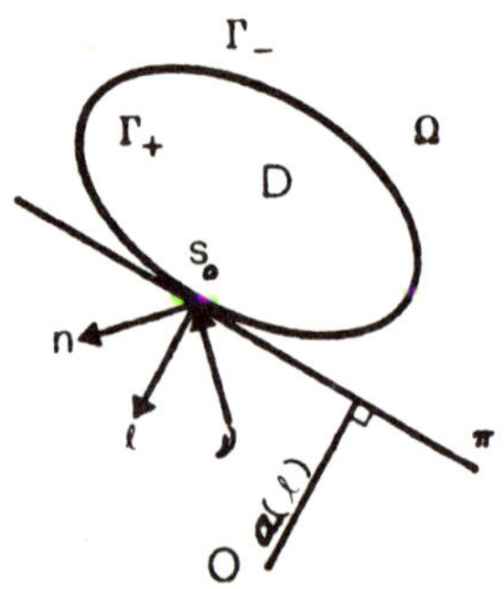

FIGURE 1

Take the origin outside of D and define the support function by formula (2). Compute integral (13) asymptotically as $k \to +\infty$ by the stationary phase method. The following lemma is well known (see e.g. Fedorjuk [1]).

LEMMA 1. *Let* $f \in C_0^\infty(D)$, $S(x) \in C^\infty(D)$, *and* $D \subset R^r$ *be a finite domain. Let* $S(x)$ *be a real-valued function which has only one stationary (critical) point* x_0 *in* D, *and assume that this point is nondegenerate (i.e.* $\mathrm{d}S(x_0) = 0$, $\mathrm{d}^2 S(x_0) \neq 0$). *Then*

$$\int_D f(x) \exp(ikS(x))\, \mathrm{d}x = \left(\frac{2\pi}{k}\right)^{r/2} \exp\left\{ikS(x_0) + \frac{i\pi}{4} \operatorname{sgn} S''(x_0)\right\} \cdot \cdot \left|\det S''(x_0)\right|^{-1/2} \left[f(x_0) + 0\left(\frac{1}{k}\right)\right], \tag{14}$$

where

$$S''(x_0) = \frac{\partial^2 S(x_0)}{\partial x_i\, \partial x_j}, \quad \operatorname{sgn} S''(x_0) = n_+ - n_-, \tag{15}$$

and n_+ (n_-) *is the number of positive (negative) eigenvalues of the matrix* $S''(x_0)$.

In order to apply this lemma let us calculate the term $S(s) = (\nu - n)s$ in (13) around the stationary point s. One has

$$(\nu - n) \cdot s = -|n - \nu| \ell \cdot s, \quad \text{where } \ell \equiv \frac{n - \nu}{|n - \nu|}. \tag{16}$$

The stationary point of the function (16) is the point $s_0 \in \Gamma_+$ at which $\ell = N_{s_0}$. Since Γ is convex there is only one stationary point on Γ_+, and

$$\ell \cdot s = \ell \cdot s_0 + \ell \cdot (s - s_0) = \ell \cdot s_0 + \tfrac{1}{2} b_{ij} u^i u^j + o(|u|^2), \tag{17}$$

where one should sum over the repeated indices, b_{ij} is the matrix of the second differential form of the surface Γ *at the point* s_0, u^i, $i = 1, 2$, are the local coordinates in the tangential plane to Γ, $s_0 = (0, 0)$ in these coordinates and the third axis is directed along $\ell - N_{s_0}$, the unit positive normal to Γ at the point s_0. Therefore

$$S''(s_0) = -|n - \nu| b_j, \quad r = 2, \tag{18}$$

$$\det(-|n - \nu| b_{ij}) = |n - \nu|^2 \det b_{ij} = |n - \nu|^2 K(s_0) \tag{19}$$

where $K = K_1 K_2 = \det b_{ij}$ is the Gaussian curvature, and the principal curvatures K_j, $j = 1, 2$, are the eigenvalues of the matrix b_{ij}. If the surface is convex then the matrix b_{ij} is positive definite, $K_j > 0$, $j = 1$,

2, $n_+ = 2$, $n_- = 0$ and sgn $b_{ij} = 2$. Thus, applying formula (14) to integral (13) yields

$$f = -\frac{ik}{2\pi} \cdot \frac{2\pi}{k} \exp\left\{ -ik|n - \nu|\, \ell \cdot s_0 + \frac{i\pi}{4} \cdot 2 \right\} \cdot$$
$$\cdot \frac{1}{|n - \nu|\{K(s_0)\}^{1/2}} \cdot \left[\ell \cdot \nu + 0\left(\frac{1}{k}\right) \right], \quad k \to \infty. \tag{20}$$

Since $|n - \nu| = 2\, n \cdot \ell$, $\ell \cdot \nu = -\ell \cdot n$ and $\ell \cdot s_0 = -a(\ell)$, one can write (20) as

$$\mathrm{f} = f(\nu, n, k) = -\frac{\exp\{i2kn \cdot \ell\, a(\ell)\}}{2K^{1/2}(s_0)} \cdot$$
$$\cdot \left[1 + 0\left(\frac{1}{k|\ell \cdot \nu|} \right) \right], \quad k \to \infty. \tag{21}$$

Let us formulate the result.

THEOREM 1. *If* Γ *is a smooth convex surface then the asymptotic behavior of the scattering amplitude as* $k \to +\infty$ *is given by formula* (21) *in which* s_0 *is the specular point; that is, the point at which the unit positive normal to* Γ_+ *is the vector* $\ell = (n - \nu)/|n - \nu|$, $K(s_0)$ *is the Gaussian curvature of* Γ *at the specular point* s_0, $a(\ell)$ *is the support function of* Γ_+ *defined by formula* (21), $n \cdot \ell > 0$ *and* $\nu \cdot \ell = -n \cdot \ell < 0$.

From formula (21) it follows that

$$|f(\nu, n, k)| \sim \frac{1}{2K^{1/2}(s_0)} \text{ as } k \to \infty. \tag{22}$$

Actually the Gaussian curvature in (22) is known as a function of the unit normal $\ell = (n - \nu)/|n - \nu|$ to Γ as follows from the derivation of formula (21). The problem of recovering a convex compact closed surface from its Gaussian curvature given as a function of the exterior normal to the surface (Minkowski's problem) has been studied extensively (see Nirenberg [1], and Hsiung [1] where further references are given). It is known (see e.g. Hsiung [1]), that this problem has at most one solution apart from the translations. Existence of a convex surface with its Gaussian curvature equal to a given function $K(\ell)$ has also been studied (see Nirenberg [1]). However, there is no constructive way to recover the surface from its Gaussian curvature: the problem was reduced to a nonlinear elliptic problem. Therefore it is important to use the phase information in the scattering amplitude. The knowledge of the phase of the scattering amplitude (21) allows one to compute the support function $\alpha(\ell)$ and to recover the surface effectively by formula (9). Actually one needs only the knowledge of the high frequency scattering amplitude for a set of n and ν such that the vector $\ell = (n - \nu)/|n - \nu|$ runs through all of S^2. For example, the backscattering amplitude $f(n, -n, k)$, $n \in S^2$ is sufficient for the effective recovery of the surface. If the high frequency scattering amplitude is known for a set of n and ν such that the vector ℓ runs through a subset $\tilde{S}^2$ then formula (9) gives the corresponding part of the surface Γ: the recovery of a part of $\tilde{\Gamma}$ of the surface by formula (9) does not require the knowledge of $\alpha(\ell)$ on all of S^2 but only on $\tilde{S}^2 \subset S^2$, where $\tilde{S}^2 = \psi(\tilde{\Gamma})$ and ψ is the spherical map of the surface defined in Section 2.1.

One sees from (21) that the knowledge of the scattering amplitude at one sufficiently high frequency and n and ν running through such a subset of $S^2 \times S^2$ that $\ell = (n - \nu)/|n - \nu|$ runs through all of S^2 is sufficient for recovering the support function $\alpha(\ell)$ and, therefore, the surface Γ itself.

2.3. The knowledge of the mean curvature of a convex closed smooth surface determines the surface uniquely up to translations. Existence of the surface with the mean curvature equal to a given function was studied in the literature (see Husing [1] and Nirenberg [1]). The problem was reduced to a nonlinear elliptic problem and there is no constructive way to recover the surface from its mean curvature.

2.4. However, there is a simple and constructive way to recover the surface from its harmonic curvature which is defined as

$$h = K_1^{-1} + K_2^{-1}. \tag{23}$$

The function $h = h(\ell)$ is assumed to be known as a function of the unit normal ℓ to the surface. Let us derive the equation of the surface from the knowledge of $h(\ell)$. Actually, the support function $\alpha(\ell)$ will be recovered from $h(\ell)$. If $a(\ell)$ is found then the equation of the surface is Equation (9).

Let us extend $a(\ell)$ and $h(\ell)$ from S^2 to R^3 and set $y = r\ell$

$$A(y) = ra(\ell), \quad h(y) = r^{-1}h(\ell). \tag{24}$$

LEMMA 2. *The following equation holds*:

$$\nabla_y^2 A = h(y). \tag{25}$$

Proof. Let $\ell = (\alpha_1, \alpha_2, \alpha_3)$ and $x = x(\alpha_1, \alpha_2, \alpha_3)$ be the radius vector of a point of the surface Γ. Let R be the radius of the curvature of the normal cut of Γ in the principal direction $(d\alpha_1, d\alpha_2, d\alpha_3)$. From the Rodrigues's formula

$$dx_j - R\, d\alpha_j = 0, 1 \leqslant j \leqslant 3, \tag{26}$$

and formula (9) it follows that

$$0 = \frac{\partial^2 a}{\partial\alpha_j\, \partial\alpha_j}\, d\alpha_i - R\, da_j = \left(\frac{\partial^2 a}{\partial\alpha_i \partial\alpha_j} - R\, \delta_{ij}\right) d\alpha_i. \tag{27}$$

Here and below one sums over the repeated indices, $\delta_{ij} = \left\{\begin{smallmatrix} 0 & i\neq j \\ 1 & i=j \end{smallmatrix}\right.$. Since the vector $(d\alpha_1, d\alpha_2, d\alpha_3) \neq 0$ one concludes from (27) that

$$\det\left(\frac{\partial^2 a}{\partial\alpha_i \partial\alpha_j} - R\delta_{ij}\right) = 0. \tag{28}$$

Furthermore,

$$\det \frac{\partial^2 a}{\partial\alpha_i \partial\alpha_j} = 0. \tag{29}$$

Indeed, differentiate (4) in α_i to obtain

$$\frac{\partial^2 a}{\partial\alpha_i \partial\alpha_j} + \frac{\partial a}{\partial\alpha_j}\, \delta_{ji} = \frac{\partial a}{\partial\alpha_i}, \quad 1 \leqslant i \leqslant 3$$

which is equivalent to

$$\frac{\partial^2 a}{\partial\alpha_j\, \partial\alpha_i}\, \alpha_j = 0 \tag{31}$$

Since $(\alpha_1, \alpha_2, \alpha_3) = \ell \neq 0$, one concludes from (31) that (29) holds.

It follows from (29) that Equation (28) has a root $R = 0$. Its two other roots R_1 and R_2 are the principal radii of the surface, $R_i = K_i^{-1}$. Since the surface is convex, $R_i > 0$, $i = 1, 2$. The trace of the matrix $\partial^2 a/(\partial\alpha_i\, \partial\alpha_j)$ is equal to $R_1 + R_2 + 0 = h(\alpha_1, \alpha_2, \alpha_3) = h(\ell)$. Since $\operatorname{tr} \partial^2 a/(\partial\alpha_i\, \partial\alpha_j) = \nabla_\alpha^2 a$ one has

$$\nabla_\alpha^2 a = h. \tag{32}$$

Because of the properties (24) Equation (32), which is Equation (25) for $|y| = 1$, implies (25).

Let us give a simple effective algorithm for solving Equation (25). Let (θ, ϕ) be the spherical coordinates of the end point of vector $\ell = (\alpha_1, \alpha_2, \alpha_3)$ and Y_{nm} be the spherical harmonics normalized in $L^2(S^2)$:

$$Y_{nm}(\theta, \phi) = \left(\frac{(2n+1)\,(n-m)!}{4\pi(n+m)!}\right)^{1/2} P_{n,m}(\cos\theta)\exp(im\phi), \tag{33}$$

$n = 0, 1, 2, \ldots, -n \leqslant m \leqslant n$. The known function $H(\ell) = h(\theta, \phi)$ can be expanded in the Fourier series

$$h = \sum_{n=0}^{\infty} \sum_{m=-n}^{n} h_{nm} Y_{nm}, \tag{34}$$

where

$$h_{nm} = \int_{S_2} h(\theta, \phi)\overline{Y}_{nm}(\theta, \phi)\, \mathrm{d}\omega, \quad \mathrm{d}\omega = \sin\theta\, \mathrm{d}\theta\, \mathrm{d}\phi, \tag{35}$$

and the bar denotes the complex conjugate. From (34) and (24) it follows that

$$h(y) = \frac{1}{r} \sum_{n=0}^{\infty} \sum_{m=-n}^{\infty} h_{nm} Y_{nm}(\theta, \phi) \tag{34'}$$

where (r, θ, ϕ) are the spherical coordinates of the point $y = r\ell$. A simple calculation shows that

$$\begin{aligned} \nabla_y^2 r Y_{nm}(\theta, \phi) &= \left(\frac{\partial^2}{\partial y_1^2} + \frac{\partial^2}{\partial y_2^2} + \frac{\partial^2}{\partial y_3^2}\right) r Y_{nm} \\ &= -\frac{(n-1)\,(n+2)}{r} Y_{nm} \end{aligned} \tag{36}$$

Therefore Equation (25) can be written as

$$-\sum_{n=0}^{\infty}\sum_{m=-n}^{n}\frac{(n-1)(n+2)}{r}A_{nm}Y_{nm}=\frac{1}{r}\sum_{h=0}^{\infty}\sum_{m=-n}^{n}h_{nm}Y_{nm} \tag{37}$$

where A_{nm} are the coefficients in the expansion

$$A(y)=r\alpha(\theta,\phi)=r\sum_{n=0}^{\infty}\sum_{m=-n}^{n}A_{nm}Y_{nm}. \tag{38}$$

From (37) it follows that

$$A_{nm}=-\frac{h_{nm}}{(n-1)(n+2)},\quad n\neq 1. \tag{39}$$

We will prove shortly that a necessary condition for a function $h(\theta,\phi)$ to be equal to R_1+R_2, where R_i, $i=1, 2$ are the principal radii of a closed smooth surface, is

$$h_{1m}=0,\quad m=0,\quad \pm 1 \qquad \square \tag{40}$$

Let us summarize the result:

THEOREM 2. *If $h(\theta,\phi)$ is the sum of the principal radii of a smooth closed convex surface then the support function $\alpha(\ell)$ of this surface is given by the formula*

$$\alpha(\theta,\phi)=-\sum_{n=0,\,n=1}^{\infty}\sum_{n=-m}^{n}\frac{h_{nm}}{(n-1)(n+2)}Y_{nm}(\theta,\phi) \tag{41}$$

where h_{nm} are given by formula (35). *The equation of the surface is given by formula* (9) *in which $(\alpha_1, \alpha_2, \alpha_3)$ are the cartesian coordinates of the point $(\theta,\phi)\in S^2$.*

It remains to be proved that (40) holds.

LEMMA 3. *If Γ is a closed compact surface (smooth enough for the Gauss formula to hold), then Equation* (40) *holds.*

Proof. Let ds be the element of the surface area of Γ, $\mathrm{d}s=R_1\,\mathrm{d}s_1\,\mathrm{d}s_2$, where $\mathrm{d}s_1\,\mathrm{d}s_2=\mathrm{d}\omega$ is the element of the surface area of S^2, $|R_j\,\mathrm{d}s_j|$, $j=1, 2$, are the elements of the lengths along the lines of principal curvatures on Γ, and $\mathrm{d}s_1\,\mathrm{d}s_2$ is the inner product of the vectors $\mathrm{d}s_1$ and

ds_2. Consider a family Γ_z of parallel surfaces for which the normals are the same as for $\Gamma = \Gamma_0$. Then

$$ds(z) = (R_1 + z)(R_2 + z)\, ds_1\, ds_2 = ds + z(R_1 + R_2)\, d\omega = z^2\, d\omega \quad (42)$$

where $ds(z)$ is the element of the surface area of the surface Γ_z. From the Gauss formula it follows that

$$\int_{\Gamma_z} N_j\, ds(z) = 0, \quad j = 1, 2, 3, \quad 0 \leqslant z \leqslant z_0, \quad (43)$$

where N_j is the j-th component of the exterior unit normal to Γ_z, and z_0 is so small that N does not depend on z. From (42) and (43) one obtains

$$\int_{S^2} (R_1 + R_2) N_j\, d\omega = 0, \quad 1 \leqslant j \leqslant 3. \quad (44)$$

Since $R_1 + R_2 = h$, $N_1 = \sin\theta \cos\phi = \frac{1}{2}(Y_{11} + Y_{1-1})$, $N_2 = \sin\theta$ $\sin\phi = (1/2i).(Y_{11} - Y_{1-1})$ and $N_3 = \cos\theta = Y_{10}$, it is clear that (44) is equivalent to (40). □

EXAMPLE. If Γ is a sphere then $R_1 = R_2 = R = \text{const}$, $h_{nm} = 0$ for $n > 0$. Therefore $A_{nm} = 0$ for $n > 0$, $a(\theta, \phi) = A_0 Y_0 = A_0/\sqrt{4\pi}$, $A_0 = h_0/2$, $h_0 = 2R\sqrt{4\pi}$. Thus, $a(\ell) = R$ and the sphere is recovered from the knowledge of the function $h = 2R = \text{const}$. □

We proved that (40) is a necessary condition for $h(\theta, \phi)$ to be the sum of the principal radii of a closed smooth surface Γ. Is this condition also sufficient?

By the algorithm given in Theorem 2 one can construct $a(\theta, \phi)$ (formula (41)) and then the surface by formula (9). It is not guaranteed, however, that this algorithm produces a convex surface.

LEMMA 3. *A necessary and sufficient condition for h to be the sum of the principal radii of a closed convex surface is*

$$(40) \text{ holds}, \quad (45)$$

and

$$\int_{S^2} \frac{(\ell \cdot \ell_2)(\nabla h(\ell) \cdot \ell_2)}{1 - (\ell \cdot \ell_1)}\, d\omega \leqslant 0, \quad \ell_2 \cdot \ell_1 = 0 \quad (46)$$

for all $\ell_2, \ell_1 \in S^2$ *such that* $\ell_2 \cdot \ell_1 = 0$, *with strict inequality in* (46) *for some* ℓ_2 *and* ℓ_1. *The symbol* $\nabla h(\ell)$ *means* $\nabla_y h(y)|_{y=\ell}$.

A proof of this result can be found in Firey [1]. Since the result is not used in the sequel its proof is not given here.

One can give another necessary and sufficient condition for h to be the sum of the principal radii of a convex closed smooth surface: (45) and

$$\text{the matrix } a_{ij}(\theta, \phi) \text{ is semi-positive definite for all } (\theta, \phi), \tag{47}$$

where

$$a_{ij}(\theta, \phi) = \frac{\partial^2 a(\ell)}{\partial \alpha_i \, \partial \alpha_j} = - \sum_{n=0,\, n\neq 1}^{\infty} \sum_{m=-n}^{n} \frac{h_{nm}}{(n-1)(n+2)} \left. \frac{\partial^2 r Y_{nm}(\theta, \phi)}{\partial y_i \, \partial y_j} \right|_{r=1}. \tag{48}$$

A matrix a_{ij} is semi-positive definite iff $\sum_{i,j=1}^{n} a_{ij} t_i \bar{t}_j \geqslant 0$ for all t_i. Condition (47) follows from (41) and from the following lemma.

LEMMA 4. *Let* $x(\ell)$, $\ell = (\alpha_1, \alpha_2, \alpha_3)$ *be the normal of the surface* Γ, *that is* $x(\ell)$ *is the point of* Γ *at which the exterior unit normal to* Γ *is* ℓ. *A continuously differentiable function* $x(\ell)$ *is the normal representation of a convex surface* Γ *iff the matrix* $\partial x_i/\partial \alpha_j$ *is a semi-positive definite matrix for all* $\ell \in S^2$.

Proof. (i) If Γ is convex then $a(\ell) = \ell \cdot x(\ell)$, $x_j = \partial a(\ell)/\partial \alpha_j$, $\partial x_i/\partial \alpha_j = \partial^2 a(\ell)/(\partial \alpha_i \, \partial \alpha_j) \equiv a_{ij}$, and this matrix is semi-positive definite if Γ is convex. This proves the necessity. (ii) Suppose now that the matrix $\partial x_i/\partial \alpha_j$ is semi-positive definite. Then so is the matrix a_{ij}. But then the support function $a(\ell)$ of Γ is convex and therefore Γ is convex. □

3. STABILITY OF THE SURFACE WITH RESPECT TO SMALL PERTURBATIONS OF THE DATA

3.1. If $k = k_0$ is so large that the reminder term in formula (2.21) $0(1/(k_0|\ell \cdot \nu|)) < 1$, then formula (2.21) shows that $f(\nu, n, k_0)/|f(n, \nu, k_0)| = -\exp\{i2k_0 n \cdot \ell \, a(\ell)\}\,(1 + 0(1/(k_0 b))$ where $|\ell \cdot \nu| \geqslant b > 0$.

Therefore

$$a(\ell) = \frac{1}{2ik_0 n \, . \, \ell} \ln \left\{ - \frac{f(\nu, n, k_0)}{|f(\nu, n, k_0|} \right\} + 0 \left(\frac{1}{(k_0 b)^2} \right). \tag{1}$$

Let us formulate the algorithm for recovering the surface from the high-frequency scattering data based on formula (1).

THEOREM 1. *If the surface is smooth, convex and closed and* k_0 *is sufficiently large, then the support function of the surface is given by formula* (1) *and the equation of the surface is given by formula* (2.9).

3.2. Suppose that the scattering data are measured with some error $\delta > 0$, so that f_δ is measured, and $|f_\delta - f| \leqslant \delta$. Assume that

$$\sup_{s \in \Gamma} K(s) \leqslant \mathrm{d}^2, \quad |\ell \cdot \nu| \geqslant b > 0. \tag{2}$$

Then

$$f_\delta = -\frac{\exp\{i2kn \cdot \ell\, \alpha(\ell)\}}{2K^{1/2}(s_0)}\left[1 + 0\left(\frac{1}{kb}\right) + 0(\delta d)\right]. \tag{3}$$

Therefore

$$\alpha(\ell) = \frac{1}{2ik_0 n \cdot \ell} \ln\left\{-\frac{f_\delta(\nu, n, k_0)}{|f_\delta(\nu, n, k_0)|}\right\} + 0\left(\frac{1}{(k_0 b)^2} + \frac{\delta d}{k_0 b}\right). \tag{4}$$

The remainder in formula (4) depends on the error δ. Therefore one sees that the error in the measurements is added to the error of the asymptotic formula. The reconstruction algorithm for noisy data is the same as in Theorem 1, but one uses formula (4) for $\alpha(\ell)$ instead of formula (1). The basic point to discuss is the ill-posedness of the inverse diffraction problem, which comes from the differentiation of the data given with some error.

Let us denote

$$\alpha_\beta(\ell) = \frac{1}{2ik_0 n \cdot \ell} \ln\left\{-\frac{f_\delta(\nu, n, k_0)}{|f_\delta(\nu, n, k_0)|}\right\}. \tag{5}$$

Then

$$|a_\beta(\ell) - \alpha(\ell)| < \beta, \quad \beta = 0\left(\frac{1}{(k_0 b)^2} + \frac{\delta d}{k_0 b}\right). \tag{6}$$

We want to estimate the derivatives $\partial \alpha(\ell)/\partial \alpha_j$, $\ell = (\alpha_1, \alpha_2, \alpha_3)$ from the knowledge of $a_\beta(\ell)$ and β. It is impossible to estimate $\partial a/\partial\alpha_j$ stably without additional information about $a(\ell)$. This additional information can be taken in the form of the apriori assumption

$$|a''(\ell)| \leqslant M. \tag{7}$$

Here a'' denotes any second derivative of $a(\ell)$.

The stable estimate $\partial\hat{a}/\partial\alpha_j$ is such estimate that

$$\varepsilon(\beta) = \left|\frac{\partial\hat{a}}{\partial\alpha_j} - \frac{\partial a}{\partial\alpha_j}\right| \to 0 \quad \text{as} \quad \beta \to 0. \tag{8}$$

Under the assumption (7) one calculates the stable estimate of the derivative $\partial\hat{a}/\partial\alpha_j$ by the formula in which $j = 1$

$$\frac{\partial\hat{a}}{\partial\alpha_1} = \frac{a(\alpha_1 + h(\beta), \alpha_2, \alpha_3) - a(\alpha_1 - h(\beta), \alpha_2, \alpha_3)}{2h(\beta)}. \tag{9}$$

Here

$$h(\beta) = \left(\frac{2\beta}{M}\right)^{1/2} \tag{10}$$

and the error of the estimate (9) is

$$\varepsilon(\beta) \leqslant (2M\beta)^{1/2}. \tag{11}$$

These results are proved in Appendix 4, where it is also proved that the estimate (9) is the best possible estimate (in a certain sense) among all linear and nonlinear estimates on the set of data specified by the number (β, M).

To conclude this section let us summarize the basic result: an algorithm is given for recovery of the surface from the scattering amplitude measured at one sufficiently high frequency and for the set of n and ν such that the vector $\ell = (n - \nu)/|n - \nu|$ runs through all of S^2. The measurements are noisy. The algorithm is stable with respect to the noise and to the error due to the asymptotic nature of the high frequency approximation of the scattering amplitude.

Remark. One can solve the inverse diffraction problem with the impedance boundary condition (1.6). The argument, based on the

Kirchhoff approximation, leads to the following formula for the leading term of the scattering amplitude for the convex impedance obstacle:

$$f_\eta = Bf, \quad \text{where} \quad B \equiv \frac{ikv \cdot \ell + \eta}{-ikv \cdot \ell + \eta} \tag{12}$$

where $\eta = \eta(s)$ is the function in the boundary condition (1.6), and $f = f(v, n, k)$ is given by (2.21). Notice that if $\eta = \infty$ (the Dirichlet boundary condition) then $B = 1$; if $\eta = 0$ (the Neumann boundary condition) then $B = -1$. If $|kv \cdot \ell| \gg |\eta|$, then $B = -1$, that is the body scatters as if it were acoustically hard ($\eta = 0$). Formulas (12) and (2.21) allow one to find both the shape of Γ and the impedance $\eta(s)$ from the scattering data $f(v, n, k)$ at a fixed large k and $v, n \in S^2$.

Chapter III

Time-Dependent Problems

1. STATEMENT OF THE PROBLEMS

1.1. Consider the problem

$$u_{tt} = \nabla^2 u \text{ in } \Omega, \quad t > 0 \tag{1}$$

$$u = 0 \text{ on } \Gamma \tag{2}$$

$$u(x, 0) = 0, \quad u_t(x, 0) = f(x). \tag{3}$$

where $f \in C_0^\infty(\Omega)$.

Define

$$v(x, k) = \int_0^\infty \exp(ikt)u(x, t)\,\mathrm{d}t. \tag{4}$$

Then

$$\nabla^2 v + k^2 v = -f \quad \text{in } \Omega, \tag{5}$$

$$v = 0 \qquad \text{on } \Gamma, \tag{6}$$

$$\int_{S_r} \left| \frac{\partial v}{\partial r} - ikv \right|^2 \mathrm{d}s \to 0, \quad r \to \infty, \tag{7}$$

and

$$u(x, t) = \frac{1}{2\pi} \int_{-\infty}^{\infty} \exp(-ikt)v(x, k)\,\mathrm{d}k. \tag{8}$$

The solution to (5)—(7) can be written as

$$v(x, k) = \int_\Omega G(x, y, k)f(y)\,\mathrm{d}y \tag{9}$$

where $G(x, y, k)$ is the Green function for problem (5)—(7). The Green function satisfies the equation

$$G = g - \int_\Gamma gh \, \mathrm{d}s, \quad h = \frac{\partial G(s, y, k)}{\partial N_s} \tag{10}$$

and h solves the equation

$$[I + A(k)]h = 2\, \frac{\partial g}{\partial N_s} \tag{11}$$

where $A(k)$ is defined by formula (I.1.4). Since the operator function $A(k)$ is compact and analytic in k on the whole complex plane k, and since $I + A(k)$ is invertible for some k (e.g. for Im $k > 0$, or for $k > 0$ such that k^2 is not an eigenvalue of the interior Neumann Laplacian), it follows from Lemma I.4.2 that h is a meromorphic function of k on the whole k- plane and the Laurent coefficients of h are finite rank (degenerate) kernels. The Green function G has the same properties. This follows from (10) since g is analytic in k. The functions G and h are analytic in k in the half-plane Im $k \geqslant 0$ since for Im $k > 0$ the operator $(I + A(k))^{-1}$ is invertible and on the real lime Im $k = 0$ the Green function has bounded limit values as follows from the limiting absorption principle (Theorem I.2.3). Therefore G is analytic in a neighborhood of the real axis. The poles of G in the half-plane Im $k < 0$ are called complex poles, resonances, and natural frequencies by mathematicians, physicists, and engineers. These poles play an important role in the study of the behavior of the solution to problem (1)—(3) for large t. In order to explain this let us assume that function (9) satisfies the estimate

$$\begin{gathered} |v(x, k)| \leqslant c(b)(1 + |k|)^{-a}, \ a > \tfrac{1}{2}, \\ |\mathrm{Re}\, k| \to \infty, \quad \mathrm{Im}\, k = b, \ -\infty < b < \infty, \end{gathered} \tag{12}$$

uniformly in x in compact domains, where b is an arbitrary constant. Estimate (12) and the meromorphic nature of v imply that

$$|\mathrm{Im}\, k_j| \to \infty \quad \text{as} \quad j \to \infty, \tag{13}$$

where k_j are the poles of v ordered so that

$$|\mathrm{Im}\, k_1| \leqslant |\mathrm{Im}\, k_2| \leqslant \ldots . \tag{14}$$

The poles of v are among the poles of G but due to some special choice of f it can happen that some of the poles of G are not poles of v. If (12) holds then one can move contour in (8) down and obtain the asymptotic expansion of $u(x, t)$ for large t:

$$u(x, t) = \sum_{j=1}^{N} c_j(x, t) \exp(-ik_j t) + 0(\exp\{-|\mathrm{Im}\, k_{N+1}|t\}), \quad t \to +\infty. \tag{15}$$

This argument is known as the asymptotic singularity expansion method (SEM). The coefficients $c_j(x, t)$ can be written as

$$c_j(x, t) = \operatorname*{Res}_{k=k_j} \{v(x, k) \exp(-ikt)\}. \tag{16}$$

If m_j is the order of the pole k_j then it follows from (16) that

$$c_j(x, t) = 0(t^{m_j - 1}), \quad t \infty +\infty. \tag{17}$$

If the pole k_j is simple, $m_j = 1$, then

$$c_j(x, t) = c_j(x). \tag{18}$$

In the literature one can see the formula

$$u(x, t) = \sum_{j=1}^{\infty} c_j(x, t) \exp(-ik_j t). \tag{19}$$

This formula is not justified and does not hold in general. It is an open problem to find out the cases when (19) holds. Convergence of the series in (19) can be understood in various ways. For example, it can be uniform convergence on compacts in $R^3 \times R_+$, $x \in R^3$, $t \in R_+$.

It is interesting to study the properties of the complex poles. Let us mention some questions of interest: (1) existence of infinitely many poles, (2) behavior of the pole k_j as $j \to \infty$, e.g. property (13), (3) estimates from below and from above of Im k_1 on the class of obstacles $D \subset B_R$, where R is a fixed number; it is possible that a bound on the curvatures of the surface Γ will enter the estimates, (4) in which cases does (13) not hold and there exists a sequence of poles k_j such that Im $k_j \to 0$, $j \to \infty$? in which cases does (13) not hold but there exists a constant $b > 0$ such that the strip $-b <$ Im $k < 0$ is free from the poles? (5) when does (12) hold and when does it not hold?

These questions are essentially open although partial answers are known and will be discussed later.

1.2. Let us write the integral equation similar to (10) in the time domain. Applying to (10) the inverse Fourier transform (8) one obtains

$$\mathscr{G}(x, y, t) = \mathscr{G}_0(x, y, t) - \int_\Gamma \frac{ds}{4\pi r_{xs}} \frac{\partial \mathscr{G}(s, y, t - r_{xs})}{\partial N_s}, \tag{20}$$

where

$$\mathscr{G}_0(x, y, t) = \frac{\delta(r_{xy} - t)}{4\pi r_{xy}}, \tag{21}$$

$\delta(x)$ is the delta function, and $\mathscr{G} = 0$ for $t < 0$.

The scattering problem in the time domain one obtains by taking the inverse Fourier transform (8) of the problem (0.1)—(0.5):

$$\nabla^2 U - U_{tt} = 0 \quad \text{in} \quad \Omega \times R_+ \tag{22}$$

$$U(x, t) = 0, \quad x \in \Gamma \tag{23}$$

$$U(x, t) = \delta(t - n \cdot x) + V(x, t), \tag{24}$$

$$r\left(\frac{\partial V}{\partial r} + \frac{\partial V}{\partial t}\right) \to 0, \quad r \to \infty. \tag{25}$$

Note that this is a problem without initial data. The integral equation (20) was not used for deriving any qualitative properties of Green's function such as its existence, uniqueness, and behavior at large t.

1.3. Consider the problem

$$u_{tt} = \nabla^2 u + f(x) \exp(-ik_0 t) \text{ in } \Omega, \quad k_0 > 0, f \in C_0^\infty(\Omega) \tag{26}$$

$$u(x, 0) = u_t(x, 0) = 0 \tag{27}$$

$$u(s, t) = 0 \text{ on } \Gamma. \tag{28}$$

The limiting amplitude principle says that the limit

$$v(x, k_0) = \lim_{T \to \infty} T^{-1} \int_0^T u(x, t) \exp(ik_0 t)\, dt \tag{29}$$

exists in an appropriate topology (e.g. uniformly in $x \in B_R$, for any fixed $R > 0$, or in $L^2(B_R)$) and solves the steady state problem (5)–(7). The validity of this principle can be established for problem (26)–(28). This is done in Section 2 in an abstract setting. Of some interest is to give conditions under which the limit (29) can be substituted by the limit

$$v(x, k_0) = \lim_{t \to \infty} u(x, t) \exp(ik_0 t). \tag{30}$$

It is interesting to estimate the rate at which $u(x, t)$ approaches the steady state $v(x, k_0) \exp(-ik_0 t)$, that is to estimate

$$u(x, t) - v(x, k_0) \exp(-ik_0 t).$$

If one uses the analytic properties of $v(x, k)$ in the half-plane $\operatorname{Im} k < 0$ then this question is a part of the singularity expansion method, and the result is of the type (15). We give also another method which is based on the spectral theorem for self-adjoint operators.

Some of the basic results and definitions from the spectral theory of self-adjoint operators are given in Appendix 2. These results and definitions are used in the rest of this chapter.

2. THE LIMITING AMPLITUDE PRINCIPLE (ABSTRACT RESULTS)

2.1. Let us develop an abstract theory. Let H be a Hilbert space, L be a linear, densely defined in H, closed, not necessarily self-adjoint operator. Let P be an orthoprojection in H onto a certain subspace $F \subset H$. In applications $F = L^2(B_R \setminus D)$, so that F is a subspace of functions with support in a certain compact domain. Let

$$u_{tt} + Lu = f \exp(-ik_0 t), \qquad k_0 > 0, \quad f \in H \tag{1}$$

$$u(0) = u_t(0) = 0. \tag{2}$$

The limiting amplitude principle says that the limit in H

$$Pv = \lim_{T \to \infty} T^{-1} \int_0^T Pu(t) \exp(ik_0 t)\, dt \tag{3}$$

exists, and v solves the problem

$$Lv - k_0^2 v = f. \tag{4}$$

In the problems for differential equations, (4) has many solutions, and the requirement that its solution is the limit (3) of the uniquely solvable problem (1)—(2) is one of the ways to select a unique solution to (4). The other way, which is called the limiting absorption principle, is to solve the problem

$$(L - (k_0 + i\varepsilon)^2)v_\varepsilon = f, \quad \varepsilon > 0 \tag{5}$$

and to pass to the limit $\varepsilon \to +0$

$$Pv = \lim Pv(k_0 + i\varepsilon), \quad \varepsilon \to +0. \tag{6}$$

If this limit exists and the element v solves in some sense Equation (4) then the limiting absorption principle holds. In the problems for differential equations the element v often does not belong to H, but it solves Equation (4) in the sense of the theory of distributions and in H^s_{loc}. The classical way to select a unique solution to (4) in the case when L is the Dirichlet or Neumann Laplacian in Ω is to use the radiation condition. The limiting amplitude and limiting absorption principles in this case will give the solution which satisfies the radiation condition. The limiting absorption and the limiting amplitude principles are more general than the radiation condition. For example, in the cylindrical domain the radiation condition (1.7) is not applicable.

Let us introduce the class of linear operators which we study. Applying the Laplace transform to (1) yields

$$(L + p^2 I)\bar{u} = (p + ik_0)^{-1} f,$$
$$\bar{u} = \int_0^\infty \exp(-pt)u(t)\,\mathrm{d}t, \quad p = \sigma + i\tau. \tag{7}$$

Let us formulate the basic assumptions.

A. The operator $R(-p^2) = (L + p^2 I)^{-1}$ exists, is bounded, and is defined on all of H for all p with $\mathrm{Re}\,p = \sigma > 0$ except, possibly, for a finite number of points $p_0, \ldots, p_M$, $\mathrm{Re}\,p_0 > \mathrm{Re}\,p_1 > \ldots$, which are poles of the operator $R(-p^2)$, and the following estimates hold

$$\|PR(-p^2)f\| \leqslant \frac{c}{1 + |p|^\gamma}, \quad \gamma > 0,$$
$$\sigma > 0, |\mathrm{Im}\,p| = |\tau| > N > 0, \quad c = c(f) \tag{8}$$

where N is sufficiently large, γ is a positive number, and the constant c does not depend on p. If $|\tau| \leqslant N$ and $\sigma = 0$ then $PR(-p^2)f$ may have a finite number of points $p = i\omega_m$, $\omega_m \neq 0$ in a neighborhood of which

$$PR(-p^2)f = \frac{q(p, f)}{(p - i\omega_m)^\delta}, \quad 0 \leqslant \delta < 2, \quad 1 \leqslant m \leqslant m_0, \quad \omega_m \neq 0 \tag{9}$$

where $\|q(p, f)\| < c$ if $|p - i\omega_m| \leqslant \varepsilon$, $\sigma > 0$, $\varepsilon > 0$ is a small number, c denotes various constants which do not depend on p, without loss of generality it is assumed that δ does not depend on m, and $\|PR(-p^2)f\| \leqslant c$ if $|p - i\omega_m| > \varepsilon$, $\sigma > 0$, $m = 1, 2, \ldots, m_0$, $|\tau| < N$ and $c = c(\varepsilon, N)$ does not depend on $\sigma > 0$. If there are no poles of $R(-p^2)$ in the half-plane $\sigma > 0$ and A holds, then we say that A_0 holds. Assume that $\{\delta = 0$ and $A_0\}$ implies (11) for $-\infty < \tau < \infty$ and Hölder continuity of $W(\tau)$ in (11).

B_τ. There exists the following limit in H:

$$\lim_{p \to 0} PR(-(p - i\tau)^2)f = W(\tau), \quad |\arg p| < \frac{\pi}{2} \tag{10}$$

where τ is a real number. If this limit exists for all $-\infty < \tau < \infty$ then we say that condition B holds.

C. Suppose that B_τ holds and

$$\|PR(-(p - i\tau)^2)f - W(\tau)\| \leqslant c(\tau)|p|^\alpha, \quad 0 < \alpha \leqslant 1, \quad |\arg p| < \frac{\pi}{2}. \tag{11}$$

Then we say that condition C_τ holds. If (11) holds for any $\tau \neq 0$ and $c(\tau) \leqslant c(a)$ if $|\tau| \geqslant a > 0$, where $a > 0$ is an arbitrary number, then we say that condition C holds.

Remark 1. The above assumptions and most of our arguments and results below remain valid for linear operators in a Banach space.

2.2. In this section a lemma will be proved. This lemma is used extensively in what follows.

LEMMA 1. *Let $F(p)$ be a function with values in H,*

$$F(p) = \int_0^\infty \exp(-pt)f(t)\,dt, \quad \sigma > \sigma_0 \geqslant 0. \tag{12}$$

Assume that

$$F(p) \text{ is analytic in the half-plane } \sigma > 0, \tag{13}$$

$$F(p) \text{ satisfies conditions (8) and (9) with } F(p) \text{ in the place of } PR(-p^2)f, \tag{14}$$

and

$$\lim_{p \to 0} p^{\alpha} F(p) = b, \ \alpha > 0, 1 + \alpha > \delta, \text{ where } \delta \text{ is defined in (9).} \tag{15}$$

Then

$$\lim_{T \to \infty} T^{-\alpha} \int_0^T f(t) \, \mathrm{d}t = \frac{b}{\Gamma(1 + \alpha)}, \tag{16}$$

where $\Gamma(t)$ *is the Gamma function. If*

$$\limsup_{p \to 0} \| p^{\alpha} F(p) \| \leqslant c, \tag{17}$$

then

$$\limsup_{T \to \infty} \left\| T^{-\alpha} \int_0^T f(t) \, \mathrm{d}t \right\| \leqslant c_1. \tag{18}$$

Remark 2. Lemma 1 is a Tauberian lemma in the sense that some conclusion about $f(t)$ is obtained from the assumption about its Laplace's transform. But it is an unusual Tauberian lemma since the known Tauberian theorems have different logical structure: one assumes something about the integral transform of $f(t)$ and about $f(t)$ itself. These additional assumptions on f are called Tauberian conditions. For example, one assumes often that $f(t) > -c$ where c is a constant (see e.g. Hardy [1]). In Lemma 1 there are no assumptions on $f(t)$. All the assumptions are on $F(p)$. This is the unusual feature of Lemma 1.

Proof of Lemma 1. If $F(p)$ satisfies conditions (13) and (8) then $F(p)$ is a Laplace tranform of a function $f(t) \in L^1(0, \infty)$, and

$$\int_0^T f \, \mathrm{d}t = \frac{1}{2\pi i} \int_{c - i\infty}^{c + i\infty} \frac{\exp(pT) - 1}{p} F(p) \, \mathrm{d}p, \quad c > 0. \tag{19}$$

Let us set $c = T^{-1}$ and $y = pT$. Then

$$I(T) \equiv \frac{1}{T^{\alpha}} \int_0^T f(t)\,\mathrm{d}t = \frac{1}{2\pi i T^{\alpha}} \int_{T^{-1}-i\infty}^{T^{-1}+i\infty} \frac{\exp(pT)-1}{p} F(p)\,\mathrm{d}p$$
$$= \frac{1}{2\pi i} \int_{1-i\infty}^{1+i\infty} \frac{\exp(y)-1}{y^{1+\alpha}} \left(\frac{y}{T}\right)^{\alpha} F\left(\frac{y}{T}\right) \mathrm{d}y. \tag{20}$$

Let $T \to +\infty$. Condition (15) says that for any fixed y the integrand in (20) tends to $((\exp(y)-1)/y^{1+\alpha})b$. We want to apply Lebesgue's dominated convergence theorem in order to pass to the limit $T \to \infty$ in (20). To check the basic condition of this theorem, assume first that $\delta = 0$ in the estimate (9) for $F(p)$, and consider the inequality.

$$\left\| \frac{\exp(y)-1}{y^{1+\alpha}} \left(\frac{y}{T}\right)^{\alpha} F\left(\frac{y}{T}\right) \right\|$$
$$= \left\| \frac{\exp(y)-1}{y^{1+\gamma_1}} \frac{1}{T^{\alpha-\gamma_1}} \left(\frac{y}{T}\right)^{\gamma_1} F\left(\frac{y}{T}\right) \right\| \leqslant \frac{c}{|y|^{1+\gamma_1}}, \tag{21}$$

where estimate (8) for $F(p)$ was used in the form

$$\|p^{\gamma_1} F(p)\| \leqslant c \tag{22}$$

and γ_1 is an arbitrary number in the interval $0 < \gamma < \min(\gamma, \alpha)$. It follows from (21) that the Lebesgue's theorem is applicable to the integral (20). This theorem yields

$$\lim_{T\to\infty} I(T) = \frac{b}{2\pi i} \int_{1-i\infty}^{1+i\infty} \frac{\exp(y)-1}{y^{1+\alpha}} = \frac{b}{\Gamma(1+\alpha)} \tag{23}$$

where $\Gamma(\alpha)$ is the Gamma-function, and the integral is computed by the residue theorem. Thus, Lemma 1 is proved under the assumption that $\delta = 0$, where δ is defined in (9). If $1 < \delta < 2$ then the estimate (22) does not hold in neighborhoods of the points $y_m = 1 + i\omega_m T$. In this case we write

$$I(T) = I_0(T, \varepsilon) + \sum_{m=1}^{m_0} I_m(T, \varepsilon), \tag{24}$$

where I_m is the part of the integral (20) taken over the contour $|T^{-1}y - i\omega_m| < \varepsilon$, $\operatorname{Re} y = 1$, and $\varepsilon > 0$ is fixed, sufficiently small number, $\varepsilon < \min_{m \neq m'} |\omega_m - \omega_{m'}|$, and $I_0(T, \varepsilon)$ is defined by (24). The integrand in $I_0(T)$ can be majorized as in (21). To estimate $I_m(T, \varepsilon)$ one writes

$$\|I_m\| = \left\| \frac{1}{2\pi i T^a} \int_{|T^{-1}+i\tau - i\omega_m| < \varepsilon} \frac{\exp(pt) - 1}{p} F(p)\, \mathrm{d}p \right\| \leqslant$$

$$\leqslant \frac{c}{T^a} \int_{-\varepsilon}^{\varepsilon} \frac{\mathrm{d}x}{(T^{-2} + x^2)^{\delta/2}} \leqslant \frac{c(\delta)}{T^{a+1-\delta}}, \quad 1 < \delta < \min(1 + a, 2), \tag{25}$$

where $c(\delta)$ is a constant which does not depend on ε and T. Therefore, if $1 < \delta < \min(1 + a, 2)$, then

$$\lim_{T\to\infty} I(T) = \lim_{T\to\infty} I_0(T, \varepsilon) + \sum_{m=1}^{m_0} \lim_{T\to\infty} I_m(T, \varepsilon) = \frac{b}{\Gamma(1 + a)}. \tag{26}$$

Notice that although I_0 and I_m depend on ε, their limits, as $T \to \infty$, do not.

Consider now the case of $0 < \delta < 1$. In this case

$$\|I_m\| < \frac{c}{T^a} \int_{-\varepsilon}^{\varepsilon} \frac{\mathrm{d}x}{(T^{-2} + x^2)^{\delta/2}} \leqslant \frac{c(\delta)}{T^a}, \quad a > 0. \tag{27}$$

Finally, if $\delta = 1$, then

$$\|I_m\| \leqslant \frac{c \ln T}{T^a}, \quad a > 0. \tag{28}$$

Given (27) or (28) one obtains (26) as before. The same argument shows that (17) implies (18). □

2.3. Let us formulate the main results which establish the relationship between the spectral properties of the operator L and the behavior of the solution to the nonstationary problem as $t \to +\infty$. In this subsection we drop the projection P in front of $w(t)$ and $\bar{w}(p, h)$ in the proof of Theorem 1 below. This is possible if the analytic properties of $\bar{w}(p, h)$ and $P\bar{w}(p, h)$ are the same. In the problems for differential equations the operator P is the operator of multiplication by a characteristic function of a compact set, $P\bar{w} = \eta\bar{w}$, $\eta = 1$ if $|x| \leqslant R$, $\eta = 0$ if $|x| > R$. Therefore, $\bar{w}(p)$ and $\eta\bar{w}(p)$ have the same analytic properties as functions of p.

THEOREM 1. *Let the assumption A hold. The operator L does not have eigenvalues in the half-plane Re* $\lambda < 0$ *iff*

$$\left\| \int_0^t w(t)\,\mathrm{d}t \right\| = 0(\exp(\varepsilon t)) \quad \text{as} \quad t \to +\infty, \quad \varepsilon > 0, \tag{29}$$

where $w(t)$ *solves the problem*

$$w_{tt} + Lw = 0, \quad t > 0,\ w(0) = 0, \quad w_t(0) = h, \tag{30}$$

h runs through a dense set S in H, and $\varepsilon > 0$ *is an arbitrarily small number.*

If $L = L^*$ *is selfadjoint and* $L \geqslant 0$, *then a sufficient condition for L not to have an eigenvalue at the point* k_0^2, $k_0 > 0$, *is*

$$\left\| \int_0^t w(t)\exp(ik_0 t)\,\mathrm{d}t \right\| = o(t), \quad \text{as } t \to +\infty. \tag{31}$$

A necessary condition for $L \geqslant 0$ *to have no positive eigenvalues is*

$$\left\| \int_0^t w(t)\,\mathrm{d}t \right\| = 0(1), \quad \text{as} \quad t \to +\infty. \tag{32}$$

Proof. (1) If h runs through a dense set of H then the element $R(-p^2)h$ and the operator-function $R(-p^2)$ have the same analytic properties as functions of p. The Laplace transform of (30) yields

$$(L + p^2 I)\bar{w} = h, \quad \bar{w} = \bar{w}(p, h) = (L + p^2 I)^{-1}h. \tag{33}$$

If λ_0 is an eigenvalue of L then the resolvent $(L - \lambda I)^{-1}$ has a pole at $\lambda = \lambda_0$. Set $\lambda = -p^2$, $p = -i\sqrt{\lambda}$, where $\sqrt{\lambda} = |\lambda|^{1/2}\exp(i \arg \lambda/2)$, $0 \leqslant \arg \lambda < 2\pi$. Suppose λ_0 is an eigenvalue of L, Re $\lambda_0 < 0$. Then Rep = Re$(-i\sqrt{\lambda_0}) > 0$, and $\bar{w}(p, h)$ has a pole at $p = p_0$ for some $h \in$

S. Suppose for simplicity that p_0 is the pole with maximal real part. Then, if $\delta > \operatorname{Re} p_0$ and $\operatorname{Re} p_1 < \operatorname{Re} p_0$, one has

$$\left\| \int_0^t w(t)\,dt \right\| = \left\| \frac{1}{2\pi i} \int_{c_0 - i\infty}^{c_0 + i\infty} \frac{\exp(pt) - 1}{p} \bar{w}(p, h)\,dp \right\|$$

$$= \left\| \frac{1}{2\pi i} \int_{c_1 - i\infty}^{c_1 + i\infty} \frac{\exp(pt) - 1}{p} \bar{w}(p, h)\,dp + \right.$$

$$\left. + \operatorname*{Res}_{p=p_0} \frac{\exp(pt) - 1}{p} \bar{w}(p, h) \right\| \geqslant \exp(t \operatorname{Re} p_0)\, d_0 + \tag{34}$$

$$+ 0(e^{t \operatorname{Re} p_1}) \text{ as } t \to +\infty, \text{ where } d_0 = \operatorname*{Res}_{p=p_0} \frac{\bar{w}(p, h)}{p},$$

$$d_0 \neq 0.$$

Therefore condition (29) does not hold if $\operatorname{Re} p_0 > 0$. This proves that condition (29) is sufficient for the absence of the eigenvalues of L in the half-plane $\operatorname{Re} \lambda < 0$. To prove the necessity of (29), let us assume that L has no eigenvalues in the half-plane $\operatorname{Re} \lambda < 0$. This is equivalent to the assumption that $\bar{w}(p, h)$ has no poles in the half-plane $\operatorname{Re} p > 0$. Therefore the number c_0 in (34) can be an arbitrarily small positive number, and, using (8), one obtains (29).

(2) If L is selfadjoint then its eigenvalues are real numbers. Suppose $\lambda_0 = k_0^2$, $k_0 > 0$, is an eigenvalue of L. Then $p = -ik_0$ is a pole of $\bar{w}(p, h)$ for some h, and the function $\overline{w(t) \exp(ik_0 t)} = \bar{w}(p - ik_0, h)$ has a pole at $p = 0$.

Since L is selfadjoint the pole $p = 0$ is simple. Therefore

$$\lim_{p \to 0} pF(p) = b, \quad F(p) = \bar{w}(p - ik_0, h) \tag{35}$$

where $b \neq 0$, $b \in H$. The assumption of Lemma 1 holds for $\bar{w}(p - ik_0, h)$ with $\sigma_0 = 0$ (since $L \geqslant 0$), $\delta = 1$, and $\alpha = 1$ (see (35)). Therefore

$$\lim_{t \to \infty} \frac{1}{t} \int_0^t w(t) \exp(ik_0 t)\,dt = b. \tag{36}$$

Thus, if (31) holds then $b = 0$ and $\lambda_0 = k_0$ is not a pole of $(L + p^2 I)^{-1}$. Therefore $\lambda_0 = k_0$ is not an eigenvalue of L.

Finally, assume that $L \geqslant 0$ has no positive eigenvalues. Then $\delta = 0$ in (9) and one has

$$\left\| \int_0^t w(t)\, \mathrm{d}t \right\| = \left\| \frac{1}{2\pi i} \int_{-i\infty}^{+i\infty} \frac{\exp(pt) - 1}{p} \bar{w}(p, h)\, \mathrm{d}p \right\| \leqslant$$

$$\leqslant c \int_{|\tau| > N} \frac{|\exp(pt) - 1|}{|p|(1 + |p|)^{\gamma}} |\mathrm{d}p| + \tag{37}$$

$$+ \frac{1}{2\pi} \left\| \int_{|\tau| < N} \frac{\exp(i\tau t) - 1}{\tau} W(\tau)\, \mathrm{d}\tau \right\| < \infty.$$

Thus (32) holds. □

2.4. Here we prove that the limiting absorption principle is equivalent to the limiting amplitude principle defined in (3). We also give an estimate of the remainder in the form

$$Pv = \frac{1}{T} \int_0^T \exp(ik_0 t) Pu(t)\, \mathrm{d}t + 0\left(\frac{1}{T^{\alpha}}\right) \tag{38}$$

where α is defined in (11).

THEOREM 2. *Let the assumption A hold. Then, for the limiting amplitude principle* (3) *to hold it is necessary and sufficient that assumptions* A_0 *and* B_{k_0} *hold. If in addition assumption* C_{k_0} *holds and* $\delta < 2 - \alpha$, *where* δ *is defined in* (9) *and* α *in* (11), *then* (38) *holds.*

Remark 3. If one defines the limiting amplitude principle by the formula

$$Pv = \exp(ik_0 t) Pu(t) + o(1), \text{ as } t \to +\infty, \tag{39}$$

then the limiting absorption principle is not equivalent to the limiting amplitude principle. This will be clear from the proof of Theorem 2, and can be illustrated as follows. Let $u = \exp(-it) + \exp(-ik_0 t)$, $k_0 \neq 1$.

Then the function $\exp(ik_0t)u = 1 + \exp\{i(k_0 - 1)t\}$ does not have a limit as $t \to +\infty$, but the function $T^{-1}\int_0^T \exp(ik_0t)u(t)\,\mathrm{d}t \to 1$ as $T \to +\infty$.

To prove Theorem 2 we need the following well known (Abelian type) lemma.

LEMMA 2. *If* $f \in L^1_{\mathrm{loc}}$, *and*

$$\lim_{T\to+\infty} T^{-1}\int_0^T f(t)\,\mathrm{d}t = b, \tag{40}$$

then

$$\lim_{p\to 0} p\int_0^\infty \exp(-pt)f(t)\,\mathrm{d}t = b, \quad |\arg p| < \frac{\pi}{2}. \tag{41}$$

Proof of Lemma 2. One has

$$\begin{aligned} p\int_0^\infty \exp(-pt)f(t)\,\mathrm{d}t &= p\exp(-pt)\int_0^t f\,\mathrm{d}t\Big|_0^\infty + \\ &\quad + p^2\int_0^\infty \exp(-pt)t\,t^{-1}\int_0^t f\,\mathrm{d}t\,\mathrm{d}t \\ &= p^2\int_0^\infty \exp(-pt)t\left(t^{-1}\int_0^t f\,\mathrm{d}t\right)\mathrm{d}t, \\ &\quad |\arg p| < \frac{\pi}{2}, \quad p \neq 0. \end{aligned} \tag{42}$$

Furthermore,

$$\lim_{p\to 0} p^2\int_0^\infty \exp(-pt)tb\,\mathrm{d}t = b. \tag{43}$$

Let $h(t) = t^{-1}\int_0^t f(t)\,\mathrm{d}t - b$. Then $\lim_{t\to+\infty} h(t) = 0$, and $h \in L^1_{\mathrm{loc}}$.

The desired conclusion (41) follows from (42), (43), and (40) if one shows that

$$\lim_{t \to +\infty} h(t) = 0 \tag{44}$$

implies

$$\lim_{p \to 0} p^2 \int_0^\infty \exp(-pt)th(t)\,\mathrm{d}t = 0, \quad |\arg p| < \frac{\pi}{2}. \tag{45}$$

Fix $t(\varepsilon)$ such that $|h(t)| < \varepsilon$ if $t \geqslant t(\varepsilon)$. Clearly

$$\lim_{p \to 0} p^2 \int_0^{t(\varepsilon)} \exp(-pt)th(t)\,\mathrm{d}t = 0. \tag{46}$$

Moreover, if $\operatorname{Re} p = \sigma = |p| \cos\beta$, $|\beta| < \pi/2$, then

$$\left| p^2 \int_{t(\varepsilon)}^\infty \exp(-pt)th(t)\,\mathrm{d}t \right| \leqslant \varepsilon |p|^2 \int_{t(\varepsilon)}^\infty \exp(-\sigma t)t\,\mathrm{d}t \leqslant$$

$$\leqslant \varepsilon |p|^2 \sigma^{-2} \leqslant \frac{\varepsilon}{\cos^2\beta}. \tag{47}$$

Since $\varepsilon > 0$ is arbitrary, (46) and (47) imply (45).

Proof of Theorem 2. (1) Necessity. Assume (3). Then, by Lemma 2, the following limit exists:

$$Pv = \lim_{p \to 0} p \int_0^\infty \exp(-pt)Pu(t)\exp(ik_0 t)\,\mathrm{d}t$$

$$= \lim_{p \to 0} Pp \cdot \frac{v(p - ik_0)}{p} = \lim_{p \to 0} Pv(p - ik_0) \tag{48}$$

Therefore condition B_{k_0} holds. Condition A_0 holds as well. Indeed, if (3) holds then

$$\left\| \int_0^t Pu(t)\exp(ik_0 t)\,\mathrm{d}t \right\| = 0(t), \quad \text{as} \quad t \to +\infty. \tag{49}$$

Moreover,

$$\overline{Pu(t)\exp(ik_0t)} = \frac{1}{p} PR(-(p - ik_0)^2)f. \tag{50}$$

If condition A_0 does not hold, then there exists at least one pole p_0, $\operatorname{Re}p_0 > 0$, of the resolvent $R(-p^2)$. Therefore, as in the proof of Theorem 1, one can show that, for some f,

$$\left\| \int_0^t Pu(t)\exp(ik_0t)\,dt \right\| \geqslant d\exp(t\operatorname{Re}p_0), \quad d > 0,$$

which contradicts (49). This contradiction proves that A_0 holds.

(2) Sufficiency. Assume A_0 and B_{k_0}. Consider (50). Condition B_{k_0} and (50) imply that the following limit exists

$$\lim_{p\to 0} p\,\overline{Pu(t)\exp(ik_0t)} = \lim_{p\to 0} PR(-(p - ik_0)^2f, \quad |\arg p| < \frac{\pi}{2}. \tag{51}$$

Equation (51) and condition A_0 show that the assumptions of Lemma 1 hold.

It follows from (51) and Lemma 1 that (3) holds.

(3) Let us prove the last conclusion of Theorem 2. Suppose that A_0, B_{k_0} and C_{k_0} hold, and $\delta < 2 - \alpha$. Then, from (50) and (11) one obtains:

$$\|\overline{Pu(t)\exp(ik_0t) - W(k_0)}\| = \left\| \frac{PR(-(p - ik_0)^2f - W(k_0)}{p} \right\| \leqslant \leqslant c(k_0)|p|^{\alpha-1}. \tag{52}$$

Therefore

$$\lim_{p\to 0} p^{1-\alpha}\|\overline{Pu(t)\exp(ik_0t) - W(k_0)}\| \leqslant c. \tag{53}$$

Inequality (53) implies condition (17) of Lemma 1 with $a = 1 - \alpha$. Inequality $\delta < 2 - \alpha$ says that condition $1 + a > \delta$ in (15) is satisfied. Thus, inequality (53) and Lemma 1 yield

$$\overline{\lim_{p\to 0}} \left\| \frac{1}{T^{1-\alpha}} \int_0^T [Pu(t)\exp(ik_0t) - W(k_0)]\,dt \right\| \leqslant c. \tag{54}$$

This inequality is equivalent to (38) since

$$W(k_0) = \lim_{p \to 0} Pv(p - ik_0) = Pv, \quad |\arg p| < \frac{\pi}{2}. \tag{55}$$

This completes the proof. □

2.4. The limiting amplitude principle for some selfadjoint operators in a Hilbert space H can be justified with the help of the spectral theorem. Let $L \geqslant 0$ be a selfadjoint operator in H, and E_λ be its resolution of identity. The solution to problem (1)—(2) can be written as

$$u = \int_0^\infty \alpha(t, \lambda)(\lambda - k_0^2)^{-1}\, \mathrm{d}E_\lambda f, \quad k_0 > 0 \tag{56}$$

where the integral is understood as the principal value and

$$\begin{aligned} \alpha(t, \lambda) &= \exp(-ik_0 t) - \cos(t\lambda^{1/2}) + ik_0\lambda^{-1/2}\sin(t\lambda^{1/2}) \\ &= \exp(-ik_0 t) + b(t, \lambda). \end{aligned} \tag{57}$$

It is clear that (56) satisfies the initial conditions (2) and Equation (1). When checking that (56) solves (1), one uses the identity $L \int_0^\infty h(\lambda)\, \mathrm{d}E_\lambda f = \int_0^\infty \lambda h(\lambda)\, \mathrm{d}E_\lambda f$. Formula (56) is the solution to (1)—(2) by Fourier's method. From the spectral theory one has $\mathrm{d}E_\lambda f = \operatorname{Im} \pi^{-1} R(\lambda + i0) f\, \mathrm{d}\lambda$. Therefore

$$\begin{aligned} P\, \mathrm{d}E_\lambda f &= \operatorname{Im} \frac{1}{\pi} PR(\lambda + i0) f\, \mathrm{d}\lambda \equiv h(\lambda, f)\, \mathrm{d}\lambda, \\ R(\lambda + i0) f &= (L - \lambda - i0)^{-1} f. \end{aligned} \tag{58}$$

where P is the orthoprojection on a subspace F of H. It has the same meaning as in Section 2.3. The function $h(\lambda, f)$, $a < \lambda < b$, is defined almost everywhere if $\|PR(\lambda \pm i\varepsilon) f\| \leqslant c$ where $c = \text{const}$ does not depend on ε, and $\lambda \varepsilon (a, b)$. Indeed, in this case, the weak limits $v_\pm = PR(\lambda \pm i0) f$ exist almost everywhere in (a, b).

Assume that

$$\int_0^\infty (1+\lambda)^{-1}\|h(\lambda, f)\|(1+\lambda^{1/2})\lambda^{-1/2}\,\mathrm{d}\lambda < \infty. \tag{59}$$

This assumption means that $\|h(\lambda)\|\lambda^{-1}$ is integrable near infinity, and $\|h(\lambda)\|\lambda^{-1/2}$ is integrable near zero. This allows one to use the Riemann—Lebesgue lemma which yields:

$$\lim_{t\to\infty}\left\|\int_{|\lambda-k_0^2|>\varepsilon} b(t,\lambda)\frac{\mathrm{d}E_\lambda}{\lambda-k_0^2}\right\| = 0. \tag{60}$$

Let us assume that

$$\|h(\lambda)-h(k_0^2)\| \leqslant c|\lambda-k_0^2|^\alpha, \quad 0<\alpha\leqslant 1. \tag{61}$$

Then

$$\int_{|\lambda-k_0^2|\leqslant\varepsilon} b(t,\lambda)\frac{\mathrm{d}E_\lambda}{\lambda-k_0^2} = \int_{|\lambda-k_0^2|\leqslant\varepsilon} b(t,\lambda)\frac{h(\lambda)-h(k_0^2)}{\lambda-k_0^2}\,\mathrm{d}\lambda + h(k_0^2)\times$$

$$\times\int_{|\lambda-k_0|\leqslant\varepsilon}\frac{b(t,\lambda)\,\mathrm{d}\lambda}{\lambda-k_0^2} = J_1+J_2. \tag{62}$$

It follows from (61) and the Riemann—Lebesgue lemma that

$$\|J_1\| \to 0 \text{ as } t\to+\infty. \tag{63}$$

One can check that

$$\lim_{t\to+\infty}\int_a^b \frac{\exp(ixt)}{x}\,\mathrm{d}x = i\pi, \quad a<0, \quad b>0. \tag{64}$$

Therefore, if $\|h(x)-h(\lambda)\| \leqslant c|\lambda-x|^\alpha, 0<\alpha\leqslant 1$, then

$$\lim_{t\to+\infty}\int_a^b \frac{\exp\{i(\lambda-x)t\}}{\lambda-x}h(x)\,\mathrm{d}x = i\pi h(\lambda), \quad a<\lambda<b. \tag{65}$$

Consider the integral

$$
\begin{aligned}
J = & \int_{|\lambda - k_0^2| \leqslant \varepsilon} \frac{b(t, \lambda)\, \mathrm{d}\lambda}{\lambda - k_0^2} = \int_{|\lambda - k_0^2| \leqslant \varepsilon} \Bigg[- \frac{\exp(it\lambda^{1/2}) + \exp(-it\lambda^{1/2})}{2} + \\
& + \frac{ik_0\lambda^{-1/2}}{2i} (\exp(it\lambda^{1/2}) - \exp(-it\lambda^{1/2})) \Bigg] \frac{\mathrm{d}\lambda}{\lambda - k_0^2} \\
= & \int_{|\lambda - k_0^2| \leqslant \varepsilon} \frac{\mathrm{d}\lambda}{\lambda - k_0^2} \Bigg[\frac{\exp(it\lambda^{1/2})}{2} (-1 + k_0\lambda^{-1/2}) - \\
& - \frac{\exp(-it\lambda^{1/2})}{2} (1 + k_0\lambda^{-1/2}) \Bigg] \mathrm{d}\lambda \\
= & \exp(-ik_0 t) \int_{|\lambda - k_0^2| \leqslant \varepsilon} \frac{\mathrm{d}\lambda}{\lambda - k_0^2} \Bigg[- \frac{\exp\{it(\lambda^{1/2} + k_0)\}}{2} (1 - k_0\lambda^{-1/2}) - \\
& - \frac{\exp\{i(k_0 - \lambda^{1/2})t\}}{2} (1 + k_0\lambda^{1/2}) \Bigg].
\end{aligned} \tag{66}
$$

It follows from (65) and (66) that

$$
\begin{aligned}
& \lim_{t \to +\infty} \{J \exp(ik_0 t)\} \\
& = \frac{1}{2k_0} \lim_{t \to +\infty} \int_{|\lambda - k_0^2| \leqslant \varepsilon} \frac{\mathrm{d}\lambda \exp\{i(k_0 - \lambda^{1/2})t\}}{k_0 - \lambda^{1/2}} = i\pi.
\end{aligned} \tag{67}
$$

Combining (56), (57), (59)–(63), (65) and (67) one obtains

$$
\begin{aligned}
\lim_{t \to +\infty} \exp(ik_0 t) Pu & = \int_0^\infty \frac{h(\lambda, f)\, \mathrm{d}\lambda}{\lambda - k_0^2} + i\pi h(k_0^2, f) \\
& = \int_0^\infty \frac{P\, \mathrm{d}E_\lambda f}{\lambda - k_0^2 - i0},
\end{aligned} \tag{68}
$$

where $h(\lambda, f)$ is defined in (58), the well known formula

$$(x - i0)^{-1} = x^{-1} + i\pi\delta(x) \tag{69}$$

is used, the distribution x^{-1} is understood as Cauchy's principal value, i.e. $\int_{-\infty}^{\infty} x^{-1}\, h(x)\, \mathrm{d}x = \lim_{\varepsilon\to 0} \int_{|x|>\varepsilon} x^{-1}\, h(x)\, \mathrm{d}x$, and $h(\lambda) = h(\lambda, f)$ is assumed to be Hölder-continuous at $\lambda = k_0^2$ (see (61)).

It follows from the spectral theorem that

$$\int_0^\infty \frac{P\,\mathrm{d}E\lambda f}{\lambda - k_0^2 - i0} = \lim_{\varepsilon\to+0} PR(k_0^2 + i\varepsilon)f = PR(k_0^2 + i0)f. \tag{70}$$

Therefore, formula (68) can be written as

$$Pu(t) = \exp(-ik_0 t)v_+ + o(1), \quad t \to +\infty \tag{71}$$

where

$$v_+ = PR(k_0^2 + i0)f \tag{72}$$

is the solution to the stationary problem

$$(L - k_0^2)v_+ = f \tag{73}$$

obtained by the limiting absorption principle. This means that we call v_+ the solution to (73) iff

$$v_+ = \lim_{\varepsilon\to+0} PR(k_0^2 + i\varepsilon)f. \tag{74}$$

Let us study the dependence of the asymptotics of $Pu(t)$ as $t \to +\infty$ on the singularity of $h(\lambda)$ at $\lambda = 0$. Condition (59) allows $\|h(\lambda)\|$ to be $0(\lambda^{-\alpha})$, $\alpha < \frac{1}{2}$.

Let us assume that in a neighborhood of $\lambda = 0$ the function $h(\lambda)$ behaves as follows:

$$h(\lambda) = h_0\lambda^{-\alpha} + h_1(\lambda), \quad 1 > \alpha \geqslant \tfrac{1}{2}, \quad 0 < \lambda \leqslant \lambda_0, \tag{75}$$

where $h_1(\lambda)$ satisfies inequality (59), λ_0 is a small number, $0 < \lambda_0 < k_0^2$, $h_0 \in H$ is a constant, i.e. an element which does not depend on λ. The arguments in Section 2.4 remain valid except for the conclusion (60). This conclusion is no longer valid because inequality (59) is not valid now. The term $h_0\lambda^{-a}$ produces the new expression for the limiting behavior of the integral

$$I(t, a) = \int_{|\lambda - k_0^2|\varepsilon} b(t, \lambda)h(\lambda)\,\mathrm{d}\lambda$$

$$= h_0 \int_0^{\lambda_0} ik_0\lambda^{-1/2-a} \sin(\lambda^{1/2}t)\,\mathrm{d}\lambda + o(1) \text{ as } \quad t \to +\infty \tag{76}$$

The term $o(1)$ in (76) is obtained as before from the Riemann–Lebesgue lemma. But, if $\frac{1}{2} \leqslant a < 1$, the integral

$$I_0(t, a) = \int_0^{\lambda_0} \lambda^{-1/2-a} \sin(\lambda^{1/2}t)\,\mathrm{d}\lambda \tag{77}$$

does not tend to zero as $t \to +\infty$. If $a = \frac{1}{2}$ then

$$I_0(t, \tfrac{1}{2}) = \int_0^{\lambda_0} \frac{\sin(\lambda^{1/2}t)}{\lambda}\,\mathrm{d}\lambda = 2\int_0^{\lambda_0^{1/2}t} \frac{\sin x}{x}\,\mathrm{d}x \underset{t\to\infty}{\to} \pi. \tag{78}$$

From (76) and (78) it follows that

$$I(t, \tfrac{1}{2}) = ik_0\pi h_0 + o(1), \quad t \to +\infty. \tag{79}$$

If $\frac{1}{2} < a < 1$, one has

$$I_0(t, a) = 2t^{2a-1}\int_0^{\lambda_0 1/2t} \frac{\sin x}{x^{2a}}\,\mathrm{d}x \sim I_0 t^{2a-1}, \quad \text{as} \quad t \to +\infty. \tag{80}$$

where (see Gradshteyn-Ryzhik [1], formula 3.761.4)

$$I_0 = 2\int_0^{\infty} \frac{\sin x}{x^{2a}}\,\mathrm{d}x = 2\Gamma(1-2a)\cos\pi a, \quad \tfrac{1}{2} < a < 1. \tag{81}$$

Therefore

$$I(t, a) = ik_0 I_0 h_0 t^{2a-1} + o(1), \quad \text{as} \quad t \to +\infty, \quad \tfrac{1}{2} < a < 1 \tag{82}$$

where the constant I_0 is given by (81).

It follows from (79) and (81) that

$$Pu(t) = \exp(-ik_0 t)v_+ + I(t, \alpha), \quad \tfrac{1}{2} \leqslant \alpha < 1 \tag{83}$$

where $I(t, \alpha)$ is given by formulas (76)—(82), and the assumption (75) holds.

Let us summarize our results.

THEOREM 3. *Let $L \geqslant 0$ be a selfadjoint operator on a Hilbert space, P be the orthoprojection on a subspace F. If assumptions* (59) *and* (61) *hold then the solution $u(t)$ to problem* (1)—(2) *has asymptotics* (71) *where v_+ is defined by* (72). *If assumptions* (75) *and* (61) *hold then $u(t)$ is given by* (83), *where the asymptotics of $I(t, \alpha)$ is given by* (76)—(82).

2.5. In the proof of Theorems 2 and 3 we used conditions (8) and (59) which give some decay as $|p| \to \infty$ and $\lambda \to +\infty$. If L is selfadjoint and the absolutely continuous subspace H_{ac} of L is the whole space $H = L^2(\Omega)$ then one can drop condition (59). It is difficult to prove an estimate of the form

$$\|h(\lambda, f)\| \leqslant c(1 + |\lambda|)^{-\gamma}, \quad \gamma > 0, \tag{84}$$

which is sufficient for (59) to hold. Since it is easy to establish absolute continuity of the spectrum of the Dirichlet or Neumann Laplacian in the exterior domain, it is important to be able to drop assumption (59) and substitute it by absolute continuity of the spectrum and the equality $H_{ac} = H = L^2(\Omega)$. Let us show that the following theorem holds.

THEOREM 4. *Let $L \geqslant 0$ be a selfadjoint operator, $H_{ac} = H$ where H_{ac} is the absolutely continuous subspace of H, and assumption* (61) *holds, then the conclusion of Theorem* 3 *holds.*

Proof. Condition (59) was used when we referred to the Riemann—Lebesgue lemma. The conclusion of this lemma holds for the integrals of the type (60) and (76). For the Riemann—Lebesgue lemma to hold it is sufficient, for example, that

$$\int_0^\infty \|h(\lambda, f)\|^2 \, d\lambda < \infty. \tag{85}$$

If $H_{ac} = H$ then

$$\int_0^\infty \|h(\lambda, f)\|^2 \, d\lambda = \int_0^\infty (dE_\lambda f, f) < \infty, \tag{86}$$

Therefore the Riemann—Lebesgue lemma is applicable. The rest of the proof is the same as in the proof of Theorem 3. □

3. THE LIMITING AMPLITUDE PRINCIPLE FOR THE LAPLACIAN IN EXTERIOR DOMAINS

3.1. Let us apply the results in Section 2 to the Dirichlet Laplacian in Ω. The assumption about Ω and notations are the same as in Chapter I. Thus, consider the problem

$$\begin{aligned} u_{tt} + Lu &= u_{tt} - \nabla^2 u \\ &= f(x)\exp(-ik_0 t) \ \text{in} \ \Omega \subset R^3, \ t > 0, \ k_0 > 0 \end{aligned} \tag{1}$$

$$u = 0 \text{ on } \Gamma \tag{2}$$

$$u(x, 0) = u_t(x, 0) = 0, \quad x \in \Omega \tag{3}$$

where L stands for the self-adjoint realization of the Dirichlet Laplacian in $L^2(\Omega)$.

Let us assume that $f(x) \in C_0^\infty(\Omega)$. A much weaker assumption on f would be sufficient, e.g.,

$$f(x) \in L^2_{\text{loc}} \quad \text{and} \quad |f(x)| \leqslant c(1 + |x|)^{-a}, \quad a > 2, c = \text{const} > 0. \tag{4}$$

In Chapter I it was proved that the resolvent kernel $G(x, y, k)$ of the operator $L \geqslant 0$ is meromorphic in k and analytic in a neighborhood of the real axis on the complex k-plane. The resolvent kernel $G(x, y, k)$ is the kernel of the operator $(L - k^2 I)^{-1}$. If $\text{Im}\, k^2 > 0$, this kernel is uniquely defined by the requirement $G(x, y, k) \in L^2(\Omega_x)$, that is, for any fixed $y \in \Omega$ the function $G(x, y, k)$ belongs to $L^2(\Omega)$. For $k > 0$ the resolvent kernel is defined as the limit $G(x, y, k + i0) = G(x, y, k)$. It was proved in Chapter I that this limit satisfies the radiation condition, and the estimate

$$|G(x, y, k)| \leqslant c|x - y|^{-1}, \quad x, y \in \Omega, \quad 0 \leqslant a \leqslant k \leqslant b \tag{5}$$

where $c = \text{const} > 0$ does not depend on x, y, and k, $a \leqslant k \leqslant b$.

Therefore the function

$$h(\lambda, f) = \frac{1}{\pi} \operatorname{Im} \int_{\Omega} G(x, y, \sqrt{\lambda + i0}) f \, dy, \quad f \in C_0^{\infty}(\Omega), \quad (6)$$

is not only Hölder-continuous in λ, $\lambda > 0$, but analytic in $k = \sqrt{\lambda}$ in a neighborhood of any point $\lambda > 0$. Therefore condition (2.61) is satisfied for $k_0 \geqslant 0$, as well as condition (2.11) with $\alpha = 1$ and P being an orthoprojection in $L^2(\Omega)$ onto $L^2(B_R \setminus D)$, where $D \subset B_R$ and $R > 0$ is an arbitrarily large fixed number.

Conditions (2.8) and (2.59) are satisfied as follows from Theorem I.6.1. Conditions (2.9) and (2.10) are obviously satisfied. Therefore Theorems 2 and 3 of Section 2 yield the following result.

THEOREM 1. *The limiting amplitude principle is valid for the solution to problem* (1)–(3) *under the assumptions of Theorem* I.6.1 *about D. More precisely, the following formulas hold*:

$$Pv = \exp(ik_0 t) Pu(x, t) + o(1), \quad t \to +\infty, \quad (7)$$

and

$$Pv = \frac{1}{T} \int_0^T \exp(ik_0 t) Pu(x, t) \, dt + 0\left(\frac{1}{T}\right), \quad T \to \infty, \quad (8)$$

where Pv is defined in (2.6), *and v is given by* (1.9).

Actually, under the assumptions of Theorem I.6.1 a stronger form of the limiting amplitude principle holds

$$u(x, t) = v(x) \exp(-ik_0 t) + 0(\exp(-at)), \quad a > 0, \quad t \to +\infty, \quad (9)$$

where 0 is uniform in $x \subset B_R \setminus D$ for any fixed R, and v solves the stationary problem (1.5)–(1.7). The proof of (9) is much more difficult than the proof of (7) and (8). The abstract results do not help. One should use analytical estimates.

Let us give a sketch of the proof. One starts with the formula

$$u(x, t) = (2\pi)^{-1} \int_{\infty}^{\infty} \exp(-ikt) V(x, k) \, dk, \quad (10)$$

where

$V(x, k)$ solves the stationary problem

$$-\nabla^2 V - k^2 V = f(x) i (k - k_0 + i0)^{-1} \tag{11}$$

$$V = 0 \text{ on } \Gamma. \tag{12}$$

Therefore

$$\begin{aligned} V &= i(k - k_0 + i0)^{-1} \int_\Omega G(x, y, k) f(y)\, \mathrm{d}y \equiv \\ &\equiv i(k - k_0 + i0)^{-1} v(x, k), \end{aligned} \tag{13}$$

and

$$u(x, t) = i(2\pi)^{-1} \int_{-\infty}^{\infty} \exp(-ikt) v(x, k)\, (k - k_0 + i0)^{-1}\, \mathrm{d}k. \tag{14}$$

If the assumption (1.12) holds then one can move the contour of integration in (14) down and obtain

$$\begin{aligned} u(x, t) &= i(2\pi)^{-1} \int_{-ib-\infty}^{-ib+\infty} \exp(-ikt) v(x, k)\, (k - k_0 + i0)^{-1}\, \mathrm{d}k + \\ &+ \exp(-ik_0 t) v(x, k_0) = \exp(-ik_0 t) v(x, k_0) + \\ &+ O(\exp(-bt)), \quad 0 < b < |\mathrm{Im}\, k_1|, \end{aligned} \tag{15}$$

where k_j are the poles of $v(x, k)$, $|\mathrm{Im} \cdot k_1| \leqslant |\mathrm{Im}\, k_2| \leqslant \ldots$. The argument is complete if (1.12) is established. Another approach is given in Section 4.

In the proof of the limiting amplitude principle and in the study of the asymptotic behavior of solutions to time-dependent problems for large t we use three different types of assumptions. The first deals with the local behavior of the limiting values of the resolvent on the continuous spectrum of L (assumptions (2.10), (2.11), (2.61)). The second deals with the behavior of the resolvent on the continuous spectrum for large λ (assumptions (2.8), (2.59), (2.84), (2.85)). The third deals with the behavior of the resolvent near its singular points (assumptions (2.9), (2.15), (2.75)).

In scattering problems assumptions (2.8) and (2.84) are often difficult to check. For example, these estimates are not obtained for the Robin and Neumann boundary conditions. If they are available then one can usually obtain the exponential decay of the remainder in the limiting amplitude principle (see (15)).

But for a selfadjoint operator with absolutely continuous spectrum the limiting amplitude principle holds in the form (2.71) without assumptions for (2.8) or (2.84) (see Theorem 2.4). Therefore we obtain the following theorem.

THEOREM 2. *For the solution to problem* (1)—(3) *the limiting amplitude principle holds in the form*

$$u(x, t) = \exp(-ik_0 t)v + o(1), \quad t \to +\infty, \quad k_0 > 0, \tag{16}$$

where v *solves the stationary problem* (1.5)—(1.7), *and* $o(1)$ *means that for any fixed* $R > 0$, *such that the ball* B_R *contains the obstacle* D, *one has*

$$\|u(x, t) - \exp(-ik_0 t)v\|_{L^2(B_R \setminus D)} \to 0 \quad as \quad t \to +\infty. \tag{17}$$

If the boundary condition (2) *is replaced by the Neumann or Robin boundary condition the limiting amplitude principle* (16) *holds provided that the function* $\eta(s)$ *in the Robin boundary condition*

$$\frac{\partial u}{\partial N} + \eta(s)u = 0 \tag{18}$$

is real-valued and the spectrum of the exterior Robin Laplacian is absolutely continuous (*which is the case e.g. if* $\eta(s) < 0$).

Proof. The conclusion of Theorem 2 follows from Theorem 2.4. The assumptions of Theorem 2.4 are satisfied for the operator L corresponding to the problem (1)—(3), that is for the exterior Dirichlet Laplacian in $H = L^2(\Omega)$.

The same argument holds for the Neumann boundary condition. If $\operatorname{Im} \eta(s) = 0$, and $\eta(s) \in C(\Gamma)$ in the Robin boundary condition, then the corresponding operator L is selfadjoint and its positive spectrum is absolutely continuous. The resolvent kernel of the Robin Laplacian in Ω solves the problem

$$(-\nabla^2 - k^2)\mathscr{G}(x, y, k) = \delta(x - y) \text{ in } \Omega, \tag{19}$$

$$\frac{\partial \mathscr{G}}{\partial N} + \eta(s)\mathscr{G} = 0 \text{ on } \Gamma, \quad \operatorname{Im} \eta = 0, \tag{20}$$

$$r\left(\frac{\partial \mathscr{G}}{\partial r} - ik\mathscr{G}\right) \to 0, \quad r = |x| \to \infty. \tag{21}$$

This kernel may have a pole $k = 0$ if $\eta > 0$. For example, let $\Gamma = S^2 = \{s : |s| = 1\}$ be a sphere, and $\eta = 1$. Then the homogeneous problem (19)–(21) with $k = 0$ has a nontrivial solution $u = r^{-1}$:

$$\nabla^2 v = 0 \quad \text{if} \quad r > 1, \frac{\partial v}{\partial r} + v = 0 \quad \text{if} \quad r = 1. \tag{22}$$

Thus the point $k = 0$ is a pole of $\mathscr{G}$ if $\eta = 1$ and $\Gamma = S^2$. Similarly one can check that $\mathscr{G}$ may have poles $k = ib$, $b > 0$, if $\eta > 0$. Indeed, if $\Gamma = S^2$ and $v = r^{-1} \exp(-br)$, $b > 0$, then

$$\begin{aligned} &-\nabla^2 v + b^2 v = 0 \quad \text{if} \quad r > 1, \\ &\frac{\partial v}{\partial r} + \eta v = 0 \quad \text{if} \quad r = 1 \quad \text{and} \quad \eta = b + 1. \end{aligned} \tag{23}$$

Thus $\mathscr{G}$ has a pole ib, $b > 0$, if $\eta = b + 1$ and $\Gamma = S^2$. In these two examples the assumptions of Theorem 2 do not hold, and the limiting amplitude principle does not hold. If $\eta \leqslant 0$ then the Robin Laplacian is a nonnegative selfadjoint operator L in $H = L^2(\Omega)$, and $H_{ac} = H$. In this case the limiting amplitude principle for L holds as follows from Theorem 2.4. Since the positive spectrum of L is absolutely continuous, we need only to show that the condition $\eta \leqslant 0$ guarantees that $L \geqslant 0$

and $k = 0$ belongs to the absolutely continuous spectrum. We have, for any v which satisfies (18) and decays rapidly at infinity,

$$(Lv, v) = \int_\Omega -\nabla^2 v \bar{v} \, \mathrm{d}x = \int_\Omega |\nabla v|^2 \, \mathrm{d}x + \int_\Gamma v_N \bar{v} \, \mathrm{d}s -$$

$$- \lim_{R \to \infty} \int_{S_R} v_r \bar{v} \, \mathrm{d}s = \int_\Omega |\nabla v|^2 \, \mathrm{d}x - \tag{24}$$

$$- \int_\Gamma \eta(s) |v|^2 \, \mathrm{d}s > 0, \quad \text{if } \eta(s) \leqslant 0.$$

Thus $L > 0$, and $k = 0$ is not a pole of $\mathscr{G}$. Therefore we proved (under the assumptions $\eta \leqslant 0$, $\eta \in C(\Gamma)$) the uniqueness theorem

$$\{Lv = 0, \quad v = 0(r^{-1}) \quad \text{as} \quad r \to \infty\} \Rightarrow v = 0. \tag{25}$$

From (25), as in Section 1.2, one can derive that the point $\lambda = 0$ belongs to the absolutely continuous spectrum of L.

It is sufficient to prove that for any $f \in C_0^\infty(\Omega)$ the limit

$$\lim_{\varepsilon \to +0} \int_\Omega \mathscr{G}(x, y, i\varepsilon) f \, \mathrm{d}y = v \tag{26}$$

exists in L^2_{loc}. This is done as in Theorem I.2.3. □

Remark 1. The limiting amplitude principle holds for the problem of the type

$$u_{tt} = \nabla^2 u \text{ in } \Omega \tag{27}$$

$$u(x, 0) = u_t(x, 0) = 0 \tag{28}$$

$$u(s, t) = f(s) \exp(-ik_0 t) \text{ on } \Gamma, \quad k_0 > 0. \tag{29}$$

In this case the function v in (16) solves the problem

$$\nabla^2 v + k_0^2 v = 0 \text{ in } \Omega \tag{30}$$

$$v = f(s) \text{ on } \Gamma \tag{31}$$

$$v \text{ satisfies the radiation condition (1.7).} \tag{32}$$

4. DECAY OF ENERGY

4.1. Consider the problem (1.1)—(1.3)

The solution to this problem can be written as

$$u = L^{-1/2}\sin(tL^{1/2})f. \tag{1}$$

This is a particular case of the formula (2.56), and L is the Dirichlet Laplacian in $H = L^2(\Omega)$. Define the energy norm by the formula

$$\begin{aligned} \|u\|_1 &= \{(u, u) + (Lu, u)\}^{1/2}, \quad u \in D(L), \\ (u, u) &= (u, u)_{L^2(\Omega)} = \|u\|_0^2. \end{aligned} \tag{2}$$

The closure of $D(L_0) = \mathring{H}^2(\Omega)$ in the norm (2) is the Sobolev space $\mathring{H}^1(\Omega)$. The norm of this space is the graph norm of the operator $L^{1/2}$, $\|\cdot\|_1 = \|\cdot\|_{\mathring{H}^1(\Omega)}$. Let E_λ be the resolution of the identity for L. Since the embedding $\mathring{H}^1(\Omega) \to L^2(B_R \setminus D)$ is compact, it is sufficient to prove that

$$(u(t), h)_1 \to 0 \quad \text{as} \quad t \to +\infty, \quad h \in \mathring{H}^1(\Omega) \tag{3}$$

in order to obtain that

$$\|u(t)\|_{L^2(B_R \setminus D)} \to 0 \quad \text{as} \quad t \to +\infty. \tag{4}$$

In this Section $R > 0$ denotes an arbitrarily large fixed number, $D \subset B_R$. From (2) it follows that

$$(u(t), h)_1 = \int_0^\infty [\lambda^{-1/2}\sin(t\lambda^{1/2}) + \lambda^{1/2}\sin(t\lambda^{1/2})]\, \mathrm{d}(E_\lambda f, h). \tag{5}$$

Since $f \in D(L^{1/2})$ one has

$$\int_0^\infty \lambda\, \mathrm{d}(E_\lambda f, f) < \infty, \tag{6}$$

and

$$\begin{aligned} \int_0^\infty \lambda^{1/2}\, \mathrm{d}(E_\lambda f, h) &\leqslant \left\{\int_0^\infty \lambda\, \mathrm{d}(E_\lambda f, f)\right\}^{1/2} \left\{\int_0^\infty \mathrm{d}(E_\lambda h, h)\right\}^{1/2} \\ &= \|L^{1/2} f\|_0 \cdot \|h\|_0 < \infty \end{aligned} \tag{7}$$

The spectrum of L is absolutely continuous. Therefore the Riemann–Lebesgue lemma is applicable to the integral (5), and this lemma yields (3). Thus we proved (4). Let us summarize the result.

LEMMA 1. *If* $f \in \mathring{H}^1(\Omega)$ *then the solution to problem* (1.1)–(1.3) *satisfies equation* (4).

One can obtain a stronger decay:

$$\|u(t)\|_{1,R}\|\dot{u}(t)\|_{1,R} \to 0 \quad as \quad t \to +\infty, \quad \dot{u} = \frac{\mathrm{d}u}{\mathrm{d}t}, \tag{8}$$

where $\|.\|_{1,R}$ *is the* $\mathring{H}^1(B_R \setminus D)$ *norm, and* $f \in \mathring{H}^1(\Omega)$.

THEOREM 1. *If* $f \in D(L) = \mathring{H}^2(\Omega)$ *then* (8) *holds.*

Proof. Let us prove that

$$\|\dot{u}\|_{1,R} \to 0 \quad \text{as} \quad t \to +\infty. \tag{9}$$

It follows from (1) that $\dot{u} = \cos(tL^{1/2})f$. As in the proof of Lemma 1, one obtains

$$(\dot{u}, h)_1 = \int_0^\infty [\cos(t\lambda^{1/2}) + \lambda \cos(t\lambda^{1/2})]\, \mathrm{d}(E_\lambda f, h). \tag{10}$$

If $f, h \in \mathring{H}^1(\Omega) = D(L^{1/2})$, then

$$\int_0^\infty \lambda |\mathrm{d}(E_\lambda f, h)| \leqslant \left\{\int_0^\infty \lambda\, \mathrm{d}(E_\lambda f, f)\right\}^{1/2} \left\{\int_0^\infty \lambda\, \mathrm{d}(E_\lambda h, h)\right\}^{1/2} < \infty. \tag{11}$$

Therefore the Riemann–Lebesgue lemma applied to (10) yields

$$\dot{u} \to 0 \quad \text{weakly in } \mathring{H}^1, \text{as } t \to +\infty. \tag{12}$$

As in the proof of Lemma 1, this and the embedding theorem imply that

$$\|\dot{u}\|_{0,R} \to 0, \quad t \to +\infty. \tag{13}$$

Furthermore $L\dot{u} = L\cos(tL^{1/2})f$, and

$$\begin{aligned}(\dot{u}, h)_L &\equiv (\dot{u}, h_0) + (L\dot{u}, Lh)_0 \\ &= \int_0^\infty [\cos(t\lambda^{1/2}) + \lambda^2\cos(t\lambda)^{1/2}]\, \mathrm{d}(E_\lambda f, h), \quad h \in D(L).\end{aligned} \tag{14}$$

Since $f \in D(L)$ one has $\int_0^\infty \lambda \, d(E_\lambda f, f) < \infty$, and the Riemann–Lebesgue lemma applied to (14) yields

$$(\dot{u}, h)_L \to 0 \quad \text{as} \quad t \to +\infty, \quad \forall h \in \mathring{H}^2(\Omega) = D(L). \tag{15}$$

From (15) and the embedding theorem Equation (9) follows. Equation (8) can be proved similarly. □

Remark 1. If $f \in D(L^n) = \mathring{H}^{2n}(\Omega)$ then

$$\begin{aligned}&((L+I)^n \dot{u}, \quad (L+I)^n h) \\ &\quad = (\dot{u}, h)_{H^{2n}(\Omega)} \to 0 \quad \text{as} \quad t \to +\infty, \quad h \in \mathring{H}^{2n}(\Omega).\end{aligned} \tag{16}$$

From (16) and the embedding theorem it follows that

$$\|\dot{u}\|_{\mathring{H}^{2n-1}(B_R \setminus D)} \to 0 \quad \text{as} \quad t \to +\infty. \tag{17}$$

4.2. If the obstacle D satisfies a non-trapping condition, e.g., D is convex or star-shaped (or even more sophisticated non-trapping conditions like in Vainberg [2] or Melrose [1]) then function (1.4) satisfies inequality (1.12) and the solution to problem (1.1)–(1.3) decays exponentially

$$|u(x, t)| \leqslant c(R, D) \exp(-b_0 t), \quad b_0 > 0, \quad |x| \leqslant R. \tag{18}$$

In Melrose [1] a similar result is mentioned for the problem with the Neumann boundary condition. We discuss the literature on this question in the Bibliographical notes. In the literature the estimate (18) is obtained in the course of a study of the time-dependent problem. If (18) is established then the function (4) is seen to be analytic in the half-plane Im $k > -a$, and by Parseval's equality,

$$\int_{\operatorname{Im} k = b} |v(x, k)|^2 |dk| < c(b, r), \quad b > -b_0, \quad |x| \leqslant R. \tag{19}$$

This estimate and the analyticity of v imply estimate (12) for $b > -b_0$. Conversely, analyticity of $v(x, k)$ in the half-plane $b > -b_0$ and the estimate (19) imply (18). A direct proof of the estimates (1.12) or (19), which does not use the time-dependent problem, is not yet found. For the Schrödinger operator in R^3 a proof of such estimates was given in Ramm [6h], [2f], and, under weaker assumptions on the potential, in Saito [3].

5. SINGULARITY AND EIGENMODE EXPANSION METHODS (SEM AND EEM)

The singularity expansion method has been explained in Section 1. The eigenmode expansion method is a method for solving linear equations using expansions in root vectors of B. Consider an example. Look for the solution to the problem

$$(\nabla^2 + k^2)w = 0 \text{ in } \Omega, \quad w = f \text{ on } \Gamma \tag{1}$$

w satisfies the radiation condition, in the form

$$w = \int_\Gamma g\sigma \, \mathrm{d}s, \; g = \frac{\exp(ik|x - s|)}{4\pi|x - s|}. \tag{2}$$

Then

$$Q\sigma = \int_\Gamma g(s, t)\sigma(t) \, \mathrm{d}t = f(s) \tag{3}$$

is an equation in $H = L^2(\Omega)$ with a linear nonselfadjoint compact operator Q. If B is a compact selfadjoint operator in H then the equation

$$B\sigma = f \tag{4}$$

is solvable in H iff

$$\sum_{n=1}^{\infty} |f_n|^2 \lambda_n^{-2} < \infty, \quad B\phi_n = \lambda_n\phi_n \tag{5}$$

where $f_n = (f, \phi_n)$, and the system $\{\phi_n\}$ of the eigenvectors of B forms an orthonormal basis of H. If (5) holds then the solution to (4) is of the form

$$\sigma = \sum_{n=1}^{\infty} \lambda_n^{-1} f_n\phi_n. \tag{6}$$

If $\lambda_j = 0$ and $f_j = 0$ then the solution to (4) exists but is not unique:

$$\sigma = \sum_{n \neq j} \lambda_n^{-1} f_n\phi_n + c_j\phi_j \tag{6}$$

where c_j are arbitrary constants.

If B is nonselfadjoint, then it may have no eigenvectors (e.g., $Bh = \int_0^x b(x, y)h(y)\,dy$, $B: L^2[0, 1] \to L^2[0, 1]$). A nonselfadjoint operator may have root vectors (generalized eigenvectors), its eigenvectors are not necessarily orthogonal and their linear span may be not the whole space. Root vectors of a compact linear opertaor B corresponding to an eigenvalue λ are the vectors which satisfy the equation

$$(B - \lambda I)^j \phi = 0, \phi \neq 0 \tag{7}$$

with some integer j. If $j = 1$ then ϕ is an eigenvector. If B is compact then one knows that $N\{(B - \lambda I)^j\} = N\{(B - \lambda I)^{j+1}\}$ if $j \geqslant r$, where $r = r(\lambda)$ is an integer which is called the Riesz number corresponding to the eigenvalue λ. The elements of $N\{(B - \lambda I)^r\}$ are called root vectors of B corresponding to the eigenvalue λ. Let us define the Jordan chain associated with a pair (λ, ϕ), $B\phi = \lambda\phi$. Consider the equation $B\phi^{(1)} - \lambda\phi^{(1)} = \phi$. If it is not solvable then we say that the Jordan chain corresponding to the pair (λ, ε) consists of one vector ϕ and has length 1. If it is solvable consider the equation $B\phi^{(2)} - \lambda\phi^{(2)} = \phi^{(1)}$. Continue this process. Since $r(\lambda) < \infty$, there is a number $r = r(\lambda, \phi)$ such that the equations

$$(B - \lambda I)\phi^{(j)} = \phi^{(j-1)}, \quad j = 1, 2, \ldots, r(\lambda, \phi), \phi^{(0)} \equiv \phi \tag{8}$$

are solvable but the equation with $j = r(\lambda, \phi) + 1$ is not solvable. Then we say that $\{\phi^{(j)}\}$, $0 \leqslant j \leqslant r(\lambda, \phi)$ is the Jordan chain associated with the pair (λ, ϕ) and $r(\lambda, \phi) + 1$ is the length of this chain. Clearly $r(\lambda) = \max r(\lambda, \phi)$, where the maximum is taken over $\phi \in N(B - \lambda I)$. The basic question concerning EEM is

when does B have a root system which forms a basis of H? (9)

It is well known (Dunford–Schwartz [1]) that if B is normal, i.e. $B^*B = BB^*$, then its eigenvectors form an orthonormal basis of H. One can check (Ramm [3]) that the operator (3) is normal if Γ is a sphere.

Before answering the basic question (9) for the operator $B = Q$, one can discuss a much simpler question:

Is the root system of Q complete in H? (10)

The answer is positive. An abstract theorem which is sufficient for our purposes can be formulated.

THEOREM 1. *Let Q be a compact linear operator on a Hilbert space H,*

$Q = Q_0 + T$, $Q_0 \geqslant 0$, $T_j = \text{Im } T = (T - T^*)/2i \geqslant 0$, *and T belongs to the trace class. Then the root system of Q is complete in H.*

Proof. If $Q_0 \geqslant 0$ and $S = S^*$ are compact then

$$N_{\lambda}^{-}(Q_0 + S) \leqslant N_{\lambda}^{-}(S), \quad \text{where} \quad N_{\lambda}^{-}(S) \equiv \sum_{\lambda_n(S) \leqslant -\lambda} 1,$$

and $\lambda_n = \lambda_n(S)$ are the eigenvalues of S, $|\lambda_1| \geqslant |\lambda_2| \geqslant \dots$.

Take $S = T_R = (T + T^*)/2$. Then

$$0 \leqslant \lim_{\lambda \to 0} \lambda N_{\lambda}^{-}(Q_0 + T_R) \leqslant \lim_{\lambda \to 0} \lambda N_{\lambda}^{-}(T_R) = 0. \tag{11}$$

The last equation holds because T and, therefore, T_R are trace-class. One can now use the following proposition (Gohberg—Krein [1], p. 292).

PROPOSITION 1. *The root system of a compact operator Q such that* Im $Q \geqslant 0$ *and* $\lim_{\lambda \to 0} \lambda N_{\lambda}^{-}(Q_R) = 0$, *is complete in H.*

Since $\text{Im } Q = T_j \geqslant 0$ by assumption, $Q_R = Q_0 + T_R$, and (10) holds, the conclusion of Theorem 1 follows from Proposition 1. □

THEOREM 2. *The root system of the operator* (3) *is complete in* $H = L^2(\Gamma)$.

Proof. Let $Q = Q(k) = Q_0 + T$, where $Q_0 = Q(0) \geqslant 0$ and $T = Q(k) - Q(0)$ is trace-class.

The conclusion of Theorem 2 follows from Theorem 1 if one notices that the operator (3) is compact, and

$$\begin{aligned}(\text{Im } Tf, f) &= (\text{Im } Qf, f) \\ &= \int_\Gamma \int_\Gamma \frac{\sin kr_{st}}{4\pi r_{st}} f(t)\overline{f(s)}\, \mathrm{d}s\, \mathrm{d}t \geqslant 0, \quad r_{st} = |s - t|.\end{aligned} \tag{12}$$

This inequality shows that $\text{Im } Q \geqslant 0$. The inequality (12) follows from the identity

$$\frac{\sin kr_{st}}{4\pi r_{st}} = \frac{k}{(4\pi)^2} \int_{S^2} \exp\{ik\omega \cdot (s - t)\}\, \mathrm{d}\omega.$$

This identity implies that

$$\int_\Gamma \int_\Gamma \frac{\sin kr_{st}}{4\pi r_{st}} f(t)\overline{f(s)}\,\mathrm{d}s\,\mathrm{d}t$$
$$= \frac{k}{(4\pi)^2} \int_{S^2} \left| \int_\Gamma f(s) \exp(ik\omega \cdot s)\,\mathrm{d}s \right|^2 \mathrm{d}\omega \geqslant 0. \tag{12}$$

Theorem 2 is proved. □

We prove that the root system of the operator (3) is not only complete in H but forms a Bari basis with brackets of H. The definition of this notion is given in Appendix 3. The basic idea of the proof is as follows. $Q = Q_R + iQ_J$. The operator Q_R is a selfadjoint compact operator of order -1, i.e.

$$\|Q_R f\|_{\tilde{H}^{q+1}} \leqslant c\|f\|_{\tilde{H}^q}, \quad c > 0. \tag{13}$$

If $k^2 \notin \sigma(-\Delta_D)$, i.e. k^2 is not an eigenvalue of the Dirichlet Laplacian in D, then Q is an isomorphism of $\tilde{H}^q$ onto $\tilde{H}^{q+1}$ (Lemma I.7.2), and so is Q_R. The operator Q_J has an analytic kernel $(4\pi r_{st})^{-1}\sin(kr_{st})$, so that Q_J: $\tilde{H}^2 \to \tilde{H}^\infty$. From the abstract results in Appendix 3, the following theorem follows.

THEOREM 3. *The root system of the operator* (3) *forms a Bari basis with brackets of* $L^2(\Gamma)$.

Proof. We assume first that $0 < k^2 \notin \sigma(-\Delta_D)$. Then Q and Q_R are invertible as operators from $\tilde{H}^0$ onto $\tilde{H}^1$, and

$$Q = Q_R(I + iQ_R^{-1}Q_J), \quad \text{ord } Q_R = -1, \quad \text{ord } Q_R^{-1}Q_J = -\infty, \tag{14}$$
$$\lambda_n(Q_R) \sim cn^{-1/2}, c = \text{const}, n \to \infty. \tag{15}$$

Formula (15) is a particular case of the formula

$$\lambda_n(L) \sim cn^{\ell d^{-1}} \quad c = \text{const}, \quad n \to \infty$$

where $\lambda_n(L)$ are the eigenvalues (counted by their multiplicities) of a pseudodifferential elliptic operator L of order ℓ on a compact manifold of dimension d. In our case $\ell = -1$ and $d = 2$. The conclusion of Theorem 3 follows from (14), (15) and Theorem 2′ in Appendix 3. In our case $p = \frac{1}{2}$, and α can be taken arbitrary negative, so that $p(1 - \alpha)$ can be made as large as one wishes. Theorem 5 in Appendix 3 additional information: the eigenvector expansion for Q_R and root vector expansion for Q are equiconvergent uniformly on bounded sets in $\tilde{H}^0$. If $\Gamma \in C^\infty$ the same is true in $\tilde{H}^q$ for any real q. □

Chapter IV

The *T*-Matrix Scheme

1. STATEMENT OF THE PROBLEM

1.1. The T-matrix scheme for solving exterior boundary value problems and scattering problems has been popular for the last 25 years. Although many papers have been published on this method and its applications to acoustical electromagnetic and elastic wave scattering, the theoretical foundations of the method were given only in 1982 (Ramm [7]).

These foundations are given in this chapter. They include analysis of the convergence of the method, stability of the numerical scheme towards small perturbations of the data and round-off errors, and estimates of the rate of convergence. Numerous practical applications of the method are not discussed. References concerning these applications can be found in Varadan and Varadan [1]. Only scalar wave scattering is discussed in detail since the basic principles and the results of the theory are fully represented in this case.

1.2. Let us describe the T-matrix scheme in a general formulation. Let $D \subset R^3$ be a bounded obstacle with boundary Γ and Ω be the exterior domain. Let N be the exterior unit normal to Γ, i.e. the normal pointing into Ω. Consider the problem

$$(\nabla^2 + k^2)u = 0 \text{ in } \Omega, \quad k > 0, \tag{1}$$

$$u = f \text{ on } \Gamma \tag{2}$$

$$r\left(\frac{\partial u}{\partial r} - iku\right) \to 0, \quad r \to \infty. \tag{3}$$

We are not concerned here with determining the minimal smoothness requirements on the surface Γ under which the theory is valid. It is certainly sufficient to assume that Γ is a Liapunov surface but this requirement can be considerably relaxed, e.g. edges and conical points are admissible.

Let $\{\psi_n\}_{\lambda=1}^{\infty}$ be a system of outgoing waves, i.e.

$$(\nabla^2 + k^2)\psi_n = 0 \text{ in } \Omega, \tag{4}$$

$$r\left(\frac{\partial \psi_n}{\partial r} - ik\psi_n\right) \to 0, \quad r \to \infty. \tag{5}$$

From Green's formula it follows that

$$\int_\Gamma \psi_n \frac{\partial u}{\partial N}\,\mathrm{d}s = \int_\Gamma u \frac{\partial \psi_n}{\partial N}\,\mathrm{d}s, \quad \forall n, \tag{6}$$

where u solves (1)—(3). Using the boundary condition (2) one obtains

$$\int_\Gamma \psi_n h\,\mathrm{d}s = f_n, \quad \forall n \tag{7}$$

where

$$f_n = \int_\Gamma f \frac{\partial \psi_n}{\partial N}\,\mathrm{d}s, \tag{8}$$

$$h = \left.\frac{\partial u}{\partial N}\right|_\Gamma. \tag{9}$$

The *T*-matrix scheme consists of the following. Let $\{\phi_j\}$ be a linearly independent, complete system of functions in $H_0 = L^2(\Gamma)$. Let

$$h_m = \sum_{j=1}^{m} c_j^{(m)} \phi_j, \tag{10}$$

where $c_j^{(m)}$, $1 \leqslant j \leqslant m$, are constant coefficients which should be determined from the linear algebraic system

$$\sum_{j=1}^{m} a_{nj} c_j^{(m)} = f_n, \quad 1 \leqslant n \leqslant m \tag{11}$$

$$a_{nj} = \int_\Gamma \psi_n \phi_j\,\mathrm{d}s = (\phi_j, \bar{\psi}_n). \tag{12}$$

One obtains this system if (10) is substituted in (7) and only the first m equalities (7) are used.

1.3. Justification of the T-matrix scheme requires positive answers to the following questions:

Q.1. Is (11) solvable for sufficiently large m? Is the solution of (11) unique?

Q.2. Does $h_m \underset{H_0}{\to} h$ as $m \to \infty$? Here h is defined in (9).

Q.3. Does $c_j^{(m)} \to c_j$ as $m \to \infty$? Is the convergence uniform in j, $1 \leqslant j < \infty$?

Q.4. Does the equality $h = \sum_{j=1}^{\infty} c_j \phi_j$ hold, where c_j are defined in Q.3 and it is assumed that the limits c_j exist?

Q.5. How does the rate of convergence depend on the choice of the systems $\{\phi_j\}$ and $\{\psi_n\}$?

Q.6. Is the numerical scheme based on Equation (11) stable with respect to small perturbations of f_n and the matrix a_{nj}?

Remark 1. The following questions are easy to answer. Is the set of Equations (7) solvable? Is the solution to (7) unique? The answer to the first questions is yes. Indeed, one knows that problem (1)–(3) is solvable. If u solves (1)–(3), then apply Green's formula to u and ψ_n and obtain (7).

The answer to the second question is yes if and only if the system $\{\psi_n\}$ is closed in $H_0 = L^2(\Gamma)$. This means that

$$\left\{\int_\Gamma h\psi_n \, ds = 0 \, \forall n\right\} \Rightarrow h = 0. \tag{13}$$

This requirement is equivalent to completeness of the system $\{\psi_n\}$ in H_0. The system $\{\psi_n\}$ is called complete in H_0 iff for any $\varepsilon > 0$ and $h \in H_0$ there exists $m(\varepsilon)$ and coefficients $c_j = c_j(\varepsilon, h)$ such that $\| h - \sum_{j=1}^{m(\varepsilon)} c_j \psi_j \| < \varepsilon$. We assume below that the system $\{\psi_j\}$ is closed in H_0.

1.4. If one takes as ψ_1 in (7) the function $g = \exp(ik|s-y|)/(4\pi|s-y|)$, $y \in D$, and does not use Equations (7) for $n > 1$, then one obtains the integral equation

$$\int_\Gamma g(s, y)h(s) \, ds = \int_\Gamma f(s) \frac{\partial g(s, y)}{\partial N} \, ds \equiv F(y), \quad y \in D. \tag{14}$$

If this equation holds for $y \in B \subset D$, where B is any ball lying strictly in D, then (14) holds in D, because both sides in (14) are solutions to Helmholtz's equation and, by the unique continuation property, they are identical in D if they are identical in B. If one lets $y \to s' \in \Gamma$ then one obtains from (14) the following equation

$$Qh = \int_\Gamma g(s, s')h(s)\,\mathrm{d}s = b(s'), \tag{15}$$

where

$$b(s') = \lim F(y), y \in D, \quad y \to s' \tag{16}$$

If one looks for a solution of (15) of the form (10) and uses a projection method for finding the coefficients $c_j^{(m)}$, then one obtains the system

$$\sum_{j=1}^{m} A_{nj} c_j^{(m)} = b_n, \quad 1 \leqslant n \leqslant m \tag{17}$$

where

$$A_{nj} = (Q\phi_j, \eta_n), \quad b_h = (b, \eta_n), \tag{18}$$

$$(f, h) \equiv \int_\Gamma f\bar{h}\,\mathrm{d}s, \tag{19}$$

and the bar denotes complex conjugation. The same questions Q.1–Q.6 as in Section 3 can be studied for the system (17), where η_n plays the role of ψ_n, but now there is no need to assume anything about the properties of η_n in Ω. The functions η_n are defined only on Γ. One can go back from the integral equation (14) to the system (7) assuming that

$$g(s, y) = \sum_{j=1}^{\infty} \phi_j(y)\psi_j(s), \quad |y| < |s|, \tag{20}$$

substituting (20) into (14) and equating coefficients in front of ψ_j. In the literature, expansion (20) is used with ψ_j being the outgoing spherical waves and ϕ_j being the regular (i.e. finite at the origin) solutions to Helmholtz's equation. Matrix (18) is identical to matrix (12) if ψ_n in (12) are chosen so that $Q\bar{\eta}_n = \psi_n$. This corresponds to a specific choice of the outgoing waves, since for any linearly independent system of functions η_n in $L^2(\Gamma)$ the system $\psi_n = Q\bar{\eta}_n$ is a system of outgoing waves whose boundary values on Γ form a linearly independent system in H_0 provided that $Q: H_0 \to H_0$ is injective, i.e. $Q\eta = 0$ implies $\eta = 0$. This will be the case iff k^2 is not an eigenvalue of the Dirichlet Laplacian in D. If k^2 is an eigenvalue of the Dirichlet Laplacian in D then one can use g_ε instead of g and the corresponding operator Q_ε has no zeros. Here g_ε is defined in Section I.1.3.

Discussion of questions Q.1—Q.6 is given in Section 2.

2. JUSTIFICATION OF THE *T*-MATRIX SCHEME

2.1. The strategy is to treat Equation (1.11) as an operator equation obtained by applying a projection method to Equation (1.7) and to use some general results about the convergence of projection methods.

Therefore some background material is needed. First, we give some definitions.

A system of elements of a Hilbert space H is called minimal if none of the elements belongs to the closure of the linear span of the others. A minimal system $\{\phi_j\}$ is called strongly minimal if $\lim_{m\to\infty} \lambda_m = \lambda > 0$, where λ_m is the minimal eigenvalue of the Gram matrix $\phi_{ij} = (\phi_i, \phi_j)$, $1 \leqslant i, j \leqslant m$. A system $\{\tilde{\phi}_j\}$ is called biorthogonal to the system $\{\phi_j\}$ if $(\tilde{\phi}_j, \phi_i) = \delta_{ij}$. The biorthogonal system $\{\tilde{\phi}_j\}$ is uniquely defined iff the system $\{\phi_j\}$ is minimal and closed in H.

A minimal system is called a Schauder basis of a Banach space X if any element $f \in X$ can be uniquely represented as $f = \sum_{j=1}^{\infty} c_j \phi_j$, where the series converges in the norm of X. Uniqueness means that if $f = \sum_{j=1}^{\infty} c_j \phi_j$ and $f = \sum_{j=1}^{\infty} c'_j \phi_j$, then $c_j = c'_j$, $\forall j$.

A minimal system $\{\phi_j\}$ in a Hilbert space H is called a Riesz basis of H if $\phi_j = Th_j$, where $\{h_j\}$ is an orthonormal basis of H and T is an isomorphism on H, i.e. T is a linear bounded injective and surjective operator defined on all of H. In this case T^{-1} has the same properties. A system $\{\tilde{\phi}_j\}$ biorthogonal to a Riesz basis is also a Riesz basis (see Appendix 3).

A complete system $\{\phi_j\}$ in H is a Riesz basis iff the matrix $\phi_{ij} = (\phi_i, \phi_j)$ generates an isomorphism on ℓ^2. Another useful criterion for a complete system $\{\phi_j\}$ to form a Riesz basis of H is the inequality

$$a_1^2 \sum_{j=1}^{m} |c_j|^2 \leqslant \left\| \sum_{j=1}^{m} c_j \phi_j \right\|^2 \leqslant a_2^2 \sum_{j=1}^{m} |c_j|^2, \quad a_1 > 0$$

which should hold for any constants c_j and for any m, and the positive constants a_1 and a_2 should not depend on m and c_j, $1 \leqslant j \leqslant m$. If $\{\phi_j\}$ is a Riesz basis then $\sup_j \|\phi_j\|^2 < a_2$, $\inf_j \|\phi_j\|^2 > a_1$ and similar inequalities hold for $\tilde{\phi}_j$. In particular, $\sup_j \|\tilde{\phi}_j\| < \infty$. Proofs of the above statements one can find in, e.g., Gohberg—Krein [1] and Young [1].

Consider an interpolation problem in the Hilbert space H. The problem consists in finding an element $h \in H$ such that

$$(h, \psi_n) = f_n, \; n = 1, 2, \ldots . \tag{1}$$

where $\{f_n\}$ is a given sequence of numbers and $\{\psi_n\}$ is a system of linearly independent elements of H, i.e. any finite number of elements ψ_n forms a linearly independent system of elements in H.

Definition 1. A system $\{\psi_n\}$ is called a Bessel system if $\sum_{n=1}^{\infty} |(h, \psi_n)|^2 < \infty$ whenever $h \in H$.

Definition 2. A system $\{\psi_n\}$ is called a Riesz—Fischer system if for any sequence $\{f_n\} \in \ell^2$ there exists an element $f \in H$ such that $(f, \psi_n) = f_n$, $\forall n$. The set of sequences $\{(h, \psi_n)\}$, $h \in H$ is called the moment space M for the system $\{\psi_n\}$. We are interested in the questions:

(1) when does a sequence $\{f_n\} \in M$. In other words, when is (1) solvable?

(2) is the solution to (1) unique?

(3) how does one construct the solution to (1)?

The answer to question (1) is as follows: for (1) to be solvable in H it is necessary and sufficient that for any m and any scalars c_j the inequality

$$\left| \sum_{j=1}^{m} c_j \bar{f}_j \right| \leqslant c \left\| \sum_{j=1}^{m} c_j \psi_j \right\|, \quad c > 0, \quad f_j = (f, \psi_j) \tag{2}$$

be valid, where c does not depend on m and c_j; if (2) holds then $\|h\| \leqslant c$. The necessity follows from the estimate

$$\left| \sum_{j=1}^{m} c_j \bar{f}_j \right| = \left| \left(h, \sum_{j=1}^{m} c_j \psi_j \right) \right| \leqslant \|h\| \left\| \sum_{j=1}^{m} c_j \psi_j \right\|.$$

To prove the sufficiency, notice that (2) and the inequality $\|h\| \leqslant c$ imply that the functional defined on the elements ψ_n by formula (1) can be extended to the linear space of these elements as a linear bounded functional by the formula $(\sum_{j=1}^m c_j\psi_j, h) = \sum_{j=1}^m c_j\bar{f}_j$. By continuity it can be uniquely extended on the closure Ψ of the linear span of the elements $\{\psi_j\}$ and then by the Hahn—Banach theorem it can be extended on all of H with the same norm.

The answer to question (2) is obvious: (1) has no more than one solution iff the system $\{\psi_j\}$ is closed.

One can prove that if (1) is solvable then the solution to (1) of minimal norm exists and is unique. This solution $h \in \Psi$. Indeed, let h_0, $\|h_0\| \leqslant c$, be a solution to (1). Consider the set S of all solutions to (1) such that $\|h\| \leqslant c$. Let $h \in S$. One can write $h = h_1 + h_2$, $h_1 \in \Psi$, $h_2 \perp \Psi$. Since $(h_2, \psi_n) = 0$, $\forall n$, the element h_1 solves (1) and $\|h_1\|^2 = \|h\|^2 - \|h_2\|^2 \leqslant \|h\|^2$. Let us show that h_1 is the solution to (1) with minimal norm and that this solution is unique. Clearly it is sufficient to prove that if h_1', $h_1 \in \Psi$ and solve (1) then $h_1' = h_1$. But if h_1' and h_1 solve (1) then $(h_1' - h_1, \psi_n) = 0$, $\forall n$. Since $h_1' - h_1 \in \Psi$, one concludes that $h_1' = h_1$. For any fixed m one can find the solution of minimal norm to the first m equations (1) by the formula

$$h = h_m \equiv -[\det(\psi_i, \psi_j)]^{-1} \begin{pmatrix} 0 & \psi_1 & \psi_m \\ f_1 & (\psi_1, \psi_1) & (\psi_1, \psi_m) \\ \vdots & \vdots & \vdots \\ f_m & (\psi_m, \psi_1) & (\psi_m, \psi_m) \end{pmatrix}. \tag{3}$$

Moreover, if (1) is solvable then its unique solution of minimal norm can be obtained by the formula

$$h = \lim h_m, \quad m \to \infty \tag{4}$$

where h_m is defined in (3).

2.2. Let (a_{ij}), $1 \leqslant i, j < \infty$ be a matrix of a linear operator on ℓ^2, i.e. $(Ax)_i = a_{ij}x_j$. Here and below one should sum over the repeated indices from 1 to ∞. When is A bounded and when is it compact in ℓ^2?

One has

$$\|Ax\|^2 = a_{ij}x_j\bar{a}_{ik}\bar{x}_k \leqslant |a_{ij}\bar{a}_{ik}| \frac{|x_k|^2 + |x_j|^2}{2} \leqslant \sup_{1 \leqslant k < \infty} \sum_{j=1}^{\infty} |a_{ij}a_{ik}| \, \|x\|^2.$$

Therefore A is bounded iff

$$\|A\| \leqslant \left(\sup_k \sum_{j=1}^{\infty} \sum_{i=1}^{\infty} |a_{ij} a_{ik}|\right)^{1/2} < \infty. \tag{5}$$

The operator A is compact if (5) holds and

$$\sup_{1 \leqslant k < \infty} \sum_{j=1}^{\infty} \sum_{i=N}^{\infty} |a_{ij} a_{ik}| \to 0 \text{ as } N \to \infty. \tag{6}$$

A useful criterion for boundedness of A gives the following lemma.

LEMMA (Schur). *Assume that* $a_{ij} = \bar{a}_{ji}$ *and* $\sup_i \sum_{j=1}^{\infty} |a_{ij}| \leqslant c$. Then $\|A\| \leqslant c, A: \ell^2 \to \ell^2$.

Proof. If A is symmetric, densely defined and $|(Ax, x)| \leqslant c\|x\|^2$, then $\|A\| \leqslant c$. One has

$$|(Ax, x)| = \left| \sum_{i,j=1}^{\infty} a_{ij} x_j \bar{x}_i \right| \leqslant \sum_{i,j=1}^{\infty} |a_{ij}| \frac{|x_i|^2 + |x_j|^2}{2} \leqslant$$

$$\leqslant \sum_{i,j=1}^{\infty} |a_{ij}| |x_i|^2 \leqslant \sup_i \sum_{j=1}^{\infty} |a_{ij}| \|x\|^2 \leqslant c\|x\|^2.$$

Therefore, $\|A\| \leqslant c$. □

2.3. Let us discuss convergence of projection methods. Let A be a linear bounded and boundedly invertible operator from a Hilbert space H onto a Hilbert space G. Let $L_m \subset L_{m+1}$, $\dim L_m = m$, be a sequence of subspaces limit dense in H, i.e. $\rho(h, L_m) \to 0$ as $m \to \infty$ for any $h \in H$, where $\rho(h, L)$ is the distance from the element h to the subspace L. Let P_m be the orthoprojection onto L_m. Let $M_m \subset M_{m+1}$ be a limit dense sequence of subspaces in G, $\dim M_m = m$, and Q_m be the orthoprojection onto M_m in G. Consider the equation

$$Ah = f \tag{7}$$

and the projection method for its approximate solution:

$$Q_m A P_m h_m = Q_m f. \tag{8}$$

Consider the following statement:

Equation (8) is uniquely solvable for all sufficiently

$$\text{large } m \text{ and } \|h_m - h\| \to 0 \text{ as } m \to \infty, \tag{9}$$

where h_m is the solution to (8). We want to give conditions for (9) to be true.

Equation (1.11) is of the form (8). Indeed, in the Equation (1.11) $H = L^2(\Gamma)$, $G = \ell^2$, L_m is the linear span of $(\phi_1, \ldots, \phi_m)$, M_m is the subspace of the sequences in ℓ^2 whose components $f = 0$ for $j > m$, the elements h_m are given by (1.10) and $Q_m f = (f_1, \ldots, f_m)$. Therefore the abstract results will be immediately applicable to Equation (1.11).

THEOREM 1. *For* (9) *to hold it is necessary and sufficient that*

$$\|Q_m A P_m h\| \geqslant c\|P_m h\|, \quad m > m_0, \quad h \in H, \quad c > 0 \tag{10}$$

where m_0 is some integer and c does not depend on m and h.

Proof. Assume (9). Then (8) is uniquely solvable for $m > m_0$ and $h_m = (Q_m A P_m)^{-1} Q_m f \to h = A^{-1} f$, $m \to \infty$, $f \in G$. Therefore $\sup_m \|(Q_m A P_m)^{-1} Q_m\| \leqslant c < \infty$. Here and below c denotes various positive constants which do not depend on m. Thus $\|P_m h\| = \|(Q_m A P_m)^{-1} Q_m Q_m A P_m h\| \leqslant c\|Q_m A P_m h\|$, i.e. (10) holds. Note that $(Q_m A P_m)^{-1} Q_m A P_m = I_m$, where I_m is the identity in $P_m H$ but not in all of H. The necessity is proved.

Assume (10). The operator $Q_m A P_m$: $P_m H \to Q_m G$, $m > m_0$, is injective because of (10). Therefore, it is surjective since $\dim P_m H = \dim Q_m G = m < \infty$. Therefore, Equation (8) is uniquely solvable for $m > m_0$. To prove that $\|h_m - h\| \to 0$, $m \to \infty$, consider the equations

$$Q_m A[P_m h + (I - P_m) h] = Q_m f$$

$$Q_m A P_m h_m = Q_m f$$

which imply that

$$Q_m A P_m (h_m - P_m h) = Q_m A (I - P_m) h. \tag{11}$$

Since the sequence of the subspaces L_m is limit dense in H one has $\|(I - P_m) h\| \to 0$, $m \to \infty$, $h \in H$. Therefore (11) and (10) imply that

$$\begin{aligned} \|h_m - P_m h\| &= \|P_m (h_m - P_m h)\| \\ &\leqslant c^{-1} \|Q_m A (I - P_m) h\| \to 0 \quad \text{as} \quad m \to \infty. \end{aligned} \tag{12}$$

Thus

$$\|h - h_m\| \leqslant \|h - P_m h\| + \|P_m h - h_m\| \to 0, \quad m \to \infty. \tag{13}$$

The proof is complete. □

Remark 1. One can give the same proof without the assumption $\dim L_m = \dim M_m = m < \infty$. In this case one should assume in addition to (10) that

$$Q_m A P_m H = Q_m G, \quad m > m_0. \tag{14}$$

If $m < \infty$ then, as was shown in the proof of Theorem 1, assumption (10) implies assumption (14).

Let us study the stability of the projection method. Suppose that (9) holds for the operator A in (7). These are the questions we are interested in:

(1) will (9) hold for $A + B$ if $\|B\| < \delta$ and $\delta > 0$ is sufficiently small?

(2) will (9) hold for $A + B$ if B is compact and $A + B$ is boundedly invertible?

THEOREM 2. *Under the above assumptions the answers to questions* (1) *and* (2) *are both yes.*

Proof. (1) Let $\delta = c - \varepsilon$, where c is the constant in (10) and ε is an arbitrary number in the interval $0 < \varepsilon < c$. Let $\|B\| < \delta$. Then

$$\|Q_m(A+B)P_m h\| \geqslant c\|P_m h\| - \delta\|P_m h\| = \varepsilon\|P_m h\|$$

and condition (10) holds with the constant $\varepsilon > 0$.

(2) If B is compact then $\|(Q_m A P_m)^{-1} Q_m B - A^{-1}B\| \to 0$, $m \to \infty$, because $(Q_m A P_m)^{-1} Q_m - A^{-1} \to 0$, $m \to \infty$ strongly. Here we use the following simple statement: if $A_m \to 0$ strongly and B is compact then $\|A_m B\| \to 0$, $m \to \infty$. If A and $A + B$ are invertible then $I + A^{-1}B$ is invertible. Thus $\|P_m h + A^{-1}BP_m h\| \geqslant c_1\|P_m h\|$, $c_1 \to 0$. Therefore

$$\begin{aligned}\|Q_m(A+B)P_m h\| &= \|Q_m A P_m [P_m h + (Q_m A P_m)^{-1} Q_m B P_m]\| \geqslant \\ &\geqslant c\|P_m h + (Q_m A P_m)^{-1} Q_m B P_m h\| \geqslant \\ &\geqslant c\|P_m h + A^{-1}BP_m h\| - c\|[(Q_m A P_m)^{-1} Q_m B - \\ &\quad - A^{-1}B]P_m h\| \geqslant cc_1\|P_m h\| - c\alpha_m\|P_m h\|, \\ &\quad \alpha_m \to 0, \quad m \to \infty.\end{aligned}$$

Therefore (10) holds for sufficiently large m with, e.g., the constant 0.5 cc_1. □

EXAMPLE 1. As an application of Theorems 1 and 2 let us study the projection method for the equation

$$(I + T)h = f \tag{15}$$

where T is a compact linear operator on a Hilbert space H. Many integral equations in Chapter I are of this type. Assume that $N(I + T) = \{0\}$, where $N(A) = \{h: Ah = 0\}$. Then, by Fredholm's alternative, the operator $I + T$ is an isomorphism of H onto H. Let $\{\phi_j\}$ be a complete linearly independent system of elements in H, $L_m = \text{span}(\phi_1, \ldots, \phi_m)$, P_m be the orthoprojection onto L_m in H. The projection method with $H = G$, $L_m = M_m$, leads to the equation

$$h_m + P_m Th_m = P_m f. \tag{16}$$

One can apply Theorem 2 with $A = I$ and $B = T$. Since T is compact, $I + T$ is an isomorphism, and for $A = I$ condition (10) is obviously satisfied with $c = 1$, one concludes that Equations (16) are uniquely solvable for $m > m_0$ and $\|h_m - h\| \to 0$, $m \to \infty$, where h is the solution to Equation (15).

2.4. Let us turn to the analysis of the T-matrix scheme, i.e. let us answer questions Q.1—Q.6 in Section 1.

Let us write equation (1.7) as an operator equation

$$\mathscr{A} h = p, \quad p = (f_1, f_2, \ldots), \tag{17}$$

where $\mathscr{A}: H_0 \to \ell^2$, $H_0 = L^2(\Gamma)$, $p \in \ell^2$,

The operator $\mathscr{A}$ is bounded iff

$$\sum_{j=1}^{\infty} |(h, \overline{\psi}_j)|^2 \leqslant c^2 \|h\|^2_{H_0}, \quad c > 0, \quad h \in H_0. \tag{18}$$

If (18) does not hold but the system $\{\psi_j\}$ forms a (Schauder) basis of H_0 then the operator $\mathscr{A}$ is densely defined on H_0 (i.e., $D(\mathscr{A})$, the domain of the definition of $\mathscr{A}$, is a dense set in H_0). Indeed, if the system $\{\psi_j\}$ is a basis then the biorthogonal system $\{\tilde{\psi}_j\}$ is also a basis of H_0 and any linear combination $h_m = \sum_{j=1}^{m} c_j \tilde{\psi}_j$ belongs to $D(\mathscr{A})$, $\mathscr{A} h_m = (c_1, c_2, \ldots, c_m, 0, 0 \ldots)$. The operator $\mathscr{A}$ transforms functions

$h \in H_0$ into sequences $\{(h, \overline{\psi}_n)\}$. The range of α, Ran α, is dense in ℓ^2 if the system $\{\psi_j\}$ is a basis of H_0. Indeed, suppose $(\alpha h, p) = 0$, $\forall h \in H_0$ and some $p \in \ell^2$. Take $h = \tilde{\psi}_i$. Then $\alpha \tilde{\psi}_i = \delta_{ij}$, $(\alpha h, p) = p_i = 0$, and, since i is arbitrary, one concludes that $p = 0$. The operator α^{-1} is bounded and defined on all of ℓ^2 iff $\|\alpha h\| \geqslant \|h\|$, $c > 0$, $h \in D(\alpha)$. The operator α is an isomorphism of H_0 onto ℓ^2 if

$$c_1\|h\| \leqslant \|\alpha h\| \leqslant c_2\|h\|, \quad c_1 > 0, \quad h \in H_0 \tag{19}$$

and they hold iff the system $\{\psi_j\}$ is a Riesz basis of H_0. Consider the projection method for solving Equation (17). Let $Q_m p = (f_1, \ldots, f_m, 0, \ldots)$ be an orthoprojection in ℓ^2 and P_m be an orthoprojection in H_0 onto the linear span of $\phi_1, \phi_2, \ldots \phi_m$. Here $\{\phi_j\}$ is a closed linearly independent system of elements of H_0. Equation (1.11) can be written as

$$Q_m \alpha P_m h_m = Q_m p \tag{21}$$

where

$$h_m = P_m h_m = \sum_{j=1}^{m} c_j^{(m)} \phi_j. \tag{22}$$

Let us write condition (10) for Equation (21):

$$c^2 \left\| \sum_{j=1}^{m} c_j \phi_j \right\|_{H_0}^2 \leqslant \left\| Q_m \alpha \left(\sum_{j=1}^{m} c_j \phi_j \right) \right\|_{\ell^2}^2 = \sum_{i=1}^{m} \left| \sum_{j=1}^{m} \alpha_{ij} c_j \right|^2, \; m > m_0, \tag{23}$$

where $c_1, \ldots c_m$ are arbitrary constants, $c > 0$,

$$\alpha_{ij} = (\phi_j, \overline{\psi}_i). \tag{24}$$

One can write (23) as

$$c^2 \sum_{j,j'=1}^{m} (\phi_j, \phi_{j'}) c_j \bar{c}_{j'} \leqslant \sum_{j,j'=1}^{m} \sum_{i=1}^{m} \alpha_{ij} \bar{\alpha}_{ij'} c_j \bar{c}_{j'}, \quad m > m_0. \tag{25}$$

Let us denote by $(\alpha)_m$ the truncated matrix:

$$(\alpha)_m = \alpha_{jj'}, \qquad \leqslant j, j' \leqslant m. \tag{26}$$

Let Φ denote the Gram matrix

$$\Phi_{jj'} = (\phi_j, \phi_{j'}). \tag{27}$$

Then (25) can be rewritten as the matrix inequality

$$c^2(\Phi)_m \leqslant (a^*a)_m, \quad c > 0, \quad \forall m > m_0, \tag{28}$$

where a^* is the matrix $a^*_{jj'} = \bar{a}_{j'j}$,

$$(a^*a)_m \equiv (a^*)_m(a)_m \tag{29}$$

and the inequality $A \leqslant B$ means that $(Ap, p) \leqslant (Bp, p), \forall p$.

Theorem 1 implies the following result.

THEOREM 3. *Assume that* $\{\psi_j\}$ *is a Riesz basis of* H_0 *and* $\{\phi_j\}$ *is a closed linearly independent system in* H_0. *The inequality* (28) *is a necessary and sufficient condition for the system* (1.11) *to be uniquely solvable for all* $m > m_0$ *and for the equation* $\|h_m - h\| \to 0,\ m \to \infty$ *to hold, where* h_m *is defined by* (1.10), $\{c_j^{(m)}\}$, $1 \leqslant j \leqslant m$, *is the solution to* (1.11), *the sequence* $\{f_n\} \in \ell^2$, *and* h *solves Equation* (1.7).

Proof. It was already established that the operator a is an isomorphism of H_0 onto ℓ^2 if the system $\{\psi_j\}$ is a Riesz basis of H_0. Therefore Equation (1.7) is uniquely solvable in H_0 and inequality (28) implies that condition (10) holds. The conclusion of Theorem 3 follows from Theorem 1.

Let us derive some sufficient conditions for the conclusion of Theorem 3 to hold. These sufficient conditions will be easy to verify in practice. Let $\lambda(A)$ ($\Lambda(A)$) denote the minimal (maximal) eigenvalue of a selfadjoint matrix A.

PROPOSITION 1. *If*

$$\inf_m \lambda((a^*a)_m) \geqslant \lambda > 0 \textit{ and} \sup_m \Lambda((\Phi)_m) \leqslant \Lambda < \infty, \tag{30}$$

then (28) *holds and therefore the conclusion of Theorem* 3 *holds.*

Proof. Indeed, $(a^*a)_m \geqslant \lambda I_m$, where I_m is the identity operator in $Q_m\ell^2$, and $c^2(\Phi)_m \leqslant c^2\Lambda I_m$. Therefore inequality (28) holds with $c^2 = \lambda\Lambda^{-1}$.

Condition (30) is a spectral condition which one can verify directly if the systems $\{\phi_j\}$ and $\{\psi_j\}$ are given. The practical conclusion is that the T-matrix scheme converges if the system $\{\psi_j\}$ is a Riesz basis of H_0, the system $\{\phi_j\}$ is a closed linearly independent system in H_0, and (30) holds.

If one takes $\phi_j = \overline{\psi_j}$ then $\alpha_{ij} = \Phi_{ij}$, $\alpha = \alpha^*$, and inequality (28) takes the form

$$c^2(\alpha)_m \leqslant (\alpha)^2_m, \quad c > 0. \tag{31}$$

Inequality (31) holds for the selfadjoint positive definite matrix $\alpha_{ij} = (\phi_i, \phi_j)$ iff the first inequality (30) holds. To see this one can use the spectral theorem for a selfadjoint operator $A \geqslant 0$ and write

$$((A^2 - c^2 A)f, f) = \int_0^\infty (t^2 - c^2 t)\, \mathrm{d}(E_t f, f) \tag{32}$$

where E_t is the resolution of the identity for A. Since $(\min_{t \geqslant 0} (t^2 - c^2 t) = -c^4/4 < 0$ one sees that a necessary and sufficient condition for the inequality $A^2 \geqslant c^2 A$ to hold with some positive c^2 is the validity of the inequality $A \geqslant bI$ with some $b > 0$. If $A \geqslant bI$, $b > 0$, then, e.g., $A^2 - b\,A \geqslant 0$. Let us formulate the result. □

PROPOSITION 2. *If* $\phi_j = \overline{\psi_j}$ *and* $\inf_m \lambda(\alpha)_m \geqslant \lambda > 0$ *then* (28) *holds and therefore the conclusion of Theorem* 3 *holds.*

2.5. Theorem 3 and Propositions 1, 2 answer questions Q.1 and Q.2 from Section 1. The rate of convergence of the T-matrix scheme with the fixed systems $\{\psi_j\}$ and $\{\phi_j\}$ depends on the rate of convergence of the quantity $\|h - P_m h\|$, in other words it depends on the rate of approximation of the function $h = \alpha^{-1} p$ by linear combinations of function $\sum_{j=1}^m c_j \phi_j$. It follows from (12) and (13) that

$$\|h - h_m\| \leqslant \|h - P_m h\|(1 + c^{-1}\|\alpha\|). \tag{33}$$

This inequality gives the rate of convergence of the T-matrix scheme. It is a problem of approximation theory to estimate

$$\|h - P_m h\| = \inf_{c_1, \ldots, c_m} \left\| h - \sum_{j=1}^m c_j \phi_j \right\|. \tag{34}$$

Estimate (33) gives an answer to question Q.5 from Section 1.

Let us assume that

$$\{\phi_j\} \text{ is a complete minimal system in } H_0. \tag{35}$$

Then there exists a unique biorthogonal system $\{\tilde{\phi}_j\}$, $(\phi_i, \tilde{\phi}_j) = \delta_{ji}$. The system (1.11) can be written as $(\mathscr{a})_m c^{(m)} = p^{(m)}$, where $c^{(m)} = (c_1^{(m)} \ldots c_m^{(m)})$, $p^{(m)} = (f_1, \ldots f_m)$, and the solution to (1.11) gives

$$h_m = \sum_{j=1}^{m} c_j^{(m)} \phi_j(x).$$

Therefore

$$c_j^{(m)} = (h_m, \tilde{\phi}_j). \tag{36}$$

If $\| h_m - h \| \to 0, m \to \infty$, then Equation (36) shows that

$$|c_j^{(m)} - c_j| \to 0, \quad m \to \infty, \quad c_j = (h, \tilde{\phi}_j). \tag{37}$$

The rate of convergence can be estimated:

$$|c_j^{(m)} - c_j| \leqslant \| h_m - h \| \, \| \tilde{\phi}_j \|. \tag{38}$$

Therefore, if

$$\sup_{1 \leqslant j < \infty} \| \tilde{\phi}_j \| \leqslant c < \infty \tag{39}$$

then

$$\sup_{1 \leqslant j < \infty} |c_j^{(m)} - c_j| \leqslant c \| h_m - h \|. \tag{40}$$

Inequality (40) implies uniform convergence (in j) of the sequence $c_j^{(m)}$ provided that $\| h_m - h \| \to 0, m \to \infty$.

Condition (39) holds e.g. if the system $\{\phi_j\}$ is a Riesz basis of H_0. Let us summarize the result.

PROPOSITION 3. *Assume that the systems* $\{\psi_j\}$ *and* $\{\phi_j\}$ *are Riesz bases of* H_0 *and the first inequality* (30) *holds. Then the T-matrix scheme converges, i.e. Equation* (1.11) *is uniquely solvable for all* $m > m_0$ *and* $\| h_m - h \| \to 0, m \to \infty$, *where* h_m *is defined by* (1.10), *and* (37) *holds uniformly in* j, $1 \leqslant j < \infty$. *Moreover, the series* $h = \sum_{j=1}^{\infty} c_j \phi_j$ *converges in* H_0 *to the solution of* (1.7).

Proof. All the statements of this proposition were proved except for the very last. To prove the last statement let us write the series for the solution to the problem (1.7): $h = \sum_{j=1}^{\infty} a_j \phi_j$. This series converges in H_0 because the system $\{\phi_j\}$ is a basis of H_0. The coefficients a_j are uniquely determined by h: apply the functional $\tilde{\phi}_j$ and use the biorthogonality $(\phi_j, \tilde{\phi}_i) = \delta_{ij}$ to obtain $(h, \tilde{\phi}_j) = a_j$. Therefore $a_j = c_j$ (see (37)).

The questions Q.1—Q.5 from Section 1 are answered. Let us discuss the last question Q.6. □

2.6. Consider the perturbed system (1.11):

$$\sum_{j=1}^{m} (a_{nj}^{(m)} + b_{nj})\tilde{c}_j^{(m)} = \tilde{f}_n, \quad 1 \leqslant n \leqslant m, \quad |\tilde{f}_n - f_n| \leqslant \varepsilon_n, \tag{41}$$

where $\tilde{f}_n$ and b_{nj} are small perturbations of the right side and of the matrix $a_{nj}^{(m)}$ respectively.

Let us assume that: the operator b in ℓ^2 generated by the matrix b_{nj} has the norm

$$\|b\| \leqslant \delta \tag{42}$$

and

$$\|\tilde{f} - f\| \leqslant \|\{\varepsilon_n\}\| < \varepsilon, \quad f = \{f_n\}, \quad \tilde{f} = \{\tilde{f}_n\}, \tag{43}$$

where $\{\varepsilon_n\}$ is an element of ℓ^2. The matrix b_{nj} can be considered as the perturbation due to round-off errors and to perturbations of the basis system $\{\psi_j\}$ and $\{\phi_j\}$.

PROPOSITION 4. *If the assumptions in Proposition* 3 *hold and* $\delta > 0$ *is sufficiently small then Equation* (41) *is uniquely solvable for all* $m > m_0$, *the function* $\tilde{h}_m = \sum_{j=1}^{m} \tilde{c}_j^{(m)}\phi_j$ *converges in* H_0 *to a function* $\tilde{h}$ *such that* $\|\tilde{h} - h\| \leqslant c(\varepsilon + \delta)$, *where* $c > 0$ *is a constant which does not depend on* ε *and* δ.

Proof. The unique solvability of system (41) for all sufficiently large m follows from Theorem 2 provided that $\|b\| < \delta$, where $\delta > 0$ is sufficiently small.

Convergence in H_0 of $\tilde{h}_m$ to an element $\tilde{h} \in H_0$ follows from the same theorem. In order to estimate $\|h - \tilde{h}\|$ one writes

$$\begin{aligned} \|h - \tilde{h}\| &= \|a^{-1}f - (a+b)^{-1}\tilde{f}\| \leqslant \|a^{-1}\|\varepsilon + \|[a^{-1} - (a+b)^{-1}]f\| \\ &= \|a^{-1}\|\varepsilon + \|(a+b)^{-1}ba^{-1}\tilde{f}\| \leqslant \|a^{-1}\|\varepsilon + \\ &\quad + \|(a+b)^{-1}\|\,\|a^{-1}\| \cdot \|\tilde{f}\| \cdot \delta \leqslant c(\varepsilon + \delta), \\ c &= \|a^{-1}\| + \|\tilde{f}\|\,\|a^{-1}\|^2(1 - \delta\|a^{-1}\|)^{-1}. \end{aligned} \tag{44}$$

Thus, if (42) and (43) hold and δ is sufficiently small the answer to the question Q.6 is yes.

Assumptions (42) and (43) mean that the perturbations decay as $n \to \infty$. Let us consider the case when the perturbations do not depend on n:

$$|f_n - \tilde{f}_n| \leqslant \varepsilon \quad \forall n, \tag{45}$$

$$|b_{nj}| < \delta \qquad \forall n, j. \tag{46}$$

In this case the number m which determines the size of the truncated matrix should be chosen as a function of ε and δ, $m = m(\varepsilon, \delta)$, so that it will not be too large or too small. The problem of how to choose this $m(\varepsilon, \delta)$ is discussed in Ramm [11c]. □

2.7. Equation (1.7) was obtained for the Dirichlet boundary condition (1.2), but similar equations can be obtained in the same way for other boundary conditions, e.g.,

$$\frac{\partial u}{\partial N} = f \text{ on } \Gamma \tag{47}$$

or

$$\frac{\partial u}{\partial N} + \eta(s)u = f \text{ on } \Gamma, \operatorname{Im} \eta \geqslant 0, \tag{48}$$

where $\eta(s) \in C(\Gamma)$ is a given function.

For example, (1.6) and (48) imply

$$\int_\Gamma u \left\{ \frac{\partial \psi_n}{\partial N} + \eta(s)\psi_n \right\} \mathrm{d}s = \int_\Gamma f\psi_n \,\mathrm{d}s. \tag{49}$$

This equation is of the form (1.7) with $\int_\Gamma f\psi_n \,\mathrm{d}s$, $u(s)$, and $\partial\psi_n/\partial N + \eta(s)\psi_n$ playing the roles of f_n, h, and ψ_n respectively.

If $\{\psi_j\}$ is the system of the outgoing spherical waves, $\psi_n = h_n^{(1)}(kr)\, Y_n(\omega)$, where $h_n^{(1)}(kr) = (\pi/2kr)^{1/2} H_{n+1/2}(kr)$ is the spherical Hankel function, $H_{n+1/2}^{(1)}(kr)$ is the Hankel function of the first kind, and $Y_n(\omega)$ are the spherical harmonics normalized in $L^2(S^2)$, $\omega = (\theta, \phi)$, then this system on Γ as well as the systems $\{\partial\psi_n/\partial N + \eta(s)\psi_n\}$ and $\{\partial\psi_n/\partial N\}$ on Γ are closed in $H_0 = L^2(\Gamma)$. In order to prove this, assume, e.g., that

$$\int_\Gamma f\psi_n \,\mathrm{d}s = 0, \forall n \tag{50}$$

The Green function g can be expanded as

$$g = \frac{\exp(ik|x - y|)}{4\pi|x - y|} \tag{51}$$

$$= \sum_n ik\psi_n^{re}(x)\psi_n(s), \quad |x| < \min_{s \in \Gamma} |s| \equiv a$$

where $\{\psi_n^{re}\}$ are regular solutions of the Helmholtz equation, $\psi_n^{re} = j_n(kr)Y_n(\omega)$, $j_n(kr) = (\pi/2kr)^{1/2}J_{n+1/2}(kr)$ is the spherical Bessel function, and the origin is inside D. From (50) and (51) one obtains

$$v(x) \equiv \int_\Gamma g(x, s)f\,\mathrm{d}s = 0, \quad |x| < a. \tag{52}$$

The function v is a solution of Helmholtz's equation in D and in Ω, $v = 0$ in D by the unique continuation theorem. Therefore $v = 0$ on Γ, and $v = 0$ in Ω by Lemma I.2.2. Therefore $f = 0$ by the jump relation (I.1.3). This argument proves the closedness of the two other systems. For example, if

$$\int_\Gamma f\left(\frac{\partial\psi_n}{\partial N} + \eta(s)\psi_n\right)\mathrm{d}s = 0, \tag{53}$$

then

$$u(x) \equiv \int_\Gamma \left\{\frac{\partial g(x, s)}{\partial N} + \eta(s)g(x, s)\right\} f\,\mathrm{d}s = 0, \quad |x| < a. \tag{54}$$

Therefore

$$(\nabla^2 + k^2)u(x) = 0 \text{ in } D \cup \Omega, \quad u \equiv 0 \text{ in } D. \tag{55}$$

Thus

$$u_+ = \frac{\partial u}{\partial N_+} = 0. \tag{56}$$

From (I.1.3) it follows that

$$u_- - u_+ = f, \quad \frac{\partial u}{\partial N_+} - \frac{\partial u}{\partial N_-} = \eta f. \tag{57}$$

From (56) and (57) one obtains

$$u_- = f, \quad \frac{\partial u}{\partial N_-} = -\eta f, \quad \frac{\partial u}{\partial N_-} + \eta u_- = 0. \tag{58}$$

Therefore $u(x)$ solves the problem

$$(\nabla^2 + k^2)u = 0 \text{ in } \Omega, \quad k > 0, \quad \frac{\partial u}{\partial N_-} + \eta(s)u = 0 \text{ on } \Gamma \tag{59}$$

u satisfies the radiation condition (0.5) (60)

The argument in Lemma I.2.2 shows that $u = 0$ in Ω. Therefore $f = u_- = 0$. Let us summarize this result.

LEMMA 1. *Let D be a bounded domain with a Liapunov boundary Γ. Then each of the systems*

$$\{\psi_n\}, \quad \left\{\frac{\partial \psi_n}{\partial N}\right\} \text{ and } \left\{\frac{\partial \psi_n}{\partial N} + \eta(s)\psi_n\right\}, \quad \text{where } \psi_n = h_n^{(1)}(kr)Y_n(\omega), \tag{61}$$

is closed in $L^2(\Gamma)$. The function $\eta(s) \in C(\Gamma)$, $\operatorname{Im} \eta \geqslant 0$.

Proof. A proof of this lemma was given above for the first and third systems. The proof for the second system is essentially the same as for the first.

Our argument holds for domains which are not necessarily connected. Actually the following stronger lemma is valid. □

LEMMA 2. *Let $\tilde{\Gamma}$ be an open set on a Liapunov surface. Then each of the systems* (61) *is closed in $L^2(\tilde{\Gamma})$.*

Proof. Assume that, e.g.,

$$\int_{\tilde{\Gamma}} f\bar{\psi}_n \, ds = 0 \, \forall n. \tag{62}$$

Then as in the proof of Lemma 1, one obtains

$$v(x) \equiv \int_{\tilde{\Gamma}} fg(x, s) \, ds = 0, \quad |x| < a. \tag{63}$$

By the unique continuation theorem one has

$$v(x) = 0 \text{ outside } \tilde{\Gamma}. \tag{64}$$

From the jump relation (I.1.3) it follows that $f = 0$. Similar argument holds for the other two systems (61). □

Remark 2. The system $\{\psi_n^{re}\}$ may not be closed in $L^2(\Gamma)$. For example, if $\Gamma = \{s: |s| = 1\}$ and $k > 0$ is a zero of the Bessel function: $J_{n+1/2}(k) = 0$ for some $n = n_0$, then the function $Y_{n_0}(\omega)$ is orthogonal in $L^2(\Gamma)$ to all of the functions ψ_n^{re} but $Y_{n_0}(\omega) \not\equiv 0$.

LEMMA 3. *The system* $\{\psi_n^{re}\}$ *is closed in* $L^2(\Gamma)$ *iff* k^2 *is not an eigenvalue of the Dirichlet Laplacian* $-\Delta_D$ *in* D. *If* k^2 *is an eigenvalue of* $-\Delta_D$ *then the system* $\{\psi_n^{re}\}$ *is not closed in* $L^2(\Gamma)$ *but it has a finite defect m where m is the multiplicity of the eigenvalue* k^2. *This means that there exists m functions* $\mu_1, \ldots, \mu_m$, *such that the system*

$$\{\mu_j\} \cup \{\psi_n^{re}\} \text{ is complete in } L^2(\Gamma).$$

Proof. Let $\int_\Gamma f\psi_n^{re}\, ds = 0$, $\forall n$. Then $v(x) \equiv \int_\Gamma fg(x, s)\, ds \equiv 0$, $x \in \Omega$. Therefore

$$(\nabla^2 + k^2)v = 0 \text{ in } D, \quad v = 0 \text{ on } \Gamma. \tag{55'}$$

The jump relation (I.1.3) shows that $f = \partial u/\partial N_+$. If there are m linearly independent (in $L^2(D)$) eigenfunctions $v_1, \ldots, v_m$ then the functions $\mu_j = \partial v_j/\partial N_+$ are linearly independent in $L^2(\Gamma)$. Indeed, suppose $\sum_{j=1}^m c_j\mu_j = 0$. Then $v(x) = \sum_{j=1}^m c_jv_j(x)$ solves (55′) and $\partial v/\partial N_+ = 0$ on Γ. This implies that $v = 0$ in D. Since the system $\{v_j\}$, $1 \leqslant j \leqslant m$, is linearly independent in $L^2(D)$, it follows that $c_j = 0$, $1 \leqslant j \leqslant m$. Thus, the system $\{\mu_j\}$, $1 \leqslant j \leqslant m$, is linearly independent in $L^2(\Gamma)$. Therefore, if $\int_\Gamma f\psi_n^{re}\, ds = 0\, \forall n$ and $\int_\Gamma f\mu_j\, ds = 0$, $1 \leqslant j \leqslant m$, then $f = 0$. □

Remark 3. The systems (61), although closed in $L^2(\Gamma)$, do not form a Riesz basis on Γ unless Γ is a sphere. This conclusion follows from the fact that the condition number of the truncated Gramians of systems (61) grows to infinity as the truncation size grows to infinity, and is illustrated by the numerical results given in Section 3 below. In general these systems do not form a Schauder basis either. This question is closely connected with the so called Rayleigh hypothesis. This hypothesis says that the series

$$u(x) = \sum_{n=0}^{\infty} c_n\psi_n(x), \quad |x| > R \tag{65}$$

where u solves problem (1.1)–(1.3) and ψ_n are given in (61), converges on the surface Γ. This hypothesis is not true in general: although the solution to problem (1.1)–(1.3) for any f can be represented by the series (65) if R is so large that $D \subset B_R = \{x: |x| \leqslant R\}$, this series diverges on Γ for certain shapes of Γ. A discussion of this question and further references can be found in Barantsev *et al.* [1], Millar [1], and van den Berg–Fokkema [1]. Examples of finite scatterers and periodic surfaces for which the Rayleigh hypothesis does not hold are given in these references.

Remark 4. The systems (61) are complete in $L^p(\Gamma)$, $p > 1$. The proof is the same as the proof of Lemma 1. The jump relation for the L^p densities, $p > 1$, is given in Fabes *et al.* [1].

Suppose that

$$\{(f, \phi_n) = 0 \,\forall n \,\&\, f \in M\} \Rightarrow f = 0 \tag{*}$$

where M is a linear dense set in the Hilbert space H, $\{\phi_n\}$ is a linearly independent system in H, and (f, ϕ) is the inner product in H. Given (*), is it true that

$$\{(f, \phi_n) = 0 \,\forall n \,\&\, f \in H\} \Rightarrow f = 0. \quad (?) \tag{**}$$

This question is of interest in the following situation.

Suppose that $H = L^2(\Gamma)$, ψ_n is the system (61), and $M = C(\Gamma)$. For the densities $f \in C(\Gamma)$ the classical potential theory is well known while for the densities $f \in L^p(\Gamma)$ the theory is more complicated and is not available in book form yet. Since $C(\Gamma)$ is dense in $H = L^2(\Gamma)$ one could avoid the reference to this more complicated theory if one could show that (*) implies (**). However, this implication does not hold in general.

Example. Let $H = \ell^2$, $\phi_1 = (1, -2, 0, 0 \ldots)$, $\phi_{n+1} = S\phi_n$, $n \geqslant 1$, where S is the shift operator defined by $Sf = (0, f_1, f_2, \ldots)$, $f = (f_1, f_2, \ldots)$, and M be the set of all compactly supported sequences $f \in \ell^2$, i.e. $f \in M$ iff $f \in \ell^2$ and $f_j = 0$ if $j > N(f)$. Suppose that $(f, \phi_n) = 0 \,\forall n$. Then $f_n - 2f_{n+1} = 0$, $f_{n+1} = f_n/2$, $f_{n+1} = f_1 2^{-n}$. If $f \in M$ then $f_1 2^{-n} = 0$ for $n > N(f)$. Therefore $f_1 = 0$, and $f = 0$. Thus, (*) holds. But (**) does not hold: take $f = (1, 2^{-1}, 2^{-2}, \ldots)$ and check that $(f, \phi_n) = 0 \,\forall n$, $f \in \ell^2$, and $f \neq 0$.

If $M = C[0, 1]$ then (*) does not imply (**). Indeed, chose a set $S = \cup_{j=1}^{\infty} I_j$, where I_j is an open interval of length $|I_j| \leqslant \varepsilon 2^{-(j+1)}$ centered at r_j, where $\{r_j\} \subset I = [0, 1]$ is the countable set of all rationals in the segment $[0, 1]$. The set S is dense in I, $|S| \leqslant \varepsilon$, $|T| \equiv |I \setminus S| \geqslant 1 - \varepsilon$, where $|S|$ is the Lebesque measure of the set S. Let ϕ_j be a nonnegative function with support I_j, $\int_0^1 \phi_j \, dx = 1$. One can assume that $\phi_j(r_j) \geqslant \phi_j(x)$, that $\phi_j(x)$ is monotone on the intervals $[0, r_j]$, $[r_j, 1]$, and that the system $\{\phi_n\}$, $n = 1, 2, 3 \ldots$ is linearly independent in $L^2[0, 1]$. If $M = C[0, 1]$ then (*) holds for the system $\{\phi_j\}$. Indeed, suppose $f \in C[0, 1]$ and $(f, \phi_n) = 0 \, \forall n$. Let us show that $f(r_j) = 0 \, \forall j$. This and the continuity of f imply that $f = 0$. Assume that $f(r_m) \neq 0$ at some rational number r_m. Without loss of generality assume $f(r_m) \geqslant \delta > 0$. Then there exists an interval $I_{j(m)}$, $r_m \in I_{j(m)}$ such that $f(x) \geqslant \delta/2$ if $x \in I_{j(m)}$. Therefore $\int_0^1 f(x)\phi_{j(m)}(x) \, dx \geqslant \delta/2 > 0$, which is a contradiction. Thus (*) holds. Let us show that (**) does not hold. Let $\varkappa_T(x)$ be the characteristic function of the set $T = I \setminus S$, $\varkappa_T(x) = 1$ if $x \in T$ and 0 otherwise. Then $0 \not\equiv \varkappa_T \in L^2[0, 1]$, and $\int_0^1 \varkappa_T \phi_n \, dx = 0 \, \forall n$. Thus (**) does not hold.

2.8. One can use the T-matrix method to solve problems of potential scattering theory.

Let

$$(-\nabla^2 + q(x) - k^2)u = 0 \text{ in } R^3, \quad k > 0, \tag{66}$$

$$q(x) = 0, \; |x| \geqslant R, \quad q \in L^2(B_R), \quad \operatorname{Im} q(x) = 0, \tag{67}$$

$$u = u_0 + v, \; u_0 = \exp(-ikn \cdot x), \; v \text{ satisfies the radiation condition.} \tag{68}$$

The integral equation for u is

$$u = u_0 - \int_{B_R} gqu \, dy = u_0 - Tu, \quad g = \frac{\exp(ik|x - y|)}{4\pi|x - y|}. \tag{69}$$

Let $\{f_j\}$, $1 \leqslant j < \infty$, be an orthonormal basis of $L^2(B_R)$, $u_{0R} = u_0$ in B_R, $u_{0R} = 0$ in $\Omega_R = R^3 \setminus B_R$,

$$u_N = \sum_{j=1}^{N} c_j f_j, \tag{70}$$

$L_N = \operatorname{span}\{f_1, \ldots f_N\}$, and P_N is the orthoprojection in $L^2(B_R) = H$ onto L_N. Consider the equation

$$u_N + P_N T u_N = P_N u_{0R}. \tag{71}$$

The problem is to prove that (71) is uniquely solvable for all sufficiently large N and $u_N \to u$ as $N \to \infty$, where u solves (69) with $u_0 = u_{0R}$.

Clearly

$$\|P_N T - T\| \to 0, \quad N \to \infty \tag{72}$$

because T is compact in H and $P_N \to I$ strongly in H, where I is the identity in H. Assume for a moment that $I + T$ is invertible. (This assumption is proved later). Then (71) is identical with (16) and the result of Example 1, p. 154, allows one to show that the following proposition holds.

PROPOSITION 5. *Under the assumption* (67) *Equation* (71) *is uniquely solvable for all sufficiently large N and* $\|u_N - u_R\| \to 0$, $N \to \infty$, *where* u_R *solves equation* (69) *with* $u_0 = u_{0R}$. *The solution* u_R *determines uniquely the solution to Equation* (69) *in* R^3: $u = u_R$ *in* B_R, $u = u_0 - Tu_R$ *in* Ω_R.

Proof. The result of Example 1 shows that Equation (71) is uniquely solvable for all sufficiently large N and $\|u_N - u_R\| \to 0$, $N \to \infty$, where u_R soles Equation (69) with $u_0 = u_{0R}$. This is true under the assumption that $I + T$ is invertible. The last statement of the proposition can be verified immediately. It remains to be proved that $I + T$ is invertible. This follows from compactness of T and lemma 4. □

LEMMA 4. *If* $(I + T)w = 0$ *then* $w = 0$.

Proof. If $(I + T)w = 0$ then w solves Equation (66) and satisfies the radiation condition. Therefore $w \equiv 0$. Indeed the radiation condition can be written as

$$\begin{aligned} 0 &= \lim_{r \to \infty} \int_{S_r} |w_r - ikw|^2 \, ds \\ &= \lim_{r \to \infty} \left\{ \int_{S_r} [|w_r|^2 + k^2 |w|^2] \, ds + ik \int_{S_r} (\bar{w}w_r - w\bar{w}_r) \, ds \right\}. \end{aligned} \tag{73}$$

One can show that (67) implies

$$\lim_{r \to \infty} \int_{S_r} (\bar{w}w_r - w\bar{w}_r) \, ds = 0. \tag{74}$$

Therefore

$$\lim_{r\to\infty} \int_{S_r} (|w_r|^2 + k^2|w|^2)\, ds = 0. \tag{75}$$

This and the remark at the end of the proof of Lemma I.2.1 imply that $w = 0$ in Ω_R, and, by the unique continuation property of solutions to elliptic equations, $w = 0$ in R^3. Let us finally check (74). Multiply Equation (66) for w by $\bar{w}$ and Equation (66) for $\bar{w}$ by w, subtract and use the Green formula. As a result one obtains (74). □

Remark 5. If Im $q \neq 0$ then Lemma 4 is not valid: there exists a potential $q(x) \in C_0^\infty$ such that Equation (66) has a nontrivial solution satisfying the radiation condition. *Example*: take $w = r^{-1} \exp(ikr)$ if $r = |x| \geqslant 1$ and continue w in B_1 as a function $w \in C^\infty$ such that $|w| \geqslant \varepsilon > 0$. This is possible since $|w| = 1$ on ∂B_1. Define

$$q(x) = 0, \quad r \geqslant 1 \tag{76}$$

$$q(x) = w^{-1}(x)\,(\nabla^2 + k^2)w \text{ in } B_1. \tag{77}$$

Then $q \in C_0^\infty$, $w \neq 0$ satisfies Equation (66) with this q and the radiation condition.

2.9. Consider the transmission problem (scattering by a penetrable body) (I.1.40)–(I.1.44). In Section I.2.6, Theorem 2, the existence and uniqueness of the solution to this problem is established. Notations introduced in Section I.2.6 are used below. Let $\{\phi_j\}$ be a complete linearly independent system of functions in H^0 and $\{\psi_j\}$ be a complete linearly independent system in $H^{-1/2}(\Gamma)$. The union of the systems $\left\{\begin{smallmatrix}\phi_j\\0\end{smallmatrix}\right\}$ and $\left\{\begin{smallmatrix}0\\\psi_j\end{smallmatrix}\right\}$ is a complete linearly independent system in $H_0 = H^0(D) \oplus H^{-1/2}(\Gamma)$. Let $\psi_j = \partial\phi_j/\partial N_+$ and $\{\phi_j\}$ be the orthonormal (in $L^2(\Delta)$) system of eigenfunctions of the Dirichlet Laplacian, where $\Delta \supset D$ is a box or a ball, so that the system, $\{\phi_j\}$ can be given explicitly. The system $\{\phi_j\}$ is complete in H^0 while the system $\{\partial\phi_j/\partial N\}$ is complete in $H^{-1/2}(\Gamma)$. Indeed, let

$$\int_\Gamma f \overline{\frac{\partial \phi_j}{\partial N}}\, ds = 0, \forall j. \tag{78}$$

Multiply (78) by $\phi_j(x)/k_j^2$, where $(\nabla^2 + k_j^2)\phi_j = 0$ in Δ, $\phi_j = 0$ on $\partial\Delta$, and sum over j. Since

$$\sum_{j=1}^{\infty} \phi_j(x)\overline{\phi_j(s)}\, k_j^{-2} = G(x, s),$$

where

$$-\nabla^2 G = \delta(x-y) \text{ in } \Delta,\ G = 0 \text{ on } \partial\Delta,$$

one obtains

$$\int_\Gamma f(s) \frac{\partial G(x, s)}{\partial N}\, ds = 0, \quad x \in \Delta. \tag{79}$$

From (79) and the jump relations for the potential of the double layer one concludes that $f = 0$.

Let

$$w_m = \begin{pmatrix} \sum_{j=1}^{m} c_j^{(m)} \phi_j \\ \sum_{j=1}^{m} d_j^{(m)} \frac{\partial \phi_j}{\partial N} \end{pmatrix}. \tag{80}$$

The projection method leads to the system

$$(w_m - Bw_m - h, \eta_j)_{H^0} = 0, \quad 1 \leqslant j \leqslant 2m, \tag{81}$$

for finding $c_j^{(m)}$ and $d_j^{(m)}$. Here B and h are given in (I.2.35) and

$$\eta_j = \begin{pmatrix} \phi_j \\ 0 \end{pmatrix}, \quad 1 \leqslant j \leqslant m, \quad \eta_j = \begin{pmatrix} 0 \\ \frac{\partial \phi_j}{\partial N} \end{pmatrix}, \quad m+1 \leqslant j \leqslant 2m. \tag{82}$$

System (81) is a linear system of $2m$ equations for $2m$ unknowns $c_j^{(m)}$ and $d_j^{(m)}$. The result in Example 1 shows that system (81) is uniquely solvable for all sufficiently large m and $w_m \to w$ in H^0 when $m \to \infty$, where w solves Equation (I.2.34).

One can reduce by half the number of unknowns if one uses Equation (I.2.32) (which is the second equation in the vector Equation (I.2.34)) in the form (I.2.29) and set

$$d_j^{(m)} = c_j^{(m)}. \tag{83}$$

In this case system (81) takes the form

$$\sum_{j=1}^{m} a_{nj} c_j^{(m)} = u_{0n}, \quad 1 \leqslant n \leqslant m, \tag{84}$$

where

$$u_{0n} = (u_0, \phi_n), \quad (f, g) \equiv (f, g)_{H_0}, \tag{85}$$

$$a_{nj} = (\phi_j, \phi_n) + \kappa(T\phi_j, \phi_n) - (1 - \rho)\left(Q\frac{\partial \phi_j}{\partial N}, \phi_n\right) \tag{86}$$

and κ, ρ, T and Q are defined by formulas (I.2.26), (I.1.43), (I.1.20), and (I.1.6) respectively. One can also obtain the system (84) by applying the projection method to the equation

$$u = F - \kappa Ty + (1 - \rho)\, Q \frac{\partial u}{\partial N_+} \tag{87}$$

which is Equation (I.2.28) with σ given by (I.2.29).

Convergence of the scheme based on Equation (84) does not follow from the result in Example 1 because the operator $Q\, \partial/\partial N_+$ in (87) is not compact in H^0. It is not known if this scheme converges. It does converge if $\rho = 1$.

2.10. One can suggest a numerical solution to exterior boundary value problems based on the completeness of the systems (61) in $L^2(\Gamma)$. Consider, for example, problem (1.1)–(1.3). If $D \subset B_R$ then the solution to problem (1.1)–(1.3) in Ω_R can be represented by the series

$$u = \sum_{n=0}^{\infty} c_n \psi_n(x), \quad x \in \Omega_R \tag{88}$$

which converges uniformly in $x \in K$, where $K \subset \Omega_R$ is any compact. Define

$$u_m(x) = \sum_{n=0}^{m} c_n(m) \psi_n(x), \quad x \in \bar{\Omega}, \tag{89}$$

where the numbers $c_n(m)$ are chosen so that

$$\left\| f - \sum_{n=0}^{m} c_n(m)\psi_n(s) \right\|_{L^2(\Gamma)} = \varepsilon(m) \to 0 \quad \text{as} \quad m \to \infty. \tag{90}$$

For example, one can choose $c_n(m)$ by solving the minimization problem

$$\varepsilon(m) = \min. \tag{91}$$

This is equivalent to finding the distance from f to the subspace span $\{\psi_0(s) \ldots \psi_m(s)\}$. Because of the completeness of the system $\{\psi_n(s)\}$ in $L^2(\Gamma)$ condition (90) can be satisfied. It will be satisfied if the numbers $c_n(m)$ are chosen so that (91) holds. Problem (91) determines $c_n(m)$ uniquely because the system $\{\psi_0 \ldots \psi_m\}$ is linearly independent in $L^2(\Gamma)$. If (90) holds then

$$|c_n - c_n(m)| \to 0 \quad \text{as} \quad m \to \infty \tag{92}$$

uniformly in n, $0 \leqslant n \leqslant n_0$, where n_0 is an arbitrary large fixed number.

To prove (92) let us use the formulas

$$u(x) = \int_\Gamma u(s) \frac{\partial G(x, s)}{\partial N} \mathrm{d}s, \quad u_m(s) = \int_\Gamma u_m(s) \frac{\partial G(x, s)}{\partial N} \mathrm{d}s \tag{93}$$

where G is the Green function corresponding to problem (1.1)–(1.3):

$$(\nabla^2 + k^2)G = -\delta(x - y) \text{ in } \Omega, \; G = 0 \text{ on } \Gamma,$$

and G satisfies the radiation condition.

It follows from (93) that

$$|u(x) - u_m(x)| \leqslant \left(\int_\Gamma |u - u_m|^2 \mathrm{d}s\right)^{1/2} \times \left(\int_\Gamma \left|\frac{\partial G(x, s)}{\partial N_s}\right|^2 \mathrm{d}s\right)^{1/2}, \quad x \in \Omega. \tag{94}$$

Therefore

$$\int_{|x|=R} |u(x) - u_m(x)|^2 \, ds \leqslant \varepsilon^2(m)c(R),$$

$$c(R) = \int_{|x|=R} ds \int_\Gamma \left| \frac{\partial G(x, s')}{\partial N} \right|^2 ds'. \tag{95}$$

From (88), (89), and (95) it follows that

$$\sum_{n=0}^{m} |c_n - c_n(m)|^2 + \sum_{n>m} |c_n|^2 \to 0 \quad \text{as} \quad m \to \infty. \tag{96}$$

The relation (92) follows from (96) immediately.

Let us summarize the results.

LEMMA 5. *If the numbers $c_n(m)$ are chosen so that* (90) *holds, then* (92) *holds uniformly in n, $0 \leqslant n \leqslant n_0$, where n_0 is an arbitrary large fixed number. The solution to problem* (1.1)–(1.3) *can be computed by formula* (88) *in which c_n are defined by* (92). *The function* (89) *is an approximate solution to problem* (1.1)–(1.3) *in the sense that*

$$\sup_{r \geqslant R} \| u(x) - u_m(x) \|_{L^2(S_r)} \to 0 \quad \text{as} \quad m \to \infty, \tag{97}$$

$$\max_{x \in \tilde{\Omega}} | D^j u(x) - D^j u_m(x) | \to 0 \quad \text{as} \quad m \to \infty, \quad j = 0, 1, 2, \ldots \tag{97'}$$

where $\tilde{\Omega} \subset \Omega$ is an arbitrary compact subdomain.

Proof. Equation (97) follows from (I.2.72). Equation (97′) for $j = 0$ follows from inequality (94), and for $j > 0$ it follows in a similar way. The rest has already been proved.

Remark 6. If f can be represented by the convergent in $L^2(\Gamma)$ series

$$f = \sum_{n=0}^{\infty} c_n' \psi_n(s), \tag{90'}$$

then, as (96) shows, $c_n' = c_n$, where c_n are the coefficients in the series (88). In this case problem (1.1)–(1.3) is solved by formula (88) if the coefficients in the series (90′) are known.

2.11. Let us discuss briefly the completeness properties of solutions to Equation (1.1) in a compact subdomain $\tilde{\Omega} \subset \Omega$. Let u be an arbitrary solution to Equation (1.1) in Ω. Consider u in $\tilde{\Omega}$. It is well known that u in $\tilde{\Omega}$ is a C^∞ function. Let us prove that the system $\{u_0(x, n, k)\}$, $n \in S^2$, $u_0 = \exp(-ikn \cdot x)$ is complete in the linear manifold of solutions to Equation (1.1). The number $k > 0$ is fixed.

LEMMA 6. *If u solves Equation* (1.1) *and*

$$\int_{\tilde{\Omega}} u(x)u_0(x, n, k)\,\mathrm{d}x = 0, \quad \forall n \in S^2 \tag{98}$$

then $u = 0$ in $\tilde{\Omega}$.

Proof. Let

$$v(x) = \int_{\tilde{\Omega}} g(x, y)u(y)\,\mathrm{d}y. \tag{99}$$

Then $v(x)$ solves Equation (1.1) in Ω_R (where R is sufficiently large so that $\tilde{\Omega} \subset B_R$) and satisfies the radiation condition. It follows from (98) that

$$v(x) = 0(|x|^{-2}) \quad \text{as} \quad |x| \to \infty. \tag{100}$$

Thus, by Lemma (I.2.1), $v = 0$ in Ω_R. By the unique continuation property of solutions to Helmholtz's equation

$$v = \frac{\partial v}{\partial N} = 0 \text{ on } \partial\tilde{\Omega}. \tag{101}$$

Moreover,

$$(\nabla^2 + k^2)v = u(x) \text{ in } \tilde{\Omega}. \tag{102}$$

From (101), (102), and Equation (1.1) for u it follows that $(\nabla^2 + k^2)v = 0$ in $\tilde{\Omega}$, and, therefore, $u = 0$ in $\tilde{\Omega}$. Indeed, from (1.1) and (102) it follows that

$$(\nabla^2 + k^2)^2 v = 0 \text{ in } \tilde{\Omega}. \tag{103}$$

Multiply (103) by $\bar{v}$ and integrate over $\tilde{\Omega}$. Apply Green's formula to the identity and use (101) to obtain

$$0 = \int_{\tilde{\Omega}} (\nabla^2 + k^2)^2 v\bar{v}\,\mathrm{d}x = \int_{\tilde{\Omega}} |(\nabla^2 + k^2)v|^2\,\mathrm{d}x. \tag{104}$$

Thus, $(\nabla^2 + k^2)v = 0$, and, by (102), $u = 0$ in $\tilde{\Omega}$. □

Remark 7. Similar result holds for solutions to the Schrödinger equation

$$(-\nabla^2 + q(x) - k^2)u = 0, \quad k > 0, \tag{105}$$

where

$$|q(x)| = 0(|x|^{-1}) \quad \text{as} \quad |x| \to \infty. \tag{106}$$

The role of u_0 in this case is played by the solutions to the scattering problem:

$$u_0(x, n, k) = \exp(-ikn \cdot x) + v, \tag{107}$$

where u_0 solves Equation (105) and v satisfies the radiation condition.

3. NUMERICAL RESULTS

3.1. The purpose of the numerical results presented here is to illustrate the properties of the commonly used complete sets of functions in $L^2(\Gamma)$ as basis systems in $H_0 = L^2(\Gamma)$. Do they form a Riesz basis of H_0? Are they good for expanding of functions $f \in H_0$? Consider first a simple example of a complete system in $L^2[0, \pi]$:

$$\phi_n(x) = (2\pi^{-1})^{1/2} \sin nx \exp(-qnx), \quad n = 1, 2 \ldots. \tag{1}$$

This set is obtained from an orthonormal basis of $L^2[0, \pi]$ consisting of the functions $\phi_{0n} = (2\pi^{-1})^{1/2} \sin nx$, $n = 1, 2, \ldots$, by multiplying ϕ_{0n} by the function $\exp(-qnx)$. Qualitatively the set $\{\phi_n\}$ is similar to the set (2.61). The constant $q > 0$ will serve in the case of the system (2.61) as a measure of the eccentricity of the obstacle D.

The Gram matrix (ϕ_m, ϕ_n) can be calculated analytically.

$$\int_0^\pi \phi_m \overline{\phi}_n \, dx = \pi^{-1} q(m+n) [1 - (-1)^{m+n} \exp(-q\pi(n+m))] \times$$

$$\times \left\{ \frac{1}{q^2(m+n)^2 + (m-n)^2} - \frac{1}{q^2(m+n)^2 + (m+n)^2} \right\}. \tag{2}$$

Let $\kappa = \|A\| \, \|A^{-1}\|$ be the condition number of a matrix A. The Gram matrix is always selfadjoint nonnegative definite, and if the system $\{\phi_j\}$ is linearly independent, then the matrix (ϕ_i, ϕ_j), $1 \leqslant i, j \leqslant m$ is positive definite. For positive definite matrices A the condition number

TABLE I
The condition number κ as a function of different truncation sizes and eccentricities for the model problem (2). The corresponding condition number κ for the normalized functions $\phi_m \cdot (\phi_m, \phi_m)^{-1/2}$ is given in parentheses.

	Truncation size							
Eccentricity	5		10		20		40	
$q = 0.1$	6	(4)	60	(40)	8×10^3	(5×10^3)	2×10^8	(1×10^8)
$q = 1/3$	1×10^2	(80)	7×10^4	(4×10^4)	3×10^{10}	(2×10^{10})	$> 10^{16}$	$(> 10^{16})$
$q = 1$	8×10^3	(5×10^3)	1×10^9	(6×10^8)	$> 10^{16}$	$(> 10^{16})$	$> 10^{16}$	$(> 10^{16})$

TABLE II

The condition number $\varkappa$ as a function of truncation sizes and ecccentricities for three different spherical waves $\{\psi_n\}$, $\{\text{Re } \psi_n\}$, $\{\chi_n\}$, and the spherical harmonics $\{Y_n\}$. The corresponding condition number for the normalized functions is given in parentheses. $m = 0$ in all cases.

	5 even	5 odd	9 even	9 odd	19 even	19 odd
$ka = 4$						
$kb = 2$						
ψ_n	30 (3)	400 (4)	2×10^7 (30)	1×10^9 (50)	$> 10^{16}$ (3×10^4)	$> 10^{16}$ (6×10^4)
Re ψ_n	30 (20)	400 (10)	5×10^6 (200)	2×10^8 (400)	$> 10^{16}$ (1×10^6)	$> 10^{16}$ (3×10^6)
χ_n	1×10^3 (5)	1×10^3 (8)	2×10^6 (50)	2×10^6 (100)	1×10^{14} (4×10^4)	1×10^{14} (1×10^5)
Y_n	3 (3)	3 (3)	4 (4)	4 (4)	4 (4)	4 (4)
$ka = 6$						
$kb = 2$						
ψ_n	30 (3)	40 (5)	2×10^7 (40)	1×10^9 (90)	$> 10^{16}$ (1×10^5)	$> 10^{16}$ (3×10^5)
Re ψ_n	400 (400)	20 (80)	4×10^4 (500)	8×10^5 (1×10^3))	$> 10^{16}$ (1×10^8)	$> 10^{16}$ (3×10^8)
χ_n	2×10^4 (6)	2×10^4 (10)	3×10^6 (100)	3×10^6 (200)	5×10^{14} (2×10^5)	4×10^{14} (4×10^5)
Y_n	7 (7)	6 (6)	9 (9)	9 (9)	10 (10)	10 (10)
$ka = 10$						
$kb = 2$						
ψ_n	30 (3)	300 (5)	2×10^7 (50)	1×10^9 (100)	$> 10^{16}$ (2×10^5)	$> 10^{16}$ (4×10^5)
Re ψ_n	30 (30)	3 (2)	60 (30)	600 (200)	5×10^{14} (8×10^8)	7×10^{15} (2×10^9)
χ_n	3×10^3 (6)	2×10^3 (10)	5×10^6 (100)	4×10^6 (300)	1×10^{14} (3×10^5)	7×10^{14} (8×10^5)
Y_n	20 (20)	10 (10)	30 (30)	30 (30)	40 (40)	40 (40)

These tables are taken from the paper Kristensson, Ramm, Ström [1].

$\kappa(A) = \Lambda\lambda^{-1}$ where $\lambda(\Lambda)$ is the minimal (maximal) eigenvalue of A. In Table I the condition number is given for the matrix (ϕ_i, ϕ_j) and (in parentheses) for the matrix (ϕ_i^0, ϕ_j^0) where $\{\phi_j^0\} \equiv \{\phi_j/\|\phi_j\|\}$ is a normalized system. It is seen that: (1) $\kappa = \kappa(m, q)$ grows rapidly as m grows, where m is the size of the matrix (ϕ_i, ϕ_j), $1 \leqslant i, j \leqslant m$; (2) $\kappa(m, q)$ increases as q increases; (3) the Gram matrix for the normalized system has a smaller condition number than that for the corresponding nonnormalized system.

The system $\{\phi_j\}$ and the biorthogonal system $\{\tilde{\phi}_j\}$ do not form a Riesz basis since the necessary conditions $\inf_m \|\phi_m\| > 0$, $\sup_m \|\tilde{\phi}_m\| < \infty$ are violated as Table I shows.

3.2. The following systems have similar properties: $\{\psi_n\}$, $\{\psi_n^{re}\}$, and $\{\varkappa_n\}$, where $\varkappa_n = (kr)^{-\ell-1} Y_n(\omega)$, $n = (\ell, m)$,

$$Y_n = \left\{ \frac{(2\ell + 1)(\ell - m)!}{4\pi(\ell + m)!} \right\}^{1/2} P_{\ell,m}(cos\,\theta) \exp(im\phi), \quad \ell = 0, 1, 2, \ldots,$$

$-\ell \leqslant m \leqslant \ell$, $P_{\ell,m}(\cos\theta)$ are the associated Legendre functions, $\psi_n(x) = h_n^{(1)}(kr) Y_n(\omega)$, $\psi_n^{re}(x) = j_n(kr) Y_n(\omega)$, $h_n^{(1)}(kr)$ is the spherical Hankel function of the first kind and $j_n(kr)$ is the spherical Bessel function. The system $\{Y_n(\omega)\}$, $n = 0, 1, 2, \ldots$ forms an orthonormal basis of $L^2(S^2)$. If $n \gg kr$ then

$$\psi_n \sim -i(2n - 1)!!\,(kr)^{-n-1} Y_n(\omega) \tag{3}$$

$$\psi_n^{re} \sim [(2n + 1)!!]^{-1}\,(kr)^n Y_n(\omega). \tag{4}$$

Therefore ψ_n and ψ_n^{re} for large n and $r = r(\omega)$, where $r = r(\omega)$ is the equation of the surface Γ, the boundary of an obstacle D, have a structure similar to that of functions (1): both systems are obtained from orthonormal systems by multiplication by functions of the form $\exp(n\rho(\omega))$, where $\rho = -qx$ in (1) and $\rho = \ln\{kr(\omega)\}$ in (3) and (4). This similarity motivated us to consider the model example (1).

Let us consider a spheroidal obstacle. The equation for its surface Γ is

$$r(\theta) = (a^{-2} \sin^2\theta + b^{-2} \cos^2\theta)^{-1/2} \tag{5}$$

where a and b are the semiaxes, the spheroid is axially symmetric and the semiaxis corresponding to b is the axis of symmetry. Since $r(\theta) =$

$r(\pi - \theta)$ the elements of the Gram matrix are zeros if $n + n'$ is odd. One can change the numeration of the columns and rows in the matrix so that it becomes a block diagonal matrix with two blocks. The size of the first block, which corresponds to the rows with even numbers, is $(n_{\max} + 1)/2$ if $n_{\max}$ is odd, and $n_{\max}/2 + 1$ if $n_{\max}$ is even. In Table II the condition numbers of the truncated Gram matrices for the systems $\{\psi_n\}$, $\{\psi_n^{re}\}$, $\{\varkappa_n\}$ and $\{Y_n\}$ on Γ are given. The numerical results in Table II indicate that the systems $\{\psi_n\}$, $\{\psi_n^{re}\}$ and $\{\varkappa_n\}$ do not form a Riesz basis of $L^2(\Gamma)$. However, the system $\{Y_n(\omega)\} = Y_n\{p^{-1}(s)\}$ on Γ does form a Riesz basis of $L^2(\Gamma)$. Here $s = p(\omega)$ is the equation of Γ and p: $S_2^2 \to \Gamma$ is assumed to be a diffeomorphism. The three properties of the system $\{\tilde{\phi}_j\}$ formulated in Section 1 hold for the system $\{\psi_n\}$, $\{\psi_n^{re}\}$ and $\{\varkappa_n\}$. One sees that the Gram matrices of the systems $\{\psi_n\}$ and $\{\varkappa_n\}$ have smaller condition numbers than that for the system $\{\psi_n^{re}\}$. The numerical results indicate that the normalized functions should be used for expansions of the field on the surface since their Gram matrix has a smaller condition number. A large condition number means that the corresponding Gram matrix is difficult to invert numerically. It means numerical instability, ill-posedness, i.e., strong dependence of the numerical results on round-off errors and errors in the data.

Let us prove that the system $Y_n(p^{-1}(s))$ forms a Riesz basis of $L^2(\Gamma)$.

LEMMA 1. *Assume that the surface Γ can be defined by the equation $s = p(\omega)$ where p: $S^2 \to \Gamma$ is a diffeomorphism. Then the system $\{Y_n(p^{-1}(s))\}$, $n = 0, 1, 2 \ldots$ forms a Riesz basis of $L^2(\Gamma)$.*

Proof. If $\mathrm{d}s$ is the element of the area of the surface Γ and $\mathrm{d}\omega$ is the corresponding element of the area of S^2 then our assumption implies that $0 < c_1 \leqslant \mathrm{d}s/\mathrm{d}\omega \leqslant c_2$. Let $q(s) = (\mathrm{d}s/\mathrm{d}\omega)^{-1/2}$ and

$$\phi_n(s) = Y_n(p^{-1}(s))q(s). \tag{6}$$

Then

$$\begin{aligned}\int_\Gamma \phi_n(s)\overline{\phi_m(s)}\,\mathrm{d}s &= \int_{S^2} Y_n(\omega)\overline{Y_m(\omega)}\,q^2\frac{\mathrm{d}s}{\mathrm{d}\omega}\,\mathrm{d}\omega \\ &= \int_{S^2} Y_n(\omega)\overline{Y_n(\omega)}\,\mathrm{d}\omega = \delta_{nm}.\end{aligned} \tag{7}$$

The operator $T: L^2(\Gamma) \to L^2(\Gamma)$ defined by the formula

$$Tf = q(s)f \tag{8}$$

is an isomorphism of $L^2(\Gamma)$ onto $L^2(\Gamma)$ because

$$c_2^{-1/2} \leqslant q(s) \leqslant c_1^{-1/2}. \tag{9}$$

From (6) and (7) it follows that the system $\{Y_n(p^{-1}(s))\}$ forms a Riesz basis of $L^2(\Gamma)$. □

Remark 1. If $p: S^2 \to \Gamma$ is a $C^{\ell+1}$ diffeomorphism then $Y_n(p^{-1}(s))$ is a Riesz basis of $H^\ell(\Gamma)$, $\ell \geqslant 0$. The proof is similar to that given above.

4. OTHER SCHEMES

4.1. Problem (1.1)—(1.3) can be reduced in many ways to various boundary integral equations. These equations were discussed in Section I.1. The classical approach is to look for the solution in the form of a double layer potential. This leads to Equation (I.1.12) which is not solvable for all $k > 0$, and therefore is not equivalent to the problem (1.1)—(1.3) at some discrete set of $k_j > 0$, namely at those values k_j^2 which are eigenvalues of the interior Neumann Laplacian. The problem (1.1)—(1.3) is uniquely solvable for all $k > 0$. Therefore Equation (I.1.12) is not convenient for numerical solution especially for large k, because for large k_j the distance between k_j^2 becomes very small: $k_j^2 \sim j^{2/3}$ as $j \to \infty$. Thus, every large k lies in a neighborhood of an exceptional value k_j at which Equation (I.1.2) is not solvable for some f. Numerically this means that the inverse of the operator $I + A'(k)$, if it exists, has a large norm.

4.2. Another integral equation, Equation (I.1.7), was derived in Section I.1 from Green's representation (I.1.5). This equation is solvable for all $k > 0$ and all f. But for the exceptional values k_j (which are sometimes called irregular frequencies) this equation has more than one solution if no other conditions on the solution are imposed. Uniqueness can be guaranteed, for example, if the solution satisfies both Equations (I.1.6) and (I.1.7). Therefore again one has difficulties in the numerical treatment.

4.3. Equation (I.1.16) is uniquely solvable for all $k > 0$ and is therefore more convenient for the numerical treatment.

But the representation (I.1.15) is not very convenient for the Neumann and Robin boundary conditions since the derivative of the double layer potential leads to a strong singularity in the kernel of the boundary integral equation.

4.4. It is not quite convenient numerically to use the so called modified Green functions of the type (I.1.26). First, if one takes a finite number of terms in the sum of the right hand side of formula (I.1.26) then only a finite interval $[0, k_0]$ is free from the irregular frequencies. Moreover, the modified Green functions are more complicated than the free space Green function.

4.5. One can use an equation of the first kind which is convenient from the following points of view: (1) it is solvable for any $k > 0$; (2) at the values of k_j at which it has more than one solution all of these solutions will produce the same (unique) solution to probelm (1.1)–(1.3). Let us discuss this in more detail. Consider Equation (I.1.8) which is an integral equation of the first kind with respect to h. If h solves this equation then the unique solution to problem (1.1)–(1.3) is given by the formula

$$u(x) = Df - Sh. \tag{1}$$

Suppose that Equation (I.1.8):

$$Qh = \frac{A'f - f}{2} = (Df)_+ \tag{2}$$

has more than one solution. Let

$$h = h_p + h_0 \tag{3}$$

be the general solution to (2), where h_0 is the general solution to the homogeneous Equation (2) and h_p is a particular solution to (2).

LEMMA 1. *The solution $u(x)$ to problem* (1.1)–(1.3) *given by formula* (1) *does not depend on the choice of the solution to Equation* (2).

Proof. It is sufficient to prove that

$$Sh_0 = 0 \text{ in } \Omega \tag{4}$$

provided that $Qh_0 = 0$. But this follows from Lemma I.2.2 since Sh_0 solves the homogeneous problem (1.1)–(1.3) and this problem has only the trivial solution. □

LEMMA 2. *Equation* (2) *is solvable for any* $k > 0$.

Proof. This follows from the fact that problem (1.1)–(1.3) is solvable (see Section I.2) and the solution can be represented by Green's formula (I.1.5).

Equation (2) is therefore convenient for numerical purposes. One can look for any solution to this equation. Practically it is useful to look for the solution with minimal norm. This solution is stable towards perturbations of f and round-off errors.

A numerical scheme for solving Equation (2) can be formulated: take a complete linearly independent system of elements $\{\phi_j\} \in H_0 = L^2(\Gamma)$, let

$$h_m = \sum_{j=1}^{m} c_j\phi_j, \; Q\phi_j = \psi_j, \; c_j = \text{const.}$$

Let

$$\left\| \sum_{j=1}^{m} c_j\psi_j - F \right\| = \min, \quad F \equiv (Df)_+, \tag{5}$$

where $\|.\|$ is the norm in H_0. Variational problem (5) leads to a linear system for the coefficients c_j. This system is uniquely solvable if the Gramian $\det(\psi_i, \psi_j) \neq 0$, $1 \leqslant i, j \leqslant m$. If $\det(\psi_i, \psi_j) = 0$ then (5) is still solvable because geometrically (5) is the problem of finding the distance from F to the finite dimensional space span $(\psi_1, \ldots, \psi_m)$. It will have many solutions in the case $\det(\psi_i, \psi_j) = 0$, and one should find among them the solution with minimal norm. Such solution exists and is unique. Indeed, for a linear equation $Ah = f$ finding the solution with minimal norm among the set of all solutions $h_p + h_0$ is equivalent to solving the problem of finding the distance from h_p to the null space $N(A)$ of the operator A. If $\dim N(A) < \infty$ this problem is uniquely solvable. In the case of the finite dimensional equation (5) this problem is always uniquely solvable. Notice that the operator Q has a finite dimensional null space, $\dim N(Q) = q < \infty$, and therefore for any m the null-space of the Gramian has dimension $\leqslant q$ and the rank of the Gramian is $\leqslant m - q$ for all m. □

4.6. We will not discuss here finite differences and finite element methods (see, e.g., Goldstein [1]).

Chapter V

Scattering by Small Bodies

1. SCATTERING BY A SINGLE SMALL BODY

1.1. If the body $\mathscr{D}$ is small in comparison with the wave length, i.e. $ka \ll 1$, where a is the characteristic dimension of $\mathscr{D}$ and k is the wave number, then one does not have to solve integral equations in order to find the scattering amplitude. The basic idea in this chapter is as follows: for small bodies the scattering amplitude depends on some functionals of the field on capacitances or polarizability tensors as we will see shortly. If $ka \ll 1$ then one can find linear algebraic equations for these quantities. Let us illustrate this by an example. Consider problem (0.1)–(0.5). For the scattered field v one has

$$(\nabla^2 + k^2)v = 0 \quad \text{in} \quad \Omega,\ k > 0,\ \Omega \subset \mathbb{R}^3, \tag{1}$$

$$v = -u_0 = -\exp(ikn \cdot s) \text{ on } \Gamma \tag{2}$$

$$\begin{aligned} &v \sim \gamma(r) f(n, \nu, k) \quad \text{as} \quad |x| = r \to \infty, \quad xr^{-1} = \nu, \\ &\gamma(r) = r^{-1}\exp(ikr). \end{aligned} \tag{3}$$

Let

$$v = \int_\Gamma g(x, t)\sigma(t)\,\mathrm{d}t, \quad g = \frac{\exp(ik|x - y|)}{4\pi|x - y|}. \tag{4}$$

Then Equation (1) is satisfied, the radiation condition (3) is satisfied, and

$$f(n, \nu, k) = \frac{1}{4\pi}\int_\Gamma \exp(-ik\nu \cdot t)\sigma(t)\,\mathrm{d}t. \tag{5}$$

The boundary condition (2) leads to the equation for σ:

$$Q\sigma \equiv \int_\Gamma g(s, t)\sigma(t)\,\mathrm{d}t = -\exp(ikn \cdot s). \tag{6}$$

So far the assumption $ka \ll 1$ was not used. For an arbitrary ka one should solve Equation (6) for σ and then compute the scattering amplitude by formula (5). If $ka \ll 1$ then Equation (5) can be written as

$$f(n, \nu, k) = \frac{1}{4\pi} \int_\Gamma \sigma(t)\, dt(1 + 0(ka)) \quad \text{if} \quad \int_\Gamma \sigma\, dt \neq 0. \tag{7}$$

Equation (6) can be written as

$$\int_\Gamma \frac{\sigma(t)\, dt}{4\pi|s - t|} = -1 + 0(ka), \quad ka \ll 1. \tag{8}$$

If one neglects the term $0(ka)$ then Equation (8) becomes

$$\int_\Gamma \frac{\sigma(t)\, dt}{4\pi|s - t|} = -1. \tag{9}$$

This equation can be interpreted as the equation for the equilibrium charge distribution on the surface Γ of the perfect conductor $\mathscr{D}$ charged to the potential $U = -1$. Formula (7) shows that the scattering amplitude is defined by one number $q = \int_\Gamma \sigma\, dt$, which has the physical meaning of the total charge on the conductor. If C is the electrical capacitance of the conductor $\mathscr{D}$, then $q = CU = -C$. Thus it follows from (9) that

$$\int_\Gamma \sigma\, dt = -C \tag{10}$$

and

$$f(n, \nu, k) = \frac{-C}{4\pi}. \tag{11}$$

The assumption $\int_\Gamma \sigma\, dt \neq 0$ is satisfied. One can see that there is no need to solve Equation (9) for σ since all one needs is the number $\int_\Gamma \sigma\, dt$. Of course, there is a question of how to compute C. This question we answer later.

1.2. Let us pass to the general method for calculating the scattering amplitude by a small body of arbitrary shape.

Let

$$(\nabla^2 + k^2)v = 0 \text{ in } \Omega,\ k > 0 \tag{12}$$

$$\frac{\partial v}{\partial N_-} + \eta v = -\left(\frac{\partial u_0}{\partial N} + \eta u_0\right) \text{ on } \Gamma, \tag{13}$$

v satisfies (3).

Substitute (4) into (13) to obtain

$$\sigma = A(k)\sigma + 2\eta Q(k)\sigma + 2\eta u_0 + 2\frac{\partial u_0}{\partial N} \tag{14}$$

where A and Q are defined in (I.1.4) and (I.1.6). Let us expand $\sigma = \sigma(k)$, $A(k)$, $2Q(k)$, and u_0 in powers of ik:

$$\sigma = \sigma_0 + ik\sigma_1 + \frac{(ik)^2}{2}\sigma_2 + \dots, \tag{15}$$

$$A(k) = A_0 + ikA_1 + \frac{(ik)^2}{2}A_2 + \dots, \tag{16}$$

$$2Q(k) = Q_0 + ikQ_1 + \frac{(ik)^2}{2}Q_2 + \dots, \tag{17}$$

$$u_0 = u_{00} + iku_{01} + \frac{(ik)^2}{2}u_{02} + \dots. \tag{18}$$

Convergence of the expansions (16)–(18) is obvious. Convergence of the expansion (15) follows from analyticity of v in k in a neighborhood of $k = 0$ and the fact that the operator $Q(k)$ in (6) is invertible and analytic in k in a neighborhood of $k = 0$.

From (15)–(18) and (14) it follows that

$$\sigma_0 = A_0\sigma_0 + \eta Q_0\sigma_0 + 2\eta u_{00} + 2\frac{\partial u_{00}}{\partial N}, \tag{19}$$

$$\sigma_1 = A_0\sigma_1 + \eta Q_0\sigma_1 + A_1\sigma_0 + \eta Q_1\sigma_0 + 2\eta u_{01} + 2\frac{\partial u_{01}}{\partial N}, \tag{20}$$

$$\begin{aligned}\sigma_2 = {} & A_0\sigma_2 + \eta Q_0\sigma_2 + A_2\sigma_0 + 2A_1\sigma_1 + \eta Q_2\sigma_0 + 2\eta Q_1\sigma_1 + \\ & + 2\eta u_{02} + 2\frac{\partial u_{02}}{\partial N}.\end{aligned} \tag{21}$$

From (5) and (15) it follows that

$$4\pi f = \int_\Gamma \left[1 - ik\nu \cdot t + \frac{(ik)^2}{2} (\nu \cdot t)^2 + \ldots \right] \times$$

$$\times \left[\sigma_0 + ik\sigma_1 + \frac{(ik)^2}{2} \sigma_2 + \ldots \right] \mathrm{d}t$$

$$= \int_\Gamma \sigma_0 \,\mathrm{d}t + ik \left[\int_\Gamma \sigma_1 \,\mathrm{d}t - \int_\Gamma \nu \cdot t\sigma_0 \,\mathrm{d}t \right] + \tag{22}$$

$$+ \frac{(ik)^2}{2} \left[\int_\Gamma \sigma_2 \,\mathrm{d}t - 2 \int_\Gamma \nu \cdot t\sigma_1 \,\mathrm{d}t + \right.$$

$$\left. + \int_\Gamma \sigma_0 (\nu \cdot t)^2 \,\mathrm{d}t \right] + \ldots .$$

We assume that $u_0 = \exp(ikn \cdot x)$ and that the origin is inside $\mathscr{D}$. Then

$$u_{00} = 1,\; u_{01} = n \cdot s,\; u_{02} = (n \cdot s)^2 \tag{23}$$

$$\frac{\partial u_{00}}{\partial N} = 0,\; \frac{\partial u_{01}}{\partial N} = n \cdot N,\; \frac{\partial u_{02}}{\partial N} = 2(n \cdot s)(n \cdot N) \tag{24}$$

where $s \in \Gamma$. We can not take $s = 0$ in (23) and (24) because we are interested in the dependence of the scattering amplitude on the characteristic size a of the body. As one can see later the formulas for the scattering amplitude will depend on the parameter ka. Let us note that

$$A_0\sigma = \int_\Gamma \mathrm{d}t\, \sigma(t) \frac{\partial}{\partial N_s} \frac{1}{2\pi r_{st}},$$

$$A_1\sigma = 0,\; r_{st} = |s - t|, \tag{25}$$

$$- \int_\Gamma A_0\sigma \,\mathrm{d}t = \int_\Gamma \sigma \,\mathrm{d}t \tag{26}$$

$$Q_0\sigma = \int_\Gamma \frac{\sigma \, \mathrm{d}t}{2\pi r_{st}},$$
$$Q_1\sigma = \frac{1}{2\pi}\int_\Gamma \sigma \, \mathrm{d}t. \tag{27}$$

Let us integrate (19) over Γ and use (23)—(27). This yields

$$\int_\Gamma \sigma_0 \, \mathrm{d}t = \int_\Gamma \eta \, \mathrm{d}t + \int_\Gamma \int_\Gamma \frac{\eta(s)\sigma_0(t)\, \mathrm{d}s \, \mathrm{d}t}{4\pi r_{st}}. \tag{28}$$

If one assumes that $\eta = \text{const}$ and uses the approximate formula

$$\int_\Gamma \frac{\mathrm{d}s}{r_{st}} \approx \frac{1}{S}\int_\Gamma \mathrm{d}t \int_\Gamma \frac{\mathrm{d}s}{r_{st}} = JS^{-1}, \; J = \int_\Gamma \int_\Gamma \frac{\mathrm{d}s \, \mathrm{d}t}{r_{st}} \tag{29}$$

where S is the area of the surface Γ, then (28) can be written as

$$\int_\Gamma \sigma_0 \, \mathrm{d}t \approx \eta S + \frac{\eta J}{4\pi S}\int_\Gamma \sigma_0 \, \mathrm{d}t. \tag{30}$$

Therefore

$$\int_\Gamma \sigma_0 \, \mathrm{d}t \approx \frac{\eta S}{1 - \eta J (4\pi S)^{-1}}, \quad \eta J(4\pi S)^{-1} \ll 1. \tag{31}$$

If one uses the approximate formula for the capacitance (Ramm [4] page 88)

$$C \approx 4\pi S^2 J^{-1}$$

then (31) can be written as

$$\int_\Gamma \sigma_0 \, \mathrm{d}t \approx \frac{\eta S}{1 - \eta S C^{-1}}. \tag{32}$$

From (22) and (32) it follows that

$$f = \frac{\eta S}{4\pi(1 - \eta S C^{-1})}, \quad \eta S C^{-1} \ll 1, \tag{33}$$

where C is the electrical capacitance of the perfect conductor with the shape and size of the scatterer $\mathscr{D}$. If $\eta \to \infty$ then the boundary condition (13) becomes condition (2), and formula (33) becomes formula (11). If $\eta \to 0$ then, up to the terms $0(ka)$ the scattering amplitude vanishes. Let us calculate the scattering amplitude for the limit case $\eta = 0$ corresponding to the acoustically hard obstacle. Notice that for the acoustically soft obstacle, that is for the boundary condition (2), the main term in the scattering amplitude for a small body does not depend on n, and k, the scattering is isotropic, $f \sim a$ since $C \sim a$.

In the next section we will see that for the acoustically hard small obstacle the main term in the scattering amplitude depends on n, ν, and k, the scattering is anisotropic, $f \sim k^2 a^3$.

1.3. Assume $\eta = 0$. Then (19) takes the form

$$\sigma_0 = A_0\sigma_0 + 2\frac{\partial u_{00}}{\partial N} = A_0\sigma_0. \tag{34}$$

Since 1 is not an eigenvalue of A_0 (Günter [1]) Equation (34) implies that $\sigma_0 = 0$. Let us prove that 1 is not an eigenvalue of A_0. Consider the potential of the single layer

$$w = \int_\Gamma \frac{1}{4\pi r_{xt}}\,\sigma_0\,\mathrm{d}t.$$

Equation (34) and formula (I.1.3) imply that $\partial w/\partial N_- = 0$. Therefore $w = 0$ in Ω and $w_- = 0$. Since $w_+ = w_-$, one has $w_+ = 0$, $\nabla^2 w = 0$ in $\mathscr{D}$. Thus $w = 0$ in $\mathscr{D}$, and $\sigma_0 = \partial w/\partial N_+ - \partial w/\partial N_- = 0$.

Consider Equation (20) for $\eta = 0$ and $\sigma_0 = 0$, and use (24) to obtain:

$$\sigma_1 = A_0\sigma_1 + 2\frac{\partial u_{01}}{\partial N} = A_0\sigma_1 + 2n \cdot N. \tag{35}$$

Integrating (35) over Γ and using (26) one obtains

$$2\int_\Gamma \sigma_1\,\mathrm{d}t = 2\int_\Gamma \frac{\partial u_{01}}{\partial N}\,\mathrm{d}t = 2n \cdot \int_\Gamma N_t\,\mathrm{d}t = 0. \tag{36}$$

Therefore formula (22) when $\eta = 0$ takes the form

$$4\pi f = -\frac{k^2}{2}\int_\Gamma \sigma_2 \, \mathrm{d}t + k^2 \int_\Gamma \nu \cdot t\sigma_1 \, \mathrm{d}t. \tag{37}$$

If $\sigma_q = A\sigma_q - 2N_q$ then the tensor

$$\beta_{pq} = V^{-1}\int_\Gamma t_p \sigma_q \, \mathrm{d}t \tag{38}$$

is the magnetic polarizability tensor of the body $\mathscr{D}$ (Ramm [4], page 58 formula (19)), and V is the volume of the body $\mathscr{D}$. From (35) and (38) it follows that

$$k^2 \nu \int_\Gamma t\sigma_1 \, \mathrm{d}t = -k^2 V \beta_{pq} n_p \nu_q. \tag{39}$$

Here and below one should sum over the repeated indices.

In order to calculate the first integral in (37) one uses Equation (21) and takes into account that $\eta = 0$, $\sigma_0 = 0$, and $A_1 \sigma = 0$:

$$\sigma_2 = A_0 \sigma_2 + 2\frac{\partial u_{02}}{\partial N} = A_0\sigma_2 + 4(n \cdot s)(n \cdot N). \tag{40}$$

Here formula (24) was used.

Integrating (40) over Γ and using (26) one obtains

$$\int_\Gamma \sigma_2 \, \mathrm{d}t = 2\int_\Gamma (n \cdot t)(n \cdot N_t) \, \mathrm{d}t = 2n \int_\Gamma N_t (n \cdot t) \, \mathrm{d}t$$
$$= 2n \int_{\mathscr{D}} \nabla(n \cdot x) \, \mathrm{d}x = 2n \cdot nV = 2V. \tag{41}$$

From (37), (39), and (41) it follows that

$$f(n, \nu, k) = -\frac{k^2 V}{4\pi}\beta_{pq} n_p \nu_q - \frac{k^2 V}{4\pi}, \quad f \sim k^2 a^3. \tag{42}$$

This formula justifies the statement in the end of Section 5.2. since $k^2 V \sim k^2 a^3$.

1.4. If the initial field u_0 is not necessarily a plane wave but any solution to Equation (1), then, for $ka \ll 1$ and $\eta = 0$, the main term in the scattering amplitude can be written as (Ramm [4], page 90, formula (39)):

$$f(\nu, k) = \frac{ikV}{4\pi} \beta_{pq} \frac{\partial u_0(0)}{\partial x_p} \nu_q + \frac{V\nabla^2 u_0(0)}{4\pi} \tag{43}$$

where $u_0(0)$ is the value of the incident field at the point 0 where the small body is placed and $\partial u_0/\partial x_p$ should be calculated at the same point, and $\nabla^2 u_0 = -k^2 u_0$. For $u_0 = \exp(ikn \cdot x)$ one has $\partial u_0(0)/\partial x_p = ikn_p$, $u_0(0) = 1$, and (43) reduces to (42).

2. SCATTERING BY MANY SMALL BODIES

2.1. Consider scattering by r small bodies. There are two aspects of this problem. The first aspect is the scattering by $r \sim 10$ bodies. In this problem one is interested in mutual interference of waves. This is a many-body problem in which the scattering amplitude again will be defined by a finite number of functionals of the scattered field. The second aspect of the problem is the scattering by $r \sim 10^{23}$ small bodies. In this problem one describes the scattering as the scattering in a continuous medium with effective parameters to be calculated from the knowledge of the distribution of the bodies (particles) and the scattering properties of a single particle.

Let us start with the statement of the scattering problem. Let

$$\mathscr{D} = \bigcup_{j=1}^{r} \mathscr{D}_j, \quad \Gamma = \bigcup_{j=1}^{r} \Gamma_j, \quad \mathscr{D}_i \cap \mathscr{D}_j = \emptyset$$
$$\text{if } i \neq j, \Omega = R^3 \setminus \mathscr{D}, \tag{1}$$

where $\emptyset$ is the empty set and Ω is the complement to $\mathscr{D}$ in R^3. Let

$$\eta = \eta_j \quad \text{on} \quad \Gamma_j, 1 \leqslant j \leqslant r \tag{2}$$

$$a = \max_{1 \leqslant j \leqslant r} a_j, \quad a_j = \text{diam } D_j, \tag{3}$$

$$d = \min d_{ij}, \quad i \neq j, \quad 1 \leqslant i, \quad j \leqslant r, d_{ij} = \text{dist}(D_i, D_j), \tag{4}$$

$$\ell = \max d_{ij}, \quad 1 \leqslant i, \quad j \leqslant r, \quad \int_\Gamma = \sum_{j=1}^{r} \int_{\Gamma_j}. \tag{5}$$

The scattering problem consists in finding the solution to problem (1.12), (1.13), (1.3). Let $s_j \in \mathscr{D}_j$, $1 \leqslant j \leqslant r$, denote the positions of the particles and $\sigma = (\sigma_1, \ldots, \sigma_r)$. We look for the solution of the form (1.4) and the scattering amplitude is given as in Section 1 by formula (1.5). The basic difference between the one- and many-body problems is that in the one-body problem $|t| \sim a$ while in the many-body problem $|t|$ can be large. Therefore, let us rewrite formula (1.5) as

$$f = \frac{1}{4\pi} \sum_{j=1}^{r} \exp(-ikv \cdot s_j) \int_{\Gamma_j} \exp\{-ikv \cdot (t - s_j)\} \sigma_j \, \mathrm{d}t. \quad (6)$$

Notice that $|t - s_j| \sim a$, $1 \leqslant j \leqslant r$, and the phase shifts between the bodies are described by the factors $\exp(-ikv \,.\, s_j)$. One obtains the integral equation for σ_j by substituting (1.4) into the boundary condition (1.13). This yields

$$\sigma_j = A_j(k)\sigma_j + 2\eta_j Q_j(k)\sigma_j + \sum{}' A_{jp}(k)\sigma_p + 2\eta_j \sum{}' Q_{jp}(k)\sigma_p + \\ + 2 \frac{\partial u_0}{\partial N(t_j)} + 2\eta_j u_0(t_j), \quad 1 \leqslant j \leqslant r, \quad (7)$$

where $\sum' = \sum_{p=1,\, p \neq j}^{r}$,

$$A_{jp}\sigma_p = 2 \int_{\Gamma_p} \frac{\partial g(t_j, t_p)}{\partial N(t_j)} \sigma_p(t_p) \, \mathrm{d}t_p \quad (8)$$

$$Q_{jp}\sigma_p = \int_{\Gamma_p} g(t_j, t_p)\sigma_p(t_p) \, \mathrm{d}t_p. \quad (9)$$

Suppose that

$$d \gg a, \quad \lambda \gg a \quad (10)$$

where λ is the wave length, $\lambda = 2\pi k^{-1}$.

This assumption does not exclude the case when there are many particles on the wavelength: the inequalities $\lambda \gg d \gg a$ are compatible. The second inequality (10) is equivalent to the basic assumption $ka \ll 1$. If (10) holds and $\eta_j = \infty$, $1 \leqslant j \leqslant r$, then Equation (7) becomes

$$Q_j(k)\sigma_j = -\sum{}' Q_{jp}\sigma_p - u_0(t_j), \quad 1 \leqslant j \leqslant r. \quad (11)$$

This equation can be written as

$$\int_{\Gamma_j} \frac{\sigma_j(t)\,\mathrm{d}t}{4\pi|t_j - t|} = -\exp(ikn \cdot s_j) - \\ - \sum_{p=1, j\neq p}^{r} \int_{\Gamma_p} g(s_j, t)\sigma_p(t)\,\mathrm{d}t + 0(ka), \quad (12)$$

$$1 \leqslant j \leqslant r.$$

It follows from (12) that

$$q_j \equiv \int_{\Gamma_j} \sigma_j\,\mathrm{d}t = -C_j \left\{ \exp(ikn \cdot s_j) + \right. \\ \left. + \sum_{p=1, j\neq p}^{r} g(s_j, s_p) q_p \right\}, \quad 1 \leqslant j \leqslant r, \quad (13)$$

where C_j is the electrical capacitance of the perfect conductor $\mathscr{D}_j$. Equation (13) is a linear system of algebraic equations for q_j, $1 \leqslant j \leqslant r$. The matrix of this system is $\delta_{jp} + C_j g(s_j, s_p)(1 - \delta_{jp})$. The system (13) can be solved by iterations if

$$\max_{1 \leqslant j \leqslant r} \sum_{p=1, p\neq j}^{r} C_j \frac{1}{4\pi|s_j - s_p|} < 1. \quad (14)$$

Condition (14) will be satisfied if the much stronger condition

$$(4\pi)^{-1} Crd^{-1} < 1, C = \max_{1 \leqslant j \leqslant r} C_j \quad (15)$$

holds. If the system (13) is solved and the numbers q_j are found then it follows from (6) that

$$f = \frac{1}{4\pi} \sum_{j=1}^{r} \exp(-ik\nu \cdot s_j) q_j + 0(rka). \quad (16)$$

The idea of this derivation is exactly the same as in Section 1.1.

2.2. Consider now the case $\eta_j \neq 0, 1 \leqslant j \leqslant r$. Let us assume that (10) holds and define

$$U_j = u_0(x) + \sum_{p \neq j}{}' \int_{\Gamma_p} g(x, t)\sigma_p(t)\,\mathrm{d}t. \tag{17}$$

Then Equation (7) can be written as

$$\sigma_j(k) = A_j(k)\sigma_j + 2\eta_j Q_j(k)\sigma_j + 2\,\frac{\partial U_j}{\partial N(t_j)} + 2\eta_j U_j(t_j). \tag{18}$$

This equation is the same as (1.14) if one substitutes U_j in the place of u_0 in (1.14). Therefore U_j defined in (17) plays the role of the effective field. One can repeat the arguments after Equation (1.14) and obtain for the scattering amplitude (6) the following approximate expressions:

$$f = \frac{1}{4\pi} \sum_{j=1}^{r} \exp(-ik\nu \cdot s_j)\,\frac{\eta_j S_j U_j}{1 - \eta_j S_j C_j^{-1}}, \quad \eta_j S_j C_j^{-1} \ll 1, \tag{19}$$

where S_j is the area of the surface Γ_j, C_j is the capacitance of $\mathscr{D}_j$ if it were a perfect conductor, and it is assumed that η_j do not vanish; and if $\eta_j = 0, 1 \leqslant j \leqslant r$, then

$$f = \frac{1}{4\pi} \sum_{j=1}^{r} \exp(-ik\nu \cdot s_j)\left(ikV_j\beta_{pq}^{(j)}\,\frac{\partial U^{(j)}(s_j)}{\partial x_p}\,\nu_q + V_j\nabla^2 U_j(s_j)\right), \tag{20}$$

where $V_j = |\mathscr{D}_j|$ is the volume of $\mathscr{D}_j$, $\beta_{pq}^{(j)}$ is the magnetic polarizability tensor of $\mathscr{D}_j$.

If $\eta_j = \infty, 1 \leqslant j \leqslant r$, and $r \gg 1$, then (16) (or (19)) can be written as

$$f = -\frac{1}{4\pi} \sum_{j=1}^{r} \exp(-ik\nu \cdot s_j) C_j U_j. \tag{21}$$

If the medium consisting of many small particles is rarefied, then one can use formulas (19)—(21) with $U_j = u_0(s_j)$. In general, in order to use formulas (19)—(21) one should calculate the effective field U.

Let us derive the equations for the effective field U. Let $N(x)\,\mathrm{d}x$ be the number of the particles in the volume $\mathrm{d}x$ around the point x, $C(x)$ and $\beta_{pq}(x)$ be the average capacitance and magnetic polarizability tensor of the particles in the volume dx. If the particles are acoustically soft, i.e. $\eta = \infty$, $1 \leqslant j \leqslant r$, then the field in the medium is

$$U(s_j) = u_0(s_j) - \sum_{p \neq j,\, p=1}^{r}{}' \; g(s_j, t_p) C_p U(t_p) V_p. \tag{22}$$

This follows from (17) if one takes into account that, for $j \neq p$,

$$\int_{\Gamma_p} g(s_j, t)\sigma_p \, \mathrm{d}t \simeq g(s_j, t_p) q_p = -g(s_j, t_p) C(t_p) U(t_p) V_p. \tag{23}$$

Indeed, the total field is $u = U + u_p$, where u_p is the field due to the p-th body $\mathscr{D}_p$. Although away from $\mathscr{D}_p$ the field u_p is negligible, but on Γ_p this field should enforce the boundary condition $U(t_p) + u_p = 0$. Thus, $U(t_p) = -u_p$ on Γ_p, and the total charge on Γ_p, i.e. q_p, is given by $q_p = -C(t_p)U(t_p)V_p$. Passing to the limit $r \to \infty$ in (22) yields

$$U(x) = u_0(x) - \int g(x, y) q(y) U(y) \, \mathrm{d}y \tag{24}$$

where

$$q(y) = N(y)C(y). \tag{25}$$

If Equation (24) is solved then the scattering amplitude is given by the formula

$$f = -\frac{1}{4\pi} \int \exp(-ik\nu \cdot y) q(y) U(y) \, \mathrm{d}y. \tag{26}$$

A similar argument gives the same Equations (24) and (25) in the case when η_j, $1 \leqslant j \leqslant r$, are finite and do not vanish.

In this case

$$q(y) = N(y)a(y), \quad \text{where} \quad a(y) = \overline{\left(\frac{\eta S}{1 - \eta S C^{-1}} \right)}, \tag{27}$$

and the bar denotes the mean value.

If $\eta_j = 0, 1 \leqslant j \leqslant r$, then the equation for the effective field is

$$U(x) = u_0(x) + \int g(x, y) \left\{ ikB_{pq}(y) \frac{\partial u(y)}{\partial y_q} \frac{x_p - y_p}{|x - y|} + \right. \tag{28}$$

$$\left. + b(y)\nabla^2 u(y) \right\} dy,$$

where

$$b(y) = \lim_{r \to 0} \frac{\Sigma V_j}{|B_r(y)|}, \quad B_{pq}(y) = \lim_{r \to 0} \frac{\Sigma V_j \beta_{pq}^{(j)}}{|B_r(y)|}, \tag{29}$$

$|B_r(y)|$ is the volume of the ball of radius r centered at the point y, the sum is taken over the particles situated in this ball, and it is assumed that the limits (29) exist. The function $b(y)$ is the average volume occupied by the particles around the point y, and $B_{pq}(y)$ is the average polarizability tensor around point y.

3. ELECTROMAGNETIC WAVE SCATTERING BY SMALL BODIES

3.1. The theory is developed in Ramm [4] and here we just formulate one of the basic results. Let $\mathscr{D}$ be a homogeneous body with characteristic dimension a. Let ε, μ and σ be its dielectric permeability, magnetic permeability, and conductivity, ε_0, μ_0 and $\sigma_0 = 0$ be the corresponding parameters of the exterior medium, $\varepsilon' = \varepsilon + i\sigma\omega^{-1}$, ω be the frequency of the initial field, λ_0 be its wavelength and $k_0 = 2\pi\lambda_0^{-1}$. Let $\delta = 1/(|\mathrm{Im}(\varepsilon'\mu)^{1/2}|\omega)$ be the depth of the skin layer in $\mathscr{D}$. If $\varepsilon \ll \sigma\omega^{-1}$ then $\delta = (2/(\sigma\mu\omega))^{1/2}$. We consider scattering under the following assumptions

$$|\varepsilon'| \gg 1, \delta \gg a \quad \text{and} \quad k_0 a \ll 1 \tag{1}$$

$$|\varepsilon'| \gg 1, \delta \ll a \quad \text{and} \quad k_0 a \ll 1 \tag{2}$$

$$\{|(\varepsilon' - \varepsilon_0)\varepsilon_0^{-1}| + |(\mu - \mu_0)\mu_0^{-1}|\}(1 + k_0^2 a^2) \ll 1. \tag{3}$$

Assumption (1) means that the body is dielectric and ε' is not very large; assumption (2) means that the body is a perfect conductor or that $|\varepsilon'|$ is very large; and assumption (3) means that the body can be of arbitrary size, not necessarily small, but it should be a small perturbation of the medium. In fact, assumption (3) is a sufficient condition for convergence of the Born series for the corresponding integro-differential equation for the electromagnetic field in the presence of the body Ramm [4].

Let us choose a Cartesian coordinate system with the origin inside $\mathcal{D}$. Let the initial field be a plane wave propagating in the positive direction e_3 of the x_3-axis, ν be the unit vector, the direction of propagation of the scattered field, θ be the angle between e_3 and ν, the angle of scattering. Let E_1 and E_2 be the components of the incident field along the x_1 and x_2 axes. The scattered field is considered in the (x_1', x_2', x_3') coordinate system. The x_3'-axis of this system is directed along the vector ν, the x_1'-axis is the same as the x_1-axis, and the planes (x_2, x_3) and (x_2', x_3') are the same. Thus, the x'-coordinate system is obtained from the X-coordinate system by a rotation around the x_1-axis over the angle θ. Let f_1 and f_2 be the components of the scattered electric field onto the axes x_1' and x_2' respectively. The (x_2, x_3) plane is called the plane of scattering.

The scattering matrix, S-matrix, is defined by the formula

$$f = SE, \quad f = \begin{pmatrix} f_2 \\ f_1 \end{pmatrix}, \quad E = \begin{pmatrix} E_2 \\ E_1 \end{pmatrix}, \quad S = \begin{pmatrix} S_2 S_3 \\ S_4 S_1 \end{pmatrix}, \tag{4}$$

or

$$\begin{pmatrix} f_2 \\ f_1 \end{pmatrix} = \begin{pmatrix} S_2 S_3 \\ S_4 S_1 \end{pmatrix} \begin{pmatrix} E_2 \\ E_1 \end{pmatrix}. \tag{5}$$

Our basic results are the following theorems.

THEOREM 1. *Under the assumptions* (1) *or* (2) *the following formula holds*

$$S = \frac{k_0^2 V}{4\pi} \times \\ \times \begin{pmatrix} \mu_0\beta_{11} + \alpha_{22}\cos\theta - \alpha_{32}\sin\theta & \alpha_{21}\cos\theta - \alpha_{31}\sin\theta - \mu_0\beta_{12} \\ \alpha_{12} - \mu_0\beta_{21}\cos\theta + \mu_0\beta_{31}\sin\theta & \alpha_{11} + \mu_0\beta_{22}\cos\theta - \mu_0\beta_{32}\sin\theta \end{pmatrix} \tag{6}$$

where α_{ij} and β_{ij} are the polarizability tensors.

This theorem is proved in Ramm [4], where the polarizability tensors $\alpha_{ij} = \alpha_{ij}(\gamma)$, $\gamma \equiv (\varepsilon' - \varepsilon_0)/(\varepsilon' + \varepsilon_0)$, and $\beta_{ij} = \alpha_{ij}(-1)$ are defined and the following formula is proved:

$$|\alpha_{ij}(\gamma) - \alpha_{ij}^{(n)}(\gamma)| \leqslant cq^n, \quad 0 < q < 1, \tag{7}$$

where q and c are constants which depend on the geometry of Γ,

$$\alpha_{ij}^{(n)}(\gamma) \equiv \frac{2}{V} \sum_{m=0}^{n} \frac{(-1)^m}{(2\pi)^m} \frac{\gamma^{n+2} - \gamma^{m+1}}{\gamma - 1} b_{ij}^{(m)}, \quad n \geqslant 1, \tag{8}$$

$$b_{ij}^{(0)} = V\delta_{ij}, \quad b_{ij}^{(1)} = \int_\Gamma \int_\Gamma \frac{N_i(t)N_j(s)}{r_{st}} \,\mathrm{d}s\,\mathrm{d}t, \tag{9}$$

$N_i(t)$ is the i-th Cartesian component of the unit normal to Γ at the point t, N is directed into Ω, $r_{st} = |s - t|$, $\mathrm{d}s$ and $\mathrm{d}t$ are the elements of the area of the surface Γ,

$$b_{ij}^{(m)} = \int_\Gamma \int_\Gamma \mathrm{d}s\,\mathrm{d}t\; N_i(t)N_j(s) \underbrace{\int_\Gamma \cdots \int_\Gamma}_{m-1} \frac{1}{r_{st_{m-1}}} \psi(t_1, t) \cdot$$
$$\cdot\, \psi(t_2, t_1) \ldots \psi(t_{m-1}, t_{m-2})\,\mathrm{d}t_1\,\mathrm{d}t_{m-1}, \tag{10}$$

and

$$\psi(t, s) \equiv \frac{\partial}{\partial N_t} \frac{1}{r_{ts}}. \tag{11}$$

In particular

$$\alpha_{ij}^{(1)}(\gamma) = 2(\gamma + \gamma^2)\delta_{ij} - \frac{\gamma^2}{\pi V} b_{ij}^{(1)}, \tag{12}$$

$$\beta_{ij}^{(1)} = \alpha_{ij}^{(1)} = -\frac{1}{\pi V} b_{ij}^{(1)}. \tag{13}$$

THEOREM 2. *If* (3) *holds then the scattering amplitude for the electric field is given by*

$$f_E = -\frac{\varepsilon' - \varepsilon_0}{4\pi\varepsilon_0} k_0^2 \nu \times [\nu \times A_E(\nu)] - \frac{k_0\omega(\mu - \mu_0)}{4\pi} \nu \times A_H(\nu), \tag{14}$$

where

$$A_E(\nu) \equiv \int_{\mathscr{D}} \exp(-ik_0\nu \cdot y)E_0(y)\,\mathrm{d}y,$$
$$A_H(\nu) = \int_{\mathscr{D}} \exp(-ik_0\nu \cdot y)H_0(y)\,\mathrm{d}y \tag{15}$$

and (E_0, H_0) *is the initial field.*

Sketch of the proof (*Ramm* [4]). The integral equations for the total field (E, H) are

$$E(x) = E_0(x) + \frac{\varepsilon' - \varepsilon_0}{\varepsilon_0} \nabla \times \nabla \times \int_{\mathscr{D}} g(x, y)E(y)\,\mathrm{d}y + {} \\ + i\omega(\mu - \mu_0)\nabla \times \int_{\mathscr{D}} g(x, y)H(y)\,\mathrm{d}y - \frac{\varepsilon' - \varepsilon_0}{\varepsilon_0} \varkappa_{\mathscr{D}} E(x), \tag{16}$$

$$H(x) = H_0(x) + \frac{\mu - \mu_0}{\mu_0} \nabla \times \nabla \times \int_{\mathscr{D}} g(x, y)H\,\mathrm{d}y - i\omega(\varepsilon' - {} \\ - \varepsilon_0)\nabla \times \int_{\mathscr{D}} g(x, y)E(y)\,\mathrm{d}y - \frac{\mu - \mu_0}{\mu_0} \varkappa_{\mathscr{D}} H(x), \tag{17}$$

where $\varkappa_{\mathscr{D}} = 1$ if $x \in \mathscr{D}$, $\varkappa_{\mathscr{D}} = 0$ if $x \in \Omega$ (see Lemma 1 below). Under the assumption (3) one can prove [Ramm 4, p. 101] that the system (16)–(17) is uniquely solvable by iterations. If one iterates once and then computes the scattering amplitude f_E, which is defined by the formula

$$E - E_0 = \frac{\exp(ik_0 r)}{r} f_E + o\left(\frac{1}{r}\right), \quad r = |x| \to \infty, \quad \frac{x}{r} = \nu, \tag{18}$$

then one obtains formula (14).

Scattering by thin wires and by small holes in a metallic screen is studied in Ramm [4]. □

LEMMA 1. *Equations* (16)—(17) *hold.*

Proof. The Maxwell equations are

$$\nabla \times E = i\omega\mu_0 H + i\omega(\mu - \mu_0)\varkappa_{\mathscr{D}} H, \tag{19}$$

$$\nabla \times H = -i\omega\varepsilon_0 E - i\omega(\varepsilon' - \varepsilon_0)\varkappa_{\mathscr{D}} E, \tag{20}$$

$$E = E_0 + E_1, \quad H = H_0 + H_1, \tag{21}$$

where (E_0, H_0) is the incident field which satisfies Equations (19)—(20) with $\mu = \mu_0$ and $\varepsilon' = \varepsilon_0$. Therefore

$$\nabla \times E_1 = i\omega\mu_0 H_1 + i\omega(\mu - \mu_0)\varkappa_{\mathscr{D}} H \tag{22}$$

$$\nabla \times H_1 = -i\omega\varepsilon_0 E_1 - i\omega(\varepsilon' - \varepsilon_0)\varkappa_{\mathscr{D}} E. \tag{23}$$

Let us denote

$$j_e = -i\omega(\varepsilon' - \varepsilon)\varkappa_{\mathscr{D}} E, \quad j_m = i\omega(\mu - \mu_0)\varkappa_{\mathscr{D}} H, \tag{24}$$

$$A_e = \int g(x, y) j_e(y)\, \mathrm{d}y, \quad A_m = \int g(x, y) j_m(y)\, \mathrm{d}y, \quad \int = \int_{\mathscr{D}}. \tag{25}$$

Then

$$E_1 = -(i\omega\varepsilon_0)^{-1}(\nabla \times \nabla \times A_e - j_e) - \nabla \times A_m \tag{26}$$

$$H_1 = -(i\omega\mu_0)^{-1}(\nabla \times \nabla \times A_m - j_m) + \nabla \times A_e. \tag{27}$$

These formulas are equivalent to (16)—(17). □

4. BEHAVIOR OF THE SOLUTIONS TO THE EXTERIOR BOUNDARY VALUE PROBLEMS AT LOW FREQUENCIES

4.1. In R^3 the Green functions of exterior Dirichlet and Neumann boundary value problems are analytic in k near any real point, in particular, near $k = 0$. In R^2 the free space Green function $i/4\ H_0^{(1)}(kr_{xy})$, $r_{xy} = |x - y|$, does not have a finite limit as $k \to 0$, which is clear from the formula

$$\begin{aligned} g &= \frac{i}{4} H_0^{(1)}(kr_{xy}) \\ &= \alpha(k) + g_0 + 0\left((kr_{xy})^2 \ln \frac{1}{kr_{xy}}\right), \quad kr_{xy} \to 0 \end{aligned} \tag{1}$$

where γ = const, $g_0 = (2\pi)^{-1} \ln r_{xy}^{-1}$, and $\alpha(k) = -(2\pi)^{-1} \ln(k\gamma) + (2\pi)^{-1} \ln 2 + \frac{i}{4}$. Therefore it is of interest to study the behavior of solutions to the exteriior boundary value problems in R^2 as $k \to 0$. This was done in the literature (see MacCamy [1], Ramm [2f], Favorin [1], Werner [4] and references therein), where integral equations of the second kind were used. In MacCamy [1], Ramm [2f], Favorin [1], the case of the boundary with one connected component was considered. In Werner [4] the general case of the boundary with m connected components was treated. In Ramm [2f] not only the Helmholtz equation but also the Schrödinger equation was treated. The basic result in MacCamy [1] is that Green's function of the exterior Dirichlet problem has a finite limit as $k \to 0$, while Green's function of the exterior Neumann problem does not. The aim of this section is to obtain this result for the more general case of the boundary with many components by new methods based on integral equations of the first kind and a priori estimates. These methods lead to the results in an easier and shorter way than the previously used methods, and the results are more complete. Below $\mathscr{D}$ is a bounded domain with boundary $\Gamma \in C^\infty$, Ω is the exterior domain, $\partial/\partial N_\pm$ are the limit values of outer normal derivatives on Γ from inside (+) and outside (−). The boundary function f (e.g. in (2a) below) is assumed to belong to $H^p(\Gamma)$ with $p = 0, 1$ or $3/2$. These assumptions can be relaxed. However, if $p < 3/2$, then the solution is not in H^2 in a neighborhood of Γ; if $p < 1$, then the boundary integral equation, e.g. (5), should be studied in a class of distributions.

4.2. The Dirichlet problem

Let

$$(\nabla^2 + k^2)u = 0 \quad \text{in} \quad \Omega \subset R^2, k > 0, u = f(s) \text{ on } \Gamma, \tag{2a}$$

$$r^{1/2}(u_r - iku) = o(1), r = |x| \to \infty. \tag{2}$$

THEOREM 1. *For any $f \in H^1(\Gamma)$ there exists*

$$\lim_{k \to 0} u(x, k) = u_0(x), \tag{3}$$

where the limit is attained uniformly on compacts, and the function $u_0(x)$ solves the problem

$$\nabla^2 u_0 = 0 \quad \text{in} \quad \Omega, \quad u_0 = f \text{ on } \Gamma, \quad u_0 = 0(1), \quad r \to \infty. \tag{4}$$

We need some lemmas for the proof of Theorem 1. These lemmas are of independent interest.

LEMMA 1. *If the equation*

$$Q_0\sigma \equiv \int_\Gamma g_0(s, t)\sigma(t)\,\mathrm{d}t = f(s) \tag{5}$$

has at most one solution then $Q_0: \tilde{H}^q \to \tilde{H}^{q+1}$ *is an isomorphism. Here* $\tilde{H}^q = H^q(\Gamma) = W^{2,q}(\Gamma)$ *is the Sobolev space, and q is a real number.*

A proof of this lemma is given in Section I.7, Lemma I.7.2.

LEMMA 2. *If* Γ *is connected and equation* $Q_0\sigma = 0$ *has a nontrivial solution* σ *then* $\dim N(Q_0) = 1$ *and* $\int_\Gamma \sigma\,\mathrm{d}t \neq 0$. *If* Γ *consists of m connected components* Γ_j, $1 \leqslant j \leqslant m$, *where* Γ_j *is a closed smooth curve, then* $\dim N(Q_0) \leqslant 1$. *If* $d > \operatorname{diam}\Gamma = \max_{s, s' \in \Gamma} |s - s'|$ *then the operator* Q_d *with the kernel* $(2\pi)^{-1} \ln (d/r_{xy})$ *is an isomorphism between* $\tilde{H}^q$ *and* $\tilde{H}^{q+1}$.

Proof. Let $m = 1$. If $Q_0\sigma = 0$, $\sigma \not\equiv 0$, then $\int_\Gamma \sigma\,\mathrm{d}t \neq 0$. Indeed, if $Q_0\sigma = 0$ and $\int_\Gamma \sigma\,\mathrm{d}t = 0$ then $\sigma = 0$. In order to prove this, consider the single layer potential

$$\begin{aligned} w(x) = S\sigma &= \frac{1}{2\pi}\int_\Gamma \ln\frac{1}{|x-t|}\,\sigma(t)\,\mathrm{d}t \\ &= \frac{1}{2\pi}\ln\frac{1}{|x|}\int_\Gamma \sigma\,\mathrm{d}t + w_1, \end{aligned} \tag{6}$$

where the origin is placed inside $\mathscr{D}$,

$$\nabla^2 w_1 = 0 \text{ in } \Omega, \quad w_1 = 0(|x|^{-1}) \quad \text{as} \quad |x| \to \infty \tag{7}$$

If $\int_\Gamma \sigma\,\mathrm{d}t = 0$ then $w = w_1 = 0(|x|^{-1})$. If $Q_0\sigma = 0$ then $w = 0$ on Γ. Thus

$$\nabla^2 w = 0 \text{ in } \Omega, \quad w = 0 \text{ on } \Gamma, \quad w = 0(|x|^{-1}). \tag{8}$$

By the maximum principle $w = 0$ in Ω. Moreover

$$\{\nabla^2 w = 0 \text{ in } \mathscr{D}, \quad w = 0 \text{ on } \Gamma\} \Rightarrow w = 0 \text{ in } \mathscr{D}. \tag{9}$$

Therefore $\sigma = (\partial w/\partial N_+) - (\partial w/\partial N_-) = 0$ by the jump relation. It is clear now that $\dim N(Q_0) = 1$ if $N(Q_0) \neq \{0\}$. Indeed, suppose $\sigma_j \in N(Q_0)$, $j = 1, 2$. Since $\int_\Gamma \sigma_j\,\mathrm{d}t \neq 0$ one can find a constant c such that

$\int_\Gamma (\sigma_1 - c\sigma_2)\,dt = 0$. This and the equation $Q_0(\sigma_1 - c\sigma_2) = 0$ imply that $\sigma_1 = c\sigma_2$, i.e. $\dim N(Q_0) = 1$.

Let us prove that $Q_d\sigma = 0$ implies $\sigma = 0$. If $Q_d\sigma = 0$ then

$$Q_0\sigma = -\frac{\ln d}{2\pi}\int_\Gamma \sigma\,dt. \tag{10}$$

The operator Q_0: $\tilde{H}^0 \to \tilde{H}^0$ is symmetric. Therefore Equation (10) is solvable only if σ_0, $Q_0\sigma_0 = 0$, $\int_\Gamma \sigma_0\,dt \neq 0$, is orthogonal to the right-hand side of (10) which is a constant. If $\ln d \neq 0$ then this orthogonality condition implies that $\int_\Gamma \sigma\,dt = 0$, $Q_0\sigma = 0$, and therefore $\sigma = 0$. If $d = 1$ then $Q_d = Q_0$, $\ln(1/r_{st}) > 0$ and equation $Q_0\sigma = 0$ implies $\sigma = 0$. Indeed, one can assume without loss of generality that $\int_\Gamma \sigma\,dt > 0$ and $\sigma = 1 + \sigma_1$, where $\int_\Gamma \sigma_1\,dt = 0$. Then

$$Q_0\sigma_1 = -Q_0 1 < 0. \tag{11}$$

A necessary condition for (11) to have a solution is the orthogonality of the right-hand side of (11) to the solution σ of the homogeneous equation (11). This condition can be written as:

$$0 = \int_\Gamma ds\,\sigma(s)Q_0 1. \tag{12}$$

One can show that

$$\sigma(s) \geqslant 0. \tag{13}$$

Therefore (12) and (11) imply that $\sigma = 0$.

Let us prove (13). If $Q_0\sigma = 0$ then the function $w(x)$, defined by (6), solves the problem

$$\nabla^2 w = 0 \text{ in } \Omega, \quad w = 0 \text{ on } \Gamma,$$

$$w = 0\left(\int_\Gamma \sigma\,dt \ln\frac{1}{|x|}\right), \quad |x| \to \infty. \tag{14}$$

Thus, $w \to -\infty$ as $|x| \to \infty$, and by the maximum principle one concludes that $w < 0$ inside Ω. Therefore $\partial w/\partial N_- \leqslant 0$ on Γ. Since $w = 0$ in $\mathscr{D}$, one has $\partial w/\partial N_+ = 0$. Thus $\sigma = \partial w/\partial N_+ - \partial w/\partial N_- = -\partial w/\partial N_- \geqslant 0$. The first part of Lemma 2 is proved.

(2) If $m > 1$ then dim $N(Q_0) \leqslant 1$ as in the case $m = 1$. Indeed, assume that

$$0 = Q_0\sigma = \sum_{j=1}^{m} \frac{1}{2\pi} \int_{\Gamma_j} \ln \frac{1}{|s_p - t|} \sigma_j(t)\,\mathrm{d}t,\ 1 \leqslant p \leqslant m, \quad s_p \in \Gamma_p.$$

Define $w(x)$ by formula (6) where now

$$\Gamma = \bigcup_{j=1}^{m} \Gamma_j, \int_\Gamma \sigma\,\mathrm{d}t = \sum_{j=1}^{m} \int_{\Gamma_j} \sigma_j(t)\,\mathrm{d}t. \tag{15}$$

The argument after formula (6) shows that

$$\int_\Gamma \sigma\,\mathrm{d}t = 0 \Rightarrow \sigma = 0, \quad \text{i.e. } \sigma_j = 0,\ 1 \leqslant j \leqslant m. \tag{16}$$

Therefore if $\sigma \in N(Q_0)$ then $\int_\Gamma \sigma\,\mathrm{d}t \neq 0$. If $0 \not\equiv \sigma \in N(Q_0)$ and $0 \not\equiv \tilde{\sigma} \in N(Q_0)$ then one can find a constant $c \neq 0$ such that $\int_\Gamma (\sigma - c\tilde{\sigma})\,\mathrm{d}t = 0$, $\sigma - c\tilde{\sigma} \in N(Q_0)$. This implies that $\sigma = c\tilde{\sigma}$, and dim $N(Q_0) = 1$. It is worthwhile to mention that in the traditional approach, when one uses the double layer potential for solving the exterior Dirichlet problem, the corresponding integral equation has the null space of dimension m, where m is the number of the connected components of Γ (see e.g. Günter [1]).

The rest of the argument is the same as in the first part of the proof. □

Proof of Theorem 1. Let

$$u(x) = \int_\Gamma g(x, t)\sigma\,\mathrm{d}t = \sum_{j=1}^{m} \int_{\Gamma_j} g(x, t)\sigma_j\,\mathrm{d}t, \tag{17}$$

Then u solves (2) iff

$$\int_\Gamma g(s, t)\sigma\,\mathrm{d}t = f, \quad s \in \Gamma. \tag{18}$$

From (1) and (18) it follows that

$$\begin{gathered}\alpha(k) \sum_{j=1}^{m} \int_{\Gamma} \sigma_j \,\mathrm{d}t + \int_{\Gamma} g_0 \sigma \,\mathrm{d}t + \varepsilon(k)\sigma = f, \\ g_0 = \frac{1}{2\pi} \ln \frac{1}{r_{st}},\end{gathered} \tag{19}$$

where $\varepsilon(k)$ is a linear operator from $\tilde{H}^q$ into $\tilde{H}^{q+1}$ and

$$\|\varepsilon(k)\|_{\tilde{H}^q \to \tilde{H}^{q+1}} \to 0 \quad \text{as} \quad k \to 0. \tag{20}$$

Let $d >$ diam Γ. Then

$$\frac{1}{2\pi} \ln \frac{d}{r_{st}} > 0. \tag{21}$$

Let us write Equation (19) as

$$\alpha(k, d)c_\sigma + Q(k, d)\sigma = f, \quad c_\sigma = \sum_{j=1}^{m} \int_{\Gamma_j} \sigma_j \,\mathrm{d}t, \tag{22}$$

where

$$\begin{gathered}Q(k, d)\sigma = Q(0, d)\sigma + \varepsilon(k)\sigma = \frac{1}{2\pi} \int_{\Gamma} \ln \frac{d}{r_{st}} \sigma \,\mathrm{d}t + \varepsilon(k)\sigma, \\ \alpha(k, d) = \alpha(k) - (2\pi)^{-1} \ln d.\end{gathered} \tag{23}$$

From Lemma 2 and (20) it follows that

$$\|Q^{-1}(k, d) - Q^{-1}(0, d)\|_{\tilde{H}^{q+1} \to \tilde{H}^q} \to 0, \quad k \to 0. \tag{24}$$

Let us write Equation (22) as

$$\sigma + \alpha(k, d)c_\sigma Q^{-1}(k, d)1 = Q^{-1}(k, d)f, \tag{25}$$

where 1 is the function equal to 1 on Γ_j, $1 \leqslant j \leqslant m$. Integrate (25) over Γ to obtain

$$\begin{gathered}c_\sigma \left(1 + \alpha(k, d) \sum_{j=1}^{m} \int_{\Gamma_j} Q^{-1}(k, d)1 \,\mathrm{d}t \right) \\ = \sum_{j=1}^{m} \int_{\Gamma_j} Q^{-1}(k, d)f \,\mathrm{d}t.\end{gathered} \tag{26}$$

It follows from (21) that

$$\int_{\Gamma_j} Q^{-1}(k, d)1 \, dt \neq 0, \quad k \to 0. \tag{27}$$

Therefore one can solve (26) for c_σ

$$c_\sigma = \sum_{j=1}^{m} \int_{\Gamma_j} Q^{-1}(k, d)f \, dt \times \\ \times \left\{ 1 + \alpha(k, d) \sum_{j=1}^{m} \int_{\Gamma_j} Q^{-1}(k, d)1 \, dt \right\}^{-1} \tag{28}$$

Since $|\alpha(k, d)| \to +\infty$ as $k \to +0$, it follows from (28), (24), and (23) that

$$\lim_{k \to +0} c_\sigma \alpha(k, d) = \sum_{j=1}^{m} \int_{\Gamma_j} Q^{-1}(0, d)f \, dt \times \\ \times \left\{ \sum_{j=1}^{m} \int_{\Gamma_j} Q^{-1}(0, d)1 \, dt \right\}^{-1} \equiv \beta. \tag{29}$$

This, (25) and (24) imply that

$$\|\sigma(k) - \sigma(0)\|_{\tilde{H}^0} \to 0 \quad \text{as} \quad k \to 0, \tag{30}$$

where

$$\sigma(0) = \sigma_0 = Q^{-1}(0, d)f - \beta Q^{-1}(0, d)1. \tag{31}$$

It follows from (17), (29), and (30) that

$$u(x, k) \to u_0(x), \quad k \to 0, \tag{32}$$

where

$$u_0(x) = \beta + \int_\Gamma g_0(x, t)\sigma_0(t) \, dt, \quad g_0 = \frac{1}{2\pi} \ln \frac{1}{r_{xt}} \tag{33}$$

Here $\sigma_0(t)$ is given by (31). If $x \in \Gamma$ then $u_0 = f$ on Γ and convergence in (32) is in $\tilde{H}^1$. If $x \in \tilde{\Omega}$, where $\tilde{\Omega} \subset \Omega$ is a compact which has no common points with Γ, then convergence in (32) is in $C^{\infty}(\tilde{\Omega})$. Finally $\lim_{|x|\to\infty} u_0(x) = \beta$ since $\int_\Gamma \sigma_0 \, dt = \lim_{k\to 0} c_\sigma = 0$. □

4.3. The Neumann problem

Let

$$(\nabla^2 + k^2)u = 0 \quad \text{in} \quad \Omega \subset R^2, k > 0, \frac{\partial u}{\partial N} = f \text{ on } \Gamma, \quad (34)$$

and u satisfies the radiation condition (2).

THEOREM 2. *The limit* (3) *exists iff*

$$\sum_{j=1}^{m} \int_{\Gamma_j} f \, dt = 0. \quad (35)$$

Proof. (1) We start with Green's formula

$$u(x, k) = \sum_{j=1}^{m} \left(\int_{\Gamma_j} u(t) \frac{\partial g(x, t)}{\partial N_t} - \int_{\Gamma_j} gf \, dt \right), \quad x \in \Omega. \quad (36)$$

Denote

$$\sigma = u \text{ on } \Gamma. \quad (37)$$

It follows from (36) that

$$\sigma_p(s) = \frac{A_p' \sigma_p + \sigma_p}{2} + \sum_{\substack{j=1 \\ j\neq p}}^{m} A_{pj}\sigma_j - F(s), \quad s \in \Gamma_p \quad (38)$$

where

$$A_{pj}\sigma_j = \int_{\Gamma_j} \sigma_j(t) \frac{\partial g(s, t)}{\partial N_t} \, dt, \quad s \in \Gamma_p, \quad (39)$$

$$A_p' \sigma_p = \int_{\Gamma_p} \sigma_p(t) \frac{\partial g(s, t)}{\partial N_t} \, dt, \quad s \in \Gamma_p \quad (40)$$

and

$$F(s) = \sum_{j=1}^{m} \int_{\Gamma_j} g(s, t) f_j(t)\, \mathrm{d}t, \quad s \in \Gamma. \tag{41}$$

Consider first the simpler case $m = 1$. Then (38) can be written as

$$\frac{A'(k)\sigma - \sigma}{2} = F. \tag{42}$$

One has

$$F = \alpha(k) \int_\Gamma f\, \mathrm{d}t + \int_\Gamma g_0 f\, \mathrm{d}t + o(1), \quad k \to 0, \tag{43}$$

and

$$\|A'(k) - A'\|_{\tilde{H}^0} \to 0, \quad k \to 0 \tag{44}$$

where $A' = A'(0)$. The operator $A' - I$ is an isomorphism of $\tilde{H}^0$ onto $\tilde{H}^0$ since 1 is not an eigenvalue of A'. This is a well known fact (Günter [1]) but we give a short proof for the convenience of the reader: if $A'\mu = \mu$ then the potential of the double layer $v = D\mu = \int_\Gamma \mathrm{d}t\, \mu(t)\partial g_0(x, t)/\partial N_t$, $v_\pm = (A' \mp I)\mu/2$, $\partial v/\partial N_+ = \partial v/\partial N_-$, satisfies the condition $v_+ = 0$ on Γ. By the uniqueness of the solution to the interior Dirichlet problem for Laplace's equation one has $v = 0$ in $\mathscr{D}$. Therefore $0 = \partial v/\partial N_+ = \partial v/\partial N_-$, and, by the uniqueness of the solution to the exterior Neumann problem for Laplace's equation, $v = 0$ in Ω. Notice, that $v(\infty) = 0$. Thus $\mu = v_- - v_+ = 0$.

If $A' - I$ is invertible and (44) holds then

$$\|(A'(k) - I)^{-1} - (A' - I)^{-1}\|_{\tilde{H}^0} \to 0, \quad k \to 0. \tag{45}$$

It follows from (42), (43), and (45) that

$$\sigma(k) = 2\alpha(k) \int_\Gamma f\, \mathrm{d}t (A' - I)^{-1} 1 + a(k) \quad \text{as} \quad k \to 0, \tag{46}$$

where $a(k)$ is continuous at $k = 0$. Clearly $(A' - I)^{-1}1 \not\equiv 0$. Therefore $\sigma(k)$ has a finite limit as $k \to +0$ iff

$$\int_\Gamma f\, \mathrm{d}t = 0. \tag{47}$$

It follows from (36) and (37) that iff $\sigma(k)$ has a finite limit, $u(x, k)$ has a finite limit as $k \to 0$. Theorem 2 is proved for $m = 1$.

(2) If $m > 1$ then the argument is practically the same. The role of (47) is played by (35). □

4.4. The Robin Problem.

Consider the Robin problem

$$(\nabla^2 + k^2)u = 0 \quad \text{in} \quad \Omega \subset R^2, \quad k > 0,$$
$$\frac{\partial u}{\partial N} + \eta(s)u = f \text{ on } \Gamma, \quad f \in \tilde{H}^0, \tag{48}$$

$$u \text{ satisfies (2)}, \eta(s) \in C(\Gamma),$$
$$\text{and } \operatorname{Im} \eta(s) \geqslant 0. \tag{49}$$

This problem is uniquely solvable for all sufficiently small $k > 0$. Assume for simplicity that $m = 1$ and look for the solution of (48) in the form (17). Then the boundary condition (48) leads to the equation

$$\frac{A(k)\sigma - \sigma}{2} + \eta(s)Q(k)\sigma = f. \tag{50}$$

This is a Fredholm integral equation in $\tilde{H}^0$. This equation is uniquely solvable for all sufficiently small $k > 0$. Indeed, it is sufficient to prove that the homogeneous equation (50) has only the trivial solution. If ν is a nontrivial solution of the homogeneous equation (50), then the single layer potential w with density ν solves problem (48)—(49) with $f = 0$. Therefore $w = 0$ in Ω, and $w = 0$ on Γ. Thus

$$(\nabla^2 + k^2)w = 0 \text{ in } \mathscr{D}, \quad w = 0 \text{ on } \Gamma. \tag{51}$$

Since $\mathscr{D}$ is a finite domain, the interval $0 \leqslant k \leqslant a$ is free from the eigenvalues of problem (51) if a is sufficiently small. For these k one has $w = 0$ in $\mathscr{D}$ and $\nu = \partial w/\partial N_+ - \partial w/\partial N_- = 0$. Equation (50) can be written as

$$\frac{A_0\sigma - \sigma}{2} + \eta(s)\alpha(k)\int_\Gamma \sigma \, ds + \eta Q_0\sigma + \varepsilon(k)\sigma = f \tag{52}$$

where $\|\varepsilon(k)\|_{\tilde{H}^q \to \tilde{H}^{q+1}} \to 0$ as $k \to 0$. Let $B\sigma = (A_0\sigma - \sigma)/2 + \eta Q_0\sigma$. If $N(B) = \{0\}$ then B is an isomorphism of $\tilde{H}^0$ onto $\tilde{H}^0$, and (52) can be written as

$$\sigma + \alpha(k)c_\sigma B^{-1}\eta + B^{-1}\varepsilon(k)\sigma = B^{-1}f,\ c_\sigma = \int_\Gamma \sigma\, ds. \tag{53}$$

Integrate (53) over Γ to obtain

$$c_\sigma\left(1 + \alpha(k)\int_\Gamma B^{-1}\eta\, ds\right) + \int_\Gamma B^{-1}\varepsilon(k)\sigma\, ds = \int_\Gamma B^{-1}f\, ds. \tag{54}$$

If

$$\int_\Gamma B^{-1}\eta\, ds \neq 0 \tag{55}$$

then, as in the proof of Theorem 1, we conclude from (54) that $\lim_{k\to 0} c_\sigma\alpha(k) = \beta$ exists and is finite, and then it follows from (53) that $\lim_{k\to 0}\sigma = \sigma_0$ does exist and is finite. Existence of these two limits implies that the solution u in the form (17) has a finite limit as $k \to 0$. This limit $u_0 = \beta + \int_\Gamma g_0(x, t)\sigma_0(t)\, dt$ solves the equation $\nabla^2 u_0 = 0$ in Ω. It satisfies the boundary condition (48) since (53) implies that

$$B\sigma_0 + \beta\eta = f. \tag{56}$$

Therefore if $N(B) = \{0\}$ and (55) holds then the solution to (48)–(49) exists, is unique, can be found in the form (17), and has a finite limit $u_0(x)$ as $k \to 0$ for any $f \in H^0$. This limit solves problem (48) with $k = 0$ and satisfies the condition $\lim_{|x|\to\infty} u_0(x) = \beta$. This argument is the same as in the proof of Theorem 1. If $N(B) = \{0\}$ but $\int_\Gamma B^{-1}\eta\, ds = 0$ then (53) and (54) imply that $\lim_{k\to 0} c_\sigma = \int_\Gamma B^{-1}f\, ds$. Thus, if $\int_\Gamma B^{-1}f\, ds \neq 0$, it follows from (53) that $\sigma = 0(\alpha(k))$ as $k \to 0$, so that $\lim_{k\to 0}\sigma$ does not exist. We proved:

LEMMA 3. *If* $N(B) = \{0\}$ *then condition* (55) *is necessary and sufficient for the existence of the limit* $\lim_{k\to 0} u(x, k)$ *for all* $f \in \tilde{H}^0$.

Assume now that $\dim N(B) > 0$. Let us show that this assumption implies $\dim N(B) = 1$, provided that

$$\eta \leqslant 0 \quad \text{if} \quad \operatorname{Im}\eta = 0. \tag{49'}$$

Suppose that $h \in N(B)$. If $\int_\Gamma h\, ds \neq 0$ for any $h \in N(B)$, $h \not\equiv 0$, then $\dim N(B) = 1$. Indeed, if $h_1 \in N(B)$, $h_2 \in N(B)$, $\int_\Gamma h_j\, ds \neq 0$, $j = 1, 2$, then $h_2 = ch_1$, $c = \int_\Gamma h_2\, ds(\int_\Gamma h_1\, ds)^{-1}$. If there exists an $h \in N(B)$, such that $\int_\Gamma h\, ds = 0$, then $h = 0$. Indeed, let

$$\frac{A_0 h - h}{2} + \eta Q_0 h = 0 \tag{57}$$

and $\int_\Gamma h\, ds = 0$. Let $w = Sh = \int_\Gamma g_0(x, s)h\, ds$ be the single layer potential with density h. Then (57) says that

$$\begin{aligned} &\nabla^2 w = 0 \quad \text{in} \quad \Omega, \ \frac{\partial w}{\partial N_-} + \eta w = 0 \text{ on } \Gamma, \\ &w = 0(|x|^{-1}) \quad \text{as} \quad |x| \to \infty. \end{aligned} \tag{58}$$

Assumptions (49)—(49′) imply that $w = 0$ in Ω. Therefore $w = 0$ on Γ, $\nabla^2 w = 0$ in $\mathscr{D}$, and so $w = 0$ in $\mathscr{D}$. From the jump formula $h = \partial w/\partial N_+ - \partial w/\partial N_- = 0$. Thus, $\int_\Gamma h\, ds \neq 0$ for all $h \in N(B)$, $h \not\equiv 0$, and $\dim N(B) = 1$.

Let us derive a necessary condition for $h \in N(B)$. One has $\int_\Gamma A_0 h\, ds = (A_0 h, 1) = (h, A_0' 1) = -(h, 1)$ where $(h, f) = \int_\Gamma h\bar{f}\, ds$, and the well known equality $A_0' 1 = -1$ was used. Thus, integrating (57) over Γ yields a necessary condition for $h \in N(B)$:

$$\int_\Gamma \eta Q_0 h\, ds = \int_\Gamma h\, ds. \tag{59}$$

Let us notice that

$$A_0' Q_0 = Q_0 A_0. \tag{60}$$

Formula (60) follows from the identity

$$\int_\Gamma ds'\, g(s, s') \frac{\partial g(s', t)}{\partial N_{s'}} = \int_\Gamma ds'\, g(t, s') \frac{\partial g(s', s)}{\partial N_{s'}} \tag{61}$$

which is a consequence of Green's formula.

Apply Q_0 to (57) and use (60) and (59) to obtain

$$\frac{A_0' - I}{2} p + Q_0 \eta p = 0, \quad p = Q_0 h, \int_\Gamma \eta p\, ds \neq 0. \tag{62}$$

Let us summarize what we have proved.

LEMMA 4. *If* dim $N(B) > 0$ *and* η *satisfies assumptions* (49)–(49′) *then* dim $N(B) = 1$. *If* $h \in N(B)$ and $h \not\equiv 0$, *then*

$$\int_\Gamma \eta p \, ds \neq 0, \quad \text{where} \quad p = Q_0 h, \ B' p = 0. \tag{63}$$

To complete our analysis let us prove Lemma 5.

LEMMA 5. *Let* dim $N(B) > 0$ *and* η *satisfy assumptions* (49)–(49′). *For the limit* $\lim_{k\to 0} u(x, k)$ *to exist for any* $f \in \tilde{H}^0$, *it is necessary and sufficient that condition* (63) *holds.*

Proof. Necessity. Suppose $\lim_{k\to 0} u(x, k)$ exists for any $f \in \tilde{H}^0$. Without loss of generality one can assume that the operator Q_0 in Equation (52) is injective. If it is not, then, as in Lemma 2, one can use an injective operator Q_d and the constant $\alpha(k, d) = \alpha(k) - (1/2\pi) \ln d$ in place of $\alpha(k)$.

If Q_0 is injective then it is an isomorphism of $\tilde{H}^0$ onto $\tilde{H}^1$. Therefore, if $\lim_{k\to 0} u(x, k) = \lim_{k\to 0} Q\sigma$ exists then the limits (29) and (30) exist, and the limit of Equation (52) is

$$B\sigma_0 + \beta\eta = \frac{A_0\sigma_0 - \sigma_0}{2} + \eta Q_0 \sigma_0 + \beta\eta = f. \tag{64}$$

This equation is solvable iff

$$\int_\Gamma fp \, ds = \beta \int_\Gamma \eta p \, ds, \quad p \in N(B'). \tag{65}$$

If $\beta \neq 0$ and condition (63) does not hold, then $\int_\Gamma fp \, ds = 0$. This cannot hold for all $f \in \tilde{H}^0$ since $p \not\equiv 0$. Therefore, if $\beta \neq 0$ then condition (63) is necessary for the limit $\lim_{k\to 0} u(x, k)$ to exist for all $f \in \tilde{H}^0$. If $\beta = 0$ then the same argument shows that the limit $\lim_{k\to 0} u(x, k)$ does not exist for some f (e.g., for $f = \bar{p}$). In the proof of sufficiency we will show that if (63) holds then $\beta \neq 0$ for some f.

Sufficiency. Suppose (63) holds, $h \in N(B)$, $p \in N(B')$, $(p, p) = 1$. The $\lim_{k\to 0} u(x, k)$ exists if $\lim_{k\to 0} \sigma = \sigma_0$ and $\lim_{k\to 0} \alpha(k) c_\sigma$ exist. In order to establish that these limits exist provided that (63) holds, let us write Equation (52) as

$$\begin{gathered} B_1\sigma + \alpha(k) c_\sigma \eta + \varepsilon(\kappa)\sigma - P\sigma = f, \\ B_1 = B + P, \ P\sigma = (\sigma, h)\eta. \end{gathered} \tag{66}$$

The operator B_1 is invertible because it is Fredholm and injective. Indeed, if $B_1\sigma = 0$ then $B\sigma = -(\sigma, h)\eta$. Therefore $(\eta, \bar{p})(\sigma, h) = 0$. Thus $(\sigma, h) = 0$, $B\sigma = 0$, and $\sigma \in N(B)$. Since $\dim N(B) = 1$ and $h \in N(B)$, one concludes that $\sigma = 0$. If B_1 is invertible, Equation (66) can be written as

$$\sigma + \alpha(k)c_\sigma B_1^{-1}(k)\eta - (\sigma, h)B_1^{-1}(k)\eta = B_1^{-1}(k)f,$$
$$B_1(k) = B_1 + \varepsilon(k). \tag{67}$$

Multiply (67) by $\bar{h}$ and integrate over Γ to obtain

$$\sigma_h + \alpha(k)c_\sigma(B_1^{-1}(k)\eta, h) - \sigma_h(B_1^{-1}(k)\eta, h) = (B_1^{-1}(k)f, h),$$
$$\sigma_h \equiv (\sigma, h), \quad \|B_1^{-1}(k) - B_1^{-1}\|_{\tilde{H}^0} \to 0, \quad k \to 0. \tag{68}$$

Integrate (67) over Γ to obtain

$$c_\sigma + \alpha(k)c_\sigma(B_1^{-1}(k)\eta, 1) - \sigma_h(B_1^{-1}(k)\eta, 1) = (B_1^{-1}(k)f, 1). \tag{69}$$

Denote

$$(B_1^{-1}(k)\eta, h) = a, \quad (B_1^{-1}(k)\eta, 1) = b. \tag{70}$$

Then (68) and (69) can be written as a linear system with respect to c_σ and σ_h:

$$(1 - a)\sigma_h + \alpha(k)ac_\sigma = (B_1^{-1}(k)f, h)$$
$$-b\sigma_h + (1 + \alpha(k)b)c_\sigma = (B_1^{-1}(k)f, 1). \tag{71}$$

The determinant of this system is $1 + \alpha b - a$. It is not zero since $\lim_{k\to 0} b = b_0 \neq 0$. Indeed, $a \to (B_1^{-1}\eta, h) = 1$, and $\alpha(k) \to \infty$. The term $(B_1^{-1}\eta, h)$ can be computed as follows. Let $\eta = B_1 q$. Then $(\eta, \bar{p}) = (Bq, \bar{p}) + (q, h)(\eta, \bar{p}) = (q, \overline{B'p}) + (q, h)(\eta, \bar{p}) = (q, h)(\eta, \bar{p})$. Since $(\eta, \bar{p}) \neq 0$ one sees that $(q, h) = 1$. But $(q, h) = (B_1^{-1}\eta, h)$. Thus $(B_1^{-1}\eta, h) = 1$. Furthermore, $b_0 = (B_1^{-1}\eta, 1) = (q, 1) \neq 0$. One has $\eta = B_1 q = Bq + (q, h)\eta = Bq + \eta$. Thus $Bq = 0$, $q \not\equiv 0$, and Equations (59) and (63) imply that $(q, 1) \neq 0$. Therefore the system (71) is uniquely solvable for all sufficiently small k. It is an easy matter to solve this system and see that the following limits exist

$$\lim_{k\to 0} \alpha(k)c_\sigma = \beta = (B_1^{-1}f, h),$$
$$\lim_{k\to 0} \sigma_h = (B_1^{-1}f, h) - b_0^{-1}(B_1^{-1}f, 1) = \gamma \tag{72}$$

where

$$b_0 = (B_1^{-1}\eta, 1). \tag{73}$$

From (72) and (67) it follows that

$$\lim_{k\to 0} \sigma(k) = (\gamma - \beta)B_1^{-1}\eta + B_1^{-1}f = \sigma_0. \tag{74}$$

From (17), (72), and (74) it follows that

$$\lim_{k\to 0} u(x, k) = \beta + \int_\Gamma g_0(x, s)\sigma_0(s)\, \mathrm{d}s. \tag{75}$$

Lemma 5 is proved.

Let us summarize our results. □

THEOREM 3. *The solution to problem* (48)–(49′) (*with* $m = 1$) *has a finite limit as* $k \to 0$ *for any* $f \in \tilde{H}^0$ *if and only if either*

$$N(B) = \{0\} \quad \text{and} \quad (55)\ \textit{holds}, \tag{76}$$

or $\dim N(B) > 0$.

$$\begin{aligned} &\textit{If} \dim N(B) > 0 \textit{ then} \dim N(B) = 1 \textit{ and} (63) \textit{ holds,} \\ &\textit{where } h \in N(B) \end{aligned} \tag{77}$$

and $B\sigma = (A_0\sigma - s)/2 + \eta(s)Q_0\sigma$.

Example. If condition (49′) does not hold then the conclusion of Theorem 3 may not hold. Indeed, consider a circle of radius 1 centered at the origin. Green's function of the problem (48) is

$$G(x, x'k) = \sum_{m=-\infty}^{\infty} \frac{\exp\{im(\phi - \phi')\}}{2\pi} G_m(r, r')$$

where, if $\eta = \text{const}$,

$$G_m(r, r') = \frac{i\pi}{2} [J_m(kr) + b_m H_m^{(1)}(kr)] H_m^{(1)}(kr'), \quad r < r'$$

and

$$b_m = -\frac{kJ'_m(k) + \eta J_m(k)}{kH_m^{(1)\prime}(k) + \eta H_m^{(1)}(k)}.$$

The solution to (48) has a finite limit as $k \to 0$ for any $f \in \tilde{H}^0$ iff $G(x, x', k)$ has a finite limit as $k \to 0$ for all $x \neq x'$, and this is so iff $b_m H_m^{(1)}(kr') H_m^{(1)}(kr)$ has a finite limit as $k \to 0$ for all $m = 0, \pm 1, \pm 2, \ldots$. A simple computation shows that $b_m H_m^{(1)}(kr') H_m^{(1)}(kr)$ has a finite limit as $k \to 0$ iff $\eta \neq m$, $m = 0, 1, 2, \ldots$. Assumption (49′) guarantees that the last condition is satisfied.

We assumed (49) and (49′). Let us drop these assumptions. Now problem (48) may have no solutions for some $k > 0$ and some f. However, the following result holds.

THEOREM 4. *The solution to problem* (48) *has a finite limit as* $k \to 0$ *for any* $f \in \tilde{H}^0$ *iff either* (76) *holds or*

$$\dim N(B) = 1 \quad \text{and} \quad (63) \textit{ holds.} \tag{78}$$

If $\dim N(B) > 1$ *then there exist* f *such that the solution to* (48) *does not have a finite limit as* $k \to 0$.

Proof. (1) If $u(x, k) \to u_0(x)$ as $k \to 0$, where $u(x, k)$ solves (48), and u_0 solves the limit problem

$$\nabla^2 u_0 = 0 \text{ in } \Omega,\ u_{0_N} + \eta u_0 = f \text{ on } \Gamma,\ |u(\infty)| < \infty \tag{79}$$

then the formula ($\varepsilon\sigma \equiv Q(k)\sigma - \alpha(k)c_\sigma - Q_0\sigma$, $c_\sigma \equiv \int_\Gamma \sigma\, \mathrm{d}t$)

$$\begin{aligned} &u(s, k) = \alpha(k)c_\sigma + (Q_0 + \varepsilon)\sigma, \\ &\sigma + \alpha(k)c_\sigma(Q_0 + \varepsilon)^{-1}1 = (Q_0 + \varepsilon)^{-1}u \end{aligned} \tag{80}$$

shows that

$$\begin{aligned} &c_\sigma = 0(\alpha^{-1}), \quad \alpha c_\sigma \to \beta, \\ &\sigma \to \sigma_0 = Q_0^{-1}u - \beta Q_0^{-1}1, \\ &u \to u_0 = \beta + \int_\Gamma g_0\sigma_0\, \mathrm{d}t, \end{aligned} \tag{81}$$

provided that $N(Q_0) = \{0\}$, so that $\int Q_0^{-1}1\, \mathrm{d}t \neq 0$. Otherwise, write Equation (80) as

$$\begin{aligned} &u(s, k) = (\alpha(k) - c)c_\sigma + (Q_1 + \varepsilon)\sigma, \\ &Q_1\sigma \equiv Q_0\sigma + cc_\sigma, \end{aligned} \tag{82}$$

where $c = \text{const} \neq 0$, and check that $N(Q_1) = \{0\}$, so that $\int_\Gamma Q_1^{-1} 1 \, \mathrm{d}t \neq 0$. To see that $N(Q_1) = \{0\}$ one writes $Q_0\sigma + cc_\sigma = 0$. This equation is solvable only if

$$cc_\sigma \int_\Gamma \sigma_0 \, \mathrm{d}t = 0, \quad \text{where} \quad Q_0\sigma_0 = 0. \tag{83}$$

Remember that $\dim N(Q_0) = 1$ and $\int_\Gamma \sigma_0 \, \mathrm{d}t \neq 0$. Thus $c_\sigma = \int_\Gamma \sigma \, \mathrm{d}t = 0$, $\sigma \in N(Q_0)$, and therefore $\sigma = 0$. So $N(Q_1) = \{0\}$.

If (81) holds then

$$B\sigma_0 + \beta\eta = f \tag{84}$$

and

$$\beta \int_\Gamma B^{-1}\eta \, \mathrm{d}t = \int_\Gamma B^{-1} f \, \mathrm{d}t. \tag{85}$$

Choose f such that $\int_\Gamma B^{-1} f \, \mathrm{d}t \neq 0$ and obtain (55). This proves necessity.

To prove sufficiency, assume (76) and repeat the argument after formula (55).

(2) Assume now that $\dim N(B) = 1$. If $u \to u_0$ and (79) holds then, as above, we derive Equation (84). This equation is solvable only if

$$\int_\Gamma fp \, \mathrm{d}t = \beta \int_\Gamma \eta p \, \mathrm{d}t, \quad p \in N(B'). \tag{86}$$

Choose f such that $\int_\Gamma fp \, \mathrm{d}t \neq 0$ and obtain (63). This proves necessity.

To prove sufficiency, assume that (78) holds. Write equation

$$B_\varepsilon \sigma + \alpha c_\sigma \eta = f, \quad B_\varepsilon = B + \varepsilon(k),$$

where $\varepsilon = \varepsilon(k)$ is defined in (52), as

$$B_1(k)\sigma + \alpha c_\sigma \eta - \sigma_h \eta = f, \quad \sigma_h \equiv \int \sigma h \, \mathrm{d}t, \quad h \in N(B). \tag{87}$$

where

$$B_1(k) = B_1 + \varepsilon, \quad B_1\sigma = B\sigma + \sigma_h \eta. \tag{88}$$

Check that $N(B_1) = \{0\}$, so that B_1 is an isomorphism. Indeed, if $0 = B_1\sigma = B\sigma + \sigma_h\eta$, then $\sigma_h \int_\Gamma \eta p \, dt = 0$. This and (78) imply that $\sigma_h = 0$. Thus $B\sigma = 0$, $\sigma = ch$, $c = \text{const}$. Since $\sigma_h = 0$ we conclude that $\sigma = 0$. So $N(B_1) = \{0\}$.

Write (87) as

$$\sigma + ac_\sigma B_1^{-1}(k)\eta - \sigma_h B_1^{-1}(k)\eta = B_1^{-1}(k)f. \tag{89}$$

The further argument is the same as after formula (67) and proves sufficiency.

(3) Assume that $\dim N(B) > 1$ and prove that there exist f such that the solution to (48) does not have a finite limit as $k \to 0$. Indeed, if this limit does exist then, as above, one obtains (86) as a necessary condition for the existence of the limit. Since $\dim N(B') = \dim N(B) > 1$, there exist at least two linearly independent elements $p_j \in N(B')$, $j = 1, 2$. If $\int_\Gamma \eta p_j \, dt = 0$ for $j = 1$ or 2, then choose f such that $\int_\Gamma f p_j \, dt \neq 0$ for this j, and condition (86) cannot hold for this f. If $\int_\Gamma \eta p_j \, dt \neq 0$, $j = 1, 2$, then take f such that $\int_\Gamma f p_1 \, dt = 1$, $\int_\Gamma f p_2 \, dt = 0$. Then $1 = \beta \int_\Gamma \eta p_1 \, dt$ and $0 = \beta \int_\Gamma \eta p_2 \, dt$, which is a contradiction if $\int_\Gamma \eta p_j \, dt \neq 0$, $j = 1, 2$. This contradiction proves our claim.

Theorem 4 is proved. □

4.5. In this section we give an alternative approach to the study of the behavior of the solutions to the problem

$$Lu + k^2 u = 0 \text{ in } \Omega, \quad k > 0, \tag{90}$$

$$u = f(a), \quad u_N = f(b), \; u_N + \eta(s)u = f(c) \text{ on } \Gamma \tag{91}$$

$$r(u_r - iku) \to 0, \quad r \to \infty, \tag{92}$$

where Ω is the exterior domain with a smooth boundary Γ, L is a general second order strongly elliptic selfadjoint differential expression which coincides with ∇^2 outside of a sphere of radius R. Let us assume that $f \in H^{3/2}(\Gamma)$ so that $u \in H^2_{\text{loc}}(\Omega)$. We wish to show that the following result holds. Our argument is valid (with obvious modifications) in R^n, $n > 2$.

THEOREM 5. *The limit*

$$\lim_{k \to 0} u(x, k) = u_0(x) \text{ in } H^2_{\text{loc}}(\Omega) \tag{93}$$

exists for any $f \in H^{3/2}(\Gamma)$ *and solves the limit problem*

$$Lu_0 = 0 \text{ in } \Omega,\ u_0 \text{ satisfies } (2), \tag{94}$$

$$u_0 = 0(1) \quad \text{as} \quad |x| \to \infty, \quad \text{if} \quad \Omega \subset R^2, \tag{95}$$

iff

$$\text{the problem (94)–(95) has at most one solution.} \tag{96}$$

This theorem reduces the problem of existence of the limit (93)–(95) *to a proof of a uniqueness theorem.*

Proof of Theorem 5. (1) Assume that (96) holds. Suppose that the solution to problem (90)–(92) satisfies the estimate

$$\sup_{0<k<k_0} \|u(x, k)\| \leqslant c, \tag{97}$$

$$\|u\|^2 = \int_\Omega |u|^2(1 + |x|)^{-a}\,\mathrm{d}x, \quad a > 2,$$

where c is a constant which does not depend on k, $r > R$ is an arbitrary large fixed number, $k_0 > 0$ is an arbitrary small fixed number. From (97) and the known estimate (see, e.g., Berezanskii [1])

$$\|u\|_{H^2(\Omega'_r)} \leqslant c(\|Lu\|_{L^2(\Omega'_{r+1})} + \|u\|_{L^2(\Omega'_{r+1})} + \|u\|_{H^{3/2}(\Gamma)}),\ \Omega'_r = \Omega \cap B_r \tag{98}$$

it follows that

$$\|u(x, k)\|_{H^2(\Omega'_r)} \leqslant c(r). \tag{99}$$

This equation and the embedding theorem imply that there exists a subsequence

$$u_n \equiv u(x, k_n) \to u_0(x) \text{ in } H^2_{\mathrm{loc}}(\Omega), \quad \text{and (94) holds.} \tag{100}$$

In $\Omega_r = R^2 \setminus B_r$ one writes ($g_{0\varepsilon} \equiv g(k) - \alpha(k)$)

$$\begin{aligned} u_n &= \int_{S_r} (g_r u_n - u_{nr} g)\,\mathrm{d}t \\ &\doteq \int_{S_r} (g_r u_n - u_{nr} g_{0\varepsilon})\,\mathrm{d}t - \alpha(k) \int_{S_r} u_{nr}\,\mathrm{d}t. \end{aligned} \tag{101}$$

If (100) holds on S_r then u_n and the first integral on the right in (101) tend to a limit as $k \to 0$. Therefore

$$\int_{S_r} u_{0r}\,\mathrm{d}t = 0, \lim_{k \to 0} \alpha(k) \int_{S_r} u_{nr}\,\mathrm{d}t = \beta. \tag{102}$$

Thus

$$u_0(x) = \int_{S_r} (g_{0r}u_0 - u_{0r}g_0)\,\mathrm{d}t - \beta \tag{103}$$

and (94)—(95) hold. We prove sufficiency of (96) if we show that (96) implies (97). If (97) does not hold, then there exists a sequence $u(x, k_n) = u_n$, $k_n \to 0$, such that

$$\|u_n\| \geqslant n. \tag{104}$$

Let $w_n = u_n/\|u_n\|$, $\|w_n\| = 1$, and w_n solves (90)—(92) with $f_n = f\|u_n\|^{-1}$ in place of f. The above argument shows that $w \to w_0$ in $H^2_{\mathrm{loc}}(\Omega)$ and w_0 solves the homogeneous problem (94)—(95). By (96) we conclude that $w_0 = 0$. On the other hand, one can pass to the limit in the equation

$$1 = \|w_n\| \to \|w_0\| = 0, \quad n \to \infty. \tag{105}$$

This is possible by the Lebesgue dominated convergence theorem, since $w_n \to w_0$ in $H^2_{\mathrm{loc}}(\Omega)$ and, from the formula for w_0, similar to (103), one obtains

$$|w_0(x)| \leqslant c, \quad x \in \Omega_r = \Omega \setminus B_r \tag{106}$$

where c does not depend on x. The contradiction (105) proves that (104) is wrong, and (97) holds. We proved that (96) implies (93)—(95).

(2) Assume that (93)—(95) hold for any $f \in H^{3/2}(\Gamma)$. Then the problem (94)—(95) is solvable for any $f \in H^{3/2}(\Gamma)$. This problem is of Fredholm's type. Therefore the homogeneous problem has only the trivial solution, so that (96) holds. To see that problem (94)—(95) is of Fredholm's type one notices that any solution to Equation (94) which satisfies (95) can be represented as $\beta + \int_\Gamma G(x, t)\sigma(t)\,\mathrm{d}t$, where $\beta =$ const and G is a fundamental solution for L. The boundary condition (94) is equivalent to a Fredholm-type equation for σ. This proves that problem (94)—(95) is of Fredholm's type.

Theorem 5 is proved. □

Chapter VI

Some Inverse Scattering Problems of Geophysics

There are many problems in geophysics, optics, nondestructive testing and elsewhere in which a finite inhomogeneity in a homogeneous space is to be recovered from the measurements of a field scattered by the inhomogeneity. The field can be measured far away from the inhomogeneity or in some region which is not too far from the inhomogeneity. Several problems of this type are solved in this chapter. The problems are formulated in Sections 1—4. Section 1 deals with recovery of the refraction index (velocity profile) from the knowledge of the field on the plane for all positions of the source and receiver. Other problems of similar nature are also considered. Section 2 deals with the two-parameter inversion: both the density and velocity are recovered from the knowledge of the field for all positions of the source and receiver on the plane and for two distinct frequencies. In Section 1 the theory is exact, in Section 2 it is based on the first Born approximation. In Section 3 the three-dimensional inverse problem of the scattering theory by a potential is solved in the Born approximation. Section 4 deals with a model problem of induction logging theory.

1. INVERSE SCATTERING FOR GEOPHYSICAL PROBLEMS

1.1. Let

$$\nabla^2 u + \omega^2 n(x) u = -\delta(x - y) \text{ in } R^3, \tag{1}$$

$$n(x) = 1 \quad \text{if} \quad |x| \geqslant R, \quad n(x) = 1 + v(x) \quad \text{if} \quad |x| \leqslant R, \tag{2}$$

$$n(x) > 0, \quad v(x) = 0 \quad \text{if } |x| \geqslant R \quad \text{or} \quad x_3 > 0. \tag{3}$$

The function u is the outgoing (i.e., u satisfies the radiation condition) scalar, e.g. acoustic, field generated by a point source located at the point y in the homogeneous space with a compactly supported inhomogeneity $v(x)$ located in the lower half-space $R^3_- = \{x\colon x_3 < 0\}$. Let $P = \{x\colon x_3 = 0\}$, $B_R = \{x\colon |x| \leqslant R\}$. The inhomogeneity is not assumed small in any sense. No symmetry assumptions of the type $v = v(x_3)$ or $v = v(\rho, x_3)$, where $\rho = (x_1^2 + x_2^2)^{1/2}$, are made. The only assumptions about $v(x)$ are:

$$v(x) = 0, \quad |x| > R \quad \text{or} \quad x_3 \geqslant 0, \quad v(x) \in L^2(R^3). \tag{4}$$

The inverse problem we are concerned with consists in finding $v(x)$ from measurements of the field $u(x, y, \omega)$ for all positions $x, y \in P$ of the receiver and source and for small ω. The notion of the smallness of ω will be specified later. This is a model seismic exploration problem in which the refraction coefficient $n(x) = c^{-2}(x)$, where $c(x)$ is the wave velocity, and it is assumed that the measured field is the acoustic scalar field. One could assume that $c = c(x, \omega)$ but in this case only $c(x, 0)$ is recovered. Our method is exact and analytic. The Born approximation is not used. The assumption $v(x) = 0$ if $|x| \geqslant R$ can be relaxed but fast decay of $v(x)$ at infinity is necessary for our method. For example, one can assume that

$$|v(x)| \leqslant c(1 + |x_1|^2 + |x_2|^2)^{-a}(1 + |x_3|)^{-a}, \quad a > 1 \tag{5}$$

but further relaxation of the assumptions on the rate of decay of v would require some new ideas.

The requirement that $u(x, y, \omega)$ is measured for $\omega \to 0$ does not restrict the resolution ability of our method as will be clear later. Therefore the usual concept, according to which the size of the details that can be resolved by testing an object with waves of length λ, is of order λ is not applicable to our scheme. The reason is that the data we use differ from the usual data.

There are quite a few practical examples in which the condition $\omega \to 0$ is satisfied. Currently the wave length λ used in seismic prospecting is in the range from 50 to 400 m. If the characteristic dimension a of the inhomogeneity $v(x)$ satisfies the condition $a \ll \lambda$ then the condition $\omega \to 0$ is satisfied.

Practical examples include: (1) metallic ore lodes ($a \sim$ 30 m), (2) caves ($a \sim$ 20—80 m), (3) old mine structures ($a \sim$ 50 m), (4) submarines ($a \sim$ 50—100 m, (5) pipelines ($a \sim$ 3 m). The last problem requires some modifications in the theory because the inhomogeneity is infinite in one direction.

1.2. The method we use for solving the inverse problem is simple. One starts with the integral equation

$$u(x, y, \omega) = g(x, y, \omega) + \\ + \omega^2 \int g(x, z, \omega) v(z) u(z, y, \omega)\, dz \equiv g + \omega^2 Bu \tag{6}$$

where $g = \exp(i\omega|x - y|)/(4\pi|x - y|)$ and the integral is taken actually over R^3_- because of (4). Equation (6) is equivalent to (1) which can be checked by differentiation, and the solution to (6) is an outgoing wave.

Under assumption (4) the linear integral operator B in (6) is compact in the space C_y of functions with a fixed singularity at the point y. The functions are of the form $u = C/|x - y| + h(x, y)$ where $h(x, y)$ is continuous, and the norm in this space is given by the equation $\|u\| = |C| + \max_{x \in R^3} |h(x, y)|$.

LEMMA 1. *For sufficiently small ω Equation (6) is uniquely solvable in C_y, its solution is analytic in ω in a neighborhood of the point $\omega = 0$ and can be obtained by iterations*

$$u_{n+1} = g + \omega^2 Bu_n, \quad u_0 = g. \tag{7}$$

Proof. The conclusion of Lemma 1 is an immediate consequence of the Banach fixed point theorem.

Let us rewrite (6) as

$$\frac{u - g}{\omega^2} = Bg + 0(\omega^2) \tag{8}$$

and, using Lemma 1, pass to the limit $\omega \to 0$ in (8). The result is

$$Mv \equiv \int \frac{v(z)\,dz}{|x - z|\,|z - y|} = f(x, y), \tag{9}$$

where

$$f(x, y) \equiv 16\pi^2 \lim_{\omega \to 0} \frac{u - g}{\omega^2}. \tag{10}$$

Let

$$x^1 = (x_1, x_2, 0), \quad y^1 = (y_1, y_2, 0). \tag{11}$$

Setting $x = x^1$ and $y = y^1$ in (9) yields

$$M_P v \equiv \int \frac{v(z)\,dz}{|x^1 - z|\,|y^1 - z|} = f(x^1, y^1), x^1, y^1 \in P. \tag{12}$$

This is our basic equation for finding $v(z)$ from the data $f(x^1, y^1)$. □

Remark 1. One can measure the data on surfaces other than P. For example, on a sphere S_{R_1} with radius $R_1 > R$, or on a cylinder $C_{R_1} = \{x: x_1^2 + x_2^2 = R_1^2\}$, $R_1 > R$. The corresponding inverse problems can be useful also. See Section 1.5 below.

1.3. Let us solve Equation (12). Taking the Fourier transform in x^1 and y^1 and using the formula

$$\frac{1}{(2\pi)^2}\int_{-\infty}^{\infty}\int \frac{\exp(\lambda\cdot x^1)}{|x^1 - z|}\,\mathrm{d}x_1\,\mathrm{d}x_2$$

$$= \frac{\exp(i\lambda\cdot z^1)}{(2\pi)^2}\int_0^{\infty}\frac{\mathrm{d}\rho\,\rho}{(\rho^2 + z_3^2)^{1/2}}\int_0^{2\pi}\exp(i|\lambda|\rho\cos\phi)\,\mathrm{d}\phi \tag{13}$$

$$= \frac{\exp(i\lambda\cdot z^1)}{2\pi}\int_0^{\infty}\frac{\mathrm{d}\rho\,\rho J_0(|\lambda|\rho)}{(\rho^2 + z_3^2)^{1/2}}$$

$$= (2\pi)^{-1}|\lambda|^{-1}\exp(i\lambda\cdot z^1)\exp(-|\lambda|\,|z_3|),$$

one obtains,

$$\frac{1}{(2\pi)^2}\int \mathrm{d}z\; v(z)\exp\{i(\lambda+\mu)\cdot z^1\}\exp\{-(|\lambda|+|\mu|)|z_3|\} = |\lambda|\,|\mu|\tilde{f}(\lambda,\mu) \tag{14}$$

where

$$\lambda = (\lambda_1, \lambda_2), \quad |\lambda| = (|\lambda_1|^2 + |\lambda_2|^2)^{1/2},$$
$$\mu = (\mu_1, \mu_2), \quad |\mu| = (|\mu_1|^2 + |\mu_2|^2)^{1/2}, \tag{15}$$

and

$$\tilde{f}(\lambda,\mu) = \frac{1}{(2\pi)^4}\iint_{-\infty}^{\infty}\iint f(x^1, y^1)\times \exp\{i(\lambda\cdot x^1 + \mu\cdot y^1)\}\,\mathrm{d}x_1\,\mathrm{d}x_2\,\mathrm{d}y_1\,\mathrm{d}y_2. \tag{16}$$

Let

$$\begin{aligned} &p = \lambda + \mu, \quad p_1 = \lambda_1 + \mu_1, \quad p_2 = \lambda_2 + \mu_2, \\ &p_3 = |\lambda|, \quad p_4 = |\mu|. \end{aligned} \tag{17}$$

Then (14) can be written as

$$\begin{aligned} &\int \mathrm{d}z\, v(z) \exp(ip \cdot z^1) \exp\{-(p_3 + p_4)|z_3|\} \\ &\quad = (2\pi)^2 p_3 p_4 F(p_1, p_2, p_3, p_4) \end{aligned} \tag{18}$$

where $F(p_1, p_2, p_3, p_4)$ is the function $\tilde{f}(\lambda, \mu)$ in the variables p_1, p_2, p_3, p_4. The Jacobian of the mapping $(\lambda_1, \lambda_2, \lambda_3, \lambda_4) \to (p_1, p_2, p_3, p_4)$ is

$$J = \begin{vmatrix} 1 & 0 & 1 & 0 \\ 0 & 1 & 0 & 1 \\ \lambda_1|\lambda|^{-1} & \lambda_2|\lambda|^{-1} & 0 & 0 \\ 0 & 0 & \mu_1|\mu|^{-1} & \mu_2|\mu|^{-1} \end{vmatrix} = \frac{\mu_1\lambda_2 - \mu_2\lambda_1}{|\mu|\,|\lambda|}, \tag{19}$$

so that $J \neq 0$ if μ and λ are linearly independent. Let

$$p_3 = p_4 = \frac{q}{2}. \tag{20}$$

Equation (20) determines a three-dimensional manifold in the four-dimensional space of variables p_1, p_2, p_3, p_4. Equation (18) reduces to

$$\begin{aligned} &\int_0^\infty \exp(-q\zeta)\left\{\int_{-\infty}^\infty \int \mathrm{d}z_1\, \mathrm{d}z_2 \exp(ip \cdot z^1) v(z_1, z_2, -\zeta)\right\} \mathrm{d}\zeta \\ &\quad = \pi^2 \phi(p_1, p_2, q) \end{aligned} \tag{21}$$

where

$$\phi(p_1, p_2, q) \equiv q^2 F\left(p_1, p_2, \frac{q}{2}, \frac{q}{2}\right), \tag{22}$$

and $\zeta = |z_3| = -z_3$.

Equation (21) can be solved analytically by taking the two-dimensional inverse Fourier transform and one-dimensional inverse Laplace transform:

$$v(z_1, z_2, z_3) = \frac{1}{4} \int_{-\infty}^{\infty} \int \mathrm{d}p_1 \, \mathrm{d}p_2 \exp(-ip \cdot z^1) \frac{1}{2\pi i} \times$$
$$\times \int_{c-i\infty}^{c+i\infty} \mathrm{d}q \, \phi(p_1, p_2, q) \exp(-qz_3) \tag{23}$$

where $p \cdot z^1 = p_1 z_1 + p_2 z_2$ and $c > 0$ is an arbitrary constant. The Mellin integral, which gives the inversion of the Laplace transform, does not depend on $c > 0$. Computational aspects of the problem will be discussed in Section 1.6. Let us summarize the result that we have just proved.

THEOREM 1. *Equation* (12) *is uniquely solvable by formula* (23) *in which* ϕ *is given in* (22) *and* $F(p_1, p_2, p_3, p_4)$ *is the function* (16) *in the variables* (17).

Remark 2. The above argument does not give any explicit characterization of the function $f(x^1, y^1)$ for which equation (12) is solvable. Equation (12) is an integral equation of the first kind since the support of $v(z)$ is a compact set. Therefore the range of the operator M_P: $L^2(B_R) \to C(P \times P)$ is not closed, small perturbations of $f(x^1, y^1)$ in the norm of $C(P \times P)$ may lead to an equation which has no solution in $L^2(B_R)$, and therefore the problem of solving Equation (12) is ill-posed. An implicit characterization of admissible data, that is of the functions $f(x^1, y^1)$ for which Equation (12) is solvable in $L^2(B_R)$, can be given as follows: $f(x^1, y^1)$ are admissible data iff formula (23) gives a function v which satisfies conditions (4).

If the data are not measured on all of P but only on a finite domain, e.g., on a disc $\mathscr{D}_{R_1}$ of radius R_1, then one can still apply the inversion formula (23), $\tilde{f}(\lambda, \mu)$ will be the Fourier transform of the data collected on $\mathscr{D}_{R_1}$, but the integral (23) will not necessarily vanish outside of B_R.

One could try a numerical method to solve Equation (12). Since this is an equation of the first kind some regularization is needed.

1.4. If the data are measured with a random error η uniformly distributed on the interval $(-\delta, \delta)$, $\delta > 0$, then one can define the right hand side of Equation (12) as the solution of a linear regression problem. One measures $u = u(x, y, \omega_j)$ at several small frequencies ω_j,

$$u_j = g_j + \omega_j^2 b + \eta_j, \quad 1 \leqslant j \leqslant n \tag{24}$$

and estimates b from these measurements. This b is then taken as the right hand side of Equation (12). The least squares method gives

$$b = \left(\sum_{j=1}^{n} \omega_j^4 \right)^{-1} \sum_{j=1}^{n} (u_j - g_j)\omega_j^2 \tag{25}$$

where $g_j = g(x, y, \omega_j)$, n is the number of various ω_j used, and $b = b(x^1, y^1)$.

1.5. Consider the case when the lower half-space which contains the inhomogeneity, has different properties from the upper half-space. Let ρ_1, α_1 and ρ_2, $\alpha_2(x)$ be density and velocity in the upper ($x_3 > 0$) and lower ($x_3 < 0$) half-spaces respectively. Let us assume that there is a compactly supported inhomogeneity in the lower half-space R^3_-, that $\rho_1 = \text{const}$, $\alpha_1 = \text{const}$, $\rho_2 = \text{const}$, $\alpha_2(x) = \alpha_2 + v(x)$, where $\alpha_2 = \text{const}$ and $v(x) = 0$ if $|x| > R$ or $x_3 \geqslant 0$. The Green function for the problem without an inhomogeneity. i.e. for $v = 0$, satisfies the equations

$$\nabla^2 \phi_i + \frac{\omega^2}{\alpha_i^2} \phi_i = -\delta(x - y) \quad i = 1, 2 \tag{26}$$

$$\frac{\partial \phi_1}{\partial x_3} = \frac{\partial \phi_2}{\partial x_3}, \ \rho_1 \phi_1 = \rho_2 \phi_2 \quad \text{at} \quad x_3 = 0. \tag{27}$$

Conditions (27) mean that the normal displacement and the pressure are continuous across thc interface. $\phi_1(\phi_2)$ in (26)—(27) denotes the value of the Green function in $R^3_+(R^3_-)$.

In Ewing *et al.* [1], p. 95—96, the Green function is calculated:

$$\phi_2 = \frac{1}{2\pi} \int_0^\infty \frac{\exp(\nu_2 x_3 - \nu_1 y_3) J_0(k\rho) k \, dk}{\delta \nu_1 + \nu_2}, \quad x_3 < 0 \tag{28}$$

where $y_3 \geqslant 0$ (i.e. the source is in R^3_+), $\nu_i = (k^2 - \omega^2 a_i^{-2})^{1/2}$, $\rho = |x^1 - y^1| = \{(x_1 - y_1)^2 + (x_2 - y_2)^2\}^{1/2}$, $\delta = \rho_2 \rho_1^{-1}$. Take $y_3 = 0$ to obtain

$$\phi_2 = \frac{1}{2\pi} \int_0^\infty \frac{J_0(k\rho)k \exp(\nu_2 x_3)\, dk}{\nu_1 \delta + \nu_2}, \quad x_3 < 0. \tag{29}$$

If $\omega \to 0$ then $\nu_i \to k$ and (29) takes the form

$$\begin{aligned}\phi_2 &= \frac{1}{2\pi(\delta + 1)} \int_0^\infty J_0(k\rho) \exp(-k|x_3|)\, dk \\ &= \frac{2\rho_1}{\rho_2 + \rho_1} \frac{1}{4\pi|x - y^1|}, \quad y^1 = (y_1, y_2, 0).\end{aligned} \tag{30}$$

Therefore the basic integral equation (12) remains essentially the same:

$$\begin{aligned}\int \frac{v(z)\, dz}{|x^1 - z|\,|y^1 - z|} &= \left(\frac{\rho_2 + \rho_1}{2\rho_1}\right)^2 f(x^1, y^1) \\ &= \frac{(1+\delta)^2}{4} f(x^1, y^1),\end{aligned} \tag{31}$$

where $\delta = \rho_2/\rho_1$, and $f(x^1, y^1)$ is given in (10) with ϕ_2 in the place of g.

1.6. Inversion of the data given on a sphere.

Consider Equation (6) in the Born approximation at an arbitrary positive frequency ω. This equation can be written as

$$\int g(x, z, \omega) v(z) g(z, y, \omega)\, dz = f(x, y), \quad x, y \in S_{R_1} \tag{32}$$

where the right-hand side is the datum

$$f(x, y) = (u - g)\omega^{-2}, \tag{33}$$

and $R_1 > R$, so that the data are collected on the sphere S_{R_1}, $\int = \int_{B_R}$.

Let us use the well-known formulas (Gradshteyn-Ryzhik [1])

$$g(x, z, \omega) = \sum_{\ell=0}^{\infty} C_\ell \psi_\ell^{re}(z)\psi_\ell(x)$$

$$= i\omega \sum_{\ell=0}^{\infty} \sum_{m=-\ell}^{\ell} j_\ell(\omega|z|)Y_{\ell m}(z^0)h_\ell^{(1)}(\omega|x|)\overline{Y_{\ell m}(x^0)}, \tag{34}$$

$$|x| > |z|,$$

$$\exp(i\omega n \cdot z) = \sum_{\ell=0}^{\infty} a_\ell \psi_\ell^{re}(z)\overline{Y_\ell(n)}$$

$$= 4\pi \sum_{\ell=0}^{\infty} \sum_{m=-\ell}^{\ell} i^\ell j_\ell(\omega|z|)Y_{\ell m}(z^0)\overline{Y_{\ell m}(n)}, \tag{35}$$

where Y_ℓ are the normalized spherical harmonics, ψ_ℓ are the outgoing waves, and ψ_ℓ^{re} are the regular waves, $j_\ell(r) = (\pi/2r)^{1/2}J_{\ell+1/2}(r)$, $h_\ell^{(1)}(r) = (\pi/2r)^{1/2}H_{\ell+1/2}^{(1)}(r)$, $\psi_\ell(x) = h_\ell^{(1)}(|x|)\overline{Y_{\ell m}(x^0)}$, $\psi_\ell^{re}(x) = j_\ell(|x|)Y_{\ell m}(x^0)$, $Y_{\ell m}(\theta, \phi) = \{(2\ell + 1)/4\pi[(\ell - m)!]/[(\ell + m)!]\}^{1/2} e^{im\phi}P_{\ell,m}(\cos\theta)$, $x^0 = (\theta, \phi)$.

Substitute (34) into (32) to obtain

$$\sum_{\ell,\ell'=0}^{\infty} C_\ell C_{\ell'} \int v(z)\psi_\ell^{re}(z)\psi_{\ell'}^{re}(z)\,dz\,\psi_\ell(x)\psi_{\ell'}(y)$$

$$= \sum_{\ell,\ell'=0}^{\infty} f_{\ell\ell'}\overline{Y_\ell(x^0)Y_{\ell'}(y^0)}, \tag{36}$$

where

$$f(x, y) = \sum_{\ell,\ell'=0}^{\infty} f_{\ell\ell'}\overline{Y_\ell(x^0)Y_{\ell'}(y^0)},$$

$$x^0 = xR_1^{-1}, \quad y^0 = yR_1^{-1} \tag{36'}$$

Thus the numbers $v_{\ell\ell'} = f_{\ell\ell'}\{C_\ell C_{\ell'} h_\ell^{(1)}(\omega R_1) h_{\ell'}^{(1)}(\omega R_1)\}^{-1}$ are known, and

$$\int v(z)\psi_\ell^{re}(z)\psi_{\ell'}^{re}(z)\,\mathrm{d}z = v_{\ell\ell'}. \tag{37}$$

Multiply (37) by $a_\ell a_{\ell'} \overline{Y_\ell(n')Y_{\ell'}(-n)}$, where n and n' are arbitrary unit vectors, and use (35) to obtain

$$\int v(z)\exp\{-i\omega(n-n')\cdot z\}\,\mathrm{d}z = \sum_{\ell,\ell'=0}^{\infty} a_\ell a_{\ell'} v_{\ell\ell'} \overline{Y_\ell(n')Y_{\ell'}(-n)}. \tag{38}$$

Denote the right-hand side of (38) by $F(\omega(n-n'))$ and set $\omega(n-n') = p$. Then

$$\int_{B_R} v(z)\exp(-ip\cdot z)\,\mathrm{d}z = F(p), \quad |p| \leqslant 2\omega. \tag{39}$$

This equation is solved in Section 3 (see Equation (3.9)).

Let us summarize the result.

THEOREM 2. *Equation* (32) *has at most one solution* $v \in L^2(B_R)$. *If it is solvable then the solution can be computed as follows*: (1) *compute numbers* $f_{\ell\ell'}$ *by formula* (36′), (2) *compute the right-hand side of* (38), (3) *solve Equation* (39) *for* $v(z)$ *by the method of Section* 3.

In a similar way one can solve the inverse problem in the case when x *and* y *run through a cylindrical surface,* $x_1^2 + x_2^2 = R_1^2$, $R_1 > R$.

If $\omega = 0$ *then the above scheme should be modified because the right-hand side of* (39) *is known only at a point* $p = 0$ *if* $\omega = 0$.

If $\omega = 0$ *then the numbers*

$$\int v(z)\,|z|^{\ell+\ell'}\,Y_\ell(z^0)Y_{\ell'}(z^0)\,\mathrm{d}z = v_{\ell\ell'}^0 \tag{40}$$

are uniquely determined by the data $f(x, y)$ *known for* $x, y \in S_{R_1}$. *These numbers define* $v(z)$ *uniquely.*

LEMMA 2. *Equation* (32) *for* $\omega = 0$ *has at most one solution.*

Proof. Suppose $f(x, y) = 0$, $x, y \in S_{R_1}$. Then the numbers $v^0_{\ell\ell'} = 0$, $\forall \ell, \ell' \geqslant 0$. We want to prove that in this case $v(z) = 0$. Let

$$v(z) = \sum_{\ell''=0}^{\infty} v_{\ell''}(r)\,\overline{Y_{\ell''}(z^0)}, \quad 0 \leqslant r = |z| \leqslant R. \tag{41}$$

Clearly, it is sufficient to prove that $v_\ell(r) = 0$, $\forall \ell \geqslant 0$. It follows from (40) and (41) that if $v_{\ell\ell'} = 0$ then

$$\sum_{\ell''=0}^{\infty} \int_0^R \mathrm{d}r\, r^{2+\ell+\ell'} v_{\ell''}(r) b(\ell, \ell', \ell'') = 0, \quad \forall \ell, \ell' \geqslant 0 \tag{42}$$

where

$$b(\ell, \ell', \ell'') = \int_{S^2} Y_\ell(z^0)\, Y_{\ell'}(z^0)\, \overline{Y}_{\ell''}(z^0)\, \mathrm{d}z^0. \tag{43}$$

Let us consider for simplicity the two-dimensional analogue of (42)–(43):

$$\sum_{\ell''=-\infty}^{\infty} \int_0^R \mathrm{d}r\, r^{1+|\ell|+|\ell'|} v_{\ell''}(r) \tilde{b}(\ell, \ell', \ell'') = 0 \quad \forall \ell, \ell' \tag{44}$$

where $\ell, \ell', \ell'' = 0, \pm 1, \pm 2, \ldots$, and

$$\begin{aligned} \tilde{b}(\ell, \ell', \ell'') &= \frac{1}{(2\pi)^{3/2}} \int_0^{2\pi} \exp\{i(\ell + \ell' - \ell'')\phi\}\, \mathrm{d}\phi \\ &= \frac{1}{(2\pi)^{1/2}}\, \delta_{0,\, \ell+\ell'-\ell''}. \end{aligned} \tag{45}$$

It follows from (44) and (45) that

$$\int_0^R \mathrm{d}r\, r^{1+|\ell|+|\ell'|} v_{\ell''}(r) = 0, \; \ell'' = \ell + \ell'. \tag{46}$$

Let ℓ'' be fixed. Then $1 + |\ell| + |\ell'|$ runs through all integers starting from $1 + |\ell''|$ when ℓ and ℓ' run through the set 0, ±1, ±2, and $\ell + \ell' = \ell''$ is fixed. Therefore

$$\int_0^R \mathrm{d}r\, r^q v_{\ell''}(r) = 0, \quad q \geqslant 1 + |\ell''|. \tag{47}$$

From Müntz's theorem (Akhiezer [1]) it follows that $v_{\ell''}(r) = 0$ for all ℓ''. Thus $v(z) = 0$.

In the three-dimensional case one uses the formula (Lubarskii [1], §54)

$$Y_{\ell_1 m_1}(\omega)\, Y_{\ell_2 m_2}(\omega) = \sum_{\ell' = |\ell_1 - \ell_2|}^{\ell_1 + \ell_2} \rho(\ell_1, \ell_2, \ell') \times$$
$$\times (\ell_1 \ell_2 m_1 m_2 | \ell' m')\, Y_{\ell' m'}(\omega),\ m' = m_1 + m_2,$$

where $(\ell_1 \ell_2 m_1 m_2 | \ell m)$ are the Clebsch–Gordon coefficients, and $\rho(\ell_1, \ell_2, \ell) = \{(2\ell_1 + 1)(2\ell_2 + 1)/(2\ell + 1)\}^{1/2} (\ell_1 \ell_2 00 | \ell 0)$. Multiply the above equation by $\overline{Y}_{\ell_3 m_3}(\omega)$ and integrate over S^2 to obtain:

$$\int_{S^2} Y_{\ell_1 m_1}(\omega)\, Y_{\ell_2 m_2}(\omega)\, \overline{Y}_{\ell_3 m_3}(\omega)\, \mathrm{d}w$$
$$= \rho(\ell_1, \ell_2, \ell_3)\, (\ell_1 \ell_2 m_1 m_2 | \ell_3 m_3) \tag{43'}$$

where the formula $\int_{S^2} Y_{\ell' m'} \overline{Y}_{\ell m}\, \mathrm{d}\omega = \delta_{\ell\ell'} \delta_{mm'}$ was used. It follows now from (43′) and (42) that

$$0 = \sum_{\ell_3 = 0}^{\infty} \sum_{m_3 = -\ell_3}^{\ell_3} \int_0^R \mathrm{d}r\, r^{2 + \ell_1 + \ell_2} v_{\ell_3 m_3}(r) \rho(\ell_1, \ell_2, \ell_3) \times$$
$$\times (\ell_1 \ell_2 m_1 m_2 | \ell_3 m_3)$$

where we wrote ℓ_1, ℓ_2 and ℓ_3 for ℓ, ℓ' and ℓ''. Multiply this equation by $(\ell_1 \ell_2 m_1, m_2 | LM)$, sum over m_1 and m_2, and use the orthogonality condition (Lubarskii [1] §54):

$$\sum_{\substack{m_1, m_2 \\ m_1 + m_2 = m}} (\ell_1 \ell_2 m_1 m_2 | \ell m)(\ell_1 \ell_2 m_1 m_2 | \ell' m') = \delta_{\ell\ell'} \delta_{mm'}$$

to obtain

$$0 = \rho(\ell_1, \ell_2, L) \int_0^R \mathrm{d}r\, r^{2+\ell_1+\ell_2} v_{LM}(r). \tag{48}$$

The coefficient $\rho(\ell_1, \ell_2, L)$ vanishes unless $\ell_1 + \ell_2 + L$ is even and $|\ell_1 - \ell_2| \leqslant L \leqslant \ell_1 + \ell_2$. Take $\ell_2 = \ell_1 + L$. Then the above conditions are met, and therefore $\rho(\ell_1, \ell_1 + L, L) \neq 0$. Thus (48) yields

$$0 = \int_0^R \mathrm{d}r\, r^{2+2\ell_1+L} v_{LM}(r), \ell_1 = 0, 1, 2, \ldots.$$

As above, Müntz's theorem yields $v_{LM}(r) = 0$. Since LM, $L = 0, 1, 2, \ldots, -L \leqslant M \leqslant L$, are arbitrary, $v(z) = 0$. □

Remark 3. The moment problem which consists in finding a function $f(r)$ from the knowledge of its moments f_m:

$$\int_0^R f(r) r^m \,\mathrm{d}r = f_m, \quad m = m_0, \quad m_0 + 1, \quad m_0 + 2, \ldots$$

was studied extensively (see e.g. Akhiezer [2]).

1.7. Computational aspects of the inversion scheme based on Equation (21).

(a) If one inverts the Laplace transform using the Mellin contour Re $q = c$, then the function

$$\psi(p_1, p_2, z_3) = (2\pi i)^{-1} \int_{c-i\infty}^{c+i\infty} \phi(p_1, p_2, q) \exp(-qz_3)\,\mathrm{d}q \tag{49}$$

is computed for some values p_1, p_2, and $z_3 < 0$. If q is a real number then, according to (17), $|p_1| \leqslant p_3 + p_4 \leqslant q$, $|p_2| \leqslant q$. In particular, $|p_1| \leqslant c$, and $|p_2| \leqslant c$. Therefore, for a fixed $c > 0$, one obtains the values of the Fourier transform of $v(z_1, z_2, z_3)$ in a disk on the real two-dimensional space of the variables p_1 and p_2:

$$\begin{aligned} &\int_{-\infty}^{\infty}\int \exp(ip \cdot z^1) v(z_1, z_2, z_3)\,\mathrm{d}z_1\,\mathrm{d}z_2 \\ &\quad = \pi^2 \psi(p_1, p_2, z_3), \quad |p_1|^2 + |p_2|^2 \leqslant c^2. \end{aligned} \tag{50}$$

Of course, the function $\psi(p_1, p_2, z_3)$ can be computed for any desired real point (p_1, p_2) by the Mellin integral provided that the parameter c in this integral is chosen so that $|p_1| \leqslant c$ and $|p_2| \leqslant c$. Therefore, the function $\psi(p_1, p_2, z_3)$ can be computed for all real p_1 and p_2.

However, it is of interest practically to invert the Fourier transform of a function v with a known compact support (say $v = 0$ for $|z| > R$) if the Fourier transform is known on a compact sets only, for example, on a disk $|p_1|^2 + |p_2|^2 \leqslant c_2$. This problem is solved in Section 3.

(b) Let us briefly review some methods for inverting the Laplace transform (see Bellman *et al.* [1], Krylov—Skoblya [1], and Davies—Martin [1]).

(1) Let

$$F(p) = Lf = \int_0^\infty \exp(-pt)f(t)\,\mathrm{d}t, \quad p = \sigma + i\lambda, \tag{51}$$

where $|f| \leqslant c(1 + t)^{-a}$, $a > 1$, $c = \text{const} > 0$. Then $F(p)$ is analytic in the half-plane $\sigma > 0$ and continuous up to the line $\sigma = 0$.

If $F(\sigma)$, $\sigma > \sigma_0 > 0$, is given, where σ_0 is an arbitrary fixed number, then

$$f(t) = \lim_{n \to \infty} \left[\frac{(-1)^n}{n!} \left(\frac{n}{t} \right)^{n+1} F^{(n)} \left(\frac{n}{t} \right) \right]. \tag{52}$$

This is a well known formula [e.g. Davies—Martin [1]). Although there are higher order derivatives appeared in (52) this formula can be used in practice, e.g., in the case when $F(p)$ is a rational function.

Example:

$$F(p) = \frac{1}{p+1}, \quad f(t) = \lim_{n \to \infty} \left[\frac{(-1)^n}{n!} \frac{n^{n+1}}{t^{n+1}} \cdot \frac{(-1)^n n!}{(n/t + 1)^{n+1}} \right]$$

$$= \lim_{n \to \infty} \frac{n^{n+1}}{(n + t)^{n+1}} = e^{-t}.$$

If we can approximate $F(p)$ by a rational function $P_\varepsilon(p)Q_\varepsilon^{-1}(p)$ so that

$$|F(p) - P_\varepsilon(p)Q_\varepsilon^{-1}(p)| \leqslant \varepsilon(1 + |p|)^{-a}, \quad a > 1, \quad \sigma > \sigma_0 \tag{53}$$

then

$$\max_{t \geqslant 0} |f(t) - f_\varepsilon(t)| \leqslant \frac{\varepsilon}{(a-1)\pi} \tag{54}$$

where $f_\varepsilon(t) = L^{-1}\{P_\varepsilon(p) Q_\varepsilon^{-1}(p)\}$. The difficulty is that the simple stability estimate (54) holds if (53) holds in the complex half-plane. It would be interesting to obtain an estimate for $|f(t) - f_\varepsilon(t)|$ if an estimate of the type (53) is known on the real axis $\sigma > \sigma_0$ only.

(2) If $F(p)$ is known in a complex half-plane $\sigma > \sigma_0 > 0$ then the Mellin inversion formula can be used:

$$f(t) = \frac{1}{2\pi i} \int_{\sigma_0 - i\infty}^{\sigma_0 + i\infty} \exp(pt) F(p)\, \mathrm{d}p. \tag{55}$$

Integral (55) can be easily reduced to the Fourier transform:

$$f(t) = \frac{1}{2\pi} \int_{-\infty}^{\infty} \exp(i\lambda t) F(\sigma_0 + i\lambda)\, \mathrm{d}\lambda \cdot \exp(\sigma_0 t) \tag{56}$$

which is convenient for numerical computation. There are a number of quadrature formulas and the tables for numerical computation of (55) and the FFT routine for numerical computation of (56).

(3) Inversion methods which use the values of $F(p)$ on the real axis only, are of particular interest. Formula (52) is very sensitive to the errors in $F(p)$ because it requires high order derivatives of $F(p)$. Therefore the following methods, which are more stable towards the errors in $F(p)$, are of interest (Krylov—Skoblya [1]).

Suppose that f is expanded in the Fourier—Laguerre series

$$f(t) = \sum_{n=0}^{\infty} A_n \frac{\exp(-t/2) L_n(t)}{n!} \tag{57}$$

where $L_n(t)$ are the Laguerre polynomials

$$L_n(t) = (-1)^n \left[t^n - \frac{n^2}{1} t^{n-1} + \right.$$
$$\left. + \frac{n^2(n-1)^2}{2!} t^{n-2} + \ldots + (-1)^n n! \right]. \tag{58}$$

Then

$$A_n = \frac{1}{n!} \int_0^\infty \exp\left(-\frac{t}{2}\right) L_n(t) f(t)\, \mathrm{d}t. \tag{59}$$

Let

$$\alpha_m = \int_0^\infty \exp\left(-\frac{t}{2}\right) t^m f\, \mathrm{d}t. \tag{60}$$

Then

$$\alpha_m = F^{(m)}\left(\frac{1}{2}\right)(-1)^m \tag{61}$$

so that

$$A_n = \sum_{m=0}^{n} \binom{n}{m} \frac{(-1)^{n-m}}{(n-m)!}\, \alpha_{n-m} \tag{62}$$

and, by (57), (61), and (62),

$$f(t) = \exp\left(-\frac{t}{2}\right) \sum_{n=0}^{\infty} \sum_{m=0}^{n} \binom{n}{m} \frac{1}{(n-m)!} F^{(n-m)}\left(\frac{1}{2}\right) \frac{L_n(t)}{n!}. \tag{63}$$

The derivatives of F are present in this formula, but if the series (57) converges rapidly, so that only a few first terms are needed for approximating $f(t)$, then (63) might work better as an inversion algorithm than (52).

(4) One can use other expansions of $f(t)$ for inverting Laplace transform.

If $\exp(-t) = x$ then (51) becomes

$$\int_0^1 x^{p-1} \phi(x)\, \mathrm{d}x = F(p), \quad \phi(x) = f(-\ell n x). \tag{64}$$

Let $p = 1, 2, 3, \ldots$. Then

$$\int_0^1 x^m \phi(x)\, \mathrm{d}x = F_m, \quad m = 0, 1, 2, \ldots. \tag{65}$$

This is a well-known moments problem. Let $P_n(x)$ be the Legendre polynomials, $\tilde{P}_n(x) = P_n(2x - 1)$ be the polynomials defined on the interval [0, 1].

Let

$$f(x) = \sum_{n=0}^{\infty} (2n+1) C_n \tilde{P}_n(x) \tag{66}$$

where

$$C_n = \int_0^1 f(x) \tilde{P}_n(x) \, dx. \tag{67}$$

Let

$$\tilde{P}_n(y) = \sum_{m=0}^{n} P_{nm} y^m. \tag{68}$$

It follows from (68) and (65) that

$$C_n = \sum_{m=0}^{n} P_{nm} F_m. \tag{69}$$

The inversion formula is (66) with C_n given by (69).

(5) One can use expansion of ϕ in the series in Chebyshev's polynomials or Fourier series (Krylov—Skoblya [1]). In the first case one sets $\exp(-\tau) = (1 + \cos\theta)/2$ and $f(t) \to \phi(\theta)$, $0 \leqslant \theta \leqslant \pi$. In the second case one sets $\exp(-\sigma t) = \cos\theta$, $\sigma > 0$, $f(t) = f(-\sigma^{-1} \ln\cos\theta) \equiv \phi(\theta)$, $0 \leqslant \theta \leqslant \pi/2$.

(6) Finally, one can consider Equation (51) with $p = \sigma > 0$ as an equation of the first kind and solve this equation for $f(t)$ numerically. This is an ill-posed problem and therefore one should use some regularization techniques (iterative methods, variational methods, singular value decomposition, or method of quasisolutions (e.g. Ivanov, *et al.* [1], Tikhonov—Arsenin [1], Ramm [3], [11c])).

1.8. Consider the two-dimensional analogue of the problem (1)—(3). The data now are the values of the field $u(x, y, k)$ on the line $L = \{x: x_2 = 0\}$ for all positions of x and y on L. The assumptions about $v(x)$ are the same as in Section 1.1. The governing equation is Equation (6) with

$$g(x, y, \omega) = \frac{i}{4} H_0^{(1)}(\omega r_{xy}), \quad r_{xy} = |x - y| \tag{70}$$

where $H_0^{(1)}$ is the Hankel function. The basic difference between the three-dimensional and two-dimensional problems is that the function (70) does not have a finite limit as $\omega \to 0$. One has

$$g(x, y) = \alpha(\omega) + g_0(x, y) + 0\left(\omega^2 r^2 \log \frac{1}{\omega r}\right) \quad \text{as } \omega \to 0. \tag{71}$$

where

$$\alpha(\omega) = -(2\pi) - 1 \ln(\gamma\omega/2) + i/4, \quad g_0 = \frac{1}{2\pi} \ln \frac{1}{r_{xy}}, \tag{72}$$

and $\gamma = 0.5772\ldots$ is Euler's constant.

Let $u - g = \psi$. Equation (6) can be written as

$$\psi(x, y) = \omega^2 \int g(x, z)v(z)\psi(z, y)\,\mathrm{d}z + \omega^2 \psi_0(x, y) \tag{73}$$

where $\int = \int_{\mathscr{D}}$, $\mathscr{D} = \{x : v(x) \neq 0\}$,

$$\psi_0(x, y) = \int g(x, z)v(z)g(z, y)\,\mathrm{d}z. \tag{74}$$

The field ψ is the scattered field. Equation (73) is uniquely solvable in $C(R^2)$ by iterations if ω is sufficiently small. It follows from (71)–(74) that

$$\psi(x, y) = \omega^2 \left\{ \alpha^2 \int v\,\mathrm{d}z + \alpha \left(\int g_0(x, z)v(z)\,\mathrm{d}z + \right.\right.$$

$$\left.\left. + \int g_0(y, z)v(z)\,\mathrm{d}z \right) + \int g_0(x, z)v(z)g_0(z, y) \right\} + \tag{75}$$

$$+ o(\omega^2), \omega \to 0.$$

Therefore

$$f_0 \equiv \lim_{\omega \to 0} (\omega a)^{-2}\psi(x, y, \omega) = \int v(z)\,\mathrm{d}z \equiv v_0, \tag{76}$$

$$f_1 \equiv \lim_{\omega \to 0} \frac{\psi - a^2\omega^2 v_0}{\omega^2 a} = \int g_0(x, z)v(z)\,\mathrm{d}z + \int g_0(y, z)v(z)\,\mathrm{d}z \tag{77}$$

and

$$f_2 \equiv \lim_{\omega \to 0} \frac{\psi - a^2\omega^2 v_0 - a\omega^2 f_1}{\omega^2} = \int g_0(x, z)v(z)g_0(z, y)\,\mathrm{d}z. \tag{78}$$

The quantities f_j, $j = 0, 1, 2$ can be measured, and they are our data. In particular, f_0 gives the number v_0. If $x = y$ then equation (77) becomes

$$\int_{\mathscr{D}} g_0(x, z)v(z)\,\mathrm{d}z = \frac{1}{2} f_1(x, x), \quad x \in L, \quad \mathscr{D} = \{x\colon v(x) \neq 0\}. \tag{79}$$

This equation is not uniquely solvable in general. Indeed, the right-hand side of (79) is the potential of the mass distribution $v(z)$. This potential is given on the line L. These data and v_0 determine the potential uniquely in the half-plane $x_2 > 0$. By the unique continuation property the potential is uniquely determined in $\Omega = R^2 \backslash \mathscr{D}$. However, there are many distributions v which produce the zero potential in Ω. For example, take the origin in $\mathscr{D}$ and let the disk $\mathscr{D}_{r_1}$ belong to $\mathscr{D}$. Choose r_2 and r_3 such that $0 < r_3 < r_2 < r_1$. Define $v = \{\pi(r_2^2 - r_3^2)\}^{-1}$ if $r_3 \leqslant r < r_2$, $v = -\{\pi(r_1^2 - r_2^2)\}^{-1}$ if $r_2 \leqslant r \leqslant r_1$, $v = 0$ if $r > r_1$ or $r < r_3$, $r = |x|$. Then the left-hand side of (79) vanishes for $r > r_1$, but $v \not\equiv 0$. Under some special assumptions on $v(z)$ Equation (79) can be uniquely solvable. For example, if $v - v(x_2)$ for $a \leqslant x_1 \leqslant b$ and $-R < x_2 < 0$ and is zero otherwise, then taking the Fourier transform of (79) yields

$$\int_{-\infty}^{0} v(z_2) \exp(|\lambda| z_2)\,\mathrm{d}z_2 = \frac{2i\pi\lambda|\lambda| F_1(\lambda)}{\exp(i\lambda b) - \exp(i\lambda a)}, \quad -\infty < \lambda < \infty \tag{80}$$

where

$$F_1(\lambda) = (2\pi)^{-1} \int_{-\infty}^{\infty} f_1(x, x) \exp(i\lambda x)\,dx. \tag{81}$$

Inverting the Laplace transform (80) yields $v(z_2)$. Thus, under the above assumption about $v(z) = v(z_2)$ an explicit analytical inversion procedure is given in terms of the back-scattered data measured on the line $x_2 = 0$.

If no special assumptions on v are made then one can use Equation (78) to recover v:

$$\int \log|x - z| \log|y - z| v(z)\,dz = f(x, y),\ x, y \in L, \quad f(x, y) = 4\pi^2 f_2(x, y) \tag{82}$$

This equation is similar to (9). Taking the Fourier transform in $x = (x, 0)$ and $y = (y, 0)$ yields

$$\int v(z) \exp\{i(\lambda + \mu)z_1 - (|\lambda| + |\mu|)|z_2|\}\,dz = 4|\lambda|\,|\mu| F(\lambda, \mu) \tag{83}$$

where

$$F(\lambda, \mu) = (2\pi)^{-2} \iint_{-\infty}^{\infty} f(x, y) \exp\{i(\lambda x + \mu y)\}\,dx\,dy, \tag{84}$$

and

$$\frac{1}{2\pi} \int_{-\infty}^{\infty} \exp(i\lambda x) \ln\sqrt{(x - z_1)^2 + z_2^2}\,dx = -\frac{\exp(i\lambda z_1 - |\lambda|\,|z_2|)}{2|\lambda|}. \tag{85}$$

Integral (85) can be calculated as follows. Notice that

$$(2\pi)^{-1} \int_{-\infty}^{\infty} dx \exp(i\lambda x) \ln\sqrt{(x - z_1)^2 + z_2^2} = \exp(i\lambda z_1)(2\pi)^{-1} \int_{-\infty}^{\infty} \exp(i\lambda y) \ln\sqrt{y^2 + z_2^2}\,dy \tag{85a}$$

and

$$\widehat{\ln y} = (2\pi)^{-1} \int_{-\infty}^{\infty} \ln|y| \exp(i\lambda y)\, \mathrm{d}y = -(2|\lambda|)^{-1}. \tag{85b}$$

Indeed,

$$i\lambda\, \widehat{\ln|y|} = (2\pi)^{-1} \int_{-\infty}^{\infty} y^{-1} \exp(i\lambda y)\, \mathrm{d}y$$

$$= (2\pi)^{-1} \cdot i\pi \operatorname{sgn} \lambda = (i \operatorname{sgn} \lambda)/2,$$

and (85b) follows. Formula (85) follows from (85a) and (85b) if one takes into account that integral (85) is a harmonic function of (z_1, z_2). Another way to derive (85) is to differentiate the integral in the right-hand side of (85) in z_2, calculate the resulting standard integral and integrate in z_2. Integral (85b) serves as an initial condition at $z_2 = 0$.

Let

$$p = \lambda + \mu \quad \text{and} \quad q = |\lambda| + |\mu|. \tag{86}$$

As in Section 1.3 one can prove that Equation (83) has at most one solution $v \in L^2$. The Jacobian of the mapping (86), $\{\lambda, \mu\} \to \{p, q\}$ is $J = \operatorname{sgn} \mu - \operatorname{sgn} \lambda \neq 0$ if $\lambda\mu < 0$. Take $\mu > 0$ and $\lambda < 0$. Then (83) becomes

$$\int v(z) \exp(ipz_1 + qz_2)\, \mathrm{d}z = |p^2 - q^2|\, \mathscr{F}(p, q), \quad z_2 < 0, \tag{87}$$

where

$$\mathscr{F}(p, q) = F\left(\frac{p-q}{2}, \frac{p+q}{2}\right). \tag{88}$$

If $\lambda < 0$ and $\mu > 0$ then the corresponding point (p, q) lies inside the wedge $0 < q < \infty, -q < p < q$.

Equation (87) can be written as

$$\int_0^{\infty} \exp(-q\zeta) \left(\int_{-\infty}^{\infty} v(z_1, -\zeta) \exp(ipz_1)\, \mathrm{d}z_1 \right) \mathrm{d}\zeta = \phi(p, q), \tag{89}$$

where

$$\phi(p, q) = |p^2 - q^2|\mathcal{F}(p, q). \tag{90}$$

Equation (89) is analogous to (21), and one can repeat the analysis given in Sections (1.3) and (1.6).

1.9. Consider the one-dimensional case.

The problem now is to find $v(x)$, $x \in R^1$, from the measurements of $u(x, y, \omega)$ for $x \geqslant 0$ and $y \geqslant 0$. We assume that $v(x) \in L^2$, $v(x) = 0$ if $x < -R$ or $x > -\varepsilon$ where ε and R are positive constants, $\varepsilon < R$. The Green function is given by

$$\begin{aligned} g &= \frac{i}{2\omega} \exp(i\omega|x-y|) \equiv \frac{i}{2\omega} + g_1 \equiv \\ &\equiv \frac{i}{2\omega} + g_0 + 0(\omega r_{xy}) \quad \text{as} \quad \omega \to 0 \end{aligned} \tag{91}$$

where $g_0 = -|x-y|/2$, $r_{xy} = |x-y|$. The integral equation (6) becomes

$$u = g(x, y) + \frac{i\omega}{2} \int \exp(i\omega|x-z|)v(z)u(z, y)\,dz. \tag{92}$$

Let $u - g = \psi$. Then

$$\psi(x, y) = \frac{i\omega}{2} \int \exp(i\omega|x-z|)v(z)\psi(z, y)\,dz + \psi_0(x, y) \tag{93}$$

where

$$\psi_0(x, y) = -\frac{1}{4} \int \exp(i\omega|x-z| + |y-z|)v(z)\,dz. \tag{94}$$

Equation (93) is uniquely solvable in $C(R^1)$ by iterations if $(\omega/2) \cdot \int |v|\,dz < 1$, in particular, for sufficiently small ω. Thus, ψ is analytic in ω in a neighborhood of the point $\omega = 0$. The function ψ is the scattered field, it is the datum.

It follows from (93) that

$$\lim_{\omega \to 0} \psi = -\frac{1}{4} \int v \, \mathrm{d}z \equiv -\frac{v_0}{4}, \quad \text{where} \quad v_n \equiv \int z^n v(z) \, \mathrm{d}z. \tag{95}$$

Let

$$\psi = \psi(x, y, \omega) = \sum_{j=0}^{\infty} C_j(x, y) \, (i\omega)^j, \tag{96}$$

and

$$\begin{aligned} \psi_0 &= -\frac{1}{4} \sum_{j=0}^{\infty} \psi_{0j} (i\omega)^j, \\ \psi_{0j} &= \frac{1}{j!} \int (|x - z| + |y - z|)^j v(z) \, \mathrm{d}z. \end{aligned} \tag{97}$$

Substitute (96) and (97) into (93) and compare the coefficients in front of $(i\omega)^j$ to obtain the recurrence formula

$$\begin{aligned} C_j(x, y) = \frac{1}{2} \sum_{m+n+1=j} \frac{1}{m!} \int |x - z|^m v(z) C_n(z, y) \, \mathrm{d}z \, - \\ - \frac{1}{4} \psi_{0j}, \quad j = 0, 1, 2 \ldots . \end{aligned} \tag{98}$$

If $j - 0$ then (98) gives

$$C_0(x, y) = -\frac{\psi_{00}}{4} = -\frac{v_0}{4}. \tag{99}$$

If $j = 1$, then

$$C_1(x, y) = -\frac{v_0^2}{8} + \frac{v_1}{2} - \frac{x + y}{4} v_0. \tag{100}$$

If $j = 2$, then

$$C_2(x, y) = -\frac{v_0^3}{16} + \frac{3v_0v_1}{8} - \frac{v_0^2(x+y)}{8} - \frac{v_0^2 x}{4} - \\ - \frac{v_0(x+y)^2}{8} + \frac{v_1(x+y)}{2} - \frac{v_2}{2}. \tag{101}$$

The coefficients C_j can be considered as data. Equations (99)–(100) determine recurrently v_0, v_1, and v_2. In principle all the moments v_n (see (95)) can be determined. It is a classical problem, Hamburger's moment problem (Akhiezer [2]), to find $v(z)$ from the knowledge of its moments v_n:

$$\int_{-R}^{0} v(z) z^n \, dz = v_n, \quad n = 0, 1, 2, \dots . \tag{102}$$

Similar problem we had in Section 1.5, Equation (48). Since the system $\{z^n\}$, $n = 0, 1, 2 \dots$ is closed in $L^2(-R, 0)$, the moment problem (102) has at most one solution. Necessary and sufficient conditions are known for the solvability of problem (102) (Akhiezer [1]). A numerical solution similar to the T-matrix approach studied in Ramm [7d] can be suggested. Namely, take a basis $\{\phi_j(z)\}$ of $L^2(-R, 0)$, look for an approximate solution in the form

$$v_m(z) = \sum_{j=0}^{m} a_j^{(m)} \phi_j(z), \tag{103}$$

substitute (103) into (102) and use the first m Equations (102) to obtain a linear system for the coefficients $a_j^{(m)}$:

$$\sum_{j=0}^{m} b_{nj} a_j^{(m)} = v_n, \quad b_{nj} = \int_{-R}^{0} z^n \phi_j(z) \, dz. \tag{104}$$

We will not go into further detail because it is probably not so easy to measure higher order coefficients $C_j(x, y)$, $j > 2$, and therefore find the higher order moments.

It is worth mentioning that an inversion procedure within the Born approximation can be constructed easily. In the Born approximation one has $\psi \simeq \psi_0$, so that

$$\int \exp\{i\omega(|x-z| + |y-z|)\} v(z)\, dz = f(x, y), x, y > 0 \quad (105)$$

where

$$f(x, y, \omega) = -4\psi(x, y) \quad (106)$$

is the data. Set $x = y$ in (105) to obtain

$$\int_{-R}^{0} \exp(-2i\omega z) v(z)\, dz = \exp(-2i\omega x) f(x, x, \omega) \equiv f(x, \omega). \quad (107)$$

Let us fix arbitrary $x \geqslant 0$ and consider the data $f(x, \omega) = \phi(\omega)$ as a function of ω. Then one can Fourier-invert the data $\phi(\omega)$ and compute $v(z)$. Since $v(z)$ is compactly supported the Fourier inversion can be done also in the case when the data $\phi(\omega)$ are given on a finite interval of frequencies, $\omega_0 \leqslant \omega \leqslant \omega_1$. This is done in Section 3 below.

In conclusion, it is worth mentioning that if the incident field is a plane wave and the measured quantity is the reflection coefficient which is known at all frequencies, then the one dimensional problem can be reduced by the Liouville transform to the problem of inverse scattering by a potential. This problem was studied in the literature (e.g. Chadan–Sabatier [1]).

1.10. Inversion of the data given on a cylinder.

Consider Equation (32) with $x, y \in C_{R_1}$, where $C_{R_1} = \{x: \rho = R_1\}$, $R_1 > R$, $\rho = (x_1^2 + x_1^2)^{1/2}$. We have

$$g(x, z) = \sum_{n=-\infty}^{\infty} (2\pi)^{-2} \exp\{in(\phi_z - \phi_x)\} \int_{-\infty}^{\infty} \exp\{i\lambda(z_3 - x_3)\} \times \times \frac{i\pi}{2} J_n(\alpha\rho_z) H_n(\alpha\rho_x)\, d\lambda, \quad \rho_z < \rho_x, \quad (108)$$

where $\alpha = (k^2 - \lambda^2)^{1/2}$, and $H_n(r) = H_n^{(1)}(r)$ is the Hankel function, $\rho_x = \rho_y = R_1, \rho_x = (x_1^2 + x_2^2)^{1/2}$,

$$\exp\{i(\lambda x_3 + \alpha\rho\cos(\phi - \phi'))\}$$
$$= \exp(i\lambda x_3) \sum_{n=-\infty}^{\infty} i^n \exp\{in(\phi - \phi')\} J_n(\alpha\rho). \tag{109}$$

Equation (32) can be written as

$$-\frac{1}{64\pi^2} \sum_{n=-\infty}^{\infty} \sum_{n'=-\infty}^{\infty} \exp\{-i(n\phi_x + n'\phi_y) -$$
$$- i(\lambda x_3 + \lambda' y_3)\} H_n(\alpha R_1) H_{n'}(\alpha' R_1) \cdot$$
$$\cdot \int \mathrm{d}z\, v(z) \exp\{i(n + n')\phi_z +$$
$$+ i(\lambda + \lambda')z_3\} J_n(\alpha\rho_z) J_{n'}(\alpha'\rho_z) \tag{110}$$
$$= \sum_{n=-\infty}^{\infty} \sum_{n'=-\infty}^{\infty} \int_{-\infty}^{\infty} \int_{-\infty}^{\infty} \mathrm{d}\lambda\, \mathrm{d}\lambda'\, f_{\lambda n \lambda' n'} \cdot$$
$$\cdot \exp\{-i(n\phi_x + n'\phi_y) - i(\lambda x_3 + \lambda' y_3)\},$$

where $\alpha' = (k^2 - \lambda'^2)^{1/2}$, and

$$f_{\lambda n \lambda' n'} = \frac{1}{(2\pi)^4} \int_0^{2\pi} \int_0^{2\pi} \int_{-\infty}^{\infty} \int_{-\infty}^{\infty} \exp\{i(n\phi_x + n'\phi_y) +$$
$$+ i(\lambda x_3 + \lambda' y_3)\} f(\phi_x, x_3, \phi_y, y_3)\, \mathrm{d}\phi_x\, \mathrm{d}\phi_y\, \mathrm{d}x_3\, \mathrm{d}y_3.$$

Here $f(\phi_x, x_3, \phi_y, y_3)$ are the data, the right-hand side of the Equation (32). From (110) it follows that

$$\int \mathrm{d}z\, v(z) \exp\{i(n + n')\phi_z\} J_n(\alpha\rho_z) J_{n'}(\alpha'\rho_z)$$
$$= -\frac{64\pi^2 f_{\lambda n \lambda' n'}}{H_n(\alpha R_1) H_{n'}(\alpha' R_1)}. \tag{111}$$

Multiply (111) by $i^{n+n'} \exp\{-i(n\phi + n'\phi')\}$, sum over n, n', and use (109), to get:

$$\int_{|z| \leq R} dz\, v(z) \exp\{ik(m + m') \cdot z\} = F(k(m + m')), \quad k > 0, \quad (112)$$

where

$$F(k(m + m')) = -64\pi^2 \sum_{n, n' = -\infty}^{\infty} \frac{f_{\lambda n \lambda' n'} i^{n+n'} \exp\{-i(n\phi + n'\phi')\}}{H_n(\alpha R_1) H_n(\alpha' R_1)},$$

$m = (\sin\theta \cos\phi, \sin\theta \sin\phi, \cos\theta)$, $\lambda = k\cos\theta$, $\alpha = k\sin\theta$

$m' = (\sin\theta' \cos\phi', \sin\theta' \sin\phi', \cos\theta')$, $\lambda' = k\cos\theta'$, $\alpha' = k\sin\theta'$.

Equation (112) is of the type (3.9) and can be solved as in Section 3.

Equation (9) with $x, y \in C_{R_1}$ is Equation (32) with $\omega = 0$. Let us prove that the homogeneous Equation (9) with $x, y \in C_{R_1}$ has only the trivial solution. If

$$w(x, y) \equiv \int_{|z| \leq R} dz\, v(z) |x - z|^{-1} |y - z|^{-1} = 0, \quad x, y \in C_{R_1},$$

then for any fixed y, $|y| > R$, and for $|x| > R$ the function $w(x, y)$ solves the problem

$$\begin{aligned} &\nabla_x^2 w(x, y) = 0, \quad \text{if} \quad |x| \geq R \quad \text{and} \quad |y| \geq R; \quad w(x, y) = 0 \\ &\qquad \text{if} \quad y \in C_{R_1} \quad \text{and} \quad x \in C_{R_1}; \quad w(\infty, y) = 0. \end{aligned} \quad (113)$$

Thus $w(x, y) = 0$ if $|x| \geq R$ and $y \in C_{R_1}$, because (113) has only the trivial solution by the maximum principle.

Fix x, $|x| \geq R$, and notice that $w(x, y)$ as a function of y solves the problem

$$\begin{aligned} &\nabla_y^2 w(x, y) = 0, \quad \text{if} \quad |x| \geq R \quad \text{and} \quad |y| \geq R; \quad w(x, y) = 0 \\ &\qquad \text{if} \quad y \in C_{R_1} \quad \text{and} \quad |x| \geq R; \quad w(x, \infty) = 0. \end{aligned}$$

This problem has only the trivial solution $w(x, y) = 0$, $y_1^2 + y_2^2 \geq R_1^2$, $|x| \geq R$. Therefore, by the unique continuation property for solutions to Laplace's equation $w(x, y) = 0$ if $|x| \geq R$ and $|y| \geq R$. In particular, $w(x, y) = 0$ if $x, y \in P$, where P is a plane which does not intersect the support of v. This implies that $v = 0$ as is proved in Theorem 1.

1.11. Inversion of the backscattering data.

Consider Equations (32)—(33) with $x = y \in P$, where $P = \{x: x_3 = 0\}$. The data $f(x^1, x^1, \omega)$, $x^1 \in P$, $0 < \omega < \infty$, are the backscattering data. In this section we assume that the data are given for all $\omega > 0$. The problem is to find $v(z)$ from Equation (32) with the right side $f(x^1, x^1, \omega)$. Let us differentiate (32) in ω, denote $2\omega = k$, and $-2\pi i\, \partial f/\partial\omega|_{\omega=k/2} = F(x^1, k)$, to obtain

$$\int g(x^1, z, k)v(z)\,\mathrm{d}z = F(x^1, k), \quad x^1 \in P, \quad 0 < k < \infty. \tag{114}$$

We solve Equation (114) for $v(z)$ analytically. Take the Fourier transform of (114) in x^1 to obtain

$$\frac{i}{2k_3(2\pi)^2}\int \mathrm{d}z\, v(z)\exp\{i(\hat{k}\cdot z^1 + |z_3|k_3)\} = \tilde{F}(\hat{k}, k) \tag{115}$$

where $k_3 = (k^2 - |\hat{k}|^2)^{1/2} \geqslant 0$ if $k \geqslant |\hat{k}|$, and k_3 is defined on the complex k-plane with the cut $(-|\hat{k}|, |\hat{k}|)$ by analytic continuation. Here we use the formula

$$\begin{aligned}\tilde{g} &= \frac{1}{(2\pi)^2}\iint_{-\infty}^{\infty}\exp(i\hat{k}\cdot x^1)g(x^1, z, k)\,\mathrm{d}x^1\\ &= \frac{i\exp\{i(\hat{k}\cdot z^1 + |z_3|k_3)\}}{2k_3\,(2\pi)^2}\end{aligned} \tag{116}$$

which follows from formula (2.18) in the next section.

Let us write (115) as

$$(2\pi)^{-3}\int \mathrm{d}z\, v(z)\exp(i\alpha\cdot z) = A(\alpha), \quad \alpha = (\hat{k}, -k_3) \tag{117}$$

where we took into account that $|z_3| = -z_3$ if $z_3 < 0$, i.e. if $v(z) \neq 0$, and define

$$A(\alpha) = -\frac{i}{\pi}k_3\tilde{F}(\hat{k}, k) = \frac{i}{\pi}\alpha_3\tilde{F}(\hat{\alpha}, (|\hat{\alpha}|^2 + \alpha_3^2)^{1/2}),$$

$$\hat{\alpha} = (\alpha_1, \alpha_2) = \hat{k}.$$

Note that $A(\alpha)$ is an entire function of α of exponential type since $v(z)$ is compactly supported. Thus a necessary and sufficient condition for $F(x', k)$ to be the exact data, i.e. for the solvability of Equation (114) in the class of compactly supported $v(z)$, is the requirements that $\alpha_3 F(\hat{\alpha}, (|\hat{\alpha}|^2 + \alpha_3^2)^{1/2})$ is an entire function of α of exponential type. If this condition is satisfied then

$$v(z) = \int A(\alpha) \exp(-i\alpha \cdot z)\, \mathrm{d}\alpha. \tag{118}$$

Let us give an example of the inversion procedure.

Example. Let $v(z) = \delta(z)$. Then

$$F(x^1, k) = \frac{\exp(ik|x^1|)}{4\pi|x^1|}, \quad \tilde{F}(\hat{k}, k) = \frac{i}{8\pi^2 k_3}, \quad \text{and} \quad A(\alpha) = \frac{1}{8\pi^3}.$$

Thus, by formula (118), one recovers $v(z) = \delta(z)$. This example shows that the inversion procedure is valid for distributions as well as for integrable functions.

Let us summarize the result.

PROPOSITION 1. *Equation* (32) *with the right side* $f(x^1, x^1, \omega)$ $x^1 \in P$, $0 < \omega < \infty$, *has at most one solution in the class of compactly supported distributions. This equation has a compactly supported solution* $v(z)$ *if the function* $\alpha_3 \tilde{F}(\hat{\alpha}, (|\hat{\alpha}|^2 + \alpha_3^2)^{1/2})$ *is an entire function of* α *of exponential type. If a compactly supported solution exists, it can be found by formula* (118).

Remark 1. Equation (114) is equivalent to the problem of finding a compactly supported function $v(z)$ from the knowledge of the integrals $\int_{S(x^1,t)} v(z)\, \mathrm{d}s$ of $v(z)$ over the spheres centered at x^1 and of radius $t > 0$, where x^1 runs through the plane P and t runs through the positive semiaxis.

Indeed, define $F(x^1, -k) = F^*(x^1, k)$, $k \geqslant 0$, where the star denotes complex conjugate. Multiply (114) by $1/2\pi \exp(-ikt)$ and integrate in k over $(-\infty, \infty)$ to obtain

$$\int \mathrm{d}z\, v(z) \frac{\delta(|x^1 - z| - t)}{4\pi|x^1 - z|} = h(x^1, t) \tag{119}$$

where $\delta(|x^1 - z| - t)$ is the delta function and

$$h = (2\pi)^{-1} \int_{-\infty}^{\infty} F(x^1, k) \exp(-ikt)\, \mathrm{d}k.$$

Note that $h = 0$ if $t < 0$. Equation (119) can be written as

$$\int_{S(x^1, t)} v(z)\, \mathrm{d}s = 4\pi t h(x^1, t). \tag{120}$$

Thus, if one can solve (114), one can solve (120). Conversely, taking the Fourier transform of (120) in t yields (114). Let us summarise the result.

PROPOSITION 2. *Let*

$$\int_{S(x^1, t)} v(z)\, \mathrm{d}s = H(x^1, t), \quad x^1 \in P, \quad t \geqslant 0. \tag{121}$$

Define

$$h(x^1, t) = (4\pi t)^{-1} H(x^1, t),\ F(x^1, k) = \int_{-\infty}^{\infty} h(x^1, t) \exp(ikt)\, \mathrm{d}t. \tag{122}$$

Then solve Equation (114) *for $v(z)$ by the procedure given in Proposition* 1. *The function $v(z)$ solves Equation* (121).

We leave the discussion of the conditions on $H(x^1, t)$, which are sufficient for $v(z)$ to have a compact support and to be a bounded function, to the reader.

In Courant, Hilbert [1] and John [1] the problem of finding $v(z)$ from the knowledge of its integrals over the spheres $S(x, 1)$, $x \in R^3$ was considered. In Lavrentiev *et al.* [1] the problem of finding $v(z)$ from the knowledge of its integrals over the spheres $S(x^1, t)$ was considered in R^2, i.e. x^1 run through the line $-\infty < x^1 < \infty$, and $S(x^1, t)$ was the circle. The solution given in Lavrentiev *et al.* [1] requires differentiation of the data infinitely many times, and therefore it is not very suitable for practice.

The problems mentioned are the problems of integral geometry. The best known problem of this type is the Radon problem of finding a function from the knowledge of its integrals over planes (see, e.g. John [1]).

Let us prove that the data $f(x, x, \omega_j)$, $|x| = R_1 > R$, $0 < \omega_j$, $1 \leqslant j \leqslant n$, do not determine $v(z)$ uniquely.

Let $v = v(r)$, $r = |z|$. One has

$$\int_{|z| \leqslant 1} \frac{\exp(2i\omega|x - z|)}{|x - z|^2} v(r)\, \mathrm{d}z = \int_0^1 \mathrm{d}r\, r^2 v(r) \cdot$$

$$\cdot \int_0^{2\pi} \mathrm{d}\phi \int_0^{\pi} \mathrm{d}\theta \sin\theta \frac{\exp[2i\omega(|x|^2 + r^2 - 2r|x|\cos\theta)^{1/2}]}{|x|^2 + r^2 - 2r|x|\cos\theta}$$

$$= \int_0^1 \mathrm{d}r\, r^2 v(r) h(r, R_1, \omega), \quad R_1 = |x|,$$

where h is the integral in the angular variables. Therefore any function $v(r)$ which is orthogonal in $L^2([0, 1])$ to $r^2 h(r, R_1, \omega)$ solves the equation

$$\int_{|z| \leqslant 1} \frac{\exp(2i\omega|x - z|)}{|x - z|^2} v(z)\, \mathrm{d}z = 0, \quad |x| = R_1. \tag{123}$$

Clearly, there are infinitely many nontrivial solutions of Equation (123) for any finite number of $\omega = \omega_j$, $1 \leqslant j \leqslant n$ and R_j, $1 \leqslant j \leqslant m$. On the other hand, the reader can prove that if the data $f(x, x, \omega_j)$, $|x| = R_1 > R$ are given on a set $\{\omega_j\}$, $1 \leqslant j < \infty$, and the set $\{\omega_j\}$ has a finite limit point ω_0, then $v(z)$ is uniquely determined. The same is true if the data $f(x, x, \omega)$ are given at a fixed $\omega > 0$ and for $|x| = R_j$, where the set $\{R_j\}$ has a limit point (not necessarily finite).

Exercise. The data $f(x^1, x^1, \omega_0)$, $x^1 \in P$, $\omega_0 > 0$ is fixed, do not determine $v(z)$ uniquely.

2. TWO PARAMETER INVERSION

2.1. In Section 1, one unknown function $n(x)$ was found from the seismic data. In this section we use the data at two frequencies to find the density of the inhomogeneity and the velocity. Consider the equation

$$\frac{\omega^2 G}{k(x)} + \nabla \cdot \left(\frac{1}{\rho(x)} \nabla G \right) = -\delta(x-y) \tag{1}$$

where $x, y \in R^3$, $\rho(x)$ is the density and $k(x)$ is the bulk modulus. Let us assume that

$$\frac{1}{k(x)} = \frac{1}{k_r} - \frac{a_1(x)}{\rho_r}, \quad \frac{1}{\rho} = \frac{1}{\rho_r} - \frac{a_2(x)}{\rho_r} \tag{2}$$

where k_r and ρ_r are positive constants, and $a_j(x)$, $j = 1, 2$, are functions with compact support, i.e. $a_j(x) = 0$ if $|x| > R$, where $R > 0$ is an arbitrarily large fixed number. We assume that $a_1 \in L^2(B_R)$, $B_R = \{x: |x| \leqslant R\}$, and $a_2 \in H^1(B_R)$.

Equation (1) can be written as

$$\left(\nabla^2 + \frac{\omega^2}{c^2} \right) G - \omega^2 a_1 G - \nabla \cdot (a_2 \nabla G) = -\rho_r \delta(x-y) \tag{3}$$

where $c^2 = k_r \rho_r^{-1}$. Let us choose units in which

$$c = 1 \quad \text{and} \quad \rho_r = 1. \tag{4}$$

This is possible: take $x = \alpha x'$, $y = \alpha y'$, $\omega = \beta \omega'$ where α and β are constants. Then (3) takes the form

$$\begin{aligned} &\left(\alpha^{-2} \nabla'^2 + \beta^2 \frac{\omega'^2}{c^2} \right) G - \beta^2 \omega^2 a_1 G - \alpha^{-2} \nabla' \cdot a_2 \nabla' G) \\ &\quad = -\rho_r^{-3} \alpha \delta(x' - y'). \end{aligned} \tag{5}$$

Let $\alpha = \rho_r$, $\beta = c\rho_r^{-1}$ and define $a_1' = c^2 a_1$ and $a_2' = a_2$. Then (5) takes the form

$$(\nabla'^2 + \omega'^2) G - \omega^2 a_1'(x) G - \nabla' \cdot (a_2' \nabla G) = -\delta(x' - y'). \tag{6}$$

Therefore one can study the equation

$$(\nabla^2 + \omega^2) G - \omega^2 a_1(x) G - \nabla \cdot (a_2 \nabla G) = -\delta(x-y). \tag{7}$$

This equation can be written as the integral equation (which is analogous to Equation (1.6)):

$$G = g - \omega^2 \int g a_1 G \, \mathrm{d}z - \int g \nabla_z \cdot (a_2(z) \nabla_z G) \, \mathrm{d}z \tag{8}$$

where $\int = \int_{R^3}$, $\mathrm{d}z = \mathrm{d}z_1 \, \mathrm{d}z_2 \, \mathrm{d}z_3$, and

$$g = \frac{\exp(i\omega|x - y|)}{4\pi|x - y|} \tag{9}$$

In operator form Equation (8) can be written as

$$G = g - \omega^2 g a_1 G - g \nabla \cdot (a_2 \nabla G) = g - gVG \tag{10}$$

where the potential V is given by

$$V = \omega^2 a_1 + \nabla \cdot a_2 \nabla \tag{11}$$

Notice that $G = g - GVg$ so that $GVg = gVG$. Let us define the T-matrix

$$T = V - VGV. \tag{12}$$

The T-matrix here is not to be confused with the T-matrix in Chap. IV. It is easy to check that

$$T = V - VgT \tag{13}$$

and

$$Tg = VG. \tag{14}$$

The operator V is an integral operator in impulse space with kernel

$$(2\pi)^{-3} \tilde{V}(k, p) = (2\pi)^{-3} \int\int \mathrm{d}x \, \mathrm{d}y \, V(x, y) \exp(ik \cdot x - p \cdot y)$$

where $V(x, y)$ is the kernel of V. This kernel may be a distribution. For V defined in (11) we have $V(x, y) = \omega^2 a_1(x)\delta(x - y) + \nabla \cdot a_2(y)\nabla\delta(x - y)$, so that

$$\tilde{V}(k, p) = \omega^2 \tilde{a}_1(k - p) - k \cdot p \tilde{a}_2(k - p) \tag{15}$$

where $\tilde{a}(k) = \int \exp(ik \cdot x)a(x)\,\mathrm{d}x$. The scattered field $G_s = G - g$ according to Equations (10) and (14) can be written as

$$G_s = -gVG = -gTg. \tag{16}$$

Let us keep the x and y variables (the positions of the receiver and the source) on the plane $x_3 = 0$ and Fourier transform (16) in $\hat{x} = (x_1, x_2)$ and $\hat{y} = (y_1, y_2)$. Let $\hat{k} \cdot x = k_1x_1 + k_2x_2$, $\mathrm{d}\hat{x} = \mathrm{d}x_1\,\mathrm{d}x_2$ etc. Then

$$\begin{aligned}
\tilde{G}_s(\hat{k}, \hat{p}, \omega) &= \int \exp(i\hat{k} \cdot \hat{x} - i\hat{p} \cdot \hat{y})G_s(\hat{x}, \hat{y}, \omega)\,\mathrm{d}\hat{x}\,\mathrm{d}\hat{y} \\
&= -\int \exp(i\hat{k} \cdot \hat{x} - i\hat{p} \cdot \hat{y})gTg\,\mathrm{d}\hat{x}\,\mathrm{d}\hat{y} \\
&= -\iint \mathrm{d}z'\,\mathrm{d}z''\ T(z', z'') \int \frac{\exp(i\omega|z' - \hat{x}| + i\hat{k}\hat{x})}{4\pi|z' - \hat{x}|}\,\mathrm{d}\hat{x}\ \cdot \\
&\quad \cdot \int \frac{\exp(i\omega|z' - \hat{y}| - i\hat{p} \cdot \hat{y})}{4\pi|z'' - \hat{y}|}\,\mathrm{d}\hat{y} \\
&= -\iint \mathrm{d}z'\,\mathrm{d}z''\ T(z', z'')\exp(i\hat{k}\hat{z}' - i\hat{p} \cdot \hat{z}'')\ \cdot \\
&\quad \cdot\ h(|z_3'|, |\hat{k}|) \cdot h(|z_3''|, |\hat{p}|),
\end{aligned} \tag{17}$$

where $T(z', z'')$ is the kernel of T,

$$h(|z_3'|, |\hat{k}|) = \begin{cases} \dfrac{i}{2}\ \dfrac{\exp\{i|z_3'|(\omega^2 - |\hat{k}|^2)^{1/2}\}}{(\omega^2 - |\hat{k}|^2)^{1/2}}, & \omega > |\hat{k}| \\[2ex] \dfrac{1}{2}\ \dfrac{\exp\{-|z_3'|(|\hat{k}|^2 - \omega^2)^{1/2}\}}{(|\hat{k}|^2 - \omega^2)^{1/2}}, & \omega < |\hat{k}| \\[2ex] \infty & \omega = |\hat{k}|. \end{cases} \tag{18}$$

Let us give the details of the computation of integral (18). One has, for example,

$$\int \frac{\exp(i\hat{k}\,.\,x + i\omega|z-\hat{x}|)}{4\pi|\hat{x}-z|}\,\mathrm{d}\hat{x}\bigg|_{\hat{x}=\hat{u}+\hat{z}} = \exp(i\hat{k}\,.\,\hat{z})\,\cdot$$

$$\cdot \int \frac{\exp\{i|\hat{k}|\,|\hat{u}|\,\cos\phi + i\omega(|\hat{u}|^2 + z_3^2)^{1/2}\}}{4\pi(|\hat{u}|^2+z_3^2)^{1/2}}\,\mathrm{d}\hat{u}$$

$$= \frac{\exp(i\hat{k}\cdot z)}{4\pi}\int_0^\infty \frac{\mathrm{d}\rho\;\rho\;\exp\{i\omega(\rho^2+z_3^2)^{1/2}\}}{(\rho^2+z_3^2)^{1/2}}\times$$

$$\times\int_0^{2\pi}\mathrm{d}\phi\;\exp(i|\hat{k}|\rho\cos\phi) = \frac{\exp(i\hat{k}\cdot\hat{z})}{4\pi}\,\frac{2\pi}{|\hat{k}|^{1/2}}\times$$

$$\times\int_0^\infty \mathrm{d}\rho\;\rho^{1/2}\exp\{i\omega(\rho^2+z_3^2)^{1/2}\}\sqrt{\rho|\hat{k}|}\;J_0(\rho|\hat{k}|)$$

$$= \frac{\exp(i\hat{k}\cdot\hat{z})}{2|\hat{k}|^{1/2}}\,\varkappa = \frac{i\,\exp\{i\hat{k}\cdot\hat{z} + |z_3|(\omega^2-|\hat{k}|^2)^{1/2}\}}{2(\omega^2-|\hat{k}|^2)^{1/2}},\;\omega > |\hat{k}|,$$

where (see Erdelyi [1], formula (8.6.21))

$$\varkappa = \begin{cases} i\left(\dfrac{\pi|z_3|}{2}\right)^{1/2}|\hat{k}|^{1/2}\,\dfrac{H^{(1)}_{-1/2}(|z_3|\sqrt{\omega^2-|\hat{k}|^2})}{(\omega^2-|\hat{k}|^2)^{1/4}}, & \omega > |\hat{k}| \\[2ex] \left(\dfrac{2|z_3|}{\pi}\right)^{1/2}|\hat{k}|^{1/2}\,\dfrac{K_{1/2}(|z_3|\sqrt{|\hat{k}|^2-\omega^2})}{(|\hat{k}|^2-\omega^2)^{1/4}}, & \omega < |\hat{k}| \end{cases}$$

and

$$H^{(1)}_{-1/2}(z) = \left(\frac{2}{\pi z}\right)^{1/2}\exp(iz),\quad K_{1/2}(z) = \left(\frac{\pi}{2z}\right)^{1/2}\exp(-z).$$

Formula (18) follows from the last three formulas. Let us define

$$k_3 = \begin{cases} (\omega^2 - |\hat{k}|^2)^{1/2}, & |\hat{k}| < \omega \\ i(|\hat{k}|^2 - \omega^2)^{1/2}, & |\hat{k}| > \omega \end{cases} \tag{19}$$

and p_3 in the same way. Then

$$|\hat{k}|^2 + k_3^2 = \omega^2, \quad p_3^2 + |\hat{p}|^2 = \omega^2. \tag{20}$$

Vectors $\hat{k}$ and $\hat{p}$ are real valued but k_3 and p_3 may be complex numbers. If V has compact support (and this was our assumption) then T has compact support as follows from (12). Therefore the Fourier transform of T is meaningful for complex p_3 and k_3. From (17) and (19) one obtains

$$\begin{aligned} \tilde{G}_s = \frac{1}{4k_3p_3} \iint \mathrm{d}z' \, \mathrm{d}z'' \; T(z', z'') \exp(i\hat{k} \cdot \hat{z}' + \\ + ik_3|z_3'| - i\hat{p} \cdot \hat{z}'' + ip_3|z_3''|). \end{aligned} \tag{21}$$

Let us assume that the support of V belongs to the half-space $z_3 < 0$. Then $|z_3| = -z_3$ and (21) takes the form

$$\begin{aligned} &\tilde{G}_s = \frac{1}{4k_3p_3} \tilde{T}(k', p), \quad \text{where} \quad |k| = |p| = \omega, \\ &k' \equiv (\hat{k}, -k_3), \quad |k'| = |k|, \end{aligned} \tag{22}$$

and

$$\tilde{T}(k, p) \equiv \iint \mathrm{d}z' \, \mathrm{d}z'' \; T(z', z'') \exp(ik \cdot z' - ip \cdot z''). \tag{23}$$

The basic problem can be formulated as follows. Given the scattering data $G(\hat{x}, x_3 = 0, \hat{y}, y_3 = 0, \omega)$ for all $\hat{x}$ and $\hat{y}$ and two distinct frequencies ω_j, $j = 1, 2$, or small ω, find $a_j(x)$, $j = 1, 2$.

Formulas (12), (15), and (22) are basic for our first inversion scheme with the data given at two distinct frequencies. Our second inversion scheme is based on the method given in Section 1.

The inversion schemes can be generalized to include the dissipative terms (corresponding to first derivative in time with a coefficient depending on x but not on t in the time-dependent problem). Our method is based on the Born approximation. Since the potential (11) contains second derivatives it does not become small even for $\omega \to 0$.

2.2. Let us describe the first inversion scheme. Formula (15) shows that under the assumption

$$|k| = |p| = \omega, \tag{24}$$

which means that the T-matrix in (22) is known on shell, $\tilde{V}$ can be written as

$$\tilde{V}(k, p) = \omega^2 W(k, p), \quad W = \tilde{a}_1(k - p) - k^0 \cdot p^0 \tilde{a}_2(k - p), \tag{25}$$

where $k^0 = k|k|^{-1}$ and $p^0 = p|p|^{-1}$ are unit vectors which do not depend on ω. Since the functions a_1 and a_2 have compact support their Fourier transforms $\tilde{a}_1$ and $\tilde{a}_2$ are entire functions of the three complex variables k_1, k_2, and k_3, which decay at infinity in the real space in R^3.

In the Born approximation $T = V$ because one neglects the quadratic terms in V in (14), and (22) becomes in this approximation

$$\tilde{G}_s = \frac{1}{4k_3p_3} \tilde{V}(k', p) = \frac{\omega^2}{4k_3p_3} W(k', p). \tag{26}$$

Thus

$$4\tilde{G}_s k_3 p_3 \omega^{-2} = W(k', p). \tag{27}$$

Let

$$k' - p = q, \quad k' + p = s \tag{28}$$

$$k' = (s + q)/2, \quad p = (s - q)/2. \tag{29}$$

Then

$$k' \cdot p = (|s|^2 - |q|^2)/4, \quad s \cdot q = 0, \tag{30}$$

$$k'^0 \cdot p^0 = (|s|^2 - |q|^2)(4\omega^2)^{-1}, \tag{31}$$

$$\left.\begin{aligned} &|q|^2 = 2\omega^2(1 - k'^0 \cdot p^0), \\ &|s|^2 = |k'|^2 + |p|^2 + 2k' \cdot p = 2\omega^2(1 + k'^0 \cdot p^0), \ \omega > 0 \end{aligned}\right. \tag{32}$$

$$|s|^2 - |q|^2 = 4\omega^2 - 2|q|^2. \tag{33}$$

Therefore

$$W \equiv W(k', p) = \tilde{a}_1(q) - \tilde{a}_2(q) \frac{|s|^2 - |q|^2}{4\omega^2}$$
$$= \tilde{a}_1(q) - \tilde{a}_2(q)\left(1 - \frac{|q|^2}{2\omega^2}\right), \quad \omega > 0. \tag{34}$$

Let us take two arbitrary frequencies ω_1 and $\omega_2 \neq \omega_1$ and solve the two equations

$$W_1 = \tilde{a}_1(q) - \tilde{a}_2(q)\left(1 - \frac{|q|^2}{2\omega_1^2}\right) \tag{35}$$

$$W_2 = \tilde{a}_1(q) - \tilde{a}_2(q)\left(1 - \frac{|q|^2}{2\omega_2^2}\right) \tag{36}$$

for $\tilde{a}_1$ and $\tilde{a}_2(q)$. The functions W_j, $j = 1, 2$, are the data according to (27). One obtains

$$\tilde{a}_2 = \frac{2(W_1 - W_2)}{|q|^2} \frac{\omega_1^2\omega_2^2}{\omega_2^2 - \omega_1^2}, \tag{37}$$

$$\tilde{a}_1 = W_1 + (W_1 - W_2) \frac{2\omega_1^2\omega_2^2}{(\omega_2^2 - \omega_1^2)|q|^2}\left(1 - \frac{|q|^2}{2\omega_1^2}\right). \tag{38}$$

Taking the inverse Fourier transform of the functions (37) and (38) one obtains $a_2(x)$ and $a_1(x)$. It is clear from (34) that the following conclusion holds.

Conclusion: One cannot find both functions a_1 and a_2 if the data are given at only one frequency. Indeed, it follows from (34) that for a fixed frequency the data depends on q only, and there is no parameter to vary in order to find both $\tilde{a}_1$ and $\tilde{a}_2$. The conclusion that one can not recover both functions a_j, $j = 1, 2$, from the data given at one frequency is not at all obvious: at first glance one can think that the two conditions $|k| = |p| = \omega$ leave 4 degrees of freedom in the 6-dimensional space $R_k^3 \times R_p^3$ which should be enough to determine two functions $\tilde{a}_j$, $j = 1, 2$, of 3 variables. This argument, however, is not valid as one saw above. The reason is that the function $W(k', p)$ has a very special structure as a function of two vectors k' and p. Our reasoning applies to the theory in the Born approximation. It does not imply that the conclusion holds in exact theory. The exact theory unfortunately is out of reach.

Another important point concerns the numerical problem of inverting the Fourier transforms of $\tilde{a}_j$, $j = 1, 2$. Let $\omega_1 < \omega_2$. It follows from (32) that the real valued vectors q run through the ball $B_{2\omega} = \{q: |q| \leqslant 2\omega\}$ when k'^0 and p^0 run through the unit sphere S^2, so that the number $k'^0 \cdot p^0$ runs through the interval $[-1, 1]$. Therefore, for real valued vectors q formulas (37) and (38) determine the Fourier transforms a_j, $j = 1, 2$, in the ball $B_{2\omega_1}$ only. Since $a_j(x)$ are compactly supported the functions $\tilde{a}_j(q)$ are entire analytic functions of q. Therefore if one knows these functions in a ball $B_{2\omega_1}$, one can uniquely determine these functions everywhere in R^3 by analytic continuation. However, the basic numerical problem is as follows:

PROBLEM: Given the Fourier transform $\tilde{a}(q)$ of a function $a(x)$, $a(x) = 0$ if $|x| \geqslant R$, in the ball $B_{2\omega} = \{q, |q| < 2\omega\}$ compute $a(x)$ in the ball $B_R = \{x: |x| \leqslant R\}$ with a prescribed accuracy.

This problem will be treated in Section 3.

2.3. Let us describe the second inversion scheme. If $\omega \to 0$ the limit of Equation (8) is

$$G(x, y) = g_0(x, y) - \int g_0 \nabla \cdot (a_2 \nabla G)\,\mathrm{d}z, \qquad g_0 - \frac{1}{4\pi|x - y|}. \tag{39}$$

The scattered field $G_s = G - g_0$ in the Born approximation for x and y on the plane $x_3 = 0$ is of the form:

$$G_s(\hat{x}, \hat{y}) = -\frac{1}{16\pi^2} \int \frac{1}{|\hat{x} - z|} \nabla \cdot \left(a_2(z) \nabla \frac{1}{|z - \hat{y}|} \right) \mathrm{d}z. \tag{40}$$

This equation can be solved analytically by the method given in Section 1. One takes the Fourier transform in $\hat{x}$ and $\hat{y}$ of Equation (40), uses formula (1.13), and obtains

$$
\begin{aligned}
f(\lambda, \mu) &\equiv \frac{1}{(2\pi)^4} \iint \mathrm{d}\hat{x}\, \mathrm{d}\hat{y} \exp(i\lambda \cdot \hat{x} + i\mu \cdot \hat{y}) (-16\pi^2 G_s(\hat{x}, \hat{y})) \\
&= \frac{1}{(2\pi)^2} \int \mathrm{d}z \, \frac{\exp(i\lambda \cdot \hat{z} - |\lambda|\,|z_3|)}{|\lambda|} \nabla_z \cdot \\
&\quad \cdot \left(a_2(z) \nabla_z \frac{\exp(i\mu \cdot \hat{z} - |\mu|\,|z_3|)}{|\mu|} \right) \\
&= \frac{-1}{(2\pi)^2 |\lambda|\,|\mu|} \int \mathrm{d}z \exp\{ i(\lambda + \mu) \cdot \hat{z} + \\
&\quad + (|\lambda| + |\mu|) z_3 \} a_2(z) (i\lambda + |\lambda| e_3) \cdot (i\mu + |\mu| e_3) \\
&= \frac{(\lambda \cdot \mu - |\lambda|\,|\mu|)}{(2\pi)^2 |\lambda|\,|\mu|} \int \mathrm{d}z \, a_2(z) \exp(ip \cdot \hat{z} + q \cdot z_3).
\end{aligned} \tag{41}
$$

Here we integrated once by parts, used the equality $|z_3| = -z_3$ and set $p = \lambda + \mu$, $q = |\lambda| + |\mu|$, (p_1, p_2), $p_3 = |\lambda|$, $p_4 = |\mu|$. These p and q should not be confused with p and q in Section 2.2. They are the same as in Section 1.

Define

$$
\begin{aligned}
F(p, q) &\equiv \frac{(2\pi)^2 |\lambda|\,|\mu| f(\lambda, \mu)}{\lambda \cdot \mu - |\lambda|\,|\mu|} = \int \mathrm{d}z \, a_2(z) \exp(ip \cdot \hat{z} + q \cdot z_3) \\
&= \int_{-\infty}^{\infty} \int \mathrm{d}p \exp(ip \cdot \hat{z}) \int_0^{\infty} \mathrm{d}\zeta \; a_2(\hat{z}, -\zeta) \exp(-q\zeta)
\end{aligned} \tag{42}
$$

as in Section 1. Here $\zeta = -z_3$. The function $F(p, q)$ is the datum. Taking the inverse Fourier and Laplace transforms of $F(p, q)$ one finds

$a_2(z) = a_2(\hat{z}, -\xi)$:

$$a_2(z) = \frac{1}{(2\pi)^2} \int_{-\infty}^{\infty} \int \mathrm{d}p \exp(-ip \cdot z) \times$$
$$\times \frac{1}{2\pi i} \int_{c-i\infty}^{c+i\infty} \mathrm{d}q\, F(p, q) \exp(-qz_3), \quad c > 0. \tag{43}$$

If $a_2(z)$ is found, then Equation (8) in the Born approximation, and for $x = (\hat{x}, 0)$, $y = (\hat{y}, 0)$, can be written as

$$-16\pi^2 \lim_{\omega \to 0} \frac{G_s - Q}{\omega^2} = \int \frac{a_1(z)\, \mathrm{d}z}{|\hat{x} - z|\, |\hat{y} - z|} \tag{44}$$

where

$$Q = -\int g(\hat{x}, z) \nabla_z \cdot (a_2(z) \nabla_z g(z, \hat{y}))\, \mathrm{d}z \tag{45}$$

and $a_2(z)$ is given by (43).

Equation (44) was solved in Section 1. Let us summarize the second inversion scheme. First, one computes a_2 by formula (43), then a_1 by solving Equation (44).

2.4. In this section we give sufficient conditions for convergence of the Born series. These conditions can be considered as conditions under which the Born approximation is valid. Error estimates will be obtained.

Consider the equation

$$u - f + \omega^2 T_1 u + T_2 u \tag{46}$$

where

$$T_1 u = \int g a_1 u\, \mathrm{d}z,$$
$$T_2 u = \int g \nabla \cdot (a_2(z) \nabla u)\, \mathrm{d}z. \tag{47}$$

The assumptions about $a_j(x)$ are:

$$\begin{aligned} &a_j(x)=0 \quad \text{if} \quad |x| \geqslant R, \\ &|a_1| \leqslant c_1, \\ &|a_2| + |\nabla a_2| \leqslant c_2. \end{aligned} \tag{48}$$

The Born series for Equation (47) is of the form $u = \sum_{n=0}^{\infty} (\omega^2 T_1 + T_2)^n f$. It converges in some space H if $\|\omega^2 T_1 + T_2\| < 1$ where $\|T\|$ is the norm of a linear operator T acting in this space. Let $H = H^2(B_R)$ be the Sobolev space, $B_R = \{x: |x| \leqslant R\}$. Since the functions $a_j(x)$ vanish outside of B_R the values of the functions $T_j f$ are determined entirely by the values of f on B_R. The following estimate

$$\|gf\|_{2,R} \leqslant c_R \|f\|_{0,R} \tag{49}$$

is known (e.g. see Kantorovich—Akilov [1]). Here $\|u\|_{m,R} = \|u\|_{H^m(B_R)}$. If $\omega \to 0$ the inequality $\|\omega^2 T_1 + T_2\| < 1$ holds if $\|T_1\| \leqslant c$ and $\|T_2\| < 1$. Here and below we denote by c various constants. One has, using (49),

$$\|T_1 f\|_{2,R} \leqslant c_R \|a_1 f\|_{0,R} \leqslant c_R c_1 \|f\|_{0,R} \leqslant c\|f\|_{2,R}. \tag{50}$$

Thus $\|T_1\| \leqslant c$ if $|a_1| \leqslant c_1$. Actually, $\|T_1\| \leqslant c$ if $\|a_1\|_{0,R} \leqslant c_1$. Furthermore

$$\|T_2 f\|_{2,R} \leqslant c_R \|\nabla \cdot (a_2 \nabla f)\|_{0,R} \leqslant c_R \cdot c_2 \|f\|_{2,R}. \tag{51}$$

Therefore $\|T_2\| < 1$ if $c_R c_2 < 1$. Let us summarize the result.

PROPOSITION 1. *If* (48) *holds and* $q \equiv \omega^2 c_1 c_R + c_2 c_R < 1$ *then the Born series* $\sum_{n=0}^{\infty} (\omega^2 T_1 + T_2)^n f$ *converges in* H_R^2 *and is majorized by the series* $\sum_{n=0}^{\infty} q^n$.

2.5. Consider the inverse scattering problem for Equation (7) with the scattering amplitude as the data. If the incident field is the plane wave $u_0 = \exp(i\omega s_0 \cdot x)$, then the scattering amplitude is given by

$$f(s_0, s, \omega) = -\frac{1}{4\pi} \int \exp(-i\omega s \cdot y)\,[a_1(y)u(y)\omega^2 + \nabla \cdot (a_2(y)\nabla u)]\,dy. \tag{52}$$

In the Born approximation $u = u_0$, and, integrating by parts the second term in (52), one obtains for the scattering amplitude

$$f = f_B(s_0, s, \omega) = -\frac{\omega^2}{4\pi} \int \exp[i\omega(s_0 - s) \cdot y] \{a_1(y) - s_0 \cdot s a_2(y)\} \, dy. \tag{53}$$

Consider $D = -4\pi\omega^{-2} f_B$ as the data. Assume that the data are known at $\omega = \omega_j$, $j = 1, 2$, $\omega_1 < \omega_2$. Then the Fourier transforms $\tilde{a}_j$ of $a_j(y)$, $j = 1, 2$, can be found from (53). Indeed, set $\omega(s_0 - s) = \omega p$, where $|p|^2 = 2 - 2\gamma$ and $\gamma = s \cdot s_0 = 1 - |p|^2/2$. Then

$$D_1 = \tilde{a}_1(\omega_1 p) - \left(1 - \frac{|p|^2}{2}\right) \tilde{a}_2(\omega_1 p) \tag{54}$$

$$D_2 = \tilde{a}_1(\omega_2 p) - \left(1 - \frac{|p|^2}{2}\right) \tilde{a}_2(\omega_2 p) \tag{55}$$

Set $q = \omega_1 p$. Then (54) and (55) can be written as

$$D_1(q) = \tilde{a}_1(q) - \left(1 - \frac{|q|^2}{2\omega_1^2}\right) \tilde{a}_2(q) \tag{56}$$

$$D_2\left(\frac{q}{\omega_2}\right) = \tilde{a}_1(q) - \left(1 - \frac{|q|^2}{2\omega_2^2}\right) \tilde{a}_2(q) \tag{57}$$

From (56) and (57) one finds $\tilde{a}_j(q)$ and, by applying the Fourier inversion, $a_j(y)$, $j = 1, 2$. Again the problem formulated at the end of subsection 2.2 arises.

3. AN INVERSION FORMULA IN SCATTERING THEORY

3.1. In this section we solve a three-dimensional inverse scattering problem in the Born approximation, and the problem in Section 2.2.

Let

$$(\nabla^2 + k^2 - q(x))u = 0 \text{ in } R^3 \tag{1}$$

$$u = \exp(iks \cdot x) + v \tag{2}$$

$$v \sim \frac{\exp(ikr)}{r} f(s, s', k) \quad \text{as} \quad r = |x| \to \infty, \quad xr^{-1} = s' \tag{3}$$

The integral equation for u is

$$u = \exp(iks \cdot x) - \int gqu \, dy, \quad g = \frac{\exp(ik|x-y|)}{4\pi|x-y|}. \tag{4}$$

Therefore

$$f(s, s', k) = -\frac{1}{4\pi} \int \exp(-iks' \cdot y) qu \, dy. \tag{5}$$

In the Born approximation ($u = \exp(iks \cdot x)$) one has

$$f = f_B = -\frac{1}{4\pi} \int \exp\{ik(s - s') \cdot y\} q(y) \, dy. \tag{6}$$

Let us assume that

$$q(x) \in L^2(B_R), \quad q(x) = 0 \quad \text{if} \quad |x| > R. \tag{7}$$

Then one obtains for $q(y)$ the integral equation

$$\int_{|y| \leqslant R} \exp\{-ik(s' - s) \cdot y\} q(y) \, dy = F(k(s' - s)), \tag{8}$$

where $F(k(s' - s)) = -4\pi f_B$. In (8) $k > 0$ is fixed and s and s' run through the unit sphere S^2. Let $p = k(s' - s)$. When s and s' run through S^2 vector p runs through the ball $B_{2k} = \{p: |p| \leqslant 2k\}$. Equation (8) can be written as

$$\int_{|y| \leqslant R} \exp(-ip \cdot y) q(y) \, dy = F(p), \quad |p| \leqslant 2k. \tag{9}$$

This equation is solvable in the class of functions q satisfying conditions

(7) iff $F(p) \in W_R$, where W_R is the set of entire functions of exponential type $\leqslant R$ square integrable over R^3, that is of entire functions such that

$$|F(p)| \leqslant c \exp(R|p|), \int_{R^3} |F(p)|^2 \, \mathrm{d}p < \infty. \tag{10}$$

This is a variant of the Paley—Wiener theorem. If

$$\tilde{q}(p) \equiv \int \exp(-ip \cdot y) q \, \mathrm{d}y = 0 \quad \text{for} \quad |p| > 2k_0 \tag{11}$$

then one can find $q(y)$, the solution of Equation (9), by taking the inverse Fourier transform of the function $F(p)$. But (11) is equivalent to the assumption that $q(y)$ is an entire function of class W_{2k_0} which is not the case in applications where assumptions (7) are more realistic. The idea of what follows is to approximate $q(x)$ satisfying conditions (7) by an entire function $q_N(x) \in W_{2k_0}$ with arbitrary accuracy on B_R: $\|q_N - q\|_R \leqslant \varepsilon(N) \to 0$, $N \to \infty$. Here $\|.\|_R$ is the norm in $L^2(B_R)$ or $C(B_R)$ depending on whether $q \in L^2(B_R)$ or $C(B_R)$. The function $q_N(x)$ will be constructed explicitly analytically from the data.

3.2. The data in (8) are given on $S^2 \times S^2$ and it is useful to have a formula which transforms the three-dimensional Fourier integral into an integral over $S^2 \times S^2$. Such a formula in R^d, $d \geqslant 2$, is known (Devaney [1], $d = 2$, Beylkin [1], $d \geqslant 2$):

$$(2\pi)^{-d} \int_{|p| \leqslant 2k} F(p) \exp(ip \cdot x) \, \mathrm{d}p$$

$$= \frac{k^d}{8\pi^d \omega_{d-1}} \int_{S^{d-1}} \int_{S^{d-1}} \times \tag{12}$$

$$\times \frac{|s - s'| \exp\{ik(s - s') \cdot x\} F(k(s - s')) \, \mathrm{d}s \, \mathrm{d}s'}{(4 - |s - s'|^2)^{(d-3)/2}}$$

where $\mathrm{d}s$ is the element of the area of the unit sphere S^{d-1}. The identity (12) holds for any $F \in L^2(B_{2k})$, S^{d-1} is the unit sphere in R^d,

ω_d is the area of S^{d-1}, $\omega_d = 2\pi^{d/2}/\Gamma(d/2)$, where $\Gamma(x)$ is the Gamma function. For $d = 3$ one has

$$(2\pi)^{-3} \int_{|p| \leqslant 2k} F(p) \exp(ip \cdot x) \, \mathrm{d}p$$

$$= \frac{k^3}{(2\pi)^4} \int_{S^2} \int_{S^2} |s - s'| \exp\{ik(s - s') \cdot x\} \times \tag{13}$$

$$\times F(k(s - s')) \, \mathrm{d}s \, \mathrm{d}s'.$$

For $d = 2$ one has

$$(2\pi)^{-2} \int_{|p| \leqslant 2k} F(p) \exp(ip \cdot x) \, \mathrm{d}p$$

$$= \frac{k^2}{16\pi^2} \int_{S^1} \int_{S^1} |s - s'|(4 - |s - s'|^2)^{1/2} \times \tag{14}$$

$$\times \exp\{ik(s - s') \cdot x\} F(k(s - s')) \, \mathrm{d}s \, \mathrm{d}s'.$$

If (11) holds then (8) is solvable by the formula

$$q(x) = (2\pi)^{-3} \int \exp(ip \cdot x) F(p) \, \mathrm{d}p$$

$$= \frac{k^3}{(2\pi)^4} \int_{S^2} \int_{S^2} |s - s'| \exp\{ik(s - s') \cdot x\} \times \tag{15}$$

$$\times F(k(s - s')) \, \mathrm{d}s \, \mathrm{d}s'.$$

where formula (13) was used.

Before we pass to the formula for q_N let us prove the following two lemmas.

LEMMA 1. *Equation* (8) *has no more than one solution.*

LEMMA 2. *Formula* (12) *holds.*

Proof of Lemma 1. Equation (8) is equivalent to (9). If $F(p) = 0$, $|p| \leqslant 2k$ then

$$\int_{|y| \leqslant R} \exp(-ip \cdot y) q(y) \, \mathrm{d}y = 0, \quad p \in R^3. \tag{16}$$

Indeed, the left-hand side in (9) is an entire function of p. If this function vanishes in the ball B_{2k} it vanishes everywhere. From (16) one concludes that $q = 0$. □

Proof of Lemma 2. In this proof we use the identity for the spherical means (see John [1], p. 81). Let us define the spherical mean for a continuous function F at the point x:

$$I(x, r) = \frac{1}{\omega_d} \int_{S^{d-1}} F(x + rs) \, \mathrm{d}s. \tag{17}$$

Clearly, if F is continuous, then

$$I(x, 0) = F(x). \tag{18}$$

Let us define the iterated spherical mean

$$M(x, \lambda, \mu) = \frac{1}{\omega_d^2} \int_{S^{d-1}} \int_{S^{d-1}} F(x + \lambda s + \mu s') \, \mathrm{d}s \, \mathrm{d}s'. \tag{19}$$

One has $M(x, \lambda, \mu) = M(x, \mu, \lambda)$, $M(x, \lambda, 0) = M(x, 0, \lambda) = I(x, \lambda)$, $M(x, 0, 0) = F(x)$. The identity which we need is of the form ($\lambda, \mu > 0$):

$$M(x, \lambda, \mu) = \frac{2\omega_{d-1}}{\omega_d (2\lambda\mu)^{d-2}} \cdot \int_{\mu-\lambda}^{\mu+\lambda} [(r + \mu - \lambda)(r + \mu + \lambda) \cdot (r - \mu + \lambda)(\lambda - r + \mu)]^{(d-3)/2} \, r I(x, r) \, \mathrm{d}r. \tag{20}$$

Set $\lambda = \mu = k$ in (20) to obtain

$$M(x, k, k) = \frac{\omega_{d-1}}{\omega_d k^{2d-4} 2^{d-3}} \cdot \int_0^{2k} (4k^2 - r^2)^{(d-3)/2} r^{d-2} I(x, r)\, \mathrm{d}r. \tag{21}$$

Let $h_y(p) = |p|(4k^2 - |p|^2)^{-(d-3)/2} \hat{F}(p) \exp(ip \cdot y)$, where $\hat{F}(p)$ is a continuous function vanishing outside B_{2k}. Taking the spherical mean of $h_y(p)$ yields:

$$I_y(0, r) = \frac{1}{\omega_d} \frac{r}{(4k^2 - r^2)^{(d-3)/2}} \cdot \int_{S^{d-1}} \hat{F}(rs) \exp(irs \cdot y)\, \mathrm{d}s. \tag{22}$$

In formula (21), the iterated spherical mean of $h_y(p)$ is

$$M(0, k, k) = \frac{\omega_{d-1}}{\omega_d^2 k^{2d-4} 2^{d-3}} \int_0^{2k} \mathrm{d}r\, r^{d-1} \cdot \int_{S^{d-1}} \hat{F}(rs) \exp(irs \cdot y)\, \mathrm{d}s \tag{23}$$

$$= \frac{\omega_{d-1}(2\pi)^d}{\omega_d^2 k^{2d-4} 2^{d-3}} \frac{1}{(2\pi)^d} \int_{B_{2k}} \mathrm{d}p \exp(ip \cdot y) \hat{F}(p).$$

Let

$$F(y) = \frac{1}{(2\pi)^d} \int_{B_{2k}} \mathrm{d}p \exp(ip \cdot y) \hat{F}(p)\, \mathrm{d}p,$$
$$\hat{F}(p) = \int_{R^3} \mathrm{d}y \exp(-ip \cdot y) F(y). \tag{24}$$

Let us write $M(0, k, k)$ for function $h_y(p)$ using definition (19) and then equate the expression to (23). This will lead to (15). Take $F(p) = h_y(p)$ in (19) to obtain

$$M(0, k, k) = \frac{1}{\omega_d^2} \int_{S^{d-1}} \int_{S^{d-1}} |k(s+s')| \cdot$$
$$\cdot (4k^2 - k^2|s+s'|^2)^{-(d-3)/2} \hat{F}(k(s+s')) \cdot \tag{25}$$
$$\cdot \exp\{ik(s+s') \cdot y\} \, ds \, ds'.$$

Compare (25) and (23) to obtain

$$\frac{1}{(2\pi)^d} \int_{B_{2k}} dp \exp(ip \cdot y)\hat{F}(p)$$
$$= \frac{k^d}{8\pi^d \omega_{d-1}} \int_{S^{d-1}} \int_{S^{d-1}} (4 - |s+s'|^2)^{-(d-3)/2} \cdot \tag{26}$$
$$\cdot |s+s'| \hat{F}(k(s+s')) \exp\{ik(s+s') \cdot y\} \, ds \, ds'.$$

If one changes s' for $-s'$ in (26) then one obtains (12). □

3.3. Let us pass to the construction of $q_N(x)$ which approximates the solution $q(y)$ of Equation (8). We assume that the data $F(k(s'-s))$ are exact, so that Equation (8) is solvable. The calculations are given for the general case of d-dimensional space.

Let us multiply (8) by $\exp\{ik(s'-s) \cdot x\} h_N(k(s'-s))$ and integrate over $S^{d-1} \times S^{d-1}$ to obtain

$$\int_{|y| \leqslant R} dy \, q(y) \int_{S^{d-1}} \int_{S^{d-1}} \exp\{ik(s'-s) \cdot$$
$$\cdot (x-y)\} h_N(k(s'-s)) \, ds \, ds'$$
$$= \int_{S^{d-1}} \int_{S^{d-1}} F(k(s'-s)) h_N(k(s'-s)) \cdot \tag{27}$$
$$\cdot \exp\{ik(s'-s) \cdot x\} \, ds \, ds'.$$

Here $h_N(p)$ is some function which will be specified later.

Our problem will be solved if the sequence defined by the formula

$$\delta_N(x) = \int_{S^{d-1}} \int_{S^{d-1}} \exp\{ik(s' - s) \cdot x\} h_N(k(s' - s)) \, \mathrm{d}s \, \mathrm{d}s' \tag{28}$$

is a delta-sequence in $L^2(B_R)$ or $C(B_R)$ in the sense that

$$\|q(x) - q_N(x)\|_R \equiv \varepsilon(N) \to 0 \quad \text{as} \quad N \to \infty. \tag{29}$$

Here

$$q_N(x) = \int_{|y| \leqslant R} \delta_N(x - y) q(y) \, \mathrm{d}y \tag{30}$$

and the norm $\| \cdot \|_R$ in (29) is the $L^2(B_R)$ ($C(B_R)$) norm if $q(x) \in L^2(B_R)$ ($q(x) \in C(B_R)$). One can estimate $\varepsilon(N)$ if some additional information about $q(x)$ is known, e.g. bounds $|q| \leqslant a$, $|\nabla q| \leqslant b$.

Let us choose h_N such that the sequence (28) becomes a delta-sequence in the sense defined above. It follows from (12) that

$$\int_{S^{d-1}} \int_{S^{d-1}} \exp\{ik(s' - s) \cdot x\} h_N(k(s' - s)) \, \mathrm{d}s \, \mathrm{d}s'$$

$$= \frac{8\pi^d \omega_{d-1}}{k^{2d-4}(2\pi)^d} \int_{B_{2k}} \mathrm{d}p \exp(ip \cdot x) \cdot \tag{31}$$

$$\cdot (4k^2 - |p|^2)^{(d-3)/2} |p|^{-1} h_N(p).$$

Let us denote

$$h_N(p)|p|^{-1}(4k^2 - |p|^2)^{(d-3)/2} \frac{8\omega_{d-1}\pi^d}{k^{2d-4}} = A_N(p). \tag{32}$$

Then, comparing (28), (31), and (32) yields

$$\delta_N(x) = \frac{1}{(2\pi)^d} \int_{B_{2k}} \mathrm{d}p \exp(ip \cdot x) A_N(p). \tag{33}$$

Therefore we need to construct a delta sequence in $L^2(B_R)$ of functions $\delta_N \in W_{2k}$. Such a sequence was constructed in Ramm [3]:

$$\delta_N(x) = \left\{ \frac{1}{|B_{2k}|} \int_{|p| \leqslant 2k} \mathrm{d}p \exp\left(\frac{ip \cdot x}{2N+d} \right) \right\}^{2N+d} \times \left(1 - \frac{|x|^2}{4R^2}\right)^N \left(\frac{N}{4\pi R^2}\right)^{d/2}. \tag{34}$$

Here $|B_k|$ is the volume of the ball of radius k, $|B_k| = k^d \pi^{d/2}/\Gamma((d+2)/2)$, and

$$\int_{|p| \leqslant 2k} \mathrm{d}p \exp\left(\frac{ip \cdot x}{2N+d} \right) = \frac{(2k)^{d/2}(2N+d)^{d/2}(2\pi)^{d/2}}{|x|^{d/2}} J_{d/2}\left(\frac{2k|x|}{2N+d} \right) \tag{35}$$

where $J_m(x)$ is the Bessel function. Thus

$$\delta_N(x) = \left(1 - \frac{|x|^2}{4R^2}\right)^N \left(\frac{N}{4\pi R^2}\right)^{d/2} \times \left\{ 2^{d/2}\Gamma\left(\frac{d}{2}+1\right) \frac{J_{d/2}\left(\frac{2k|x|}{2N+d}\right)}{\left(\frac{2k|x|}{2N+d}\right)^{d/2}} \right\}^{2N+d}, \tag{36}$$

where $\Gamma(x)$ is the Gamma function.

In particular, for $d = 3$ one has

$$\delta_N(x) = \left(1 - \frac{|x|^2}{4R^2}\right)^N \left(\frac{N}{4\pi R^2}\right)^{3/2} \left(\frac{\sin\lambda - \lambda\cos\lambda}{\lambda^3/3}\right)^{2N+3}, \quad \lambda = \frac{2k|x|}{2N+3}. \tag{37}$$

Notice that the sequence (34) is a delta sequence on $L^2(B_R)$ but not on $L^2(R^d)$. One can not construct a delta-sequence $\delta_N \in W_{2k}$ on $L^2(R^d)$ because the Fourier transform $\tilde{\delta}_N(p)$ of a function $\delta_N \in W_{2k}$ should have its support in the ball B_{2k} while the support of the Fourier transform of a delta sequence on $L^2(R^d)$ cannot stay as $N \to \infty$ within a fixed compact. It follows from (32) and (33) that

$$A_N(p) = \int_{R^d} \delta_N(x) \exp(-ip \cdot x) \, \mathrm{d}x \tag{38}$$

and

$$h_N(p) = |p|(4k^2 - |p|^2)^{-(d-3)/2} A_N(p) k^{2d-4} (8\pi^d \omega_{d-1})^{-1}. \tag{39}$$

It follows from (27), (28), and (30) that

$$\begin{aligned} q_N(x) = \int_{S^{d-1}} \int_{S^{d-1}} & F(k(s' - s)) h_N(k(s' - s)) \cdot \\ & \cdot \exp\{ik(s' - s) \cdot x\} \, \mathrm{d}s \, \mathrm{d}s'. \end{aligned} \tag{40}$$

Let us formulate the result.

THEOREM 1. *If Equation* (8) *is solvable then formula* (40) *gives an approximate analytical solution to Equation* (8) *in the sense that Equation* (29) *holds. The sequence* $h_N(x)$ *in* (40) *is given by formulas* (39), (38), *and* (36).

Remark 1. If

$$|q| \leqslant a, \quad \text{and} \quad |\nabla q| \leqslant b \tag{41}$$

then, in the norm of $C(B_R)$, the estimate for $\varepsilon(N)$ in (29) is of the form:

$$\varepsilon(N) \leqslant \Gamma\left(\frac{d+1}{2}\right)\left\{\Gamma\left(\frac{d}{2}\right)\right\}^{-1} bRN^{-1/2} + 0(N^{-1}), \quad N \to \infty. \tag{42}$$

This is proved in Ramm [3], p. 211.

3.4. Let $d = 3$. Suppose that the data are given with some error η, so that $F_\eta(k|s' - s))$ is known such that $|F - F_\eta| < \eta$.

If one applies the inversion formula (40) to the noisy data F_η then one obtains a function $q_{N,\eta}$ such that

$$\|q_{N,\eta} - q\|_R \leqslant \|q_{N,\eta} - q_N\|_R + \|q_N - q\|_R \leqslant \eta b_N + \varepsilon(N) \tag{43}$$

where the norm is taken in $C(B_R)$, $\varepsilon(N)$, given by (29), is the error of the inversion formula (40) in the absence of noise, and

$$b_N = \int_{S^2}\int_{S^2} |h_N(k(s' - s))| \, ds \, ds', \tag{44}$$

where, for $d = 3$, from (39) one has $h_N(p) = |p| k^2 A_N(p) (16\pi^4)^{-1}$, $A_N(p)$ is given by (38) and δ_N is given by (37). For a fixed $\eta > 0$ one can find $N = N(\eta)$ for which

$$\alpha(N, \eta) \equiv \eta b_N + \varepsilon(N) = \min. \tag{45}$$

Such $N(\eta)$ does exist because $\varepsilon(N) \to 0$ and $b_N \to \infty$ as $N \to \infty$. For this $N(\eta)$ the error $\alpha(\eta) \equiv \alpha(N(\eta), \eta) \to 0$ as $\eta \to 0$. Therefore

$$\| q_{N(\eta), \eta}(x) - q(x) \|_R \to 0 \quad \text{as} \quad \eta \to 0 \tag{46}$$

so that $q_{N(\eta), \eta}(x)$ is a stable approximation of $q(x)$ on B_R.

For computing b_N the following lemma is useful.

LEMMA 3.

$$\begin{aligned} &\int_{S^{d-1}}\int_{S^{d-1}} h(k(s' - s)) \, ds \, ds' \\ &= \frac{\omega_{d-1}}{2^{d-3} k^{2d-4}} \int_{|p| \leqslant 2k} dp \, h(p) |p|^{-1} (4k^2 - |p|^2)^{(d-3)/2}. \end{aligned} \tag{47}$$

Proof. The proof is similar to the proof of Lemma 2 and therefore is left to the reader.

From (44), and (47) it follows that

$$\begin{aligned} b_N &= \frac{2\pi}{k^2} \int_{|p| \leqslant 2k} dp \, |h_N(p)| \, |p|^{-1} \\ &= \frac{1}{8\pi^3} \int_{|p| \leqslant 2k} |A_N(p)| \, dp. \end{aligned} \tag{48}$$

A rough estimate for b_N, which follows from (48), is

$$b_N \leqslant \frac{|B_{2k}|}{8\pi^3} \cdot \max_{|p| \leqslant 2k} |A_N(p)| \leqslant \frac{4k^3}{3\pi^2} \int_{R^3} |\delta_N(x)| \, \mathrm{d}x. \tag{49}$$

Another estimate for b_N follows directly from (44):

$$b_N \leqslant (4\pi)^2 \max_{|p| \leqslant 2k} |h_N(p)|. \tag{50}$$

Estimate (50) can also be obtained from the rough estimate of the first integral in (48). □

3.5. In this section we solve the following problems which are of interest in many applications.

PROBLEM 1. Given the Fourier transform $F(\omega)$ of a function $f(x) \in L^2$, $f(x) = 0$ if $|x| > R$, on a finite interval $-\omega_0 \leqslant \omega \leqslant \omega_0$, find $f(x)$. We have

$$(2\pi)^{-1} \int_{-R}^{R} f(t) \exp(-i\omega t) \, \mathrm{d}t = F(\omega), -\omega_0 \leqslant \omega \leqslant \omega_0. \tag{51}$$

The data are

$$F(\omega), -\omega_0 \leqslant \omega \leqslant \omega_0, \quad \text{and} \quad R > 0. \tag{52}$$

PROBLEM 2. Let

$$\int_0^R f(t) \exp(-pt) \, \mathrm{d}t = F(p), \quad 0 \leqslant p \leqslant b \tag{53}$$

given

$$R \text{ and } F(p), \quad 0 \leqslant p \leqslant b \tag{54}$$

find $f(t)$.

One can treat Equations (51) and (53) as integral equations of the first kind and solve them numerically using some kind of regularization procedure.

We give some explicit analytical formulas for calculating $f_N(t)$ such that

$$\|f_N(t) - f(t)\| \leqslant \varepsilon(N) \to 0 \quad \text{as } N \to \infty, \tag{55}$$

where $\|\cdot\|$ is the norm in $L^2(-R, R)$.

Solution to Problem 1:

$$f_N(t) = \int_{-\omega_0}^{\omega_0} \mathrm{F}(\omega)\, h_N(\omega) \exp(i\omega t)\, \mathrm{d}\omega \tag{56}$$

where

$$h_N(\omega) = \int_{-\infty}^{\infty} \delta_N(t) \exp(-i\omega t)\, \mathrm{d}t \tag{57}$$

and

$$\delta_N(t) = \left[\frac{N}{4\pi R^2}\right]^{1/2} \left[1 - \frac{t^2}{4R^2}\right]^N \times \left\{\frac{\sin\,[\omega_0 t(2N+q)^{-1}]}{\omega_0 t(2N+q)^{-1}}\right\}^{2N+q}, \quad q \geqslant 1. \tag{58}$$

Formula

$$F_N(\omega) = (2\pi)^{-1} \int_{-R}^{R} f_N(t) \exp(-i\omega t)\, \mathrm{d}t \tag{59}$$

solves the problem of spectral extrapolation, of extrapolation of $F(\omega)$ in the region $|\omega| > \omega_0$.

Let us prove that the function (56) satisfies (55) and thus solves Problem 1. Multiply (51) by $\exp(i\omega t)\ h_N(\omega)$ and integrate over $(-\omega_0, \omega_0)$ to obtain

$$\int_{-R}^{R} \mathrm{d}t'\ f(t')\, \delta_N(t - t') = f_N(t), \tag{60}$$

where

$$\delta_N(t) = (2\pi)^{-1} \int_{-\omega_0}^{\omega_0} h_N(\omega) \exp(i\omega t)\, \mathrm{d}\omega \tag{61}$$

so that formula (57) holds. Formula (55) is valid if $\delta_N(t)$ is a delta sequence in $L^2(-R, R)$. One can easily check that formula (58) defines a delta sequence in $L^2(-R, R)$ which is a sequence of entire functions of t of exponential type $\leqslant \omega_0$, so that by Paley-Wiener theorem the function $h_N(\omega)$ defined by (57) vanishes if $|\omega| > \omega_0$, Im $\omega = 0$. Details can be found in Ramm [3], pp. 211–215, where the estimate $\varepsilon(N) \leqslant AN^{-1/2} + 0(N^{-1})$ is obtained under the assumption $|f| + |f'| \leqslant c$, where c = const and A is an explicitly written constant proportional to c (see formula (42)). □

Solution to problem 2:

$$f_N(t) = \int_0^b F(p)\, H_N\left[p - \frac{b}{2}\right] \exp(pt)\, \mathrm{d}p \tag{62}$$

where

$$H_N(p) = (2\pi)^{-1} \int_{-\infty}^{\infty} \Delta_N(iy) \exp(-iyp)\, \mathrm{d}y, \tag{63}$$

$$H_N(p) = 0 \quad \text{if} \quad |p| > \frac{b}{2}, \tag{64}$$

and

$$\Delta_N(iy) = \left[\frac{N}{4\pi R^2}\right]^{1/2} \left[1 + \frac{y^2}{4R^2}\right]^{N} \times$$

$$\times \left[\frac{\sin \dfrac{yb}{2(2N+q)}}{\dfrac{yb}{2(2N+q)}}\right]^{2N+q}, \quad q \geqslant 1. \tag{65}$$

Let us prove that function (62) satisfies (55) and thus solves Problem 2. Multiply (53) by

$$H_N\left[p - \frac{b}{2}\right] \exp(pt)$$

and integrate in p over $(0, b)$ to get

$$\int_0^R \mathrm{d}t' \; f(t') \, \mathrm{d}_N(t - t') = f_N(t) \tag{66}$$

where

$$\begin{aligned} \mathrm{d}_N(t) &= \int_0^b H_N\left[p - \frac{b}{2}\right] \exp(pt)\, \mathrm{d}p \\ &= \int_{-b/2}^{b/2} H_N(p) \exp(pt)\, \mathrm{d}p \exp\left[\frac{b}{2}\, t\right]. \end{aligned} \tag{67}$$

Let us show that the sequence

$$a_N(t) = \int_{-b/2}^{b/2} H_N(p) \exp(pt)\, \mathrm{d}p \tag{68}$$

is a delta sequence in $L^2(-R, R)$. Then

$$\mathrm{d}_N(t) = a_N(t) \exp\left[\frac{b}{2}\, t\right]$$

is also a delta sequence in $L^2(-R, R)$. It follows from (68) and (63) that

$$\begin{aligned} &a_N(t) = \Delta_N(t), \quad t = iy, \\ &\Delta_N(t) = \left(\frac{N}{4\pi R^2}\right)^{1/2} \left(1 - \frac{t^2}{4R^2}\right)^N \left\{\frac{\sinh\left[tb/(4N + 2q)\right]}{tb/(4N + 2q)}\right\}^{2N+q}, \; q \geqslant 1 \end{aligned} \tag{69}$$

The sequencc $\Delta_N(t)$, defined by formula (69), is a delta-sequence in $L^2(-R, R)$. This is checked in the same way as for the sequence (58). Therefore $a_N(t)$ and $\mathrm{d}_N(t)$ are delta sequences in $L^2(-R, R)$, and the proof is complete. □

Formulas (56) and (62) give an analytical solution to Problems 1 and 2 respectively if the data are exact. If the data are noisy one can use formulas (56) and (62) in order to obtain a stable approximation of $f(t)$. For example, suppose $F_\eta(\omega)$ is given such that $|F_\eta(\omega) - F(\omega)| \leqslant$

η, $-\omega_0 \leqslant \omega \leqslant \omega_0$, then one can find $N(\eta)$ such that formula (56) with $F_\eta(\omega)$ in place of $F(\omega)$ and $N = N(\eta)$ will give $f_\eta(t)$ such that

$$\|f_\eta(t) - f(t)\| \to 0 \quad \text{as} \quad \eta \to 0. \tag{70}$$

Indeed

$$\begin{aligned}\|f_\eta(t) - f(t)\| &\leqslant \|f_\eta - f_{N(\eta)}\| + \|f_{N(\eta)} - f\| \leqslant \\ &\leqslant \left\| \int_{-\omega_0}^{\omega_0} (F_\eta - F) h_{N(\eta)}(\omega) \exp(i\omega t)\, \mathrm{d}\omega \right\| + \varepsilon(N(\eta)) \leqslant \\ &\leqslant \eta c(N(\eta)) + \varepsilon(N(\eta)), \quad c(N) = \int_{-\omega_0}^{\omega_0} |h_N(\omega)|\, \mathrm{d}\omega.\end{aligned} \tag{71}$$

For a fixed η one can find $N = N(\eta)$ for which

$$\eta c(N) + \varepsilon(N) = \min \equiv \alpha(\eta). \tag{72}$$

Since $\varepsilon(N) \to 0$ as $N \to \infty$, and $c(N) \to \infty$, one concludes that if $N(\eta)$ solves problem (72) then $N(\eta) \to \infty$ as $\eta \to 0$, and $\alpha(\eta) \to 0$ as $\eta \to 0$. This argument is similar to the one given in Section 3.4.

Let us summarize the result:

If the data are given with an error η, $|F_\eta - F| \leqslant \eta$, then the function

$$f_\eta(t) = \int_{-\omega_0}^{\omega_0} F_\eta(\omega) h_{N(\eta)}(\omega) \exp(i\omega t)\, \mathrm{d}\omega$$

satisfies (70)*. Here $N(\eta)$ solves problem* (72).

Practically, Problems 1 and 2 are ill posed and the given above stable approximation to $f(t)$, the function $f_\eta(t)$, may be of limited value because the function $c(N)$ grows fast and in order for $\alpha(\eta)$ to become sufficiently small one may have to choose η very small. From the practical point of view this is not always possible since the precision of the measurements may be limited.

4. A MODEL INVERSE PROBLEM OF INDUCTION LOGGING

4.1. Consider a medium properties of which we want to find from the knowledge of the field on the ray $x_3 < 0$. One can think about a borehole, a cylindrical hole in the medium; along the x_3-axis the source and receiver can be moved. They can be kept either at a fixed distance

d, or their positions can vary arbitrarily. The problem is to recover the properties of the medium from the measurements recorded by the receiver.

In induction logging the medium is characterized by the dielectric and magnetic parameters $\varepsilon = 1$, $\mu = \text{const}$, and by the conductivity $\sigma = \sigma_0 + v(X)$, where $\sigma_0 = \text{const}$ and $v(X)$ is the inhomogeneous part of σ. The displacement currents $\partial\mathscr{D}/\partial t$ are negligible. The vector potential A satisfies the equation $(\nabla^2 + k^2)A = -j$, $k^2 = i\omega\mu\sigma(X)$, and j is the source current. Let $X = (\rho, \phi, \zeta)$ be the cylindrical coordinates of the point x. In induction logging it is always assumed that j is a current in a circular loop of wire. The loop is perpendicular to the x_3-axis, it has a small radius a, and its center is at a point $(0, 0, \zeta)$. The vector potential A under these assumptions and under the additional assumption $v = v(\rho, z)$, has only one nonzero component, A_ϕ, in the cylindrical coordinates, $\nabla \cdot A = 0$, $E = i\omega\mu A$, $H = \nabla \times A$.

We consider a model scalar acoustical problem in which the field u satisfies the equation

$$\nabla^2 u + k^2(X)u = -\delta(X - Y), \quad X, Y \in R^3$$

$k^2(X) = i\omega\sigma(x)$, $\sigma(X) = \sigma_0 + v(X)$, $\sigma_0 = \text{const} > 0$, $\omega > 0$, $v(X) = 0$ if $x_3 > 0$ or $|X| > R$, where $R > 0$ is a fixed (arbitrary large) number, $v \in L^2$, $\text{Im}\, k(X) > 0$, $k_0 = (i\omega\sigma_0)^{1/2}$. The integral equation for u is

$$u(X, Y, \omega) = g(X, Y) + i\omega \int g(X, Z)v(Z)u(Z, Y)\,dZ, \tag{0}$$

where

$$g = \frac{\exp(ik_0|X - Y|)}{4\pi|X - Y|}.$$

If ω is sufficiently small then the integral equation is uniquely solvable by iterations and u is analytic in the variable $p = \omega^{1/2}$ in a neighborhood of the point $p = 0$.

Therefore the following limit exists

$$f(X, Y) \equiv \lim_{\omega \to 0} 16\pi^2 \frac{u - g}{i\omega} = \int \frac{v(Z)\,dZ}{|X - Z|\,|Y - Z|}. \tag{1}$$

The function $f(X, Y)$ is the datum. Let $X = (0, 0, x)$, $Y = (0, 0, y)$.

Under various assumptions on $v(Z)$ we give inversion algorithms for recovering $v(Z)$ from the data $f(x, y)$ measured on the x_3-axis.

4.2. Inversion in the one-dimensional case.

Assume that $v(Z) = v(z_1, z_2, z_3) = v(z_3)$, $|z_3| \leqslant R$, $r \leqslant R$, and $v = 0$ otherwise. Here $r = (z_1^2 + z_2^2)^{1/2}$. Let $X = (0, 0, x)$, $Y = (0, 0, y)$, $z_3 = \zeta$, $y = x - d$, $d = \text{const}$. Then (1) takes the form

$$\int_{-\infty}^{\infty} \mathrm{d}\zeta \; v(\zeta) h(x - \zeta) = \psi(x), \quad -\infty < x < \infty \tag{2}$$

where

$$\psi(x) \equiv (2\pi)^{-1} f(x, x - d),$$

$$h(x - \zeta) \equiv \int_0^R \frac{\mathrm{d}r\; r}{[(\zeta - x)^2 + r^2]^{1/2}[(\zeta - x + d)^2 + r^2]^{1/2}}. \tag{3}$$

Equation (2) is of convolution type and can be solved by Fourier's transform. If the data is noisy, i.e. ψ_δ is known instead of ψ, $|\psi_\delta - \psi| \leqslant \delta$ then a regularizing algorithm should be used for numerical solution of (2). For example, one can solve (2) with ψ_δ in the right-hand side by the formula

$$v_\delta(\zeta) = (2\pi)^{-1} \int_{-\infty}^{\infty} \mathrm{d}\lambda \; \exp(i\lambda\zeta) \tilde{\psi}_\delta(\lambda) \tilde{h}^{-1}(\lambda) \exp(-\alpha(\delta)|\lambda|)$$

where $\alpha(\delta) \to 0$ and $\delta\alpha^{-1}(\delta) \to 0$ (see Ramm [3], p. 207 for details and for an optimal choice of $\alpha(\delta)$). One can solve also the Wiener–Hopf type equation

$$\int_0^{\infty} \mathrm{d}\zeta \; v(\zeta) h(x - \zeta) = \psi(x), \quad 0 \leqslant x < \infty \tag{2$'$}$$

which models the case when measurements are taken strictly inside the borehole.

The same method is valid for inversion of the data at a fixed frequency $\omega > 0$ in the Born approximation. The function analogous to $h(x - \zeta)$ in (3) will be $h_\omega(x - \zeta) = \int_0^R \mathrm{d}r \; r \exp(i\omega\beta)\alpha^{-1}$, where $\alpha \equiv [(\zeta - x)^2 + r^2]^{1/2} \cdot [(\zeta - x + d)^2 + r^2]^{1/2} \equiv \alpha_1\alpha_2$, $\beta = \alpha_1 + \alpha_2$.

4.3. The two-dimensional case.

(1) The results of this section show that the data in the Born approximation

$$F(X, Y) \equiv \frac{16\pi^2(u-g)}{i\omega\mu} = \int \frac{\exp\{ik_0(|X-Z| + |Z-Y|)\}v(Z)}{|X-Z|\,|Z-Y|}\,dZ \tag{4}$$

measured for all X, Y belonging to the ζ-axis and at a fixed frequency $\omega > 0$ does determine $v(r, z)$ in some cases. Assume that $v = v(r, z) = 0$ if $r \geqslant R$ or $|z| \geqslant R$.

Let us use the spheroidal coordinates as in Ramm [8]: $z_3 = \ell st + (x_3 + y_3)/2$, $z_2 = \ell(s^2-1)^{1/2}(1-t^2)^{1/2}\cos\theta + (x_2+y_2)/2$, $z_1 = \ell(s^2-1)^{1/2}(1-t^2)^{1/2}\sin\theta + (x_1+y_1)/2$, where $\ell = |X-Y|/2$, $|X-Z| + |Z-Y| = 2\ell s$, $1 \leqslant s < \infty$, $|X-Z| - |Z-Y| = 2\ell t$, $-1 \leqslant t \leqslant 1$, θ is the angle between the plane spanned by $X - Z$ and $Y - Z$ and a fixed plane containing X and Y. The Jacobian of the transformation $Z \to (s, t, \theta)$ is $J = \ell^3(s^2 - t^2)$. In our case X and Y belong to the ζ axis. As the fixed plane we take the plane $z_1 \cdot z_3$, $z_3 = \zeta$, so that θ is the angular variable in the cylindrical coordinate system. in the new variables (4) takes the form:

$$2\pi\ell \int_1^\infty ds\,\exp(2ik_0\ell s)\int_{-1}^1 dt\; v\left(\ell(s^2-1)^{1/2}(1-t^2)^{1/2},\; \ell st + \frac{x+y}{2}\right) = F(x, y) \tag{5}$$

where x and y denote the ζ components of the vectors X and Y, $(x-y)/2 = \ell$ if $x \geqslant y$. From (5) it is clear that under our assumption $v = v(r, z)$ the data depends on $m = (x+y)/2$ and $\ell = |x-y|/2$.

(2) Let us assume that F is given for all frequencies ω i.e. for all k_0, $k_0 = [(1+i)/\sqrt{2}\,]\,\sigma_0^{1/2}\,\omega^{1/2}$. Then the function $p(s, \ell, m) \equiv \int_{-1}^1 dt\; v(\ell(s^2-1)^{1/2}(1-t^2)^{1/2}, \ell st + m)$ can be found from (5) by

taking the inverse Fourier transform in k_0. Therefore one obtains the equation for $v(r, \zeta)$

$$\int_{-1}^{1} v(\ell(s^2-1)^{1/2}(1-t^2)^{1/2}, \ell st + m)\,\mathrm{d}t = p(s, \ell, m), \quad s \geqslant 1, \quad \ell \geqslant 0, \tag{6}$$

where $p(s, \ell, m)$ is known.

Let $\xi = \ell(s^2-1)^{1/2}(1-t^2)^{1/2}$, $\eta = \ell st + m$. Consider the ellipse $E(s, \ell)$ on the plane (ξ, η) given by the equation $\xi^2(s^2-1)^{-1} + (\eta - m)^2 s^{-2} = \ell^2$. Equation (6) can be written as

$$\int_{E(s,\ell) = \{\xi,\eta:\, \xi^2(s^2-1)^{-1} + (\eta-m)^2 s^{-2} = \ell^2\}} w(\xi, \eta)\,\mathrm{d}S = p(s, \ell, m), \quad s \geqslant 1, \quad \ell \geqslant 0. \tag{7}$$

Here $\mathrm{d}S$ is the arc length of the ellipse, $\mathrm{d}S = \ell((s^2-t^2)/(1-t^2))^{1/2}\,\mathrm{d}t$, $w(\xi, \eta) = v(\xi, \eta)((1-t^2)/(s^2-t^2))^{1/2}\, 1/\ell \equiv v(\xi, \eta)a(\xi, \eta)$, and $a(\xi, \eta)$ is the function $((1-t^2)/(s^2-t^2))^{1/2}\, 1/\ell$ expressed in the variables ξ, η, $s^2 + t^2 = \xi^2\ell^{-2} + (\eta - m)^2\ell^{-2} + 1$, $s^2t^2 = (\eta - m)^2\ell^{-2}$. Problem (7) is an integral geometry problem. If w is found from (7) then v is found since $a(\xi, \eta)$ is known. It seems that Equation (7) was not studied in the literature. A similar equation in which the ellipses had one focus at the origin and the other running through one of the coordinate axes was studied in Lavrentiev *et al.* [1], [2], but their results are not applicable to Equation (7) because the families of ellipses are different.

(3) Consider the inversion scheme for the case when $v(r, z) = v(z)$. If $v = v(z)$ does not depend on r then (6) takes the form

$$\int_{-1}^{1} \mathrm{d}t\; v(\ell st + m) = p(s, \ell, m), \quad s \geqslant 1, \quad \ell \geqslant 0. \tag{8}$$

Take $s = m/\ell$ and $m(t+1) = \tau$. Then (8) can be written as

$$\int_0^{2m} v(\tau)\,\mathrm{d}\tau = mp\left(\frac{m}{\ell}, \ell, m\right) \equiv h(m, \ell), \quad \ell \geqslant 0, \quad m \geqslant \ell. \tag{9}$$

From (9) one finds by differentiation that

$$v(2m) = \frac{1}{2}\frac{\mathrm{d}h(m, \ell)}{\mathrm{d}m}, \quad v(z) = \frac{1}{2}\left(\frac{\mathrm{d}h}{\mathrm{d}m}\right)\Big|_{m=z/2}, \quad m \geqslant \ell. \tag{10}$$

Thus, the inversion scheme in this subsection can be summarized as follows. First, take the inverse Fourier transform of the data $F(x, y, k_0)$ in k_0 variable (see Equation (5)0 to obtain (6). Then, take $s = m\ell^{-1}$ in (6) and find $v(z)$ by formula (10). In this scheme the data is assumed to be three-dimensional (a function of ℓ, m and k_0), while the conductivity is assumed to be a function of z only, i.e. a function of one variable. The differentiation in (10) makes the inversion an ill-posed problem. In Ramm [11] an algorithm for stable differentiation of noisy data is given. There is considerable freedom in the choice of the method for solving (8). For example, one could set $m = \ell$ and $s = 1$ and solve (8) as in Section 6 below.

(4) Consider the two-dimensional inhomogeneity $v(r, z)$ and assume that the dependence on r is linear near the borehole, that is $v(r, z) = v(0, z) + (\partial v(0, z))/\partial r\, r \equiv v(z) + w(z)r$. Then (6) can be written as

$$\int_{-1}^{1} v(\ell st + m)\,\mathrm{d}t + \ell \int_{-1}^{1} w(\ell st + m)\,(1 - t^2)^{1/2}\,\mathrm{d}t(s^2 - 1)^{1/2} = p(s, \ell, m), \quad s \geqslant 1, \quad \ell \geqslant 0. \tag{11}$$

Take $s = 1$ in (11) and solve (11) for v. (See Section (6) below.) If $v(z)$ is known then (11) can be written as an equation for w:

$$\int_{-1}^{1} w(\ell st + m)\,(1 - t^2)^{1/2}\,\mathrm{d}t = p_1(s, \ell, m)$$
$$p_1 \equiv [p(s, \ell, m) - \int_{-1}^{1} v(\ell st + m)\,\mathrm{d}t]\,(s^2 - 1)^{-1/2}\ell^{-1}. \tag{12}$$

Equation (12) can be solved for w. Notice, that the function $p_1(s, \ell, m)$ is continuous as $s \to 1$ because of the choice of v. One has, if $m = \ell$, $p_1(1, \ell) \equiv p_1(1, \ell, \ell)$,

$$\int_{0}^{2} w(\ell\tau)\,(1 - (\tau - 1)^2)^{1/2}\,\mathrm{d}\tau = p_1(1, \ell), \ \ell \geqslant 0. \tag{13}$$

Take the Mellin transform of (13) to obtain

$$\hat{w}(u)a(u) = b(u), \quad \hat{w}(u) = b(u)a^{-1}(u) \tag{14}$$

where $\hat{w}(u) \equiv \int_0^\infty w(x)x^{u-1}\,\mathrm{d}x$, $b(u) = \hat{p}_1(1, \ell)$, $a(u) \equiv \int_0^2 \tau^{-u}(1 - (\tau - 1)^2)^{1/2}\,\mathrm{d}\tau$, and the formula $\hat{w}(x\tau) = \hat{w}(u)\tau^{-u}$ was used. To find $w(z)$ take the inverse Mellin transform of $a^{-1}(u)b(u)$.

(5) A general idea for inversion. Let $A = A_0 + i\omega\mu \int gvA\,\mathrm{d}Z$ be Equation (0) corresponding to the source which produces potential A_0 in the absence of the inhomogeneity v. $A - A_0 \equiv A'$ is the scattered field. In the Born approximation one has $A' = i\omega\mu \int gvA_0\,\mathrm{d}Z(*)$. One can try to find such A_0 that (*) can be solved for v.

(6) A way to solve the equation

$$\int_{-1}^{1} v(\ell t + m)\,\mathrm{d}t = p(1, \ell, m), \quad \ell \geqslant 0,$$

is to take $m = \ell$, set $\ell(t+1) = \tau$, write the equation as

$$\int_0^{2\ell} v(\tau)\,\mathrm{d}\tau = \ell p(1, \ell, \ell) \equiv h(\ell),$$

and find

$$v(2\ell) = \frac{1}{2}\frac{\mathrm{d}h(\ell)}{\mathrm{d}\ell}, \quad v(z) = \frac{1}{2}\frac{\mathrm{d}h(\ell)}{\mathrm{d}\ell}\bigg|_{\ell = z/2}.$$

(7) Assume $v = v(r)$. Then (6) becomes

$$\int_{-1}^{1} v(\ell[(s^2 - 1)(1 - t^2)]^{1/2})\,\mathrm{d}t = p(s, \ell), \quad s \geqslant 1, \quad \ell \geqslant 0 \tag{15}$$

(p does not depend on m). This equation can be solved analytically even in the case when s (or ℓ) is fixed. Indeed, let $\ell(s^2 - 1)^{1/2} = z$. Then (15) becomes

$$\int_0^1 v(z(1 - t^2)^{1/2})\,\mathrm{d}t = f(z), \quad f(z) \equiv 0.5\, p(s, z(s^2 - 1)^{-1/2}, \tag{16}$$

$s > 1$ if fixed.

Let $(1-t^2)^{1/2}=y$, $zy=x$. Then (16) becomes

$$\int_0^z v(x)x(z^2-x^2)^{-1/2}\,dx = zf(z), \quad z \geqslant 0. \tag{17}$$

This equation can be solved by the formula

$$v(z)=\frac{1}{z}\,\frac{2}{\pi}\,\frac{d}{dz}\int_0^z \frac{x^2 f(x)\,dx}{\sqrt{z^2-x^2}}. \tag{18}$$

To derive (18) multiply (17) by $2z(y^2-z^2)^{-1/2}$, integrate in z over $(0, y)$, and use the formula $\int_x^y (2z\,dz/[(y^2-z^2)(z^2-x^2)]^{1/2})=\pi$. This is a variant of Abel's equation.

(8) Assume $v(r, z)=v_0(z)+v_1(z)h(r_0-r)$ where $v_0(z)$, $v_1(z)$ are to be found, $r_0>0$ is to be found, $h(x)=1$, $x\geqslant 0$, $h(x)=0$, $x<0$. This model is used in geophysics. Then (6) becomes

$$\int_{-1}^1 v_0(\ell st+m)\,dt + \\ +\int_{-1}^1 v_1(\ell st+km)h(r_0-\ell(s^2-1)^{1/2}(1-t^2)^{1/2})\,dt \tag{19} \\ = p(s,\ell,m).$$

Set $s=1$ and find v_0+v_1 as in Section 6. Take $\ell(s^2-1)^{1/2}$ large, then the second integral in (19) is small and (19) reduces to (8) with $v=v_0$. Solve for v_0. Since v_0+v_1 and v_0 are found both functions v_0 and v_1 are found. Equation (19) can be written as

$$\int_{1\geqslant|t|>\{1-r_0^2\ell^{-2}(s^2-1)^{-1}\}^{1/2}} v_1(\ell st+m)\,dt = p_1(s,\ell,m)\equiv \\ \equiv p(s,\ell,m)-\int_{-1}^1 v_0(\ell st+m)\,dt, \tag{20}$$

where $p_1(s, \ell, m)$ is known. Let $\ell^{-2}(s^2 - 1)^{-1} = \beta$. Then (20) becomes

$$V(m + \ell s\sqrt{1 - r_0^2\beta}) - V(m - \ell s\sqrt{1 - r_0^2\beta}) = p_2(s, \beta, m) \tag{21}$$

where $p_2 = -[p_1(s, \ell, m)\ell s - V(m + \ell s) + V(m - \ell s)]$ and V is an antiderivative of v_1, $V' = v_1$. Choose $\ell s(1 - r_0^2\beta)^{1/2} \ll 1$. This is possible: $\ell s = \ell(1 + \ell^{-2}\beta^{-1})^{1/2}$ and one can take ℓ small and $\beta \sim r_0^{-2}$. Then Taylor's formula and (21) yield

$$2\ell s\sqrt{1 - r_0^2\beta}\, v_1(m) = p_2(s, \beta, m), \ \ell^{-2}(s^2 - 1)^{-1} = \beta. \tag{22}$$

Therefore

$$r_0 = \beta^{-1/2}\left\{1 - \left(\frac{p_2(s, \beta, m)}{2\ell s\, v_1(m)}\right)^2\right\}^{1/2}, \quad \begin{matrix} \ell s(1 - r_0^2\beta)^{1/2} \ll 1. \\ \ell^{-2}(s^2 - 1)^{-1} = \beta. \end{matrix} \tag{23}$$

Let us give an illustrative example of one of the inversion procedures. In this example all the computations are analytical.

Example. Assume that $v(X) = 1$ if $r \leqslant 1, -1 < x_3 < 0$ $v(X) = 0$ otherwise, $\mu = \sigma_0 = 1$, and $d = 0$, so that the positions of the source and the receiver are the same. The inversion procedure from Section 4.2 is based on Equation (2). Under our assumptions this equation is

$$\begin{aligned} &2\pi \int_{-\infty}^{\infty} dz\ v(z) \int_0^1 \frac{r\,dr}{(z - x)^2 + r^2} \\ &= \pi \int_{-\infty}^{\infty} dz\ v(z) \ln\left(1 + \frac{1}{(z - x)^2}\right) = \psi(x). \end{aligned} \tag{24}$$

Let $\tilde{\psi}(y) = \int_{-\infty}^{\infty} \psi(x) \exp(ixy)\, dx$. Taking the Fourier transform of (24) one obtains

$$\begin{aligned} &\pi\tilde{v}(y) \int_{-\infty}^{\infty} \exp(iyx) \ln\left(1 + \frac{1}{x^2}\right) dx \\ &= \pi\tilde{v}(y) 2\pi |y|^{-1}[1 - \exp(-|y|)] = \tilde{\psi}(y) \end{aligned} \tag{25}$$

where formula (1.5.10) from Erdelyi [1] was used. Therefore

$$\begin{aligned}\tilde{v}(y) &= |y|\,\tilde{\psi}(y)\{2\pi^2[1-\exp(-|y|)]\}^{-1} \\ &\equiv \tilde{\psi}/\tilde{h},\ \tilde{h} \equiv |y|^{-1}2\pi^2[1-\exp(-|y|)].\end{aligned} \tag{26}$$

In particular, under the above assumptions about v one has

$$\psi(x) = \int_{-1}^{0} dz \ln\left(1 + \frac{1}{(z-x)^2}\right),$$

$$\tilde{\psi}(y) = \pi \frac{1-\exp(-iy)}{iy}\, 2\pi|y|^{-1}[1-\exp(-|y|)].$$

From (26) and the last formula one finds

$$\tilde{v}(y) = \frac{1-\exp(-iy)}{iy},$$

$$v(x) = (2\pi)^{-1}\int_{-\infty}^{\infty} \exp(-ixy)\tilde{v}(y)\, dy = \begin{cases} 1, -1 < x < 0 \\ 0, \text{otherwise.}\end{cases}$$

Therefore the inhomogeneity is recovered from the exact data analytically. Suppose now that the data are noisy: ψ_δ is given instead of ψ, $\|\psi_\delta - \psi\| \leqslant \delta$, where $\delta > 0$ is known, and $\|\psi\| = (\int_{-\infty}^{\infty} |\psi|^2\, dx)^{1/2}$. Let

$$v_\delta(x) = (2\pi)^{-1}\int_{-\infty}^{\infty} \exp(-iyx)\tilde{\psi}_\delta(y)\frac{|y|\exp(-\alpha(\delta)|y|)}{2\pi^2(1 \quad \exp(-|y|))}\, dy. \tag{27}$$

We will show that under an appropriate choice of $\alpha(\delta)$ the function $v_\delta(x)$ gives a stable approximation to the solution $v(x)$ (corresponding to the exact data) in the sense that $\|v_\delta(x) - v(x)\| \to 0$ as $\delta \to 0$. The regularization parameter $\alpha(\delta)$ will be chosen optimally. Let us assume that the unknown solution $v(x)$ satisfies the inequality $\int_{-\infty}^{\infty} |\tilde{v}|^2(1 + y^2)^b\, dy \leqslant M^2$, where b and M are given positive constants. This assumption means that the solution lies in a certain compact set in L^2. Assume that b > 0. It will be clear soon how this assumption is used. Parseval's equality yields

$$\|v_\delta - v\| = (2\pi)^{-1/2}\|\tilde{v}_\delta - \tilde{v}\| \leqslant$$

$$\leqslant (2\pi)^{-1/2}\left(\|\tilde{\psi}_\delta - \tilde{\psi}\| \max_{y \geqslant 0} \frac{\exp(-\alpha|y|)|y|}{1 - \exp(-|y|)} + \right.$$

$$\left. + \left\| \frac{\tilde{\psi}}{\tilde{h}} (1 + y^2)^{b/2} \right\| \max_{y \geqslant 0} \frac{1 - \exp(-\alpha|y|)}{(1 + y^2)^{b/2}} \right) \leqslant$$

$$\leqslant c(\delta\alpha^{-1} + M\alpha^{\gamma}),\ c = \text{const} > 0,\ \gamma = \max(1, b),\ b > 0.$$

The estimate $(1 - \exp(-\alpha|y|))/(1 + y^2)^{b/2} \leqslant c\alpha^{\gamma}$ holds if $b > 0$. Let us find $\alpha = \alpha(\delta)$ such that $\delta\alpha^{-1} + M\alpha^{\gamma} = \min$. One has $\alpha(\delta) = c_1\delta^{1/(\gamma+1)}$, $c_1 = (\gamma M)^{-1/(\gamma+1)}$. For this optimal $\alpha(\delta)$ the error of the approximate solution is $\|v_\delta - v\| \leqslant \text{const}\ \delta^{\gamma/(\gamma+1)}$, as $\delta \to 0$.

In Coen and Yu [1], Minerbo [1] and Oristaglio [1] one can find many references on the work done in induction logging theory.

Chapter VII

Scattering by Obstacles with Infinite Boundaries

1. STATEMENT OF THE PROBLEM

Let $D \subset R^3$ be an infinite domain with infinite (non-compact) boundary Γ. Consider the problem

$$[\Delta + (k+i\varepsilon)^2]u_\varepsilon = -f \text{ in } D, \quad k>0, \quad \varepsilon>0, \quad f \in C_0^\infty(D) \tag{1}$$

$$u_\varepsilon = 0 \text{ on } \Gamma \tag{2}$$

$$u_\varepsilon \in L^2(D). \tag{3}$$

We prove that the weak solution to problem (1)—(3) exists and is unique for an arbitrary domain D. The weak solution to problem (1)—(3) is a function $u \in L^2(D)$ such that

$$\int \nabla u \nabla\bar{\eta}\, dx - (k+i\varepsilon)^2 \int u\bar{\eta}\, dx = \int f\bar{\eta}\, dx, \quad \forall \eta \in C_0^\infty(D), \quad \int = \int_D. \tag{4}$$

Under some assumptions about D we prove that the limit

$$\lim u_\varepsilon = u, \quad \varepsilon \to +0 \tag{5}$$

exists in an appropriate sense, the function u solves the equation

$$(\nabla^2 + k^2)u = -f \text{ in } D \tag{6}$$

and satisfies condition (2). This statement is the limiting absorption principle for the Dirichlet Laplacian in D.

For domains with boundaries which approach asymptotically the boundary of a cone we prove the eigenfunction expansion theorem similar to the Theorem I.3.1. It is assumed that the boundary of D is starshaped near infinity (i.e., in the region $|x| > R$, where $R > 0$ is an arbitrarily large fixed number) with respect to a point $0 \in \Omega = R^3 \backslash D$. For domains with infinite boundaries we study the theory of compact and smooth perturbations. The problem is to show that if the Dirichlet Laplacian in D has absolutely continuous spectrum and the eigenfunction expansion theorem holds, then this is true for the domain D_1 which has the same boundary as D in the region $B_R = R^3 \backslash B_R$, $B_R = \{x: |x| \leqslant R\}$. We also study the spectral properties of the Dirichlet Laplacian and give criteria for its essential selfadjointness and for the discreteness and continuity of its spectrum.

2. SPECTRAL PROPERTIES OF THE LAPLACIANS

2.1. Let $L_0 = -\Delta$ be the Laplacian defined on $D(L_0) = C_0^\infty(D)$. Clearly L_0 is symmetric and nonnegative on $D(L_0)$. Therefore there exists a self-adjoint nonnegative extension L of L_0 (Friedrichs extension).

The operator L_0 is not essentially selfadjoint if $D \neq R^3$, i.e. its closure $\overline{L}_0$ is not selfadjoint. Indeed, the closure of $C_0^\infty(D)$ in the graph norm $\|L_0 u\| + \|u\|$ consists of the functions which are in $H^2_{\text{loc}}(D)$, while, e.g., in the case of a finite domain D the domain of definition of the Friedrichs extension of the Dirichlet Laplacian is $\mathring{H}^1(D) \cap H^2(D)$. The index of defect of $\overline{L}_0$, $D(\overline{L}_0) = \dot{H}^2(D)$, in the case of a finite domain D is infinite.

For an arbitrary infinite domain D with a smooth boundary Γ, $\Gamma \in C^2_{\text{loc}}$ we prove that $-\Delta$, defined on the set M_0 of functions which vanish in a neighbourhood of infinity and belong to $\mathring{H}^1(D) \cap H^2_{\text{loc}}(\overline{D})$, is essentially selfadjoint.

First we prove that if $D = R^3$, then $L_0 = -\Delta$, $D(L_0) = C_0^\infty(R^3)$, is essentially selfadjoint. Indeed, this will be the case iff the equation

$$(L_0^* + 1)u = 0 \text{ in } R^3, \quad u \in D(L_0^*) \tag{7}$$

has only the trivial solution. Equation (7) means that

$$\int u\overline{(-\Delta h + h)}\, \mathrm{d}x = 0, \quad \forall h \in D(L_0), \quad u \in L^2(R^3), \quad \int = \int_{R^3}. \tag{8}$$

LEMMA 1. *Let* (8) *hold and* $D = R^3$. *Then* $u = 0$ *in* R^3.

Proof. Let $u_\varepsilon = up(\varepsilon x)$ where $p \in C_0^\infty(R^3)$, $0 \leqslant p \leqslant 1$, $p(x) = p(|x|) = \left\{\begin{smallmatrix} 0 \text{ if } |x| \geqslant 2a \\ 1 \text{ if } |x| \leqslant a, \end{smallmatrix}\right. p(r)$ decreases, monotonically on the interval $[a, 2a]$ and $B_{2a} = \{x: |x| \leqslant 2a\}$. Clearly $u_\varepsilon \in D(L_0)$ since any u which satisfies (8) is locally C^∞. One can assume that u and p_ε are real-valued functions. Then

$$\int (up_\varepsilon)^2\,dx \leqslant \int [-\Delta(up_\varepsilon) + up_\varepsilon]up_\varepsilon\,dx$$
$$= \int (-2\nabla u \nabla p_\varepsilon u p_\varepsilon - u^2 \Delta p_\varepsilon \cdot p_\varepsilon)\,dx = \int u^2 (\nabla p_\varepsilon)^2\,dx, \tag{9}$$

where we integrated by parts:

$$-2\int \nabla u \nabla p_\varepsilon u p_\varepsilon\,dx = \int u^2 [p_\varepsilon \Delta p_\varepsilon + (\nabla p_\varepsilon)^2]\,dx.$$

It follows from (9) that

$$\int_{|x| < a\varepsilon^{-1}} u^2\,dx < c^2 \varepsilon^2 \int u^2\,dx \to 0, \quad \text{as} \quad \varepsilon \to 0.$$

Here $|\nabla p(x)| \leqslant c$. It follows from (10) that $u = 0$ in R^3. □

COROLLARY 1. *The operator* $L_0 = -\Delta$ *is essentially selfadjoint on* $C_0^\infty(R^3)$.

Now we turn to the proof of the general statement mentioned above. Assume that Γ is the smooth infinite boundary of an infinite domain $D \subset R^3$, e.g., Γ belongs to C^a_{loc}, $a \geqslant 2$. Let M_0 denote that set of $H^2_{\text{loc}} \cap \mathring{H}^1$ functions vanishing outside of $D_R = D \cap B_R$, where $H^m = H^m(D)$, and $R > 0$ is a number which depends on the function. Let L_1 denote the symmetric operator defined in $H = L^2(D)$ as $L_1 u = (-\Delta + 1)u$ on $D(L_1) = M_0$.

LEMMA 1′. *The operator* L_1 *is essentially selfadjoint.*

Proof. It is sufficient to prove that the equation

$$L_1^* u = 0 \quad \text{or} \quad \int_D u(-\Delta + 1)\phi\,dx = 0, \quad \forall \phi \in M_0, \tag{7′}$$

has only the trivial solution, since in this case the defect index of a positive definite operator L_1 is $(0, 0)$. Since $C_0^\infty(D) \subset M_0$, one concludes from (7′) that $u \in H^2_{\text{loc}}(\bar{D})$, and $(-\Delta + 1)u = 0$ in D (see Berezanskii [1], p. 178, Theorem 4.6). Therefore, taking $\phi \in M_0$, and integrating by parts, one obtains

$$0 = \int_D u(-\Delta + 1)\phi \, dx = \int_D (-\Delta u + u)\phi \, dx - \int_\Gamma u \frac{\partial \phi}{\partial N} \, ds$$
$$= -\int_\Gamma u \frac{\partial \phi}{\partial N} \, ds. \tag{8'}$$

Here we used the assumption $\phi \in \mathring{H}^1$ and the equation $-\Delta u + u = 0$. Actually the integral over Γ is taken over a compact part of Γ on which $\partial\phi/\partial N$ does not vanish. We claim that the set of $\partial\phi/\partial N$ is a dense set in $L^2(\Gamma)$ if ϕ runs through M_0. Indeed, $\partial\phi/\partial N$ can be an arbitrary compactly supported smooth function h on Γ. To see this, one solves the problem

$$L_1^2\phi = 0, \quad \phi = 0 \text{ on } \Gamma_R \cup S_R, \quad \frac{\partial \phi}{\partial N} = h \text{ on } \Gamma_R,$$
$$\frac{\partial \phi}{\partial N} = 0 \text{ on } S_R,$$

where $S_R = \{x: |x| = R, x \in D\}$, and h is smooth on $\Gamma_R \cup S_R$. This Dirichlet boundary value problem is uniquely solvable. This proves the claim. It follows from (8′) that $u = 0$ on Γ. The rest of the proof is the same as the proof of Lemma 1. □

2.2. Let $L = -\Delta$ denote Friedrichs's extension of L_0, that is the Dirichlet Laplacian in D. The following is a simple sufficient condition on D for the spectrum of L to be discrete.

LEMMA 2. *The operator L has a discrete spectrum if $\lambda_1(b) \to \infty$ as $b \to \infty$. Here $\lambda_1(b)$ is the smallest eigenvalue of the Dirichlet Laplacian in the domain $d(b)$ which is the crossection of D by the plane $x_3 = b$. We assume for simplicity that $d(b)$ is a finite domain for any b.*

Proof. By Lemma 5, Appendix 2, L has a discrete spectrum iff the embedding

$$i : \mathring{H}^1(D) \to H^0(D) \tag{11}$$

is compact. Since the embedding $i : \mathring{H}^1(D) \to H^0(D_R)$ is compact for any $R > 0$, $D_R = D \cap B_R$, it is sufficient to show that $\int_{D \setminus D_R} |u|^2 \, dx \to 0$, $R \to \infty$, uniformly with respect to u: $\|u\|_{H^{01}(D)} \leqslant 1$.

Since

$$\int_{x_3 \geqslant b} |\nabla u|^2 \, dx$$

$$= \int_b^\infty dx_3 \int_{d(x_3)} |\nabla u|^2 \, dx_1 \, dx_2 \geqslant \int_b^\infty dx_3 \, \lambda_1(x_3) \int_{d(x_3)} |u|^2 \, dx_1 \, dx_2 \geqslant$$

$$\geqslant \min_{x_3 \geqslant b} \lambda_1(x_3) \int_{D \setminus D_b} |u|^2 \, dx, \text{ and } \min_{x_3 \geqslant b} \lambda_1(x_3) \to +\infty, \text{ as } b \to \infty,$$

the embedding (11) is compact. Lemma 2 is proved. □

Remark 1. A necessary and sufficient condition for L to have a discrete spectrum, i.e. for the compactness of the embedding $i : \mathring{H}^1(D) \to H^0(D)$ is known: the domain D should contain only a finite number of nonintersecting balls of radius r for any $r > 0$ (Molchanov [1]). In applications the result in Lemma 2 is sufficient.

2.3. The following is a simple condition on D for the spectrum $\sigma(L)$ of L to be continuous.

LEMMA 3. *If D contains balls of arbitrary large radii then the spectrum of L is the semiaxis* $[0, \infty)$.

Proof. It is known (Lemma 3, Appendix 2) that λ is a point of continuous spectrum $\sigma_c(L)$ of L if there exists a sequence u_m, which is not compact and for which

$$\|Lu_m - \lambda u_m\| \to 0, \quad m \to \infty, \quad \|u_m\| = 1. \tag{12}$$

As u_m one can take $u_m = \exp[i\sqrt{\lambda}(x - x_m) \cdot n]\varepsilon_m^{3/2} p(\varepsilon_m(x - x_m))$, where n is a unit vector, x_m is the center of the ball $B_m = \{x: \varepsilon_m|x - x_m|$

$\leqslant 1\}$, $\varepsilon_m \to 0$, $B_m \subset D$, and $p(x) \in C_0^\infty$ is a fixed function, $0 \leqslant p(x) \leqslant 1$, $p(x) = 1$ if $|x| \leqslant 1/2$, $p(x) = 0$ if $|x| \geqslant 1$, $\int |p|^2 \, dx = 1$. One has $\int |u_m|^2 \, dx = \varepsilon_m^3 \int |p(\varepsilon_m(x - x_m)|^2 \, dx = \int |p(z)|^2 \, dz = 1$.

Furthermore,

$$(\lambda + \Delta)u_m = 2\nabla \exp[i\sqrt{\lambda}(x - x_m) \cdot n]\nabla p(\varepsilon_m(x - x_m))\varepsilon_m^{3/2} + \\ + \exp[i\sqrt{\lambda}(x - x_m) \cdot n]\Delta p(\varepsilon_m(x - x_m))\varepsilon_m^{3/2},$$

so that (12) holds. □

The following is a simple condition on D for L to have no positive eigenvalues.

LEMMA 4. *Let N be the normal to Γ pointing out of D, and e_3 be the unit vector of the x_3 axis. Assume that D lies in a half-space $x_3 > 0$ and*

$$N \cdot e_3 \leqslant 0 \text{ on } \Gamma. \tag{13}$$

Then L does not have positive eigenvalues.

Proof. Let $u \not\equiv 0$,

$$\Delta u + \lambda u = 0 \text{ in } D, \quad u = 0 \text{ on } \Gamma, \quad u \in L^2(D), \quad \lambda > 0. \tag{14}$$

One can assume that u is real-valued. Integrate the identity

$$0 = (\Delta u + \lambda u) \frac{\partial u}{\partial x_3}$$

$$= \frac{\lambda}{2} \frac{\partial u^2}{\partial x_3} - \frac{1}{2} \frac{\partial}{\partial x_3} (\nabla u)^2 + \sum_{j=1}^{3} \frac{\partial}{\partial x_j} \left(\frac{\partial u}{\partial x_j} \frac{\partial u}{\partial x_3} \right)$$

over the part D_{ob} of D between $x_3 = 0$ and $x_3 = b$ and then use the Gauss formula to obtain

$$0 = \int_{\Gamma_b \cup P_b} \left(\frac{\partial u}{\partial x_3} \frac{\partial u}{\partial N} - \frac{1}{2} (\nabla u)^2 N \cdot e_3 + \frac{\lambda}{2} u^2 N \cdot e_3 \right) ds \tag{15}$$

where Γ_b is the part of Γ between $x_3 = 0$ and $x_3 = b$, and P_b is the part of the plane $x_3 = b$ which lies in D.

Since $u = 0$ on Γ, one has $\nabla u = (\partial u/\partial N)N$ on Γ. Therefore (15) can be written as

$$0 = \int_{\Gamma_b} \left(\frac{\partial u}{\partial N} \right)^2 N \cdot e_3 \, \mathrm{d}s + \int_{P_b} \left[\lambda u^2 + \left(\frac{\partial u}{\partial x_3} \right)^2 - (\hat{\nabla} u)^2 \right] \mathrm{d}s \qquad (16)$$

where $(\hat{\nabla} u)^2 = (\partial u/\partial x_1)^2 + (\partial u/\partial x_2)^2$.

It follows from (13) that

$$-\int_{\Gamma_b} \left(\frac{\partial u}{\partial N} \right)^2 N \cdot e_3 \, \mathrm{d}s \geqslant -\int_{\Gamma_{b_0}} \left(\frac{\partial u}{\partial N} \right)^2 N \cdot e_3 \geqslant d > 0 \quad \text{if} \quad b \geqslant b_0. \qquad (17)$$

Therefore (16) and (17) imply

$$\int_{P_b} \left[\lambda u^2 + \left(\frac{\partial u}{\partial x_3} \right)^2 - (\hat{\nabla} u)^2 \right] \mathrm{d}s \geqslant d > 0. \qquad (18)$$

It follows from (18) that

$$\int_{P_b} [\lambda u^2 + (\nabla u)^2] \, \mathrm{d}s \geqslant d > 0. \qquad (19)$$

Integrate (19) in x_3 to obtain

$$\int_{D_{b_0 b}} [\lambda u^2 + (\nabla u)^2] \, \mathrm{d}x \geqslant \mathrm{d}(b - b_0), \quad b \geqslant b_0, \qquad (20)$$

where $D_{b_0 b} = \{x \colon x \in D,\ b_0 \leqslant x_3 \leqslant b\}$. It follows from (20) that $u \notin L^2(D)$. To see that $u \notin L^2(D)$ if (20) holds, note that if $u \in L^2(D)$ and u solves (14) then $J(b) = \int_{P_b} u^2 \, \mathrm{d}s$ cannot increase for all $b \geqslant b_0$, where $b_0 > 0$ is arbitrary.

Therefore there exists a sequence $b_n \to \infty$, such that

$$J'(b_n) = 2 \int_{P_{b_n}} u \frac{\partial u}{\partial x_3} \, \mathrm{d}s \leqslant 0, \quad b_n \to +\infty. \qquad (21)$$

Since

$$0 = \int_{D_{ob_n}} (\Delta u + \lambda u) u \, \mathrm{d}x$$
$$= \int_{D_{ob_n}} [\lambda u^2 - (\nabla u)^2] \, \mathrm{d}x + \int_{P_{b_n}} u \frac{\partial u}{\partial x_3} \, \mathrm{d}s, \tag{22}$$

it follows from (21) and (22) that

$$\int_{D_{ob_n}} (\nabla u)^2 \, \mathrm{d}x \leqslant \lambda \int_{D_{ob_n}} u^2 \, \mathrm{d}x < \infty. \tag{23}$$

Therefore

$$\int_D [\lambda u^2 + |\nabla u|^2] \, \mathrm{d}x < \infty. \tag{24}$$

Lemma 4 is proved. □

Remark 2. If Γ is a periodic surface with the equation $x_3 = f(x_1, x_2)$, $f(x_1 + T_1, x_2 + T_2) = f(x_1, x_2)$, $|f| \leqslant c$, then condition (13) holds and L does not have positive eigenvalues in the domain $D = \{x\colon x_3 \geqslant f(x_1, x_2)\}$.

LEMMA 5. *Suppose that the problem*

$$\Delta u + k^2 u = 0 \text{ in } D,\ u = 0 \text{ on } \Gamma,\ k > 0 \tag{26}$$

$$\lim_{R \to \infty} \int_{S_R(D)} (|u_r|^2 + k^2 |u|^2) \, \mathrm{d}s = 0,\ S_R(D) = \{s\colon |s| = R, \quad s \in D\} \tag{27}$$

has only the trivial solution. Let one of the following conditions

$$\lim_{R \to \infty} \int_{S_R(D)} |u_r - iku|^2 \, \mathrm{d}s = 0, \tag{28}$$

or

$$\int_D |u_r - iku|^2 (1 + |x|)^{-1} \, dx < \infty, \tag{29}$$

hold. Then the problems (26), (28), *and* (26), (29) *have only the trivial solution.*

Proof. It follows from (28) that

$$\lim_{R\to\infty} \left\{ \int_{S_R(D)} (|u_r|^2 + k^2 |u|^2) \, ds + ik \int_{S_R(D)} (u_r \bar{u} - \bar{u}_r u) \, ds \right\} = 0. \tag{30}$$

It follows from (26) that

$$\lim_{R\to\infty} \int_{S_R(D)} (u_r \bar{u} - \bar{u}_r u) \, ds = 0. \tag{31}$$

Equations (30) and (31) imply (27). Therefore $u = 0$ in D. If (29) holds then

$$\int_0^\infty ds (1 + r)^{-1} \int_{S_r(D)} |u_r - iku|^2 \, ds < \infty.$$

Thus

$$\underline{\lim}_{R\to\infty} \int_{S_R(D)} |u_r - iku|^2 \, ds = 0. \tag{32}$$

Therefore $u \equiv 0$. □

3. SPECTRAL PROPERTIES OF THE DIRICHLET LAPLACIAN IN SEMI-INFINITE TUBES

3.1. Let $D \subset R^3$ be a cylinder if $x_3 \geqslant R$, (with the axis along the x_3-axis) and a bounded domain with a smooth boundary if $x_3 \leqslant R$. Let F_h denote the crossection of D by the plane $x_3 = h$, $F_h = F$ if $h \geqslant R$. Let $D_h(D'_h)$ denote the part of D in the halfspace $x_3 \leqslant h (x_3 \geqslant h)$, $\Gamma_h = \partial D_h$, $\Gamma'_h = \partial D'_h$. We wish to prove two theorems.

THEOREM 1. *The spectrum of the Dirichlet Laplacian* $-\Delta_D$ *in the above*

THEOREM 1. *The spectrum of the Dirichlet Laplacian* $-\Delta_D$ *in the above domain D, the semi-infinite tube, is continuous on the ray* $[\alpha_1, \infty)$ *and discrete on the segment* $[0, \alpha_1)$. *Here* α_1 *is the lowest eigenvalue of the Dirichlet Laplacian in F.*

Let

$$d = \inf\left\{\int_D |\nabla u|^2 \,\mathrm{d}x \left(\int_D |u|^2 \,\mathrm{d}x\right)^{-1}\right\} \tag{1}$$

where inf is taken over the set $\mathring{H}^1(D)$. Assume that

$$\alpha_1 \geqslant d + q, \quad q > 0. \tag{2}$$

THEOREM 2. *If* (2) *holds then* $d > 0$ *is an eigenvalue of* $-\Delta_D$.

3.2. *Proof of Theorem 1.* The boundary ∂D is a compact perturbation of the boundary $\partial D'_R$. By separation of variables one can check that $-\Delta_D$ in D'_R has a continuous spectrum $[\alpha_1, \infty)$. Compact perturbation of the boundary $\partial D'_R$ does not change the continuous spectrum but may introduce a discrete spectrum below α_1 and, possibly, on the ray $[\alpha_1, \infty)$, i.e. isolated eigenvalues of finite multiplicity. Let us prove these statements. First, let us prove that there are at most a finite number of eigenvalues of finite multiplicity of $-\Delta_D$ on any interval $(-\infty, \alpha)$. Suppose there are infinitely many eigenvalues λ_j on this interval, and let u_j be the corresponding normalized eigenfunctions. If $x_3 \geqslant R$ one can write

$$u_j = \sum_{m=1}^{\infty} c_{mj}(x_3) w_m(x_1, x_2), \quad x_3 \geqslant R \tag{3}$$

where $\{w_m\}$ is the complete orthonormal in $L^2(F)$ system of eigenfunctions of the Dirichlet Laplacian $\hat{\Delta}_D$ in F, and $c_{mj}(x_3) = \int_F u_j \bar{w}_m \,\mathrm{d}x_1 \,\mathrm{d}x_2 = (u_j, w_m)_F$ are the Fourier coefficients. Substitute (3) into the equation

$$(\nabla^2 + \lambda_j)u_j = 0, \quad u_j \in L^2(D), \quad \|u_j\|_{L^2(D)} = 1, \tag{4}$$

to get

$$c''_{mj} + (\lambda_j - \alpha_m)c_{mj} = 0, \tag{5}$$

where α_m are the eigenvalues of $\hat{\Delta}_D$. The Parseval equality and the inclusion (4) yield

$$\sup_{1 \leqslant j < \infty} \int_R^\infty \mathrm{d}x_3 \sum_{m=1}^\infty |c_{mj}|^2 \leqslant c < \infty. \tag{6}$$

It follows from (5) that

$$c_{mj}(x_3) = a_{mj} \exp\{(\alpha_m - \lambda_j)^{1/2} x_3\} + b_{mj} \exp\{-(\alpha_m - \lambda_j)^{1/2} x_3\}, \tag{7}$$

where a_{mj} and b_{mj} are constants. Thus, if, say, $\alpha = \alpha_n$, and $\lambda_j \geqslant \alpha_n$, then $c_{mj} = 0$ for $1 \leqslant m \leqslant n$, and if $\lambda_j < \alpha_n$, then $a_{mj} = 0$, $m \geqslant n$. Therefore $\nabla u_j \in L^2(D)$, and

$$\sup_{1 \leqslant j < \infty} \sum_{m=1}^\infty |b_{mj}|^2 (1 + \alpha_m - \lambda_j) \leqslant c < \infty. \tag{8}$$

Inequality (8) and the fact that $\alpha_m \to +\infty$ imply that the set $\{b_j\}$ of the elements of ℓ^2 is compact. Here b_j is the sequence b_{jm}, $m = 1, 2, 3 \ldots$. Therefore the set $\{u_j\}$ is compact in $L^2(D)$ which is a contradiction since $\{u_j\}$ is an orthonormal set in $L^2(D)$. This contradiction shows that on any interval $(-\infty, \alpha)$ there can be only a finite number of eigenvalues of $-\Delta_D$ in D. A similar argument and the result hold for $-\Delta_N$ in D. □

Proof of Theorem 2. Inequality (2) holds if D_R has sufficiently large volume. Indeed, in this case

$$\mathrm{d}_R = \inf_{u \in \mathring{H}^1(D_R)} \left\{ \int |\nabla u|^2 \,\mathrm{d}x \left(\int_{D_R} |u|^2 \,\mathrm{d}x \right)^{-1} \right\} \geqslant d, \tag{9}$$

because any $u \in \mathring{H}^1(D_R)$ can be extended by setting $u = 0$ in D'_R, and the extended $u \in \mathring{H}^1(D)$. If D_R has sufficiently large volume $d_R > 0$ can be as small as we wish, so that $\alpha_1 \geqslant d_R + q \geqslant d + q$, $q > 0$. By Theorem 1, any point of the spectrum of $-\Delta_D$ which is smaller than α_1 is an eigenvalue of finite multiplicity. Therefore we should prove that d is a point of the spectrum of $-\Delta_D$. But this follows from the following lemma.

LEMMA 1. *Let A be a selfadjoint operator in a Hilbert space H, and $d = \inf(Au, u)$, $\|u\| = 1$, $d > -\infty$. Then $d \in \sigma(A)$, where $\sigma(A)$ is the spectrum of A.*

Proof. Without loss of generality assume that $d = 0$. Otherwise, consider the operator $A - dI$ if $d > -\infty$. Assume that $0 \notin \sigma(A)$. Then there is a neighborhood $|\lambda| < \varepsilon$, where $\varepsilon > 0$ is sufficiently small, which does not contain any points of $\sigma(A)$. Therefore $A \geqslant \varepsilon I$ and $\inf(Au, u) \geqslant \varepsilon > 0$, $\|u\| = 1$. This is a contradiction since $\inf(Au, u) = 0$, $\|u\| = 1$. □

Remark 1. It is not known if there can be eigenvalues of $-\Delta_D$ on the ray (α_1, ∞). If the boundary condition is the Robin condition at least on a part of the boundary, then eigenvalues on the continuous spectrum of the corresponding Laplacian can occur.

EXAMPLE. Let $D = \{(x_1, x_2), 0 \leqslant x_1 < \infty, 0 \leqslant x_2 \leqslant \pi\}$, and the boundary condition is

$$u = 0 \text{ if } x_2 = 0 \text{ or } x_2 = \pi, -\frac{\partial u}{\partial x_1} = \alpha u \text{ if } x_1 = 0, \alpha > 0. \tag{10}$$

Here $\alpha_1 = 1$. Let $u = \exp(-\alpha x_1) \sin nx_2$. Then

$$\Delta u + (n^2 - \alpha^2)u = 0 \text{ in } D.$$

The boundary condition (10) is satisfied. Clearly $u \in L^2(D)$ if $\alpha > 0$. Thus $n^2 - \alpha^2$ is an eigenvalue of the Laplacian in D defined by the boundary condition (10). Therefore any real point λ can be an eigenvalue of this Laplacian with an appropriate α.

4. ABSENCE OF POSITIVE EIGENVALUES FOR THE DIRICHLET LAPLACIAN UNDER LOCAL ASSUMPTIONS AT INFINITY

4.1. The aim of this section is to establish that under certain assumptions on the boundary Γ'_R the Dirichlet Laplacian L in D has no positive eigenvalues. The number R can be arbitrarily large. Therefore only local assumptions at infinity are imposed on Γ. Any compact part of Γ is an arbitrary sufficiently smooth (say Liapunov type) surface. The local smoothness is not necessary but we will not go into details.

Let us first assume that

$$x \cdot N \leqslant 0 \text{ on } \Gamma'_R = \{x : x \in \Gamma, |x| \geqslant R\}. \tag{1}$$

THEOREM 1. *If* (1) *holds than L has no positive eigenvalues.*

Theorem 1 holds for general second-order elliptic operators (see

Bibliographical notes). Geometrically condition (1) means that every ray which crosses Γ'_R remains in D. This means that Γ'_R is illuminated by a point source situated at the origin.

In order to prove Theorem 1 we need some lemmas.

Let

$$(\nabla^2 + k^2)u = 0 \text{ in } D, \quad k^2 > 0, \quad u = 0 \text{ on } \Gamma'_R. \tag{2}$$

In the spherical coordinates Equation (2) takes the form

$$u'' + \frac{2}{r} u' - \frac{Au}{r^2} + k^2 u = 0, \quad u' = \frac{\partial u}{\partial r}, \tag{3}$$

where $A = -\Delta^*$ is the angular part of the Laplacian,

$$A = -(\sin\theta)^{-1} \frac{\partial}{\partial\theta} \sin\theta \frac{\partial}{\partial\theta} - (\sin\theta)^{-2} \frac{\partial^2}{\partial\phi^2}$$

Let $v = ru$. Then (3) becomes

$$v'' - r^{-2}Av + k^2 v = 0 \tag{4}$$

Let $v_m = r^m v$. Then

$$v''_m - 2mr^{-1}v'_m + [m(m+1) - A]r^{-2}v_m + k^2 v_m = 0. \tag{5}$$

Let

$$F(m, t, r) = \|v'_m\|^2 + (k^2 - tr^{-1} + m(m+1)r^{-2})\|v_m\|^2 - \\ - r^{-2}(Av_m, v_m) \tag{6}$$

where m and t are positive parameters,

$$\|v\| = \|v\|_r = (v, v)_r^{1/2}, \quad (v, f)_r = (v, f) = \int_{S_r^2} v\bar{f}\, ds. \tag{7}$$

Here S_r^2 is an open set on the unit sphere S^2 which is the central projection of $S_r(D) = S_r \cap D$ onto S^2. Condition (1) implies that

$$S_{r_2}^2 \supseteq S_{r_1}^2 \quad \text{if} \quad r_2 \geqslant r_1 \geqslant R. \tag{8}$$

Therefore the family of Hilbert spaces $L^2(S_r) = H_r$ depends on the parameter r and $H_{r_2} \supseteq H_{r_1}$ if $r_2 \geqslant r_1 \geqslant R$. The operator A is symmetric on the spaces $\mathring{H}^2(S_r^2)$.

Let $0 \leqslant s \leqslant 2$ be a real number. Let us compute the derivative

$$\begin{aligned}\frac{\mathrm{d}}{\mathrm{d}r}(r^s F) = sr^{s-1}F + r^s\{&2\operatorname{Re}(v''_m, v'_m) + \\ &+ 2(k^2 - tr^{-1} + m(m+1)r^{-2})\operatorname{Re}(v'_m, v_m) + \\ &+ (tr^{-2} - 2m(m+1)r^{-3})\|v_m\|^2 + 2r^{-3}(Av_m, v_m) - \\ &- r^{-2}(Av'_m, v_m) - r^{-2}(Av_m, v'_m)\} + J,\end{aligned} \tag{9}$$

where

$$\begin{aligned}J = \lim_{\Delta r \to 0} \frac{1}{\Delta r}\int_{\Delta S_r^2} &\{|v'_m|^2 + (k^2 - tr^{-1} + \\ &+ m(m+1)r^{-2})|v_m|^2 - r^{-2}Av_m\bar{v}_m\}\,\mathrm{d}s \end{aligned} \tag{10}$$

$$= \lim_{\Delta r \to 0}\frac{1}{\Delta r}\int_{\Delta S_r^2} |v'_m|^2\,\mathrm{d}s \geqslant 0, \quad r \geqslant R.$$

Here $\Delta S_r^2 = S_{r+\Delta r}^2 \backslash S_r^2$, and the limit of the last two terms is zero because $v_m = 0$ on Γ. Substituting v''_m from (5) into (9) yields

$$\frac{\mathrm{d}}{\mathrm{d}r}(r^s F) = F_1 + F_2 + J, \tag{11}$$

where

$$\begin{aligned}F_1 = r^{s-1}\{&(4m+s)\|v'_m\|^2 + (sk^2 + \\ &+ (1-s)r^{-1}t + (s-2)m(m+1)r^{-2})\|v_m\|^2 - \\ &- 2t\operatorname{Re}(v'_m, v_m) + (2-s)r^{-2}(Av_m, v_m)\},\end{aligned} \tag{12}$$

and

$$F_2 = r^{s-2}[(v'_m, Av_m) - (Av'_m, v_m)]. \tag{13}$$

Integrating by parts yields

$$\begin{aligned}F_2 &= -r^{s-2}\int_{\partial S_r^2} \frac{\partial v_m}{\partial n}\, v'_m\,\mathrm{d}\ell \\ &= -r^{s-2}\int_{\partial S_r^2} |\nabla v_m|^2 (n \cdot N)(N \cdot x^0)\,\mathrm{d}\ell.\end{aligned} \tag{14}$$

Here $x^0 = x|x|^{-1}$, n is the normal to $\partial S_r(D)$ pointing out of $S_r(D)$, N is the normal to Γ pointing out of D, ∇v_m on Γ is directed either along N or in the opposite direction because $v_m = 0$ on Γ. Since N enters twice in the integrand in (14) it does not matter whether $\nabla v_m = N|\nabla v_m|$ on Γ or $\nabla v_m = -N|\nabla v_m|$ on Γ.

Any condition on Γ'_R which implies that

$$(N \cdot n)(N \cdot x^0) \leqslant 0 \tag{15}$$

is sufficient for the inequality

$$F_2 \geqslant 0 \tag{16}$$

to hold. In particular if (1) holds, then $n \cdot N \geqslant 0$ and (16) holds. Let $m \geqslant m_0$, $r \geqslant r_0$, and $s = 2$, where m_0 and r_0 are sufficiently large. Then

$$F_1 \geqslant 0 \quad \text{for all} \quad m \geqslant m_0, \quad r \geqslant r_0, \quad s = 2. \tag{17}$$

Indeed, $(Av_m, v_m)(2 - s)r^{-2} = 0$ if $s = 2$, and the remaining part of F_1 is nonnegative if

$$(4m + s) \cdot (sk^2 + (1 - s)r^{-1}t + (s - 2)m(m + 1)r^{-2}) \geqslant t^2 \tag{18}$$

and

$$4m + s > 0, \quad sk^2 + (1 - s)r^{-1}t + (s - 2)m(m + 1)r^{-2} > 0. \tag{19}$$

Since $s = 2$, $k > 0$, $m \geqslant m_0$, $r \geqslant r_0$, and m_0, r_0 are sufficiently large, inequalities (18) and (19) hold. Thus (17) holds. We have proved the following lemma.

LEMMA 1. *If* (1) *holds and* $t = t_0 > 0$ *is fixed then*

$$\frac{\mathrm{d}}{\mathrm{d}r}(r^2 F) \geqslant 0 \quad \text{for all} \quad m \geqslant m_0, \quad r \geqslant r_0. \tag{20}$$

Let us prove

LEMMA 2. *There exist* m_0, t_0 *and* r_0 *such that*

$$F(m_0, t_0, r) > 0 \quad \textit{for all} \quad r \geqslant r_0. \tag{21}$$

Proof. We have

$$F(m, t, r) = r^{2m}\{\|v' + mr^{-1}v\|^2 + \\ + (k^2 - tr^{-1} + m(m + 1)r^{-2})\|v\|^2 - r^{-2}(Av, v)\}. \tag{22}$$

Since $v \not\equiv 0$, there is a $r_1 > r_0$ such that $\|v(r_1)\| > 0$. Therefore $F(m, t, r_1) > 0$ for all sufficiently large m. We can increase m_0, if necessary, and then $F(m, t, r_1) > 0$ for all $m \geqslant m_0$. By Lemma 1 this implies that $r^2 F(m, t_0, r) \geqslant r_1^2 F(m, t_0, r_1) > 0$ for all $r \geqslant r_1$. Lemma 2 is proved. □

Now we can prove

LEMMA 3. *If $\|v(r)\|_r^2$ does not increase monotonically for $r \geqslant R$, where $R > 0$ is arbitrarily large fixed number, then there exists a sequence $r_n \to +\infty$ such that*

$$F(0, 0, r_n) > 0. \tag{23}$$

Proof. Since $\|v(r)\|_r^2$ does not increase monotonically for $r \geqslant R$, there exists a sequence $r_n \to +\infty$ such that

$$\frac{\mathrm{d}}{\mathrm{d}r} \|v(r)\|_r^2 = 2\,\mathrm{Re}(v', v)_r \leqslant 0 \quad \text{at} \quad r = r_n. \tag{24}$$

It follows from (24) that

$$\|v' + mr^{-1}v\|^2 = \|v'\|^2 + m^2 r^{-2}\|v\|^2 + 2mr^{-1}\,\mathrm{Re}(v', v) \leqslant \\ \leqslant \|v'\|^2 + m^2 r^{-2}\|v\|^2 \quad \text{at} \quad r = r_n. \tag{25}$$

Thus

$$F(m, t, r) \leqslant r^{2m}\{\|v'\|^2 + (k^2 - tr^{-1} + \\ + m(2m+1)r^{-2})\|v\|^2 - r^{-2}(Av, v)\}, \quad r = r_n. \tag{26}$$

This and (21) imply

$$0 < r^{2m_0}\{\|v'\|^2 + k^2\|v\|^2 - r^{-2}(Av, v)\} = r^{2m_0}F(0, 0, r), \quad r = r_n \tag{27}$$

provided that r_n is sufficiently large so that $-tr_n^{-1} + m_0(2m_0 + 1)r_n^{-2} < 0$.

Inequality (23) follows immediately from (27). □

LEMMA 4. *For $\varepsilon > 0$, however small, one has*

$$\lim_{r \to \infty} r^{\varepsilon}(\|v'\|^2 + k^2\|v\|^2) = \infty. \tag{28}$$

Moreover

$$\liminf_{r \to \infty} (\|v'\|^2 + k^2\|v\|^2) > 0 \tag{29}$$

provided that $v \not\equiv 0$.

Proof. Since $(Av, v) \geqslant 0$ one has

$$F(0, 0, r) \leqslant \|v'\|^2 + k^2\|v\|^2. \tag{30}$$

One can assume that $\|v\|^2 = \|v(r)\|_r^2$ does not increase monotonically for $r \geqslant R$ (since if it does then (28) and (29) are obvious). Therefore, by Lemma 3, there exists $r = r_1$, as large as we wish, such that $F(0, 0, r_1) > 0$. Let $s > 0$ be arbitrarily small.

Inequalities (18) and (19) hold for $m = t = 0$ provided that r is sufficiently large. Therefore

$$\frac{\mathrm{d}}{\mathrm{d}r} r^s F(0, 0, r) \geqslant 0, \quad r \geqslant r_1. \tag{31}$$

Thus

$$r^s F(0, 0, r) \geqslant r_1^s F(0, 0, r_1) > 0, \quad r \geqslant r_1. \tag{32}$$

Inequalities (32) and (30) imply (28). To prove (29) set $s = 0$ in (11). Then (10) and (16) are still valid, and $F_1 = 2r^{-2}(Av, v) \geqslant 0$. Thus

$$\frac{\mathrm{d}}{\mathrm{d}r} F(0, 0, r) \geqslant 0. \tag{33}$$

From (33), (23), and (30) the inequality (29) follows. □

There are at least two ways to proceed with the proof of Theorem 1. One way is to prove the following lemma, which is of independent interest.

LEMMA 5. *If* $u \in L^2(D)$ *and conditions* (1), (2) *hold, then*

$$\nabla u \in L^2(D). \tag{34}$$

The second way is based on

LEMMA 6. *If* $v \not\equiv 0$ *solves* (4) *and* (1) *holds then*

$$\lim_{R \to \infty} \int_R^{R+b} \|v\|^2 \,\mathrm{d}r > 0 \tag{35}$$

for any $b > 0$.

Let us first prove Theorem 1 assuming either Lemma 5 or Lemma 6.

Proof of Theorem 1. (a) If u solves (2), (1) holds, and $u \in L^2(D)$, then, by Lemma 5, $\nabla u \in L^2$. Thus

$$\int^{\infty} \mathrm{d}r\, r^2(\|u\|^2 + \|u'\|^2) < \infty. \tag{36}$$

Therefore

$$\int^{\infty} \mathrm{d}r(\|v\|^2 + \|v'\|^2) < \infty. \tag{37}$$

Indeed, $v = ru$, $u' = r^{-1}v' - r^{-2}v$, $v' = ru' + r^{-1}v$, $\|v'\| \leqslant r\|u'\| + r^{-1}\|v\|$, so that (36) implies (37).

If (37) holds and $v \not\equiv 0$ then

$$\liminf_{r \to \infty} (\|v\|^2 + \|v'\|^2) = 0. \tag{38}$$

This is impossible because of (29). Therefore $u \in L^2(D)$ implies that $u \equiv 0$. Theorem 1 is proved.

(b) If u solves (2), (1) holds, and $u \in L^2(D)$, then $\int^{\infty} \mathrm{d}r\|v\|^2 < \infty$, and therefore

$$\liminf_{r \to \infty} \int_r^{r+b} \|v\|^2\, \mathrm{d}r = 0, \quad b > 0. \tag{39}$$

This is impossible because of (35). Theorem 1 is proved.

We now prove Lemma 5. □

Proof of Lemma 5. Multiply (2) by $\bar{u}$ and integrate over D_r to obtain

$$\int_{D_r} |\nabla u|^2\, \mathrm{d}x = k^2 \int_{D_r} |u|^2\, \mathrm{d}x + \int_{\Gamma_R} \bar{u} \frac{\partial u}{\partial N}\, \mathrm{d}s + \int_{S_r(D)} \bar{u} \frac{\partial u}{\partial r}\, \mathrm{d}s. \tag{40}$$

Since $u = 0$ on Γ'_R one can take the second integral on the right side of (40) over Γ_R rather than over Γ_r. It follows from (40) that in order to prove (34) it is sufficient to prove that

$$\liminf_{r \to \infty} \operatorname{Re} \int_{S_r(D)} \bar{u} \frac{\partial u}{\partial r}\, \mathrm{d}s < \infty. \tag{41}$$

Let

$$f(r) = \operatorname{Re} \int_{S_r(D)} \bar{u} \frac{\partial u}{\partial r} \, ds$$
$$= \frac{1}{2} \left(\frac{d}{dr} \int_{S_r(D)} |u|^2 \, ds - 2r \int_{S_r^2} |u|^2 \, d\omega \right). \tag{42}$$

Here $ds = r^2 \, d\omega$ and $d\omega$ is an element of the area of S_r^2. Thus

$$\frac{1}{2} \frac{d}{dr} \int_{S_r(D)} |u|^2 \, ds = f(r) + r \int_{S_r^2} |u|^2 \, d\omega \geqslant f(r). \tag{43}$$

Therefore (41) will be proved if we show that

$$\liminf_{r \to \infty} \frac{d}{dr} \int_{S_r(D)} |u|^2 \, ds < \infty. \tag{44}$$

Denote $h(r) = \int_{S_r(D)} |u|^2 \, ds$. The assumption $u \in L^2(D)$ implies that

$$\int^{\infty} dr h(r) < \infty. \tag{45}$$

If (44) does not hold then

$$\frac{dh}{dr} = \frac{d}{dr} \int_{S_r(D)} |u|^2 \, ds \geqslant \alpha > 0, \quad r \geqslant r_0 \tag{46}$$

where α is a constant and r_0 is a sufficiently large positive number. Inequality (46) cannot hold because of (45). Therefore (44) holds. Thus, (41) is valid, and Lemma 5 is proved. □

Remark 1. Lemma 5 remains valid if the boundary condition $u = 0$ on Γ'_R is replaced by either

$$\frac{\partial u}{\partial N} = 0 \text{ on } \Gamma'_R, \tag{47}$$

or

$$\frac{\partial u}{\partial N} + \eta u = 0 \text{ on } \Gamma'_R, \quad \eta \geqslant 0. \tag{48}$$

The proof is essentially the same, but Equation (43) should be modified:

$$\frac{\mathrm{d}h}{\mathrm{d}r} = 2f + 2r\int_{S_r^2} |u|^2 \,\mathrm{d}\omega + \lim_{\Delta r \to \infty} \frac{1}{\Delta r}\int_{\Delta S_r(D)} |u|^2 \,\mathrm{d}s \tag{49}$$

where $\Delta S_r(D) = S_{r+\Delta r}(D)\backslash S_r(D)$. If (1) holds then the last term in (49) is nonnegative, and the proof of Lemma 5 remains valid. If condition (48) holds then the integral

$$\int_{\Gamma_r} \bar{u}\frac{\partial u}{\partial N}\,\mathrm{d}s = \int_{\Gamma_R} \bar{u}\frac{\partial u}{\partial N}\,\mathrm{d}s - \int_{\Gamma_r\backslash\Gamma_R} \eta|u|^2\,\mathrm{d}s \leqslant \int_{\Gamma_R} \bar{u}\frac{\partial u}{\partial N}\,\mathrm{d}s, \tag{50}$$

so that the argument in the proof of Lemma 5 is valid for the boundary condition (48) as well.

It is more difficult to prove Lemma 6.

Proof of Lemma 6. One has

$$\begin{aligned}\frac{\mathrm{d}^2\|v\|^2}{\mathrm{d}r^2} &= 2\,\mathrm{Re}\,\frac{\mathrm{d}}{\mathrm{d}r}(v', v) = 2\,\mathrm{Re}[(v'', v) + (v', v')] \\ &= 2\|v'\|^2 + 2r^{-2}(Av, v) - 2k^2\|v\|^2.\end{aligned} \tag{51}$$

Therefore

$$\begin{aligned}\frac{\mathrm{d}^2\|v\|^2}{\mathrm{d}r^2} + 4k^2\|v\|^2 &= 2\,(\|v'\|^2 + k^2\|v\|^2 + r^{-2}(Av, v)) \geqslant \\ &\geqslant c > 0.\end{aligned} \tag{52}$$

The last inequality follows from (29) and the inequality $(Av, v) \geqslant 0$. The conclusion of Lemma 6 follows immediately from (52) and the following lemma. □

LEMMA 7. *Let* $y(t) \geqslant 0$ *and*

$$y'' + k^2 y \geqslant c > 0, \quad t > t_0 \tag{53}$$

where $t_0 \geqslant 0$ *is a number. Then*

$$\liminf_{R \to \infty} \int_R^{R+b} y\,\mathrm{d}t > 0 \tag{54}$$

for any $b > 0$.

The main difficulty in the proof of Lemma 6 is the proof of Lemma 7. If one takes $y = \|v\|^2$, then (52) and Lemma 7 imply (35).

Proof of Lemma 7. Let $z(t) = y(t) - k^{-2}c \geqslant -k^{-2}c$.

Then

$$z'' + k^2 z \geqslant 0. \tag{55}$$

The graph of any solution to this inequality consists of an alternating succession of negative and nonnegative parts. The length of the interval corresponding to a negative (nonnegative) part is not larger (smaller) than πk^{-1}. The area under a negative part is not larger than the area under each of its two neighboring positive parts. These properties we will establish later. Given these properties one sees that the area under any negative part of the graph of z is not larger than $\pi k^{-1} \cdot k^{-2}c = \pi c k^{-3}$, and

$$\int_R^{R+b} z(t)\,\mathrm{d}t \geqslant -\pi c k^{-3} \quad \text{for any} \quad b > 0. \tag{56}$$

Therefore

$$\int_R^{R+b} y(t)\,\mathrm{d}t \geqslant k^{-2}cb - \pi c k^{-3} = k^{-2}c(b - \pi k^{-1}). \tag{57}$$

If $y \geqslant 0$ and (53) holds with some $k > 0$ then it certainly holds for any larger value of k, say $k_1 > k$, and the inequality (57) holds with k_1 in place of k.

Fix any $b > 0$ and choose $k_1 > \pi b^{-1}$. Then inequality (57) (with k_1 in place of k) yields

$$\int_R^{R+b} y(t)\,\mathrm{d}t \geqslant k_1^{-2}c(b - \pi k_1^{-1}) > 0. \tag{58}$$

Since the middle term in (58) increases when b increases, one obtains (54) for any $b > 0$. To complete the proof let us establish the properties of the graph of $z(t)$ used in the proof.

Assume that the graph of $z(t)$ has a negative part. Otherwise inequality (56) becomes $\int_R^{R+b} z(t)\,\mathrm{d}t \geqslant 0$ and the argument after (56) becomes even simpler. The negative part must have at least one finite end, say t_0. Otherwise $z'' \geqslant -k^2 z \geqslant 0$, $-\infty < t < \infty$, so that z is convex and therefore $z > 0$ for sufficiently large t. Without loss of generality one can assume that (55) holds for $-\infty < t < \infty$. Otherwise one could extend the graph so that (55) would hold for $-\infty < t < \infty$. One can also assume without loss of generality that the finite end of the negative part is its right end. Therefore $z(t_0) = 0$, $z'(t_0) \geqslant 0$.

Let

$$x(t) = \frac{kz(t)}{\sin k(t - t_0)}, \quad -k^{-1}\pi \leqslant t - t_0 \leqslant k^{-1}\pi. \tag{59}$$

Then $x(t)$ is continuous on the interval (59) $x(t_0) = z'(t_0) \geqslant 0$, and

$$\begin{aligned} x'(t) &= k \frac{z'(t) \sin k(t - t_0) - kz(t) \cos k(t - t_0)}{\sin^2 k(t - t_0)} \equiv \\ &\equiv k \frac{h(t)}{\sin^2 k(t - t_0)}. \end{aligned} \tag{60}$$

Clearly

$$h' = (z'' + k^2 z) \sin k(t - t_0).$$

Thus

$$\begin{aligned} &h'(t) \geqslant 0 \quad \text{if} \quad 0 \leqslant t - t_0 \leqslant \pi k^{-1}, \quad \text{and} \\ &h'(t) \leqslant 0 \quad \text{if} \quad -\pi k^{-1} \leqslant t - t_0 \leqslant 0. \end{aligned} \tag{61}$$

One has $h(t_0) = 0$. Therefore (61) implies that $h(t) \geqslant 0$ on the interval (59) and $x'(t) \geqslant 0$ on this interval. Thus

$$\begin{aligned} &x(t) \geqslant x(t_0) \quad \text{if} \quad 0 \leqslant t - t_0 \leqslant \pi k^{-1} \\ &x(t) \leqslant x(t_0) \quad \text{if} \quad -\pi k^{-1} \leqslant t - t_0 \leqslant 0. \end{aligned} \tag{62}$$

From (59) and (62) it follows that

$$\begin{aligned} z(t) &= k^{-1}x(t) \sin k(t - t_0) \geqslant k^{-1}x(t_0) \sin k(t - t_0) \\ &= k^{-1}z'(t_0) \sin k(t - t_0). \end{aligned} \tag{63}$$

Therefore the graph of $z(t)$ lies above the graph $k^{-1}z'(t_0) \sin k(t - t_0)$ on the interval (59). This implies the properties of the graph $z(t)$ used in the proof. □

Remark 2. Absence of positive eigenvalues was established for general elliptic operators of the second order in R^N not necessarily selfadjoint

$$\ell u = -\sum_{p,j=1}^{N} D_p(a_{pj}(x)D_j u) + \sum_{j=1}^{N} b_j(x)D_j u + Q(x)u,$$
$$D_j = \frac{\partial}{\partial x_j}, \tag{64}$$

under the following assumptions (Mishnaevskii [1], [2]), (1) $a_{pj}(x)$ and $b_j(x)$ are real-valued functions, $a_{pj} \in C^2(\bar{D})$, $b_j \in C^1(\bar{D})$, $a_{pj} = a_{jp}$, the matrix $a_{ij}(x)$ is positive definite uniformly in $\bar{D}$, $d_1|t|^2 \leqslant a_{pj}t_j\bar{t}_p \leqslant d_2|t|^2$, where $d_2 \geqslant d_1 > 0$, $|t|^2 = |t_1|^2 + \ldots |t_N|^2$, and one sums over the repeated indices; $a_{pj} - \delta_{pj} = 0(|x|^{-a})$, $|\nabla a_{pj}| = 0(|x|^{-1-a})$, $|(\partial^2 a_{pj})/(\partial x_m \partial x_n)| = 0\,(|x|^{-2-a})$, $|x| \gg 1$, $a > 0$; $Q(x) \in L^2_{\text{loc}}(D)$, $|Q(x)| = 0\,(|x|^{-1-a})$, $|x| \gg 1$;

(2) there exists an m, $2 \leqslant m \leqslant N+1$, such that

$$D \cap \left\{x\colon \sum_{j=1}^{m-1} x_j^2 \geqslant R^2\right\} \subseteq \left\{x\colon \sum_{j=m}^{N} x_j^2 \leqslant \mathrm{d}\left(\sum_{j=1}^{m-1} x_j^2\right)^b\right\} \tag{65}$$

where $0 < b \leqslant 1$.

Condition (65) allows the domain D to be a paraboloid, in which case condition 1 is not valid for sufficiently large x.

Remark 3. In the case when condition (1) is replaced by the local condition (2.13):

$$N \cdot e_3 \leqslant 0 \text{ on } \Gamma'_R \tag{66}$$

and $u = 0$ on Γ, it is not known if there are eigenvalues of $-\Delta_D$ embedded in its continuous spectrum.

Remark 4. Zero can be an eigenvalue of the operator

$$-\nabla^2 + Q(x) \text{ in } L^2(R^3) \quad \text{even if} \quad Q \in C_0^\infty(R^3).$$

EXAMPLE 1. Let $u = r^{-2}\cos\theta$ if $r \geqslant 1$. Then $\nabla^2 u = 0$ in $\Omega_1 = \{x\colon |x| \geqslant 1\}$, $|x| = r$. Let us extend u into B_1 smoothly and so that $Q \equiv \nabla^2 u/u$ is smooth. For example, set $u = f(r)\cos\theta$, $r \leqslant 1$. Then $\nabla^2 u = (f'' + 2r^{-1}f'(r) - 2fr^{-2})\cos\theta$, $Q = f^{-1}(f'' + 2r^{-1}f'(r) - 2fr^{-2})$, $r \leqslant 1$, $Q = 0$,

$r \geqslant 1$. One can choose f such that $Q \in C^\infty$. Take $f = r(1 + h(r))$, $h(r))$, $h^{(n)}(R) = 0(r^{p-n})$, $r \to \infty$, $n \leqslant p$, $p \geqslant 2$. Then $Q \in C^\infty$ in a neighborhood of the point $r = 0$. On the interval $(0, 1)$ one can choose $f > 0$, $f \in C^\infty$, and such that $Q \in C^\infty$ in a neighborhood of the point $r = 1$. Thus

$$\nabla^2 u - Qu = 0 \text{ in } R^3, \quad u \in L^2(R^3), \quad Q \in C^\infty, \quad Q = 0 \text{ if } r \geqslant 1. \tag{67}$$

A similar construction gives a solution u to Equation (67) which decays as $0(r^{-n-1})$ with any given $n = 0, 1, 2, \ldots$, $u = r^{-(n+1)} Y_{nm}(\omega)$ if $r \geqslant 1$, where Y_{nm} are the spherical harmonics. If $n = 0$ then $u = r^{-1}$ in Ω_1, so that $u \notin L^2(R^3)$.

4.2. Let us show that the spectrum of L is continuous if condition (1) holds.

LEMMA 8. *If* (1) *holds then the spectrum of L fills the ray* $[0, \infty)$ *and does not contain eigenvalues.*

Proof. If (1) holds then D contains a cone. Therefore the conclusion of Lemma 8 follows from Lemma 2.3. and Theorem 1. □

We are now interested in the following questions. (1) is the spectrum of L absolutely continous? (2) is the limiting absorption principle for the operator L valid for all $k^2 > 0$?

If the limiting absorption principle for the operator L is valid for all $k > 0$ then the positive spectrum of L is absolutely continuous. In the next section we will prove that the limiting absorption principle for the operator L is valid for all $k^2 > 0$ provided that assumption (1) and some additional assumptions concerning Γ'_R are satisfied.

Remark 5. If D is a finite domain then any solution $u \in L^2(\Omega)$ of the Helmholtz equation

$$(\nabla^2 + k^2)u = 0 \text{ in } \Omega, \quad k > 0, \quad u \in L^2(\Omega)$$

is $u \equiv 0$ in Ω (Lemma I.2.1). Here no boundary condition u on Γ is imposed. If Γ is infinite, one needs, in general, a boundary condition on u in order to prove that L has no positive eigenvalues. However, if D contains a half-space, say R^3_+, then again no boundary condition is necessary in order to prove that L has no positive eigenvalues.

5. THE LIMITING ABSORPTION PRINCIPLE AND COMPACT PERTURBATIONS OF THE BOUNDARY

5.1. In this section we always assume that the condition (4.1) holds so that the Dirichlet Laplacian L in D has continuous spectrum $[0, \infty)$, does not have positive eigenvalues, and the radiation condition guarantees uniqueness of the solution to the Dirichlet problem in D.

By a canonical domain we mean a domain in which the resolvent kernel of L can be studied (e.g. by separation of variables). For instance a cone is a canonical domain.

Let $G(x, y, k)$ be the resolvent kernel of L, that is the kernel of the operator $(L - k^2 - io)^{-1}$, $k > 0$, in D. This kernel solves the problem

$$(-\nabla_x^2 - k^2)G(x, y, k) = \delta(x - y) \text{ in } D, \tag{1}$$

$$G(x, y, k) = 0, \quad x \in \Gamma, \tag{2}$$

$$G(x, y, \sqrt{k^2 + i\varepsilon}) \in L^2(D) \text{ if } \varepsilon > 0, \tag{3}$$

$$G(x, y, k) = \lim_{\varepsilon \to +0} G(x, y, \sqrt{k^2 + i\varepsilon}). \tag{4}$$

where the limit is attained uniformaly on compact sets $|x - y| \geqslant a > 0$, $k \geqslant b > 0$, a and b are arbitrarily small fixed numbers.

By Lemma 2.5 the radiation condition

$$\lim_{r \to \infty} \int_{S_r(D)} \left| \frac{\partial G}{\partial r} - ikG \right|^2 ds = 0 \tag{5}$$

guarantees uniqueness of the solution to problem (1), (2). Let us assume that (5) holds and

$$|G(x, y, k)| \leqslant \frac{c}{|x - y|}, \quad |\nabla G| \leqslant \frac{c}{|x - y|^2}, \quad |x - y| \leqslant 1 \tag{6}$$

$$|G| + |\nabla G| + |D^2 G| \leqslant \frac{c}{|x - y|}, \quad |x - y| \geqslant 1 \tag{7}$$

where $D^2 G$ stands for an arbitrary second derivative of G.

The local estimates for G are well known and easy to establish. The estimates for $|x - y| \geqslant 1$ are more difficult to prove. The estimates (6) and (7) are uniform with respect to k, $0 < b \leqslant k \leqslant d$, where $d > b$ is any number.

We first prove the following general result.

THEOREM 1. *Let $\tilde{D}$ denote any domain which is obtained from D by a compact perturbation, so that $\tilde{\Gamma}'_R = \Gamma'_R$ for some R, where $\tilde{\Gamma} = \partial\tilde{D}$. Assume that the resolvent kernel of L in D satisfies estimates* (5)–(7) *and the radiation condition guarantees uniqueness of the solution to the Dirichlet problem in $\tilde{D}$. Then the operator L in $\tilde{D}$ has absolutely continuous spectrum and its resolvent kernel $\tilde{G}$ satisfies estimates* (6) *and* (7).

Let us notice that if $\partial\tilde{D}_R \subset D$ then one can follow the scheme of Chapter I and prove the existence of $\tilde{G}$ by the integral equation method. If some part of $\partial\tilde{D}_R$ does not lie in D then it is not clear if the integral equation method can be used. The reason is that the kernel $G(x, y, k)$ is defined only for $x, y \in D$. In some special cases, e.g. if $D = R^3_+$, one can construct G which is defined for all $x, y, x \neq y$. In these cases the integral equation method is convenient.

The method we wish to use in the general case is based on some *a priori* estimates. It is similar to the method used in the proof of Theorem I.2.3. The proof of Theorem 1 consists of three steps.

Step 1. First we notice that it is sufficient to prove that the boundary value problem

$$(-\nabla^2 - k^2)u = f \text{ in } \tilde{D}, \quad f \in C_0^\infty(\tilde{D}) \tag{8}$$

$$u = 0 \text{ on } \tilde{\Gamma} \tag{9}$$

$$\lim_{r\to\infty} \int_{S_r(\tilde{D})} \left| \frac{\partial u}{\partial r} - iku \right|^2 \mathrm{d}s = 0 \tag{10}$$

is uniquely solvable, and its solution can be obtained by the limiting absorption principle:

$$u(x, k) = \lim_{\varepsilon\to+0} u(x, \sqrt{k^2 + i\varepsilon}) \tag{11}$$

where the limit is attained uniformly with respect to $x \in \tilde{D}_r$ and $b \leqslant k \leqslant d$, i.e. uniformly on compact sets in $\tilde{D} \times R_+$, $x \in \tilde{D}$, $k \in R_+$. Indeed, since

$$u(x, k) = \int_{\tilde{D}} \tilde{G}(x, y, k) f(y)\, \mathrm{d}y, \tag{12}$$

the knowledge of $u(x, k)$ for all $f \in C_0^\infty(\tilde{D})$ defines the kernel $\tilde{G}(x, y, k)$ uniquely as a distribution depending on a parameter x. The differential equation (8) and boundary condition (9) imply that

$$(-\nabla_x^2 - k^2)\tilde{G}(x, y) = \delta(x - y) \text{ in } \tilde{D}, \quad \tilde{G} = 0 \text{ on } \Gamma. \tag{13}$$

The standard regularity properties of the solution to Equation (13) imply that $\tilde{G}(x, y, k) \in C^\infty(\tilde{D}\backslash y)$ and satisfies estimates (6). The radiation condition allows one to use the standard proof of the symmetry of $\tilde{G}$:

$$\tilde{G}(x, y, k) = \tilde{G}(y, x, k). \tag{14}$$

Namely, in the sense of distributions one has

$$\int_{\tilde{D}} \tilde{G}(z, y)(\nabla_z^2 + k^2)\tilde{G}(z, x)\, dz = \int_{\tilde{D}} (\nabla_z^2 + k^2)\tilde{G}(z, y) \cdot \tilde{G}(z, x)\, dz. \tag{15}$$

Because of (13), Equation (15) reduces to (14). The same proof in classical terms (without using the theory of distributions) requires an application of Green's formula to $\tilde{G}(z, y)$ and $\tilde{G}(z, x)$ in the domain $\tilde{D}_{x,y} = \tilde{D}\backslash(B_\varepsilon(x) \cup B_\varepsilon(y))$, where $B_\varepsilon(x) = \{z: |z - x| < \varepsilon\}$, $\varepsilon > 0$, and $B_\varepsilon(x) \subset \tilde{D}$, $B_\varepsilon(y) \subset \tilde{D}$.

The radiation condition for $\tilde{G}$ is easy to establish as follows. Write Green's formula in $D'_R = \tilde{D}'_R$ for $G(x, y, \sqrt{k^2 + i\varepsilon})$, $\varepsilon > 0$, and then pass to the limit $\varepsilon \to +0$. This yields

$$\tilde{G}(x, y) = \alpha(y)G(x, y) + \int_{S_R(D)} \left(G(s, x)\frac{\partial \tilde{G}(s, y)}{\partial N_s} - \tilde{G}(s, y)\frac{\partial G(s, x)}{\partial N_s} \right) ds \tag{16}$$

where

$$\alpha(y) = 1 \text{ if } y \in D'_R, \quad \alpha(y) = 0 \quad \text{if} \quad y \notin D'_R. \tag{17}$$

The right side of (16) satisfies the radiation condition as a function of x because the integral is taken over a compact set and $G(s, x)$ satisfies the radiation condition (5) by assumption.

Step 2. Secondly we prove that problem (8)—(10) is uniquely solvable and the solution can be obtained by the limiting absorption principle (11). Consider the problem

$$(-\nabla^2 - k^2 - i\varepsilon)u_\varepsilon = f \text{ in } \tilde{D},\ \varepsilon > 0,\ u_\varepsilon = 0 \text{ on } \tilde{\Gamma} \tag{18}$$

$$u_\varepsilon \in L^2(\tilde{D}). \tag{19}$$

This problem is uniquely solvable in $L^2(\tilde{D})$ since L is selfadjoint in $L^2(\tilde{D})$ (see Section 2.1). Suppose that the following *a priori* estimate holds

$$N^2(u_\varepsilon) \equiv \int_{\tilde{D}} (1 + |x|)^{-1-a} |u_\varepsilon|^2 \, dx \leqslant c, \tag{20}$$

where $c > 0$ is a constant which does not depend on ε, and $a > 0$. Then it is easy to prove that

$$u_\varepsilon \underset{H^2_{\mathrm{loc}}(\tilde{D})}{\rightarrow} u \quad \text{as} \quad \varepsilon \rightarrow +0. \tag{21}$$

If (21) holds then one can pass to the limit $\varepsilon \rightarrow +0$ in (18) and see that the function u, defined in (21), satisfies Equation (8) and (9). Applying Green's formula to $G_\varepsilon \equiv G(x, y, \sqrt{k^2 + i\varepsilon})$ and u_ε in D'_R yields:

$$u_\varepsilon(x, k) = \int_{D'_R} G_\varepsilon(x, y) f \, dy + \int_{S_R(D)} \left(u_\varepsilon \frac{\partial G_\varepsilon}{\partial N} - G_\varepsilon \frac{\partial u_\varepsilon}{\partial N} \right) ds. \tag{22}$$

Passing to the limit $\varepsilon \rightarrow +0$ in (22) yields

$$u(x, k) = \int_{D'_R} Gf \, dy + \int_{S_R(D)} \left(u \frac{\partial G}{\partial N} - G \frac{\partial u}{\partial N} \right) ds. \tag{23}$$

We used the continuity of the imbedding i: $H^2(D_R) \rightarrow H^1(S_R(D))$ to pass to the limit in the last integral in (23). Since the integrals in (23) are taken over compact sets and $G(x, y, k)$ satisfies the radiation condition (5), the function $u(x, k)$ satisfies the radiation condition (10). To complete the argument in Step 2 we should prove (21) and (20). Equation (21) follows from the standard inequality

$$\|u\|_{H^2(D_1)} \leqslant c(\|\nabla^2 u\|_{L^2(D_2)} + \|u\|_{L^2(D_2)}), \quad D_1 \subset D_2 \tag{24}$$

where $c = c(D_1, D_2) > 0$, D_1 and D_2 are arbitrary compact domains, and $D_1 \subset D_2$ is a strictly inner subdomain of D_2. If $u = 0$ on ∂D_2 then

$D_1 = D_2$ is allowed in (24). If $u = 0$ on a part of ∂D_2 then part of ∂D_1 is allowed to belong to this part of ∂D_2. If (20) holds then

$$\|u_\varepsilon\|_{L^2(\tilde{D}_r)} + \|\nabla^2 u_\varepsilon\|_{L^2(\tilde{D}_r)} \leqslant c \tag{25}$$

where c denotes different positive constants, $r > 0$ is an arbitrary number, and we used Equation (18). From (25) it follows that

$$\|u_\varepsilon\|_{L^2(\tilde{D}_r)} \leqslant c. \tag{26}$$

From (26) and the compactness of the imbedding $i: H^2(\tilde{D}_r) \to H^0(\tilde{D}_r)$ it follows that the set of functions u_ε, $0 < \varepsilon \leqslant 1$ contains a subsequence which converges in $H^0(\tilde{D}_r) = L^2(\tilde{D}_r)$ as $\varepsilon \to +0$. This subsequence we denote u_{ε_n}. Equation (18) shows that $\nabla^2 u_{\varepsilon_n}$ converges in $L^2(\tilde{D}_r)$ as $\varepsilon_n \to 0$. This and inequality (24) imply that u_{ε_n} converges in $H^2(\tilde{D}_r)$ as $\varepsilon_n \to +0$. Since the limit function $u(x, k)$ solves problem (8)—(10), as we have already proved, and since, by assumption, the solution to problem (8)—(10) is unique, one concludes that not only the subsequence u_{ε_n} but the function u_ε itself converges in H^2_{loc} as $\varepsilon \to +0$. Finally, estimate (20) should be proved.

Suppose (20) is not valid. Then there exists a sequence $\varepsilon_n \to +0$ such that $N(u_n) \geqslant n$, $n = 1, 2, \ldots, u_n = u_{\varepsilon_n}$.

Let $u_n/N(u_n) = v_n$. Then

$$N(v_n) = 1 \tag{27}$$

$$(-\nabla^2 - k^2 - i\varepsilon_n)v_n = f_n, \quad f_n = \frac{f}{N(u_n)} \to 0 \text{ as } n \to \infty \tag{28}$$

$$v_n = 0 \text{ on } \tilde{\Gamma}. \tag{29}$$

Our argument above shows that

$$v_n \underset{H^2_{\mathrm{loc}}(\tilde{D})}{\to} v, \quad n \to \infty \tag{30}$$

and v solves the problem

$$(-\nabla^2 - k^2)v = 0 \text{ in } \tilde{D}, \; v = 0 \text{ on } \tilde{\Gamma} \tag{31}$$

$$\lim_{r\to\infty} \int_{S_R(\tilde{D})} \left| \frac{\partial v}{\partial r} - ikv \right|^2 ds = 0. \tag{32}$$

Therefore

$$v = 0 \text{ in } \tilde{D}. \tag{33}$$

If we show that

$$\lim_{n\to\infty} N(v_n) = 0 \tag{34}$$

then (33) and (27) cannot hold simultaneously. This contradiction proves estimate (20). To prove (34), one writes Green's formula for v_n

$$v_n = \int_{D'_R} G_{\varepsilon_n}(x, y) f_n \, dy + \int_{S_R(D)} \left(v_n \frac{\partial G_{\varepsilon_n}}{\partial N} - G_{\varepsilon_n} \frac{\partial v_n}{\partial N} \right) ds. \tag{35}$$

Each of the terms in the right side of (35) can be estimated in the same way. The estimate is possible because the integrals in (35) are taken over compact sets and because $G_{\varepsilon_n}(x, y)$ satisfies the inequality

$$|G_{\varepsilon_n}(x, y)| \leqslant \frac{c}{|x - y|}, \quad |x - y| \geqslant 1. \tag{36}$$

This inequality we need only for y running through compact sets. Let us estimate, for example, the first integral in (35). We have

$$N\left(\int_{D'_R} G_{\varepsilon_n}(x, y) f_n \, dy\right) \leqslant c \|f_n\|_{L^2(D'_R)} \left(\int_{\tilde{D}} \int_{D'_R} \frac{dx\, dy}{(1 + |x|)^{1+\alpha} |x - y|^2} \right)^{1/2}$$
$$\leqslant c_1 \|f_n\|_{L^2(D'_R)} \to 0, \quad n \to \infty. \tag{37}$$

The last relation follows from the second equation (28). It follows from (30) and (33) that

$$v_n \underset{H^1(S_r(D))}{\to} 0, \quad n \to \infty \tag{38}$$

so that the other two integrals in (35) can be estimated similarly.

Step 3. To finish the proof of Theorem 1 we should prove that $\tilde{G}$ satisfies inequality (7). This inequality is assumed to be valid for G and therefore formula (16) implies inequality (7) for $\tilde{G}(x, y)$ when y runs through compact sets. In order to prove that

$$|\tilde{G}(x, y)| \leqslant \frac{c}{|x - y|}, \quad |x - y| \geqslant 1, \quad x, y \in \tilde{D} \tag{39}$$

we assume that (39) is not valid. In this case there exist sequences x_n, $y_n \in \tilde{D}$, such that

$$|\tilde{G}(x_n, y_n)| \geqslant \frac{n}{|x_n - y_n|}, \quad |x_n - y_n| \geqslant 1 \tag{40}$$

Without loss of generality one can assume that $|x_n - y_n| \to \infty$. By assumption

$$|G(x_n, y_n)| \; |x_n - y_n| \leqslant c. \tag{41}$$

It follows from (16) and (7) that

$$|\tilde{G}(x, y)| \leqslant \frac{c}{|x|}, \quad |x| \to \infty, \quad |y| \leqslant c_1. \tag{42}$$

Therefore one can assume that

$$|x_n| \to \infty \quad \text{and} \quad |y_n| \to \infty. \tag{43}$$

Multiply equation (16) by $|x_n - y_n|$, take $x = x_n$, $y = y_n$, use (40), and notice that $\alpha(y_n) = 1$. This yields

$$\begin{aligned} n &\leqslant |x_n - y_n| \; |\tilde{G}(x_n, y_n)| \leqslant \\ &\leqslant |x_n - y_n| \; |G(x_n, y_n)| + c|x_n - y_n| \int_{S_R(D)} \frac{\mathrm{d}s}{|s - x_n| \; |s - y_n|}. \end{aligned} \tag{45}$$

The right side of (45) is bounded and we have a contradiction. This proves (39). The other estimates (7) can be proved similarly. Theorem 1 is proved.

Remark 1. In the proof of Theorem 1, we used estimate (7) only for y running through compact sets.

5.2. Suppose now that the perturbation of Γ is not compact, i.e. there is no such $R > 0$ that $\Gamma'_R = \tilde{\Gamma}'_R$. Let us assume that

$$\tilde{\Gamma}'_R \in D'_R, \text{and } \rho(r) \leqslant c(1 + r)^{-a}, \quad a > 2, \tag{46}$$

where $\rho(r)$ is the distance between $\tilde{\Gamma}'_r \cap S_r$ and $\Gamma'_r \cap S_r$. Let us also assume that condition (4.1) holds for Γ'_R and $\tilde{\Gamma}'_R$. Without loss of generality we can assume that $\tilde{D} \subset D$. Otherwise we can compactly perturb D so that the above condition is satisfied. Let us denote $D \backslash \tilde{D}$

by d. The spectrum of the Laplacian $-\Delta_d$ defined in d by the boundary conditions $u = 0$ on Γ, $\partial u/\partial N = 0$ on $\tilde{\Gamma}$, where N is the normal on $\tilde{\Gamma}$ pointing into $\tilde{D}$, is discrete because of (46), and without loss of generality we assume that k^2 is not an eigenvalue of $-\Delta_d$. Otherwise we can compactly perturb D so that k^2 will not be an eigenvalue of $-\Delta_d$. Green's formula for G and $\tilde{G}$ in $\tilde{D}$ yields

$$\tilde{G}(x, y) = G(x, y) - \int_{\tilde{\Gamma}} G(x, s)\mu(s, y)\, ds, \; \mu \equiv \frac{\partial \tilde{G}(s, y)}{\partial N_s}. \tag{47}$$

Take the normal derivative and use the formula for the normal derivative of the single layer potential to obtain

$$\mu + A\mu = 2\frac{\partial G}{\partial N}, \qquad A\mu = \int_{\tilde{\Gamma}} \frac{\partial G(s, t)}{\partial N_s} \mu(t)\, dt. \tag{48}$$

Under the assumption (46) one can check that A is compact in $C(\tilde{\Gamma})$ and $L^2(\tilde{\Gamma})$. The homogeneous equation

$$\mu + A\mu = 0 \tag{49}$$

has only the trivial solution $\mu = 0$. Otherwise k^2 is an eigenvalue of $-\Delta_d$. Therefore, by Fredholm's alternative, Equation (48) is uniquely solvable and the kernel $\tilde{G}$ of the Dirichlet Laplacian in $\tilde{D}$ is given by formula (47). To justify this argument completely, one should establish the following lemma.

LEMMA 1. *The operator* (48) *is compact in* $C(\tilde{\Gamma})$ (*and in* $L^2(\tilde{\Gamma})$) *provided that* (46) *holds.*

The proof of Lemma 1 is not difficult but relatively long. The reader can find it in Ramm [1b]. In this proof the following estimate is essential

$$\frac{\partial G(s, t)}{\partial N_s} \leqslant \begin{cases} c|s-t|^{2-\lambda}, \lambda > 0, |s-t| \leqslant 1 \\ c|s-t|^{-1}\rho(t, \Gamma), |s-t| \geqslant 1 \end{cases} \tag{50}$$

where $\rho(t, \Gamma)$ is the distance from the point $t \in \tilde{\Gamma}$ to Γ, $\rho(t, \Gamma) = \inf_{x \in \Gamma}|t - x|$,

$$\rho(t, \Gamma) \leqslant c(1 + |t|)^{-a}, \quad a > 2 \tag{51}$$

(see (46)). The constant λ in (50) depends on the smoothness of $\tilde{\Gamma}$: if the equation of $\tilde{\Gamma}$ in the local coordinates is $s_3 = f(s_1, s_2)$ and $f \in C^{1,\lambda}$, then this λ is used in (50). We assume that λ can be chosen uniformly on $\tilde{\Gamma}$. The inequality (50) for $|s-t| \leqslant 1$ is standard in potential theory, and for $|s-t| \geqslant 1$ can be obtained as follows. Let $\rho(t', t) = \rho(t, \Gamma)$, $t' \in \Gamma$. By the mean value (Lagrange) formula one has

$$\frac{\partial G(s,t)}{\partial N_s} = \frac{\partial G(s,t')}{\partial N_s} + \nabla_t \frac{\partial G(s,z)}{\partial N_s} \cdot (t-t') \tag{52}$$

where z is the "mean point" in the mean value formula. Since $t' \in \Gamma$, the first term in the right side of (52) is zero. Estimating the second term and using (7) one obtains inequality (50) for $|s-t| \geqslant 1$.

6. EIGENFUNCTION EXPANSION IN CANONICAL DOMAINS

In this section we sketch the eigenfunction expansion theory similar to the one presented in Section I.3. Let us assume that D is a cone K, which plays the role of a canonical domain. Then $G(x, y, k)$ can be constructed explicitly by separation of variables and its asymptotics is

$$G(x,y,k) = \gamma(|y|)u_0(x,n,k)(1+o(1)), |y| \to \infty, \frac{y}{|y|} = n \tag{1}$$

where $\gamma(r) = (4\pi r)^{-1}\exp(ikr)$ and n is a unit vector in the cone K. This formula is analogous to (I.3.7). From (1) and (5.47) it follows that

$$\tilde{G}(x,y,k) = \gamma(|y|)u(x,n,k)(1+o(1)), \quad |y| \to \infty, \quad \frac{y}{|y|} = n \tag{2}$$

where

$$\begin{aligned} u(x,n,k) &= u_0(x,n,k) - \int_{\tilde{\Gamma}} u_0(s,n,k)\mu(s,x,k)\,\mathrm{d}s, \\ \mu &= \frac{\partial \tilde{G}(s,x,k)}{\partial N_s}. \end{aligned} \tag{3}$$

As in Section I.3. the function $u(x, n, k)$ is the solution to the scattering problem in $\tilde{D}$:

$$(\nabla^2 + k^2)u = 0 \text{ in } \tilde{D}, \quad u = 0 \text{ on } \tilde{\Gamma}, \tag{4}$$

$$u = u_0 + v \tag{5}$$

where v satisfies the radiation condition

$$\lim_{r \to \infty} \int_{S_r(\tilde{D})} \left| \frac{\partial v}{\partial r} - ikv \right|^2 \mathrm{d}s = 0. \tag{6}$$

As in Section I.3 one can prove the eigenfunction expansion theorem. This theorem is analogous to Theorem I.3.1.

THEOREM 1. *Let $f \in L^2(\tilde{D})$. Define*

$$Ff = \tilde{f}(\kappa) = \frac{1}{(2\pi)^{3/2}} \int_{\tilde{D}} f(x) \overline{u(x, \kappa)} \, \mathrm{d}x \kappa = kn \tag{7}$$

$u(x, \kappa) = u(x, k, n)$. Then

$$F^{-1}f = f(x) = \frac{1}{(2\pi)^{3/2}} \int_K \tilde{f}(\kappa) u(x, \kappa) \, \mathrm{d}\kappa. \tag{8}$$

The operator F: $L^2(\tilde{D}) \to L^2(K)$ is an isometric isomorphism, $F^{-1} = F^*$, and

$$F(-\Delta_D(\tilde{D}))F^{-1} = k^2 \tag{9}$$

where $-\Delta_D(\tilde{D})$ is the Dirichlet Laplacian in $\tilde{D}$.

The proof of Theorem 1 is similar to that of Theorem I.3.1 and therefore omitted.

Remark 1. One can construct the wave operators and the S-matrix as in Section I.5, formulas (I.5.5); (I.5.6). The role of F_0 in (I.5.5) is played by the Fourier transform in D, that is the operator

$$F_0 f = \frac{1}{(2\pi)^{3/2}} \int_K f(x) \overline{u_0(x, \kappa)} \, \mathrm{d}x. \tag{10}$$

Appendix 1. Summary of Some Results in Potential Theory and Embedding Theorems

1. Let $D \subset R^n$ be a compact domain with a closed boundary $\partial D = \Gamma$. We discuss the basic facts from potential theory first under the classical assumptions: Γ is Liapunov-type, and the densities of the potentials of the single and double layers are Hölder continuous or continuous, and afterwards for not so smooth boundaries and densities: the boundaries of class C^1 and densities of class $L^p(\Gamma)$, $p > 1$.

1. We write $\Gamma \in C^{1,\alpha}$ iff in the local coordinates $x_3 = F(x_1, x_2)$ the function $F \in C^{1,\alpha}$, that is F has first derivatives which are Hölder-continuous. A function $h(x)$, $x \in R^n$ is Hölder-continuous $h \in C^{0,\alpha}(D)$ in an open set $D \subset R^n$ iff $|h(x) - h(y)| \leqslant c|x - y|^\alpha$, $x, y \in D$, $c = c(D, h, \alpha)$, $0 < \alpha \leqslant 1$. We write $C(D)$ for $C^0(D)$, and $C^1(D)$ for $C^{1,0}(D)$. Since the normal to Γ is defined by the formula $N(s) = (e_1F_{x_1} + e_2F_{x_2} - e_3)/\sqrt{|\nabla F|^2 + 1}$, where e_ℓ are the unit vectors of the local cartesian coordinate system and $s \in \Gamma$, one sees that $N(s)$ is Hölder-continuous if $\Gamma \in C^{1,\alpha}$. A compact closed surface $\Gamma \in C^{1,\alpha}$, $\alpha > 0$ is called a Liapunov surface.

Consider an integral operator

$$Kh = \int_D K(s, t)h(t)\,\mathrm{d}t, \tag{1}$$

where $D \subset R^n$ is a bounded domain. Let $s \in D' \subset R^m$. We write a.e. for almost every, $|D|$ for meas D, i.e. for volume of $D \subset R^n$.

LEMMA 1. *Let*

$$\int_D |K(s, t)|^r \,\mathrm{d}t \leqslant c_1^r \text{ for a.e. } s \in D', \int_{D'} |K(s, t)|^\sigma \,\mathrm{d}s \leqslant c_2^\sigma \text{ for a.e. } t \in D,$$

$$q \geqslant p, q \geqslant \sigma, (1 - \sigma q^{-1})p' \leqslant r, p, q \geqslant 1, p' = p(p-1)^{-1}.$$

Then $K: L^p(D) \to L^q(D')$ *is bounded,* $\|K\| \leqslant c_1^{1-\sigma q^{-1}} c_2^{\sigma q^{-1}} c(|D|)$. *Here* c_1 *and* c_2 *are given positive constants and* $c(|D|)$ *is a positive constant which depends only on* D.

LEMMA 2. *Let $K(s, t) = B(s, t)/|s - t|^{\lambda}$, where $B(s, t)$ is a continuous function on $D' \times D \subset R^m \times R^n$. Then K is bounded as an opperator from $L^p(D)$ into $L^q(D')$ provided that*

$$q < \frac{mp}{n - (n - \lambda)p}, \quad m > n - (n - \lambda)p. \tag{2}$$

LEMMA 3. *Let $n = m$, $D = D'$, and $K(s, t)$ is defined as in Lemma 2. Then the operator K: $L^p(D) \to C(D)$ is bounded if $\lambda < n(1 - p^{-1})$. The operator K: $L^p(D) \to C^{\alpha}(D)$ is bounded, where $\alpha = n - np^{-1} - \lambda$.*

2. In this section, the operators S, A, and A' are studied in the Hölder spaces, $\Gamma \in C^{1,\alpha}$.

LEMMA 4. *Let $n = 3$, $g(x, y)$ is given by* (I.1.1.), $\Gamma \in C^{1,\alpha}$, *and $S\sigma$, the single layer potential, be defined by* (I.1.4). Then S: $C(\Gamma) \to C^{\alpha}(R^3)$ *is a bounded operator. The operators D: $C(\Gamma) \to C(D_{\pm})$, defined by* (I.1.1.) *(with $D_+ = D$, $D_- = \Omega$) are bounded and formulas (3a) hold. The operators A', A: $C(\Gamma) \to C^{\alpha}(\Gamma)$, defined by* (I.1.4) *and* (I.1.2) *are bounded, the operator S: $C^{\beta}(\Gamma) \to C^{1,\beta}(D_{\pm})$ is bounded, and formula (3b) hold. Formulas (3b) hold even if $\sigma \in C(\Gamma)$, moreover, the limits are attained uniformly on Γ.*

Let

$$\begin{aligned} &\mu \in C^{1,\alpha}(\Gamma), \quad B\mu = \frac{\partial A'\mu}{\partial N} = \lim_{\varepsilon \to \infty} \frac{\partial}{\partial N_s}(D\mu), \\ &x = s + \varepsilon N_s, \quad s \in \Gamma, \quad \varepsilon \to \pm 0. \end{aligned} \tag{3}$$

The Liapunov—Tauber theorem says that the limits as $\varepsilon \to +0$ and, $\varepsilon \to -0$ are equal. Notice that according to the definitions (I.1.1.)—(I.1.4) *the operators A, A', and B depend on the parameter k.*

LEMMA 5. *The operator $B = B(k)$: $C^{1,\alpha}(\Gamma) \to C^{\alpha}(\Gamma)$ is bounded. The operator $B(k_1) - B(k_2)$ is compact in $C^{1,\alpha}(\Gamma)$. If k^2 is neither an eigenvalue of the Dirichlet Laplacian nor Neumann Laplacian in D, then $B^{-1}(k) = [(A' + I)/2]^{-1}Q[(A - I)/2]^{-1}$, where Q is defined by* (I.1.8).

Let us explain the last formula. Suppose $B(k)\mu = f$. Then $\partial v/\partial N = f$, where $v = D\mu$ is defined by (I.1.1.). *If this v can be written in the form $v = S\sigma$, where S is defined by* (I.1.4), *then $(A - I)\sigma/2 = f$. If k^2 is not an eigenvalue of the Dirichlet Laplacian in D, then the operator $A - I$*

is invertible and $\sigma = [(A - I)/2]^{-1}f$. *We have* $Q\sigma = v_- = (A' + I)\mu/2$. *The operator* $A' + I$ *is invertible if* k^2 *is not an eigenvalue of the Neumann Laplacian in D. Thus* $\mu = [(A' + I)/2]^{-1}Q[(A - I)/2]^{-1}f = B^{-1}(k)f$. *If* k^2 *is not an eigenvalue of the Dirichlet laplacian in D the operator* Q: $C^{a}(\Gamma) \to C^{1,a}(\Gamma)$ *is invertible. Thus*

$$B(k) = \frac{A - I}{2}\, Q^{-1}\, \frac{A' + I}{2}. \tag{4}$$

3. In this section the operators S, A and A' are studied in the Lebesgue spaces. In Lemmas 6 and 7 below we assume that

$$\Gamma \subset C^1, \quad 1 < p < \infty. \tag{5}$$

Let

$$A'_\varepsilon\mu = 2\int_{\Gamma_\varepsilon} \frac{\partial g(s, t)}{\partial N_t}\,\mu(t)\,\mathrm{d}t, \quad \Gamma_\varepsilon = \{t: |t - s| \geqslant \varepsilon > 0, \quad t \in \Gamma\} \tag{6}$$

and

$$A'\mu = \lim_{\varepsilon \to \infty} A'_\varepsilon\mu \text{ (a)}, \quad A'_*\mu = \sup_{\varepsilon > \infty} |A'_\varepsilon\mu|, \text{(b)} \tag{7}$$

The operator A' is defined by (7) coincides on $C(\Gamma)$, $\Gamma \in C^{1,a}$, with the operator A' defined by (I.1.2) as an improper integral.

LEMMA 6. *The operator* A'_*: $L^p(\Gamma) \to L^p(\Gamma)$ *is bounded. For any* $\mu \in L^p(\Gamma)$ *the limit* (7a) *exists in* $L^p(\Gamma)$ *and pointwise a.e. The operator* A': $L^p(\Gamma) \to L^p(\Gamma)$ *is compact. Let* $v(x) = D\mu$, $\mu \in L^p(\Gamma)$, $x \in D$, $0 < a < 1$. *Then there exists a* $\delta = \delta(a, D) > 0$ *such that the function*

$$v_*(t) = \sup\{|v(x)|\, x \in D, \quad |x - t| < \delta, \quad (x - t)\cdot N_t < -a|x - t|\} \tag{8}$$

belongs to $L^p(\Gamma)$,

$$\|v_*\|_{L^p(\Gamma)} \leqslant c\|\mu\|_{L^p(\Gamma)}, \tag{9}$$

and for almost every $t \in \Gamma$ *the pointwise limit exists:*

$$\lim v(x) = \frac{A'\mu + \mu}{2} \quad as \quad x \to t, \quad x \in D, \quad (x - t)\cdot N_t < -a|x - t|. \tag{10}$$

Let $\mu \in W^{1,p}(\Gamma)$, $1 < p < \infty$, *where* $\|\mu\|_{W^{1,p}(\Gamma)} = \|\mu\|_{L^p(\Gamma)} + \|\nabla_t\mu\|_{L^p(\Gamma)}$, *and* $\nabla_t \mu$ *is the surface gradient of* μ *on* Γ. *The operator* A': $W^{1,p}(\Gamma) \to W^{1,p}(\Gamma)$ *is compact. For any* α, $0 < \alpha < 1$, *there exists a* $\delta = \delta(\alpha, D)$, *such that the function*

$$(\nabla v)_*(t) = \sup\{|\nabla v| : x \in D, |x - t| < \delta, (x - t) \cdot N_t < -\alpha|x - t|\} \tag{11}$$

belongs to $L^p(\Gamma)$ *and*

$$\|\nabla v_*(t)\|_{L^p(\Gamma)} \leqslant c\|\mu\|_{L^p(\Gamma)}. \tag{12}$$

LEMMA 7. *Let*

$$A_\varepsilon \sigma = 2 \int_{|s-t|>\varepsilon} \sigma(t) \frac{\partial g(s,t)}{\partial N_s} \, dt, \quad s \in \Gamma. \tag{13}$$

Then the limit $\lim_{\varepsilon \to 0} A_\varepsilon \sigma = A\sigma$ *exists in* $L^p(\Gamma)$ (*and pointwise a.e.*), *and* A: $L^p(\Gamma) \to L^p(\Gamma)$ *is compact. Let* $w(x)$ *be defined by* (I.1.4), $x \in D_\pm$, $\sigma \in L^p(\Gamma)$, *and* $0 < \alpha < 1$. *Then there exists a* $\delta = \delta(\alpha, D)$ *such that the functions*

$$(\nabla w)_+(t) \tag{14}$$

$$= \sup\{|\nabla w| : x \,\varepsilon\, D_+, |s - t| < \delta, (x - t) \cdot N_t < -\alpha|x - t|\} \tag{15}$$

and

$$(\nabla w)_-(t) = \sup\{|\nabla w| : x \in D_-, |x - t| < \delta, (x - t) \cdot N_t > \alpha|x - t|\} \tag{16}$$

belong to $L^p(\Gamma)$

$$\|(\nabla w)_\pm\|_{L^p(\Gamma)} < c\|\sigma\|_{L^p(\Gamma)}, \tag{17}$$

and the limit

$$\lim_{x \to t^\pm} \frac{\partial w(x)}{\partial N_t} = \lim \nabla w \cdot N_t = \frac{A\sigma \pm \sigma}{2} \tag{18}$$

exists pointwise a.e. Here $x \to t^+ (x \to t^-)$ *means that* $x \to t$ *and* x *belongs to the domain indicated by the inequality in* (15) ((16)). *The operator* $I + A$ *is a continuous bijection* $L^p(\Gamma)$ *onto* $L^p(\Gamma)$, *and* $W^{1,p}(\Gamma)$. *The operators* A: $L^p(\Gamma) \to L^p(\Gamma)$ *and* A: $W^{1,p}(\Gamma) \to W^{1,p}(\Gamma)$ *are compact.*

4. In this section the operators *S*, *A* and *A′* are studied in the Sobolev spaces $H^q(\Gamma)$.

LEMMA 8. *The operator* Q: $H^q(\Gamma) \to H^{q+1}(\Gamma)$ *is Fredholm-type with index* 0. *Iff* k^2 *is not an eigenvalue of the Dirichlet Laplacian in D, the operator Q is an isomorphism of* $H^q(\Gamma)$ *onto* $H^{q+1}(\Gamma)$. *The operators A and A′ are continuous from* $H^q(\Gamma)$ *into* $H^{q+1}(\Gamma)$.

A linear operator A: $B_1 \to B_2$ from a Banach space B_1 into a Banach space B_2 is called Fredholm-type iff im A is closed, $n = n(A) = \dim \ker A < \infty$, and $n^* = n^*(A) = \dim \operatorname{coker} A < \infty$. Here im A is the range of A', ker A is the null space of A, i.e. $\ker A = \{f: Af = 0\}$, and $\operatorname{coker} A = B_2/\operatorname{im} A$, i.e. the quotient space. The index of A, ind A or $i(A)$, is defined by the formula $i(A) = n(A) - n^*(A)$.

5. Consider a compact set $D \subset R^n$, $n \geqslant 2$, and the set $M(D)$ of measures μ whose support lies in D. Let

$$u_\mu(x) = \int g_0(x, y)\, \mathrm{d}\mu, \quad g_0(x, y) = (\omega_n(n-2)|x-y|^{n-2})^{-1}, \tag{19}$$

where ω_n is the surface area of the unit sphere in R^n, $n > 2$, $g_0 = -(2\pi)^{-1} \ln|x-y|$ if $n = 2$.

The Wiener capacitance of D is defined by the formula

$$\operatorname{cap} D = C(D) = \sup \mu(D), \quad \mu \in M(D), \quad u_\mu(x) \leqslant 1, \quad x \in R^n. \tag{20}$$

The capacitance has the following properties: $C(D) \geqslant 0$, $C(D)$ is invariant under translations and rotations of D, $C(D_1) \leqslant C(D_2)$ if $D_1 \subseteq D_2$, $C(D) \leqslant \sum_{j=1}^{\infty} C(D_j)$ if $D = \bigcup_{j=1}^{\infty} D_j$. If $\mu \in M(D)$, $u_\mu(x) \leqslant m_2$ if $x \in R^n$, and $u_\mu(x) \geqslant m_1$ if $x \in D$, then $m_1 C(D) \leqslant \mu(D) \leqslant m_2 C(D)$. One can prove that

$$\operatorname{cap} D = \left(\inf_{\mu \in M_1(D)} \int_D \int_D g_0(x, y)\, \mathrm{d}\mu(x)\, \mathrm{d}\mu(y) \right)^{-1},$$

$$M_1(D) = \left\{ \mu : \mu \in M(D), \int_D \mathrm{d}\mu = 1 \right\} \tag{21}$$

and

$$\operatorname{cap} D = \inf \int_{R^n} |\nabla u|^2 \, dx, \quad u \in C_0^\infty(R^n), \quad u = 1 \text{ in } D. \tag{22}$$

The supremum in (20) is attained at a measure γ which is called the equilibrium measure of the compact D, and $u_\gamma(x) = 1$ a.e. in D. Formula (22) is easily understood if one recalls that $CU^2 = 2E$, where C is the capacitance of the perfect conductor D, U is the potential of this conductor, and E is the electrostatic energy of the field around D. If $U = 1$ (as is seen from (22)) then $C = 2E$ and the equilibrium is attained if $E = \min$. Since $2E = \int_{R^n} |\nabla u|^2 \, dx$ where u is the potential of a charge distribution in D, one obtains (22). Formula (21) is easily understood if one recalls that $Q^2/2C = E$, where Q is the total charge of the conductor D, $Q = \int_D d\mu$. According to (21), $Q = 1$, and the equilibrium distribution of the charge in D is attained when $E = \min$. Since $2E = \int_D \int_D g_0(x, y) \, d\mu(x) \, d\mu(y)$, one obtains (21).

Let $G \subset R^n$ be an open set, E be a compact subset of G,

$$M(E, G) = \{u : u \in C_0^\infty(G), \quad 0 \leqslant u \leqslant 1, \quad u = 1 \text{ on } \tilde{E} \supset E\} \tag{23}$$

where $\tilde{E}$ is an open set containing E, $\tilde{E}$ depends on u. Let $M(E) = M(E, R^n)$. If $p > 1$ one defines

$$(p, \ell) \operatorname{cap}(E, G) = \inf \left\{ \int_G \sum_{|\alpha| = \ell} |D^\alpha u|^p \, dx, \quad u \in M(E, G) \right\}. \tag{24}$$

If $\ell = 1$ then we write $(p, 1) - \operatorname{cap}(E, G) = p - \operatorname{cap}(E, G)$, $p - \operatorname{cap}(E, G) =$

$$\inf \left\{ \int_G |\nabla u|^p \, dx, \quad u \in M(E, G) \right\}. \tag{25}$$

Clearly 2-cap$(E, R^n) = \operatorname{cap} E$ (see (22)).

6. *Embedding Theorems*

THEOREM 1. *Let $D \subset R^n$ be a bounded domain with a smooth boundary Γ, $1 < p < \infty$, and $0 < \ell < s - np^{-1}$. Then the embedding operator i: $W^{s,p}(D) \to C^\ell(D)$ is compact. If $\ell \geqslant s - np^{-1}$,*

$m > n - (s - \ell)p$, and $1 < q < mp/(n - (s - \ell)p)$ then the embedding operator i: $W^{s,p}(D) \to W^{\ell,q}(\Gamma_m)$ is compact. Here Γ_m is any m-dimensional C^δ manifold in D, $\|u\|_{W^{s,p}(D)} \geqslant c\|u\|_{W^{\ell,q}(\Gamma_m)}$, where here and below $c > 0$ denotes various constants which do not depend on u $\in W^{s,p}(D)$. The trace operator r: $W^{s,p}(D) \to W^{s-p^{-1},p}(\Gamma)$ is continuous. For any given $u \in W^{s-p^{-1},p}(\Gamma)$ there exists a $v \in W^{s,p}(D)$ such that $v = u$ on Γ, and $\|v\|_{W^{s,p}(D)} \leqslant c\|u\|_{W^{s-p^{-1},p}(\Gamma)}$, so that an extension operator from $W^{s-p^{-1},p}(\Gamma)$ into $W^{s,p}(D)$ is bounded. In particular, if $p = 2$ then $W^{s,2}(D) = H^s(D)$ and the operator r: $H^s(D) \to H^{s-1/2}(\Gamma)$ is bounded.

7. *Potential theory and boundary value problems in domains with nonsmooth boundaries.*

If the boundary condition is homogeneous then the Dirichlet problem in Ω is formulated as

$$\int_\Omega (\nabla u \nabla h - k^2 uh)\,\mathrm{d}x = \int_\Omega fh\,\mathrm{d}x, \quad \forall h \in \mathring{H}^1(\Omega), \quad k > 0, \quad u \in \mathring{H}^1(\Omega), \tag{26}$$

while the Neumann problem is

$$\int_\Omega (\nabla u \nabla h - k^2 uh)\,\mathrm{d}x = \int_\Omega fh\,\mathrm{d}x, \quad \forall h \in H^1(\Omega), \quad k > 0, \quad u \in H^1(\Omega), \tag{27}$$

and in both problems u should satisfy the radiation condition (0.5). If the boundary Γ is not sufficiently smooth, so that Green's formula may be not applicable, then the usual proof of the uniqueness theorem, based on Green's formula, should be modified slightly. Namely, if $f = 0$ then (26) with $h = \bar{u}\alpha((R - |x|)/\delta)$, $\delta > 0$, $D \subset B_R$, yields

$$\int_\Omega |\nabla u|^2 \alpha\,\mathrm{d}x + \int_\Omega \nabla u \cdot \bar{u} \nabla \alpha\,\mathrm{d}x - k^2 \int_\Omega |u|^2 \alpha\,\mathrm{d}x = 0 \tag{28}$$

where $\alpha(x) = \alpha(|x|) \in C^\infty$, $0 \leqslant \alpha \leqslant 1$, $\alpha(r) = 0$, $r \leqslant 0$, $\alpha(r) = 1$, $r \geqslant 1$, $\alpha' \geqslant 0$. Also

$$\int_{\Omega_R} (|\nabla u|^2 - k^2 |u|^2)\alpha \, dx + \int_\Omega \overline{\nabla u} \cdot u \nabla \alpha \, dx = 0. \tag{29}$$

Thus

$$\int_\Omega (\nabla u \, \bar{u} - \overline{\nabla u} \, u) \nabla \alpha \, dx = 0. \tag{30}$$

Let $\delta \to 0$. Then (30) becomes

$$\int_{S_R} \left(\frac{\partial u}{\partial r} \bar{u} - u \frac{\partial \bar{u}}{\partial r} \right) ds = 0. \tag{31}$$

The radiation condition and (31) yield

$$\lim_{R \to \infty} \int_{S_R} |u|^2 \, ds = 0. \tag{32}$$

Equation (32) and Lemma I.2.1 imply that $u \equiv 0$ in Ω. This argument is applicable as well for the Neumann boundary value problem. It does not require any smoothness from the boundary. Existence of the solution to the Dirichlet boundary value problem does not require any smoothness of Γ, but there are some requirements on Γ in the case of the Neumann boundary condition. This can be explained as follows. In Section I.2.2 we showed how to reduce a boundary value problem in Ω to a boundary value problem for a compact domain. Therefore we can discuss the role of the smoothness of Γ assuming that the domain B, in which the boundary value problem is being studied, is compact. In this case the existence of the solution to the problem

$$-\nabla^2 u = f \text{ in } B, \quad u = 0 \text{ on } \partial B, \quad f \in L^2(B), \tag{33}$$

in $H^{01}(B)$ is equivalent to the validity of the inequality

$$\|h\|_{L^2(B)} \leqslant c \|\nabla h\|_{L^2(B)}, \quad \forall h \in \mathring{H}^1(B), \tag{34}$$

where $c > 0$ does not depend on h. Indeed, the problem (33) in the generalized formulation is equivalent to the validity of the following integral identity:

$$\int_B \nabla u \nabla h \, \mathrm{d}x = \int_B fh \, \mathrm{d}x, \quad \forall h \in \mathring{H}^1(B). \tag{35}$$

The right side of (35) is a linear bounded functional $F(h)$ in $\mathring{H}^1$ for any $f \in L^2$ iff (34) holds. The left side of (35) is a linear bounded functional in $\mathring{H}^1(B)$. Therefore the problem (33) is uniquely solvable in $\mathring{H}^1(B)$ for any $f \in L^2(B)$ iff (34) holds. But (34) holds for any bounded open set $B \subset R^3$ and in this case the embedding $i: \mathring{H}^1(B) \to L^2(B)$ is compact (see e.g. Schulze—Wildenhain [1], p. 331). The spectrum of the Dirichlet Laplacian in $L^2(B)$ is therefore discrete for any bounded open set B.

A similar argument shows that the Neumann boundary value problem in B is solvable in $H^1(B)$ iff the following inequality holds

$$\inf_{a \in R^1} \{\|h - a\|_{L^2(B)}\} \leqslant c\|\nabla h\|_{L^2(B)}, \quad \forall h \in H^1(B) \tag{36}$$

where $c > 0$ does not depend on h. This is Poincare's inequality. A necessary and sufficient condition on B for (36) to hold is known. A necessary and sufficient condition on B for the compactness of the embedding $i: H^1(B) \to L^2(B)$ is known.

Let us formulate these conditions. First we should introduce some definitions. Let F be a closed set, G be an open set, $F \subset G \subset B$. The set $G \backslash F = M$ is called the conductor. Let $V(M) = \{f: f \in C^\infty(B), f = 1$ on $F, f = 0$ on $B \backslash G\}$. Define the conductivity of M by the formula

$$\operatorname{cond} M = \inf\left\{\int_M |\nabla f|^2 \, \mathrm{d}x, \quad f \in V(M)\right\}. \tag{37}$$

Let $\{M\}$ be the set of the conductors for which 2 mes $G \leqslant$ mes B, where mes B is the Lebesgue measure of B. Then (36) holds iff

$$\sup_{M \in \{M\}} \frac{\operatorname{mes} F}{\operatorname{cond} M} < \infty. \tag{38}$$

The embedding $i: H^1(B) \to L^2(B)$ is compact iff

$$\lim_{c \to 0} \left\{ \sup_{\operatorname{mes} G \leqslant c} \frac{\operatorname{mes} F}{\operatorname{cond} M} \right\} = 0. \tag{39}$$

Condition (39) is necessary and sufficient for the spectrum of the Neumann Laplacian in $L^2(B)$ to be discrete. These results are formulated in Schulze—Wildenhain [1], where one can find numerous references to the original papers. An often used sufficient condition for compactness of the embedding i: $H^1(B) \to L^2(B)$ for a bounded domain B is the so-called "finite tiling property" which is defined in Wilcox [1].

In Kral [1] and Burago—Maz'ya [1] the integral equations of potential theory are studied for the surfaces which are not smooth. In this case the operators A' and A defined by (I.1.2) and (I.1.4) are not necessarily compact. If one defines the essential norm of a linear operator T: $B \to B$ on a Banach space B by the formula

$$|T|_{\text{ess}} = \inf_{Q \in K} \|T - Q\| \tag{40}$$

where K is the set of all compact linear operators on B, then the equation

$$(I + T)h = f \tag{41}$$

in B is of Fredholm's type if $|T|_{\text{ess}} < 1$. Therefore if the boundary Γ is not smooth but $|A|_{\text{ess}} < 1$, $|A'|_{\text{ess}} < 1$, then th usual potential theory is still valid. The essential norm of the operators A and A' is calculated in Kral [1], and Burago—Maz'ya [1].

Appendix 2. Summary of Some Results in Operator Theory

1. Let A be a linear operator on a Hilbert space H. This operator is called densely defined if its domain of definition $D(A)$ is dense in H. The adjoint operator A^* is defined by $A^*h = h^*$, where $D(A^*)$ consists of all h such that $(Af, h) = (f, h^*)$, $f \in D(A)$. If $(Af, h) = (f, Ah)$, $\forall f, h \in D(A)$ then A is called symmetric on $D(A)$. If $A = A^*$ then A is called selfadjoint. A linear operator is called closed if $h_n \to h$ and $Ah_n \to f$ imply that $h \in D(A)$ and $Ah = f$. This means that $D(A)$ is complete in the graph norm $\|h\|_{\Gamma(A)} \equiv \|h\| + \|Ah\|$. If $R_\lambda(A) = (A - \lambda I)^{-1} = (A - \lambda)^{-1}$ is a bounded operator defined on all of H then λ is called a regular point of A and one writes $\lambda \in \rho(A)$. Otherwise λ belongs to the spectrum of A, and one writes $\lambda \in \sigma(A)$. The operator $R_\lambda(A)$ is called the resolvent of A. The set of regular points is open, $\sigma(A) = C \backslash \rho(A)$, where C is the set of all complex numbers. If $\lambda \in \sigma(A)$ and $Af = \lambda f$, $f \in H$, $f \neq 0$, then λ is an eigenvalue of A, f is the eigenvector, $N_\lambda = N(A - \lambda I) = \{f: Af = \lambda f\}$ is the eigenspace of A corresponding to λ.

The geometric multiplicity of λ is dim $N(A - \lambda I)$. The set $M_\lambda = \bigcup_{s=1}^{\infty} N\{(A - \lambda I)^s\}$ is the rootspace of A corresponding to λ, dim M_λ is the algebraic multiplicity of λ. If $A = A^*$ then $M_\lambda = N_\lambda$. An eigenvalue λ is called normal if (i) dim $M_\lambda < \infty$ and (ii) $H = M_\lambda \oplus M'_\lambda$, where $\oplus$ is the sign of the direct sum and $A - \lambda I$ is invertible on its invariant subspace M'_λ. The dimension of A is the number $r(A) = \dim \overline{R(A)}$, where the bar denotes closure and $R(A)$ is the range of A; $r(A) = r(A^*) = r\{(A^*A)^{1/2}\} = r\{(AA^*)^{1/2}\}$. If A is a linear bounded operator with $r(A) < \infty$, then A is called finite-dimensional. In this case $Af = \sum_{j=1}^{r(A)} (f, h_j)f_j$, where $\{f_j\}$ is a basis of $R(A)$ and $\{h_j\}$ is a biorthogonal to the $\{f_j\}$ system.

2. If $A = A^*$ then there is a uniquely determined family of orthogonal projections E_λ associated with A, the resolution of the identity for A, such that $F(A) = \int_{-\infty}^{\infty} F(\lambda)\, dE_\lambda$, the integral converges strongly on $D(F(A)) = \{h: \int_{-\infty}^{\infty} |F(\lambda)|^2 \, d(E_\lambda h, h) < \infty\}$, $E(\lambda) \leqslant E(\lambda')$ if $\lambda \leqslant \lambda'$, $E(-\infty) = 0$, $E_\lambda E_\mu = E_{\min(\lambda,\mu)}$, $E(+\infty) = I$, $E_\lambda F(A) = F(A)E_\lambda$, $E_\mu \underset{s}{\to}$

E_λ as $\mu \to \lambda + 0$, where $\underset{s}{\to}$ means strong convergence. One has

$$((E_{\lambda_2} - E_{\lambda_1})h, f) = \lim_{\delta \to +0} \lim_{\varepsilon \to +0} \frac{1}{2\pi i} \int_{\lambda_1 + \delta}^{\lambda_2 + \delta} ((R_{\lambda + i\varepsilon} - R_{\lambda - i\varepsilon})h, f)\, \mathrm{d}\lambda \tag{1}$$

for any $h, f \in H$. If the weak derivative $\mathrm{d}E_\lambda/\mathrm{d}\lambda$ exists and is an integral operator then its kernel is

$$\frac{\mathrm{d}E_\lambda(x, y)}{\mathrm{d}\lambda} = \frac{1}{\pi} \operatorname{Im} R_{\lambda + i0}(x, y) \tag{2}$$

where $\operatorname{Im} R_{\lambda+i0}(x, y) = \lim_{\varepsilon \to +0} \operatorname{Im} R_{\lambda+i\varepsilon}(x, y)$, and $R_{\lambda+i\varepsilon}(x, y)$ is the kernel of the operator $R_{\lambda+i\varepsilon} = (A - \lambda - i\varepsilon)^{-1}$. Let $A = A^*$. We write $A \geqslant m$ if $(Af, f) \geqslant m(f, f)$, $f \in D(A)$, and call A semibounded from below. One defines the multiplicity $N(\lambda_1, \lambda_2)$ of the spectrum of $A = A^*$ in the interval $(\lambda_1, \lambda_2]$ as $\dim(E_{\lambda_2} - E_{\lambda_1})H$.

LEMMA 1. *Let A be semibounded from below and $N(\lambda) = \sum_{\lambda_n \leqslant \lambda} 1$, where λ_n are the eigenvalues of A counted according to their multiplicities, $\lambda_1 \leqslant \lambda_2 \leqslant \ldots$. Then*

$$N(\lambda) = \sup_M \dim M, \quad M \subseteq \{f : f \in D(A), \quad (Af, f) \leqslant \lambda(f, f)\} \tag{3}$$

where M are linear manifolds. More generally, if $N(\lambda_1, \lambda_2)$ is the multiplicity of the spectrum of A in the interval $(\lambda_1, \lambda_2]$ then

$$\begin{gathered} N(\lambda_1, \lambda_2) = \sup_M \dim M, \\ M \subseteq \{f : f \in D(A), \lambda_1(f, f) < (Af, f) \leqslant \lambda_2(f, f)\}. \end{gathered} \tag{4}$$

Sketch of the Proof. We prove (4) since (3) is a particular case of (4). $N(\lambda_1, \lambda_2) = \infty$ iff there exists a linear manifold $M \subset D(A)$ of arbitrary dimension d, such that $\lambda_1(f, f) < (Af, f) \leqslant \lambda_2(f, f)$, $f \in M$. For example, M can be constructed as a linear span of d linearly independent vectors from $(E_{\lambda_2} - E_{\lambda_1})H \equiv E_{\lambda_1\lambda_2}H$. If $N(\lambda_1, \lambda_2) = d < \infty$ then there exist d linearly independent vectors f_j in $E_{\lambda_1\lambda_2}H$. One has $(Af_j, f_j) = (AE_{\lambda_1\lambda_2}f_j, E_{\lambda_1\lambda_2}f_j) = \int_{\lambda_1}^{\lambda_2} \lambda d(E_\lambda f_j, f_j)$. Thus $\lambda_1 \|f_j\|^2 < (Af_j, f_j) \leqslant \lambda_2 \|f_j\|^2$. Conversely, if there is an M_0, such that $\lambda_1 \|f\|^2 < (Af, f) \leqslant \lambda_2 \|f\|^2$, $f \in M_0$, $\dim M_0 = d = \sup \dim M$, where M is defined in (4), then $\dim E_{\lambda_1\lambda_2}H = d$. Indeed, $\dim E_{\lambda_1\lambda_2}H \geqslant d$ because $M_0 \subset E_{\lambda_1\lambda_2}H$, (if $M_0 \not\subset E_{\lambda_1\lambda_2}H$ then there is an $f \in M_0$, $f \perp E_{\lambda_1\lambda_2}H$, $\|f\| = 1$, for which $(Af, f) \notin (\lambda_1, \lambda_2])$; and $\dim E_{\lambda_1\lambda_2}H \leqslant d$, because otherwise one could find at least $d + 1$ linearly independent vectors f_j such that $\lambda_1\|f_j\|^2 < (Af_j, f_j) \leqslant \lambda_2\|f_j\|^2$, $1 \leqslant j \leqslant d+1$, which is a contradiction. □

A point λ is called a point of the continuous spectrum of a selfadjoint operator A, $\lambda \in \sigma_c(A)$ iff $N(\lambda - \varepsilon, \lambda + \varepsilon) = \infty$ for any $\varepsilon > 0$. If $0 < N(\lambda - \varepsilon, \lambda + \varepsilon) < \infty$ for some $\varepsilon > 0$ then λ is a point of the discrete spectrum of A, it is an isolated eigenvalue of finite multiplicity. If A is not necessarily selfadjoint, then define continuous spectrum of A, $\sigma_c(A)$, by the formula: $\sigma_c(A) = \{\lambda\colon \exists f_n \in D(A), \|Af_n - \lambda f_n\| \to 0$ as $n \to \infty$, $\{f_n\}$ is a bounded noncompact sequence$\}$. Some authors call $\sigma_c(A)$ essential spectrum of A.

LEMMA 2. *$N(\lambda - \varepsilon, \lambda + \varepsilon) < \infty$, $\varepsilon > 0$, iff there exists a subspace F, $\operatorname{codim} F < \infty$, such that $\|(A - \lambda)f\| \geqslant \varepsilon\|f\|$, $f \in F \cap D(A)$.*

LEMMA 3. *If $A = A^*$, then $\lambda \in \sigma_c(A)$ iff there exists a noncompact sequence f_n, $\|f_n\| = 1$, such that $\|(A - \lambda)f_n\| \to 0$, $n \to \infty$.*

The set of $h \in D(A)$ for which the function $(E_\lambda h, h)$ is absolutely continuous in λ is an invariant subspace of A, $H_{ac}(A)$. The restriction A_{ac} of A onto $H_{ac}(A)$ is the absolutely continuous part of A, $\sigma(A_{ac}) \equiv \sigma_{ac}(A)$, its spectrum is the absolutely continuous spectrum of A, $\sigma_{ac}(A) \subset \sigma_c(A)$. One says that $\tilde{A}$ is a closed extension of A if $D(\tilde{A}) \supset D(A)$ and $\tilde{A} = A$ on $D(A)$. If $\dim D(A) < \infty$ (mod $D(A)$)) then the extension is called finite-dimensional.

Here $\dim M(\text{mod } L) = \dim M/L$, where $M \supset L$ are linear manifolds, and M/L is the quotient set. One has $\sigma_c(\tilde{A}) \supset \sigma_c(A)$ and

LEMMA 4. *If $\tilde{A}$ is a finite-dimensional extension of A then $\sigma_c(\tilde{A}) = \sigma_c(A)$. Let M_1 and M_2 be orthogonal linear manifolds invariant with respect to A. If A_j is the restriction of A onto M_j then one defines $A_1 \oplus A_2(f_1 + f_2) = A_1f_1 \oplus A_2f_2$, where $f_j \in M_j$, $D(A_1 \oplus A_2) = M_1 \oplus M_2$. Therefore A is an extension of $A_1 \oplus A_2$, $\sigma_c(A) \supset \{\upsilon_c(A) \cup \sigma_c(A_2)\}$.*

3. Let A be a linear symmetric closed operator on H, λ be a complex number, $\operatorname{Im}\lambda \neq 0$, $N_{\bar{\lambda}}^* = N(A^* - \bar{\lambda})$, the bar denotes complex conjugate. One has $D(A^*) = D(A) \dotplus N_{\bar{\lambda}}^* \dotplus N_{\lambda}^*$, where $\dotplus$ is the symbol of the direct sum. This means that any vector from $D(A^*)$ can be represented as a sum of the vectors from $D(A)$, $N_{\bar{\lambda}}^*$, and N_{λ}^*, and this representation is unique. The numbers $\dim N_{\bar{\lambda}}^* = n(\lambda)$ and $\dim N_{\bar{\lambda}}^* =$

$n(\bar{\lambda})$ are called the defect numbers of A, the pair $(n(\lambda), n(\bar{\lambda}))$ is the defect index of A. These numbers do not depend on λ, $\operatorname{Im} \lambda \neq 0$. A closed symmetric operator A is selfadjoint iff $n(\lambda) = n(\bar{\lambda}) = 0$. A closed symmetric operator has selfadjoint extensions iff $n(\lambda) = n(\bar{\lambda})$. If A is symmetric and its closure $\bar{A}$ is selfadjoint then A is called essentially selfadjoint. If A is symmetric and bounded from below then $\bar{A}$ has equal defect numbers and it has selfadjoint extensions. If $A > m$ then $A - m > 0$. Therefore for a study of symmetric extensions of A one can assume that A is positive definite, i.e. $A \geqslant m > 0$. In this case the closure of $D(A)$ in the norm $\|f\|_A = (Af, f)^{1/2}$ is a Hilbert space $H_A \subset H$. One can prove that on $D(A^*) \cap H_A$ the operator A^* is selfadjoint and is, therefore, a selfadjoint extension $\tilde{A}_m$ of A. This extension is called Friedrichs's extension. It is the minimal selfadjoint extension of A in the sense that $\tilde{A}_m \geqslant m$ and $(\tilde{A}_m^{-1}f, f) \leqslant (\tilde{A}^{-1}f, f)$ for any selfadjoint positive definite extension of A, so that $B = \tilde{A}^{-1} - \tilde{A}_m^{-1} \geqslant 0$. The operator $B = 0$ on $R(A)$, so one can consider B as an operator on the orthogonal complement to $\overline{R(A)}$, i.e. on $N(A^*)$. If $\tilde{A}$ is an arbitrary selfadjoint positive definite extension of A, then $D(\tilde{A}) = D(A) \oplus (\tilde{A}_m^{-1} + B)N(A^*)$, where $B \geqslant 0$ is a selfadjoint bounded operator on $N(A^*)$, $D(\tilde{A}^{1/2}) = D(A_m^{1/2}) \oplus R(B^{1/2})$, $D(\tilde{A}_m^{1/2}) = H_A$.

3. If A is a linear closed operator and K is a linear compact operator, then $\sigma_c(A) = \sigma_c(A + K)$.

If $A \geqslant m > 0$, K is symmetric, $D(K) \supset D(A)$, and $\tilde{A}_m^{-1}K$ is compact in H_A then K is called relatively compact in H_A with respect to A. If $K \geqslant 0$ then K is relatively compact in H_A with respect to A iff the embedding $i\colon H_A \to H_K$ is compact. If $A \geqslant m > 0$ and K is symmetric, relatively compact in H_A with respect to A, then $\sigma_c(A) = \sigma_c(A + K)$.

If A is selfadjoint and $D(A) \subset D(T)$, where T is a linear operator, then T is called relatively compact with respect to A iff $T(A + i)^{-1}$ is compact. If A is a closed linear densely defined operator in H and $\lambda \notin \sigma(A)$, then $(A - \lambda)^{-1}$ is compact iff $(I + A^*A)^{-1}$ is compact.

4. If λ is an isolated eigenvalue of finite algebraic multiplicity r of a linear operator A and Γ is a simple closed curve around λ, then the operator

$$P = -(2\pi i)^{-1} \int_\Gamma (A - zI)^{-1} \, dz \tag{5}$$

is a projection, $P^2 = P$. If there are no other points of the spectrum of A inside Γ, then $\dim PH = r$.

If L is a selfadjoint operator with a discrete spectrum, $0 < \lambda_1 \leqslant \lambda_2 < \ldots, \lambda_n \to +\infty$, then $\|P_n f - f\| \to 0$, $n \to \infty$, for any $f \in H$, P_n is the orthoprojection on the linear span of the eigenvectors of L corresponding to the eigenvalues $\lambda_1 \ldots \lambda_n$.

5. LEMMA 5. *If $A > mI$ is a selfadjoint, bounded from below, operator in a Hilbert space H, and $H_A \subset H$ is the Hilbert space which is the completion of $D(A)$ in the norm generated by the inner product $((A - mI + I)f, f)$, then the spectrum of A is discrete iff the embedding i: $H_A \to H$ is compact. The spectrum of A is called discrete if it consists of isolated eigenvalues of finite multiplicity with the only possible accumulation point at infinity.*

Proof. Without loss of generality assume that $A \geqslant I$, so that A^{-1} is bounded and defined on all of H. Then A has a discrete spectrum iff A^{-1} is compact. Since A^{-1} is selfadjoint, it is compact iff $A^{-1/2}$ is. If $A > 0$, then $(Au, u) = \|A^{1/2}u\|^2$. Therefore, compactness of the embedding i: $H_A \to H$ is equivalent to compactness of $A^{-1/2}$, i.e. to the discreteness of the spectrum of A. □

Appendix 3. Eigenfunction Expansions and Asymptotics of the s-Values for Some Nonselfadjoint Operators

1. In this section we introduce the notions of the Riesz basis and Riesz basis with brackets, and give conditions for a root system of an operator to form a Riesz basis with brackets.

A system $\{f_j\}$ of the elements of a Hilbert space H is called a Riesz basis of H, and we write $\{f_j\} \in R$ or $\{f_j\} \in R(H)$ iff $f_j = Th_j$, where T is a linear isomorphism of H onto H, and $\{h_j\}$ is an orthonormal basis of H. Let P_F denote an orthoprojection onto a subspace F.

A system $\{f_j\}$ is called a Riesz basis of H with brackets, $\{f_j\} \in R_b$, iff there exists a sequence of integers $n_1 \leqslant n_2 \leqslant \ldots \to \infty$, and a linear isomorphism T of H onto H such that $TF_j = H_j$, $j = 1, 2, \ldots$. Here F_j is the linear span of the elements $f_{n_j}, f_{n_j+1}, \ldots f_{n_{j+1}-1}$, $H_j = \operatorname{span}\{h_{n_j}, \ldots h_{n_{j+1}-1}\}$ and $\{h_j\}$ is an orthonormal basis of H. The sequence $n_1, n_2, \ldots$ defines the bracketing rule. If $\{f_j\} \in R_b$ and $\sum_{j=1}^{\infty} \|P_{F_j} - P_{H_j}\| < \infty$ then we write $\{f_j\} \in B_b$, and call $\{f_j\}$ a Bari basis with brackets.

THEOREM 1. *A system $\{f_j\} \in R(H)$ iff the system $\{f_j\}$ is complete in H and one of the following conditions holds:*

(1) *There exist $a_1 > 0$ and $a_2 \geqslant a_1$, such that for any n and any c_j one has*

$$a_1 \sum_{j=1}^{n} |c_j|^2 \leqslant \left\| \sum_{j=1}^{n} c_j f_j \right\|^2 \leqslant a_2 \sum_{j=1}^{n} |c_j|^2; \tag{1}$$

(2) *the Gram matrix (f_i, f_j), $1 \leqslant i, j < \infty$, generates an isomorphism of ℓ^2 onto ℓ^2.*

(3) *for any $f \in H$ one has*

$$\sum_{j=1}^{\infty} |(f, f_j)|^2 < \infty \quad \text{and} \quad \sum_{j=1}^{\infty} |(f, \tilde{f}_j)|^2 < \infty \tag{2}$$

where $\{\tilde{f}_j\}$ is the biorthogonal system, i.e. $(f_i, \tilde{f}_j) = \delta_{ij}$;

(4) *the system $\{f_j\}$ is an unconditional basis of H and*

$$0 < m \leqslant \|f_j\| < M < \infty.$$

A system $\{f_j\}$ is a basis (Schauder's basis) of M iff any $f \in H$ can be represented by a convergent series $f = \sum_{j=1}^{\infty} c_j f_j$, and this representation is unique. A basis is unconditional iff it remains a basis after an arbitrary transposition of its terms, i.e. $\{f_{n_j}\}$ is a basis for any transposition $(n_1, n_2, \ldots)$ of the sequence $(1, 2, 3 \ldots)$.

We write $L \in R_b(H)(L \in B_b(H))$ if the root vectors of the operator L form a Riesz (Bari) basis with brackets.

2. Let $L \geqslant m > 0$ be a selfadjoint operator on H with a discrete spectrum $0 < \lambda_1 \leqslant \lambda_2 \leqslant \ldots,$

$$\lambda_n = cn^p(1 + 0(n^{-p_1})), \quad n \to \infty, p_1 > 0, \quad p > 0, \quad c > 0, \tag{4}$$

$$M = L + T, \quad D(T) \supset D(L), \tag{5}$$

$$|Tf| \leqslant c|L^a f|, \quad a < 1, \quad D(T) \supset D(L^a). \tag{6}$$

By c here and below one denotes various positive constants, $L^a = \int_m^\infty \lambda^a \, dE_\lambda$, where E_λ is the resolution of the identity for L.

THEOREM 2. *Let* (4)–(6) *hold. If* $p(1 - a) = 1$ *then* $M \in R_b(H)$. *If* $p(1 - a) > 1$ *then* $M \in B_b(H)$. *The spectrum of* M *is discrete and* $\sigma(M) \subset \bigcup_{j=1}^{\infty} C_j$, $C_j = \{\lambda : |\lambda - \lambda_j| \leqslant c_0 \lambda_j^a\}$, $c_0 > 0$.

It is often convenient to reduce the problem of basisness for the operator M to the same problem for a compact operator. We say that an operator M has a basisness property if $M \in R_b(H)$. One writes $M = (I + TL^{-1})L$ and

$$M^{-1} = L^{-1}(I + TL^{-1})^{-1} = A(I + Q), \quad A = L^{-1}. \tag{7}$$

Here we assume that M is invertible, and use the formula

$$(I + S)^{-1} = I + Q, \quad Q = -(I + S)^{-1}S, \quad S = TL^{-1} \tag{8}$$

which holds if $I + S$ is invertible.

We assume that

$$\|TL^{-a}\| \leqslant c, a < 1. \tag{9}$$

In this case the operator $S = TL^{-a}L^{-1+a}$ is compact since TL^{-a} is bounded and L^{-1+a} is compact. The operator L^{-b}, $b > 0$, is compact because of (4). Therefore the basisness property for M holds iff it holds for the operator $B = A(I + Q)$. Here

$$A = A^* > 0 \text{ is compact, } \lambda_n(A) \sim c^{-1} n^{-p}(1 + 0(n^{-p_1})), \quad p > 0, \quad p_1 > 0 \tag{10}$$

$$Q \text{ is compact, } I + Q \text{ is invertible, } \|Qf\| \leqslant c\|A^{1-a}f\|, \quad f \in D(L^{1-a}) \tag{11}$$

where $a < 1$ is the same as in (6).

Theorem 2 can be reformulated as

THEOREM 2′. *Let* $B = A(I + Q)$ *and conditions* (10) *and* (11) *hold. If* $p(1 - a) = 1$ *then* $B \in R_b(H)$, *and if* $p(1 - a) > 1$ *then* $B \in B_b(H)$.

Completeness of the root system of an operator $B = A(I + Q)$ can be checked as follows. Let A be a compact linear operator in H, $s_n(A) = \lambda_n\{(A^*A)^{1/2}\}$ be its s-values. We say that $A \in \sigma_p$ if $\sum_{n=1}^{\infty} |s_n(A)|^p < \infty$ (*).

Let $p(A) = \inf p$, where the infimum is taken over all p for which (*) holds. If $p(A) = 1$ then A is called trace-class operator, if $p(A) = 2$ then A is called Hilbert-Schmidt class operator. We write $A \in \sigma_\infty$ if A is compact.

THEOREM 3. *Let* $B = A(I + Q)$, $A = A^*$, $p(A) < \infty$, $Q \in \sigma_\infty$ *and* $N(B) = \{0\}$. *Then the root systems of* B *and* B^* *and complete in* H.

THEOREM 4. *Let* $M = L + T$, $L = L^*$, $L^{-1} \in \sigma_\infty$, $p(L^{-1}TL^{-1}) < \infty$. *Then the root system of* M *is complete in* H.

It is interesting to known when the root vector expansion for the operator L in Theorem 2 converges or diverges simultaneously with that for the operator M, in other words, when the root vector expansions for L and M are equiconvergent. The equiconvergence is uniform on bounded sets if for any $R > 0$ the estimate (45) (see below) holds.

THEOREM 5. *The root vector expansions for* L *and* M *are equiconvergent uniformly on bounded sets if the assumptions* (4)–(6) *hold,* $p(1 - a) > 2$, *and the proper bracketing is used.*

Proofs of the Theorems 1, 3, 4 can be found in Gohberg–Krein [1]. The results in Theorems 2 and 2′ were obtained in Marcus [1] and Kacnelson [1] (see also Agranovich [1]). We prove Theorem 5 and recover part of the result of Theorem 2 in Section 4 below. The proof of Theorem 5 follows closely the arguments in Ramm [3] p. 306. Lemma 6 in Section 4 was used in Kacnelson [1].

3. In this section we prove some theorems about asymptotics of the s-values of perturbed compact operators.

We will need the following lemma.

LEMMA 1. *A linear operator Q on H is compact iff* $\sup_{h \perp L_n} \|Qh\|/\|h\| \to 0$, *as* $n \to \infty$, *where L_n is an n-dimensional subspace of H, $L_n \subset L_{n+1}$ and the sequence L_n is limit dense, i.e. $\rho(f, L_n) \to 0$, $n \to \infty$ for any $f \in H$, where $\rho(f, L_n)$ is the distance from f to L_n.*

THEOREM 6. *Let $A \in \sigma_\infty$, $Q \in \sigma_\infty$, $N(I+Q) = \{0\}$, and $B = A(I+Q)$. Then*

$$\lim_{n \to \infty} s_n(B) s_n^{-1}(A) = 1. \tag{12}$$

Proof. One has

$$\begin{aligned} s_{n+1}(B) &= \min_{L_n} \max_{f \perp L_n} \frac{\|Bf\|}{\|f\|} \leqslant \\ &\leqslant \max_{(I+Q)f \perp L_n(A)} \left\{ \frac{\|A(I+Q)f\|}{\|(I+Q)f\|} \frac{\|(I+Q)f\|}{\|f\|} \right\} \leqslant \\ &\leqslant s_{n+1}(A) \left(1 + \max_{(I+Q)f \perp L_n(A)} \frac{\|Qf\|}{\|f\|} \right) \leqslant \\ &\leqslant s_{n+1}(A)(1+\varepsilon_n), \quad \varepsilon_n \to 0. \end{aligned} \tag{13}$$

Here L_n is an n-dimensional subspace, $L_n(A)$ is the linear span of the first n eigenvectors (counted according to their multiplicities) of the operator A^*A, so that

$$s_{n+1}(A) = \max_{h \perp L_n(A)} \frac{\|Ah\|}{\|h\|}, \tag{14}$$

and Lemma 1 was used at the last step in (13). The invertibility of $I + Q$ guarantees that the sequence $(I + Q^*)L_n(A)$ is limit dense in H.

Since $A = B(I+S)$, where $I+S = (I+Q)^{-1}$, and S is compact if Q is, one concludes as above that

$$s_{n+1}(A) \leqslant s_{n+1}(B)(1+\varepsilon_n'), \quad \varepsilon_n' \to 0, \quad n \to \infty. \tag{15}$$

Inequalities (13) and (15) yield (12).

One can estimate the remainder under some additional assumptions. □

THEOREM 7. *Let* $B = A(I+Q)$, $A \in \sigma_\infty$, $Q \in \sigma_\infty$, $N(I+Q) = \{0\}$,

$$s_n(A) = cn^{-p}(1 + 0(n^{-p_1})), \quad p > 0, \quad p_1 > 0, \tag{16}$$

and

$$\|Qf\| \leqslant c\|Af\|^a\|f\|^{1-a}, \quad a > 0 \tag{17}$$

Then

$$s_n(B) = s_n(A)(1 + 0(n^{-q})), \quad q = \min\left(p_1, \frac{pa}{1+pa}\right). \tag{18}$$

The estimate of the remainder in (18) *is sharp in the sense that it is attained for some* Q.

Proof. One has

$$\begin{aligned} s_{n+m+1}(B) &\leqslant \max_{\substack{(I+Q)f \perp L_n(A)\\ f \perp L_m(A)}} \frac{\|A(I+Q)f\|}{\|(I+Q)f\|} \cdot \left(1 + \max_{\substack{(I+Q)f \perp L_n(A)\\ f \perp L_m(A)}} \frac{\|Qf\|}{\|f\|}\right) \leqslant \\ &\leqslant s_{n+1}(A)\left(1 + c\max_{f \perp L_m(A)} \frac{\|Af\|^a}{\|f\|^a}\right) \leqslant \\ &\leqslant s_{n+1}(A)(1 + cs_{m+1}^a(A)),\ n, m \to \infty. \end{aligned} \tag{19}$$

The symmetry argument used in the proof of Theorem 5 should be modified. We have $A = B(I+S)$, $S = -(I+Q)^{-1}Q$

$$\begin{aligned} s_{n+m+1}(A) &\leqslant \max_{\substack{(I+S)f \perp L_n(B)\\ f \perp L_m(A)}} \frac{\|B(I+S)f\|}{\|(I+S)f\|} \cdot \left(1 + \max_{f \perp L_m(A)} \frac{\|Sf\|}{\|f\|}\right) \leqslant \\ &\leqslant s_{n+1}(B)\left(1 + c_1\max_{f \perp L_m(A)} \frac{\|Af\|^a}{\|f\|^a}\right) \leqslant \\ &\leqslant s_{n+1}(B)(1 + c_1 s_{m+1}^a(A)). \end{aligned} \tag{20}$$

Therefore (19) and (20) imply

$$\begin{aligned} s_{n+2m+1}(A) &\leqslant s_{n+m+1}(B)(1+cs^{\alpha}_{m+1}(A)) \leqslant \\ &\leqslant s_{n+1}(A)(1+c_1 s^{\alpha}_{m+1}(A)). \end{aligned} \tag{21}$$

Thus

$$\begin{aligned} \frac{s_{n+2m+1}(A)}{s_{n+m+1}(A)}(1+0(s^{\alpha}_{m+1}(A))) &\leqslant \frac{s_{n+m+1}(B)}{s_{n+m+1}(A)} \leqslant \\ &\leqslant \frac{s_{n+1}(A)}{s_{n+m+1}(A)}(1+0(s^{\alpha}_{m}(A))). \end{aligned} \tag{22}$$

Let $m = n^{1-x}$, $x > 0$. It follows from (16) that

$$\begin{aligned} \frac{s_{n+m}(A)}{s_n(A)} &= \left(\frac{n+m}{n}\right)^{-p}(1+0(n+m)^{-p_1}+0(n^{-p_1})) \\ &= 1+0(n^{-p_1})+0\left(\frac{m}{n}\right) = 1+0(n^{-p_1})+0(n^{-x}). \end{aligned} \tag{23}$$

It follows from (22) and (23) that

$$\frac{s_{n+m+1}(B)}{s_{n+m+1}(A)} = 1+0(n^{-p_1})+0(n^{-x})+0(n^{-(1-x)p\alpha}). \tag{24}$$

Since $\min(x,(1-x)p\alpha) = p\alpha(1+p\alpha)^{-1}$ one has

$$q = \min(p_1, x, (1-x)p\alpha) = \min(p_1, p\alpha(1+p\alpha)^{-1}). \tag{25}$$

This and (24) imply (18). The estimate of the remainder is sharp: it is attained if $p_1 > p\alpha(1+p\alpha)^{-1}$, $A > 0$ and $Q = A^{\alpha}$. □

4. Here we prove Theorem 5.

LEMMA 2. *If* (4) *holds then*

$$N(\lambda) = \sum_{\lambda_n < \lambda} 1 = (\lambda c^{-1})^{1/p}(1+0(\lambda^{-p_1/p})). \tag{26}$$

Proof. The functions λ_N and $N(\lambda)$ are reciprocal, thus (4) implies that

$$(\lambda c^{-1})^{1/p} = n(1+0(n^{-p_1})), \quad n = (\lambda c^{-1})^{1/p}(1+0(\lambda^{-p_1/p})). \qquad \square$$

LEMMA 3. *If* (4) *holds then, for sufficiently large* n *and* m, $0 < q_1 \leqslant nm^{-1} \leqslant q_2 < 1$, *there exist eigenvalues* $\lambda^{(1)}$ *and* $\lambda^{(2)}$ *such that the interval* $(\lambda^{(1)}, \lambda^{(2)})$ *is free from the eigenvalues,* $\lambda_n \leqslant \lambda^{(1)} < \lambda^{(2)} \leqslant \lambda_m$, *and* $\lambda^{(2)} - \lambda^{(1)} \geqslant c_1 m^{p-1}$.

Proof. There are $m - n$ eigenvalues (counting multiplicities) on the segment $(\lambda_n, \lambda_m]$. So, there exists at least two eigenvalues $\lambda_n \leqslant \lambda^{(1)} < \lambda^{(2)} \leqslant \lambda_m$ such that

$$\lambda^{(2)} - \lambda^{(1)} \geqslant \frac{\lambda_m - \lambda_n}{m - n} \geqslant c\frac{m^p - n^p}{m - n} - 0\left(\frac{m^{p-p_1}}{m - n}\right) \geqslant c_1 m^{p-1}. \qquad \square$$

LEMMA 4. *Let* $m = m(n)$, $1 - b_2(n) \leqslant m^{-1}(n)n \leqslant 1 - b_1(n)$, $b_1(n)b_2^{-1}(n) \geqslant c_2$, $n^{p_1}b_1(n) \to \infty$, *and* $b_2(n) \to 0$, *as* $n \to \infty$. *Then the conclusion of Lemma 3 holds.*

Proof. It is similar to that of Lemma 3. One has

$$\begin{aligned}\frac{m^p - n^p - 0(m^{p-p_1})}{m - n} &= m^{p-1}\frac{1 - (m^{-1}(n)n)^p - 0(m^{-p_1})}{1 - m^{-1}(n)n} \geqslant \\ &\geqslant m^{p-1}\frac{1 - (1 - b_1(n))^p - 0(m^{-p_1})}{b_2(n)} \geqslant \\ &\geqslant m^{p-1}\frac{b_1(n)\{1 - 0(m^{-p_1}b_1^{-1}(n))\}}{b_2(n)} \geqslant \\ &\geqslant c_2 m^{p-1}(1 - o(1)) \geqslant c_1 m^{p-1}. \qquad \square\end{aligned}$$

Remark 1. If $b_1(n) = b_2(n) = n^{-p_2}$, $p_2 < p_1$ then the conditions of Lemma 4 are satisfied.

LEMMA 5. *If* (4) *holds,* $R_\lambda^0 = (L - \lambda I)^{-1}$ *and* $\lambda^{(m)} = \frac{1}{2}(\lambda^{(2)} - \lambda^{(1)})$, *where* $\lambda^{(1)}$ *and* $\lambda^{(2)}$ *are defined in Lemma 3, then*

$$\| R^0_{\lambda^{(m)}} L^a \| \leqslant c m^{1-p(1-a)}. \tag{27}$$

Proof. One has

$$\| R^0_{\lambda^{(m)}} L^a \| = \sup_{\lambda_j} \frac{|\lambda_j|^a}{|\lambda^{(m)} - \lambda_j|} \leqslant c\frac{\lambda_m^a}{m^{p-1}} \leqslant c m^{1-p(1-a)},$$

where Lemma 3 was used.

It follows from (27) that

$$\|R^0_{\lambda^{(m)}}L^\alpha\| \to 0, \quad m \to \infty, \quad \text{if} \quad p(1-\alpha) > 1. \tag{28}$$

Furthermore

$$\|R^0_{\lambda^{(m)}}\| \leqslant \sup_j \frac{1}{|\lambda^{(m)} - \lambda_j|} \leqslant \frac{c}{m^{p-1}}. \tag{29}$$

For large m one has

$$\begin{aligned} R_{\lambda^{(m)}} &= (M - \lambda^{(m)} I)^{-1} \\ &= R^0_{\lambda^{(m)}}(I + TR^0_{\lambda^{(m)}})^{-1}, \quad \|TR^0_{\lambda^{(m)}}\| \leqslant \|R^0_{\lambda^{(m)}}L^\alpha\| \cdot \|TL^{-\alpha}\|. \end{aligned} \tag{30}$$

It follows from (30), (27), and (6) that

$$\|R_{\lambda^{(m)}}\| \leqslant c\|R^0_{\lambda^{(m)}}\| \leqslant cm^{1-p}. \tag{31}$$

Furthermore (27) and (31) imply that

$$\begin{aligned} \|R_{\lambda^{(m)}} - R^0_{\lambda^{(m)}}\| &= \|R_{\lambda^{(m)}}TR^0_{\lambda^{(m)}}\| \leqslant \\ &\leqslant c\|R_{\lambda^{(m)}}\| \cdot \|R^0_{\lambda^{(m)}}L^\alpha\| \leqslant cm^{2-p(2-\alpha)} \end{aligned} \tag{32}$$

$$\begin{aligned} \|R_{\lambda^{(m)}}T\| &\leqslant \|(R_{\lambda^{(m)}} - R^0_{\lambda^{(m)}})T\| + \|R^0_{\lambda^{(m)}}T\| \leqslant \\ &\leqslant cm^{1-p(1-\alpha)}, \quad \text{if} \quad 1 < p(1-\alpha) \end{aligned} \tag{33}$$

and

$$\|R_{\lambda^{(m)}} - R^0_{\lambda^{(m)}})L\| = \|R_{\lambda^{(m)}}TR^0_{\lambda^{(m)}}L\| \leqslant cm^{2-p(1-\alpha)}. \tag{34}$$

□

Proof of Theorem 5. Let $m(n) = n/(1 - n^{-p_2})$, $p_2 < p_1$. Then the conditions of Lemma 4 are satisfied. Let the sequence $m(n)$ define the bracketing rule, and $\lambda^{(n)} = (\lambda^{(1)} + \lambda^{(2)})/2$, where $\lambda^{(1)}$ and $\lambda^{(2)}$ are defined in Lemma 3 with $\lambda_n = \lambda_{m(n)}$ and $\lambda_m = \lambda_{m(n+1)}$. Let $C_n = \{\lambda: |\lambda| = \lambda^{(n)}\}$. Integrating the identity

$$\frac{1}{2\pi i}\lambda^{-1}f = (-R_\lambda f + \lambda^{-1}R_\lambda Lf + \lambda^{-1}R_\lambda Tf)\frac{1}{2\pi i} \tag{35}$$

over C_n yields

$$f = f_n + \alpha_n(f) + b_n(f) \tag{36}$$

where

$$f_n = -\frac{1}{2\pi i}\int_{C_n} R_\lambda f\,\mathrm{d}\lambda, \quad \alpha_n(f) = \frac{1}{2\pi i}\int_{C_n} \lambda^{-1} R_\lambda L f\,\mathrm{d}\lambda,$$
$$b_n(f) = \frac{1}{2\pi i}\int_{C_n} \lambda^{-1} R_\lambda T f\,\mathrm{d}\lambda. \tag{37}$$

In a similar way

$$f = f_n^0 + \alpha_n^0(f), \quad f_n^0 = -\frac{1}{2\pi i}\int_{C_n} R_\lambda^0 f\,\mathrm{d}\lambda,$$
$$\alpha_n(f) = \frac{1}{2\pi i}\int_{C_n} \lambda^{-1} R_\lambda^0 L f\,\mathrm{d}\lambda. \tag{38}$$

According to formula (A2.5), $f_n^0 = P_n^0 f$, where P_n^0 is the orthoprojection in H onto the linear span of eigenvectors of L corresponding to the eigenvalues lying in the circle $|\lambda| < \lambda^{(n)}$. The spectral theorem for selfadjoint operators yields $\|P_n^0 f - f\| \to 0$,

$$\|\alpha_n^0(f)\| \to 0,\ n \to \infty, \quad \text{for any} \quad f \in H,$$
$$\|P_n^0\| \leqslant 1, \|\alpha_n^0\| \leqslant \|I - P_n^0\| \leqslant 2. \tag{39}$$

We have $f_n = P_n f$, where P_n is a projection on the linear span of the root vectors of M corresponding to the eigenvalues of M lying in the circle $|\lambda| < \lambda^{(n)}$. Thus (36) and (38) yield

$$\|P_n f - P_n^0 f\| \leqslant \|(\alpha_n - \alpha_n^0) f\| + \|b_n(f)\|. \tag{40}$$

To show that

$$\|P_n f - P_n^0 f\| \to 0, \quad n \to \infty, \quad f \in H \tag{41}$$

it is sufficient to show that each term in the right hand side of (40) goes to zero as $n \to \infty$.

It follows from (4), (33), and (37) that

$$\|b_n(f)\| \leqslant \|R_{\lambda^{(n)}} T f\| \leqslant$$
$$\leqslant c n^{1-p(1-\alpha)} \|f\| \quad \text{if} \quad p(1-\alpha) > 1. \tag{42}$$

Furthermore

$$\|\alpha_n(f) - \alpha_n^0(f)\| = \left\| \frac{1}{2\pi i} \int_{C_n} \lambda^{-1}(R_\lambda - R_\lambda^0)Lf \, \mathrm{d}\lambda \right\| \leqslant$$
$$\leqslant \sup_{|\lambda| = \lambda^{(m)}} \|(R_\lambda - R_\lambda^0)Lf\lambda\| \leqslant cn^{2-p(1-\alpha)}\|f\|, \tag{43}$$

where we use (34). It follows from (42), (43), and (40) that

$$\|P_n f - P_n^0 f\| \leqslant \varepsilon(n)\|f\|, \quad \varepsilon(n) = c(n^{1-p(1-\alpha)} + n^{2-p(1-\alpha)}). \tag{44}$$

If $p(1 - \alpha) > 2$ then $\varepsilon(n) \to 0$. Therefore, the partial sums $P_n f$ and $P_n^0 f$ of the root vector expansion for M and eigenvector expansion for L are not only equiconvergent at any $f \in H$ but this equiconvergence is uniform on bounded sets $\|f\| \leqslant$ const in the sense that

$$\sup_{\|f\| \leqslant R} \|P_n f - P_n^0 f\| \to 0 \quad \text{as} \quad n \to \infty, \quad \text{if} \quad p(1 - \alpha) > 2. \tag{45}$$

□

Remark 2. Assume that (4) holds with $p_1 > 1$. Then

$$\lambda_{n+1} - \lambda_n \geqslant c_1 n^{p-1} \tag{46}$$

and one can take $m(n) = n$. In this case the Riesz basis with brackets becomes a Riesz basis ([3], p. 306).

LEMMA 6. *The last statement of Theorem 2 holds.*

Proof. It follows from (30) that $R_{\lambda(m)}$ is compact (since $R_{\lambda(m)}^0$ is compact and $(I + R_{\lambda(m)}^0 T)^{-1}$ is bounded). Therefore the spectrum of M is discrete. Furthermore, it follows from (30) that if $\lambda \notin \sigma(L)$ and $\|TR_\lambda^0\| \leqslant q < 1$ then $\lambda \notin \sigma(M)$. One has $\|TR_\lambda^0\| \leqslant \|TL^{-\alpha}\| \, \|L^\alpha R_\lambda^0\| \leqslant c \max |\lambda_j^\alpha|/|\lambda_j - \lambda| \leqslant q < 1$ provided that $\max_{\lambda_j} |\lambda_j^\alpha|/|\lambda_j - \lambda| \leqslant qc^{-1}$. Therefore, if $|\lambda_j - \lambda| \geqslant c_0|\lambda_j|^\alpha$, $c_0 = cq^{-1}$, then $\lambda \notin \sigma(M)$. This is equivalent to the last statement of Theorem 2. □

Appendix 4. Stable Numerical Differentiation

If one looks for an optimal (by criterion of minimal variance) linear estimate of $s'(t)$ from the observation of $u(t) = s(t) + n(t)$, where $n(t)$ is noise and $s(t)$ is useful signal, then one can derive an integral equation for the weight function of optimal estimate. This integral equation is often difficult to solve and, even if one can solve it. It is difficult to construct the corresponding filter. In this paper an optimal estimate of s' on a subset of all linear estimates is sought and it is shown that this quasioptimal estimate is easy to calculate, the corresponding filter is easy to construct, and the error of this estimate differs little from the error of optimal estimates. It is also shown that among all estimates (linear and nonlinear) of s' for $|n| \leqslant \delta$ and $|s''| \leqslant M$ the best estimate is given by $\Delta_h u = (2h)^{-1}[u(t+h) - u(t-h)]$ with $h = (2\delta/M)^{1/2}$.

1. DETERMINISTIC CASE

Let $u = s + n$, $|n| \leqslant \delta$, $|s''| \leqslant M$. Let us assume that $u(t)$ and $s(t)$ are continuous functions defined on $\mathbb{R}^1$ and $\|\cdot\|$ denotes the norm in $C(\mathbb{R}^1)$. Let

$$\Delta_h u = (2h)^{-1}[u(t+h) - u(t-h)], \tag{1}$$

$$h(\delta) = (2\delta/M)^{1/2}, \quad \varepsilon(\delta) = (2M\delta)^{1/2}. \tag{2}$$

Let A denote the set of all operators $T: C(\mathbb{R}^1) \to C(\mathbb{R}^1)$.

THEOREM 1. *Under the above assumptions the following estimates hold*

$$\|\Delta_{h(\delta)} u - s'\| \leqslant \varepsilon(\delta), \tag{3}$$

$$\inf_{T \in A} \sup_{\substack{|s''| \leqslant M \\ |n| \leqslant \delta}} \|Tu - s'\| = \varepsilon(\delta), \tag{4}$$

and infimum in (4) *is attained on* $T = \Delta_{h(\delta)}$. *Here and below* $\varepsilon(\delta)$ *and* $h(\delta)$ *are defined in* (2).

Remark 1. If $u(t)$ and $s(t)$ and defined on a bounded domain, e.g., on $I=[0, 1]$, and if

$$\begin{aligned}
\tilde{\Delta}_h u &= h^{-1}[u(x+h)-u(x)], && 0 \leqslant x \leqslant \frac{h}{2}, \\
&= h^{-1}\left[u\left(x+\frac{h}{2}\right)-u\left(x-\frac{h}{2}\right)\right], && \frac{h}{2} \leqslant x \leqslant 1-\frac{h}{2}, \\
&= h^{-1}[u(x)-u(x-h)], && 1-\frac{h}{2} \leqslant x \leqslant 1,
\end{aligned} \tag{5}$$

then

$$|\tilde{\Delta}_{h_1(\delta)} u - s'| \leqslant 2(M\delta)^{1/2} = \sqrt{2}\varepsilon(\delta), \tag{6}$$

where

$$h_1(\delta) = 2(\delta M^{-1})^{1/2} = h(\delta)/\sqrt{2}. \tag{7}$$

Remark 2. The function $n(t)$ can be considered as a random function uniformly distributed on the segment $[-\delta, \delta]$.

Remark 3. Theorem 1 and Remark 1 combine the results of Ramm [11a] and Stechkin, Subbotin [1].

Proof of Theorem 1. We have

$$|\Delta_h u - s'| \leqslant |\Delta_h(u-s)| + |\Delta_h s - s'| \leqslant \frac{\delta}{h} + \frac{Mh}{2}. \tag{8}$$

The second term in the right-hand side of (8) is obtained by expansion of $\Delta_h s$ in Taylor series in powers of h, while the first term follows from the obvious estimates $\|\Delta_h\| \leqslant h^{-1}$, where $\|\cdot\|$ is the norm of an operator in $C(\mathbb{R}^1)$. Minimizing in h the right-hand side in (8) we obtain (3) with $h_{\min} = h(\delta)$. Let us prove (4). We have

$$\begin{aligned}
\sup_{\substack{|s''| \leqslant M \\ |n| \leqslant \delta}} \|Tu - s'\| &= \sup_{\substack{s_j,\, j=1,2 \\ |s_j''| \leqslant M \\ |n| \leqslant \delta}} \|Tu - s_j'\| \geqslant \sup_{\substack{s_j,\, j=1,2 \\ |s_j''| \leqslant M \\ |n| \leqslant \delta}} |Tu(0) - s_j'(0)| \geqslant \\
&\geqslant \inf_a \sup_{\substack{s_j,\, j=1,2 \\ |s_j''| \leqslant M \\ |n| \leqslant \delta}} |a - s_j'(0)| \geqslant \tfrac{1}{2}|s_2'(0) - s_1'(0)| \geqslant \varepsilon(\delta).
\end{aligned} \tag{9}$$

The last inequality is valid if we make the following choice of s_j:

$$s_1(t) = -\frac{M}{2}\, t(t - 2h(\delta)),$$
$$s_2(t) = -s_1(t) \quad \text{for} \quad 0 \leqslant t \leqslant 2h(\delta), \tag{10}$$

and define $s_j(t)$, $j = 1, 2$ outside the segment $[0, 2h(\delta)]$ so that the condition, $|s_j''| \leqslant M$, $|s_j| \leqslant \delta$ will be satisfied. It is easy to check this for $0 \leqslant t \leqslant 2h(\delta)$, $|s_j''| \leqslant M$, $|s_j| \leqslant \delta$, $j = 1, 2$. Thus both functions s_1 and s_2 are of the form $u - n$ with $|n| \leqslant \delta$ and $u = 0$. Taking the infimum over $T \in A$ we conclude that (4) holds with $\geqslant$ and from (3) it follows that for $T = \Delta_{h(\delta)}$ we have equality. This completes the proof of Theorem 1. The proof of Remark 1 is similar (see Ramm [11c]). □

2. STOCHASTIC CASE

Assume that $u(t) = s(t) + n(t)$, $-\infty < t < \infty$, s is a useful signal and n is noise. For simplicity assume that s and n are uncorrelated and n has a zero mean value. Let D denote variance. Assume that the random functions s and n are stationary processes, $R_s(t)$ is the covariance function of s and $n = \delta v$, where $D[v] = 1$, $\delta > 0$, $\delta^2 = D[n]$, $R_v(t)$ is the covariance function of v. The problem is to find a linear estimate of s' from the observed signal u. The optimal estimate is the solution of the variational problem

$$D[Lu - s'] = \min. \tag{10}$$

The solution of this problem was obtained in Wiener [1] for the case of infinite time of observation. Even for simple covariance functions his solution is often difficult to calculate (one has to solve some integral equation) and difficult to construct a filter which implements the estimate. That is why it is natural to look for a quasioptimal estimate which satisfies the following requirements: (1) it is easy to calculate and to implement; (2) the error of the quasioptimal estimate is approximately the same as the error of optimal estimate. The idea to look for a quasioptimal estimate in the problem of differentiation of random functions was suggested probably for the first time in Ramm [11]

Let

$$D[s^{(m)}] \leqslant M_m, \tag{11}$$

$$\Delta_h^{(Q)} s \equiv h^{-1} \sum_{k=-Q}^{Q} A_k^{(Q)} s\left(t + \frac{kh}{Q}\right). \tag{12}$$

If $m = 2q$ or $2q + 1$ let us take $Q = q$ in (12). Expanding (12) in powers of h are requiring that the order of $\Delta_h^{(Q)} s - s'$ be minimal we obtained the following system for the coefficients $A_k^{(Q)}$

$$\begin{aligned} &\sum_{k=-Q}^{Q} \left(\frac{k}{Q}\right)^j A_k^{(Q)} = \phi_{1j}, \quad 0 \leqslant j \leqslant 2Q, \\ &\delta_{1j} = 0, \quad j \neq 1, \\ &\quad\;\; = 1, \quad j = 1. \end{aligned} \tag{13}$$

The system (13) is uniquely solvable since its matrix is nondegenerate (Vandermonde determinant). We have (see Ramm [11c])

$$\begin{aligned} &Q = 1\colon A_0^{(1)} = 0, A_{\pm 1}^{(1)} = \pm\tfrac{1}{2} \\ &Q = 2\colon A_0^{(2)} = 0, A_{\pm 1}^{(2)} = \pm\tfrac{4}{3}, A^{(2)}{}_{\pm 2} = \mp\tfrac{1}{6} \\ &Q = 3\colon A_0^{(3)} = 0, A_{\pm 1}^{(3)} = \pm\tfrac{9}{4}, A_{\pm 2}^{(3)} = \mp\tfrac{9}{20}, A_{\pm 3}^{(3)} = \pm\tfrac{1}{20} \\ &Q = 4\colon A_0^{(4)} = 0, A_{\pm 1}^{(4)} = \pm\tfrac{16}{5}, A_{\pm 2}^{(4)} = \mp\tfrac{4}{5}, A_{\pm 3}^{(4)} = \pm\tfrac{16}{105}, A_{\pm 4}^{(4)} = \mp\tfrac{1}{70}. \end{aligned} \tag{13'}$$

LEMMA 1. *Let $m = 2q + 1$. We have*

$$D[\Delta_h^{(Q)} s - s'] \leqslant C_m M_m h^{2m-2} \equiv \gamma_m h^{2m-2} \tag{14}$$

if the coefficients $A_k^{(Q)}$ in (12) *satisfy* (13) (*with* $0 \leqslant j \leqslant 2q$) *and*

$$C_m = \frac{(2Q+1)}{(\mathrm{m}!)^2 Q^{2m}} \sum_{k=-Q}^{Q} |A_k^{(Q)}|^2 k^{2m}, \quad m = 2q + 1, Q = q. \tag{15}$$

In order to prove Lemma 1 we need

LEMMA 2. *Let g_j be some random variables and a_j be numbers, $D[g_j] \leqslant M$. Then*

$$D\left[\sum_{k=1}^{n} c_k g_k\right] \leqslant nM \sum_{k=1}^{n} |c_k|^2. \tag{16}$$

Proof of Lemma 2. Since $D[cg] = |c|^2 D[g]$ we can prove (16) for $c_k = 1$. If $b_k \geqslant 0$ then

$$\left(\sum_{k=1}^{n} b_k\right)^2 \leqslant n \sum_{k=1}^{n} b_k^2. \tag{17}$$

This elementary inequality follows immediately from the Cauchy inequality. Now, if $f(x_1, \ldots, x_n)$ is the probability density of the joint distribution of random variables g_k, $1 \leqslant k \leqslant n$, and if we assume (without any loss of generality) that $\bar{g}_j = 0$, and set $\mathrm{d}x = \mathrm{d}x_1, \ldots, \mathrm{d}x_n$, $\int = \int_{\mathbb{R}^n}$, $x = (x_1, \ldots, x_n)$, then

$$D\left[\sum_{k=1}^{n} g_k\right] = \int \left|\sum_{k=1}^{n} x_k\right|^2 f(x)\,\mathrm{d}x \leqslant$$

$$\leqslant n \int \sum_{k=1}^{n} |x_k|^2 f(x)\,\mathrm{d}x = n \sum_{k=1}^{n} D|g_k|.$$

Lemma 2 is proved. □

Proof of Lemma 1. We have

$$\Delta_h^{(Q)} s - s' = h^{-1} \sum_{k=-Q}^{Q} A_k^{(Q)} \frac{h^m k^m}{m! Q^m} s^{(m)}(t_k),$$

where t_k are some points in the remainders of the Taylor formula. Now we apply Lemma 2 with $g_k = s^{(m)}(t_k)$, $D[g_k] \leqslant M_m$.

LEMMA 3. *We have*

$$D[\Delta_h^{(Q)} u - s']$$

$$\leqslant \gamma_m h^{2m-2} + h^{-2}\delta^2 \sum_{k,j=-Q}^{Q} A_k^{(Q)} A_j^{(Q)} R\left(\frac{k-j}{Q} h\right) \equiv \phi(h),$$

where $m = 2Q + 1$, *and*

$$R(t - \tau) = \overline{v^*(t)v(\tau)} \tag{19}$$

is the covariance function of $v(t)$, $n(t) = \delta v(t)$, $D[v] = 1$.

THEOREM 2. *The quasioptimal estimate of* s' *has the form*

$$\hat{s}' = \Delta_h^{(Q)} u, \tag{20}$$

where the hat is the symbol of estimate (so that the right-hand side of (20) *is just a definition of* $\hat{s}'$), *the operator* $\Delta_h^{(Q)}$ *is defined in* (12), *the coefficients* $A_k^{(Q)}$ *are uniquely defined from the system* (13) *and* $h = h_m > 0$ *is defined as a minimizer of the right-hand side of* (18):

$$\phi(h) = \phi_{\min} \quad \text{for} \quad h = h_m. \tag{21}$$

In order to calculate h_m explicitly we need some assumptions about $R(t)$ or its spectral density (Fourier transform).

EXAMPLE 1. Let $m = 2$. From (13′)–(15) we obtain

$$\Delta_h^{(1)} u = \frac{u(t+h) - u(t-h)}{2h}, \quad C_2 = \tfrac{3}{8}.$$

$D[\Delta_h^{(1)} s - s'] \leqslant \frac{3}{8} M_2 h^2$, where $D[s''] \leqslant M_2$. Let $R(t) = \exp(-a|t|)$, $a > 0$. Then from (18) we obtain

$$\phi(h) = \tfrac{3}{8} M_2 h^2 + \frac{\delta^2 h^{-2}}{2} (1 - \exp(-ah)).$$

Assuming $h \ll 1$ we have

$$\phi(h) = bh^2 + \mathrm{d}h^{-1}, \quad b = \tfrac{3}{8} M_2, \quad d = \frac{\delta^2 a}{2}.$$

Thus

$$h_m = (d/2b)^{1/3}, \tag{22}$$

$$\begin{aligned} \phi_{\min} &= (2^{-2/3} + 2^{1/3})\, d^{2/3} b^{1/3} \\ &= (2^{-2/3} + 2^{1/3})(\tfrac{3}{8})^{1/3}(\tfrac{1}{2})^{2/3} M_2^{1/3} \delta^{4/3} a^{2/3}. \end{aligned} \tag{23}$$

Remark 1. Calculations similar to those give in Example 1 can be carried through in most practical cases. Note that the behavior of the spectral density $\tilde{R}(\lambda)$ at infinity $\lambda \to \infty$ defines the behavior of $R(t)$

near $t = 0$. Since only the behavior of R near $t = 0$ is essential for $h \ll 1$ for calculation of h_m and $\phi_{\min}$ we can make the following conclusion:

The above recipe for calculating $h_{\min}$ is stable towards perturbations of the spectral density $\tilde{R}(\lambda)$ preserving its asymptotics at infinity.

Remark 2. If we assume that $D[s^{(m)}] \leqslant M_m$, $m > 2$ and find h_m and $\phi_{\min}$, then we can see that for large m the error $\phi_{\min}$ of the quasioptimal estimate (12) is of the same order in δ as $\delta \to 0$ as the error of optimal Wiener estimate Wiener [1]. It is clear the estimate (12) is easy to implement as a filter: only arithmetic operations and delay lines are used. It is easy also to find the quasioptimal estimate, i.e., to find h_m. We only need to know the asymptotics of the spectral density of noise and the estimation constant M_m.

3. SEARCH OF EXTREMUM OF A RANDOM FUNCTION

In applications (e.g., in navigation theory) the following problem is of interest. A function $u = s + n$ is given, where $s(t)$ is a univalent smooth deterministic unknown function defined on the segment $I = [0, 1]$ (univalent means that $s(t)$ has exactly one maximum on $[0, 1]$), and $n(t)$ is noise, $|n| \leqslant \delta$. The problem is to find the point t^*, where $s(t)$ is maximal. We assume that $|s''| \leqslant M$ and M is known. To solve this problem let us find $h = h_1(\delta)$ (see (7)), let us divide the segment by points $t_k = kh$, and calculate $\hat{u}_k = (2h^{-1}[u(t_k + h) - u(t_k - h)]$. If $\hat{u}_k \hat{u}_{k+1} < 0$ and $|\hat{u}_j| > 2\,(M\delta)^{1/2} \forall j$, then we conclude with the probability 1 that $t_k \leqslant t^* \leqslant t_k + h$. Indeed from (6) it follows that $\hat{u}_k = 2(M\delta)^{1/2} < s'(t_k) < \hat{u}_k + 2(M\delta)^{1/2}$. If $|\hat{u}_j| > 2(M\delta)^{1/2}$ then $s'(t_k)$ has the same sign as $\hat{u}_k$. Therefore from $\hat{u}_k \hat{u}_{k+1} < 0$ it follows that s' changes sign on the segment $[t_k, t_{k+1}]$. Since $s(t)$ is univalent the point t^* lies in the segment.

Appendix 5. Limit of the Spectra of the Interior Neumann Problems when a Solid Domain Shrinks to a Plane One

Let $(\nabla^2 + \lambda)u = 0$ in $\mathscr{D} \subset \mathbb{R}^d$, $\partial u/\partial N = 0$ on $\partial\mathscr{D}$. How do the eigenvalues λ_j behave when $\mathscr{D}$ shrinks to a domain $\Delta \subset R^{d-1}$? The answer depends not only on Δ but on the way $\mathscr{D}$ shrinks to Δ. The limit of λ_j is found. Examples are given.

I. INTRODUCTION

Consider the problem

$$\nabla^2 u + \lambda u = 0 \text{ in } \mathscr{D} \subset R^2, \frac{\partial u}{\partial N} = 0 \text{ on } \Gamma. \tag{1}$$

Here $\mathscr{D} = \mathscr{D}(\varepsilon)$ is a bounded domain with a piecewise smooth boundary Γ, N is the unit outer normal to Γ, ∇^2 is the Laplacian. Let us assume that the equation of Γ is of the form

$$y = \begin{cases} \varepsilon f(x), \ a \leqslant x \leqslant b, \ y > 0, \\ \varepsilon g(x), \qquad\qquad y < 0, \end{cases} \tag{2}$$

where $\varepsilon > 0$ is a parameter, f and g are piecewise smooth functions. If $f(a) > 0$, $g(a) < 0$ then the segment $x = a$, $g(a) \leqslant y \leqslant f(a)$ belongs to Γ, and the same holds for $x = b$. Let $\lambda_j(\varepsilon)$ be the eigenvalues of (1)

$$0 = \lambda_0 \leqslant \lambda_1(\varepsilon) \leqslant \lambda_2(\varepsilon) \leqslant \dots .$$

PROBLEM. How do $\lambda_j(\varepsilon)$ behave when $\varepsilon \to 0$?

The limit of $\mathscr{D}(\varepsilon)$ is the segment Δ: $a \leqslant x \leqslant b$, $\mathscr{D}(\varepsilon) \subset R^2$, $\Delta \subset R^1$. The same problem can be formulated for $\mathscr{D}(\varepsilon) \subset R^d$, $\Delta \subset R^{d-1} d > 2$. One can also consider a general second order selfadjoint elliptic operator $\mathscr{L}$ instead of Δ. From the results of this paper it follows that

$$\lim_{\varepsilon \to \infty} \lambda_j(\varepsilon) = \lambda_j, \tag{3}$$

where λ_j are the eigenvalues of a differential operator on $L^2(\Delta) = L^2([a, b])$ with coefficients which depend on f and g, and with boundary conditions which are $u'(a) = u'(b) = 0$ if $f(a) - g(a) \neq 0$, $f(b) - g(b) \neq 0$ and $|u(a)| < \infty$ if, for example, $f(a) - g(a) = 0$. Similar results hold for the case $\mathscr{D}(\varepsilon) \subset R^d$, $d > 2$, $\Delta \subset R^{d-1}$.

Examples. 1. Let $\mathscr{D}(\varepsilon)$ be a rectangle: $g = 0$, $f = 1$. Then $\lambda_{n_1 n_2} = \pi^2(n_1^2/(b-a)^2 + n_2^2/\varepsilon^2)$. Therefore $\lambda_{n_1 n_2}(\varepsilon) \underset{\varepsilon \to 0}{\to} \infty$ if $n_2 \neq 0$, $\lambda_{n_j 0} \equiv \lambda_{n_1}(\varepsilon) = (\pi^2/(b-a)^2)n_1^2$ does not depend on ε, $\lim_{\varepsilon \to 0} \lambda_n(\varepsilon) = \pi^2 n^2/(b-a)^2 \equiv \lambda_n$, where the set $\{\lambda_n\}$ is the spectrum of the limit problem in Δ:

$$u'' + \lambda u = 0, \quad a \leqslant x \leqslant b, \quad u'(a) = u'(b) = 0.$$

2. Let $f = \varepsilon\sqrt{1 - x^2}$, $g = -f$, Γ is the ellipse $x^2 + y^2/\varepsilon^2 = 1$. By separation of variables one can find $\lambda_j(\varepsilon)$ and check that $\lim_{\varepsilon \to 0} \lambda_j(\varepsilon) \neq \pi^2 n^2/(b-a)^2$.

3. Let $\mathscr{D}(\varepsilon)$ be the sector: $g = 0$, $0 \leqslant x \leqslant 1$, $a = 0$, $b = 1$, $f = f = \begin{cases} x \tan \varepsilon, & 0 \leqslant x \leqslant \cos \varepsilon \\ \sqrt{1 - x^2}, & \cos \varepsilon \leqslant x \leqslant 1 \end{cases}$. The spectrum $\lambda_j(\varepsilon)$ one can calculate easily and and find that $\lim_{\varepsilon \to 0} \lambda_j(\varepsilon) = x_{0j}^2$, where x_{0j} are the roots of $J_0'(x) = 0$, $x_{00} = 0 < x_{01} < \ldots$.

From these examples one can see that the limits of $\lambda_j(\varepsilon)$, $\lambda_1(\varepsilon) \leqslant \lambda_2(\varepsilon) \leqslant \ldots$ as $\varepsilon \to 0$ depend on f and g. In Section II it will be explained how to calculate these limits. Notice that for the Dirichlet boundary condition and for the third boundary condition $\partial u/\partial N + hu|_\Gamma = 0$, $h > 0$ the limits $\lim_{\varepsilon \to 0} \lambda_j(\varepsilon) = \infty$.

II. CALCULATION OF THE LIMIT SPECTRA

1. The starting point is the variational definition of $\lambda_j(\varepsilon)$ as the consecutive minima of the ratio

$$J(\varepsilon) = \frac{\int_{\mathscr{D}(\varepsilon)} |\nabla u|^2 \, dx \, dy}{\int_{\mathscr{D}(\varepsilon)} |u|^2 \, dx \, dy} = \frac{\int_a^b dx \int_{\varepsilon g(x)}^{\varepsilon f(x)} (|u_x^2| + |u_y^2|) \, dy}{\int_a^b dx \int_{\varepsilon g(x)}^{\varepsilon f(x)} |u|^2 \, dy}. \tag{4}$$

One has $\int_{\varepsilon g}^{\varepsilon f} |u|^2 \, dy = \varepsilon(f - g)|u^2(x, \xi)|$, $\varepsilon g < \xi < \varepsilon f$, $\xi \to 0$ when $\varepsilon \to 0$. Since $\partial u/\partial N \to \partial u/\partial y$ as $\varepsilon \to 0$ (because $N_x =$

$-\varepsilon f'/\sqrt{1+\varepsilon^2|f'|^2} \to 0$ as $\varepsilon \to 0$) one passes to the limit $\varepsilon \to 0$ in (4) and obtains

$$J = \frac{\int_a^b \mathrm{d}x(f(x) - g(x))|v'^2|}{\int_a^b \mathrm{d}x(f(x) - g(x))|v|^2} = \frac{\int_a^b h|v'|^2\,\mathrm{d}x}{\int_a^b h|v|^2\,\mathrm{d}x}, \tag{5}$$

where

$$f(x) - g(x) \equiv h(x), \quad v = u(x, 0), \quad v' = \frac{\mathrm{d}v}{\mathrm{d}x}. \tag{6}$$

The spectrum corresponding to the variational problem (5) is the spectrum of the problem

$$\begin{cases} (h(x)v')' + \lambda hv = 0, \quad a \leqslant x \leqslant b, & (7) \\ h(a)v'(a) = 0, \quad h(b)v'(b) = 0. & (8) \end{cases}$$

The boundary conditions (8) are the natural boundary conditions for the variational problem (5). If $h(a) \neq 0$, $h(b) \neq 0$ then (8) is equivalent to the Neumann boundary conditions for the limit problem: $v'(a) = v'(b) = 0$. If, e.g., $h(a) = 0$, then Equation (7), which is of the form

$$v'' + \lambda v + \frac{h'(x)}{h(x)}\, v = 0, \quad a \leqslant x \leqslant b \tag{9}$$

has an unbounded coefficient at $x = a$ and the boundary condition at $x = a$ is $|v(a)| < \infty$. One can assume that the parameter ε enters non-linearly in the equation of Γ, e.g. $y = f(x, \varepsilon)$, $y > 0$, $y = g(x, \varepsilon)$, $y < 0$. In this case the role of $h(x)$ is played by $\partial f(x,0)/\partial\varepsilon - \partial q(x,0)/\partial\varepsilon$.

2. If $\mathscr{D}(\varepsilon) \subset R^d$, Γ is given by the equations

$x_d = \varepsilon f(x_1, \ldots, x_{d-1})$, $x_d > 0$, $x_d = \varepsilon g(x_1, \ldots, x_{d-1})$, $x_d < 0$,

and $\mathscr{D}(0) = \Delta \subset R^{d-1}$, then the argument in Section II.1 shows that $\lim_{\varepsilon\to 0} \lambda_j(\varepsilon) = \lambda_j$, where λ_j, $0 = \lambda_0 < \lambda_1 \leqslant \lambda_2 \leqslant \ldots$ are the eigenvalues of the problem

$$\begin{cases} \operatorname{div}(h(p)\nabla v) + \lambda h(p)v = 0 \text{ in } \Delta,\ p = (x_1, \ldots, x_{d-1}), & (10) \\ h\dfrac{\partial v}{\partial \nu} = 0 \text{ on } \partial\Delta,\ h \equiv f(p) - g(p). & (11) \end{cases}$$

Here $\partial\Delta$ is the boundary of Δ, ν is the unit outer normal to $\partial\Delta$, $\nabla v = \sum_{j=1}^{d-1} (\partial v/\partial x_j)e_j$, e_j are the unit vectors of the orthonormal cartesian basis in R^{d-1}, $\operatorname{div} a = \sum_{j=1}^{d-1} \partial a_j/\partial x_j$. If $h \neq 0$ on $\partial\Delta$, then (11) is the Neumann boundary condition on $\partial\Delta$. If $h = 0$ on $\partial\Delta$, then the problem (10) is degenerate near $\partial\Delta$.

Examples. 4. In example 2 one has $h = 2\sqrt{1 - x^2}$. The problem (7)—(8) is of the form

$$\begin{cases} (\sqrt{1-x^2}\, v')' + \lambda\sqrt{1-x^2}\, v = 0, -1 \leqslant x \leqslant 1, & (12) \\ |v(\pm 1)| < \infty. & (13) \end{cases}$$

One can write (12) as

$$v'' + \lambda v - \frac{x}{1-x^2}\, v = 0, -1 \leqslant x \leqslant 1. \tag{14}$$

5. In example 3 one has $h = \partial f(x, 0)/\partial\varepsilon = x$, $h'/h = 1/x$, and Equation (9) is of the form

$$v'' + \lambda v + \frac{v}{x} = 0. \tag{15}$$

The boundary condition at $x = 0$ is $|v(0)| < \infty$. The boundary condition at $x = 1$ is $v'(1) = 0$, because $h(1) \neq 0$. (Note also that for any $\varepsilon > 0$ the normal to the line $y = \sqrt{1 - x^2}$ at $x = 1$ is directed along the x-axis). The solution of (15) bounded at $x = 0$ is $J_0(\sqrt{\lambda} x)$, where $J_0(x)$ is the Bessel function. The eigenvalues λ_j are to be found from the equation $J_0'(\sqrt{\lambda}) = 0$, which follows from the boundary condition $v'(1) = 0$. Thus, $\lambda_j = x_{0j}^2$ where x_{0j} are zeros of the equation $J_0'(x) = 0$.

6. Let $\mathscr{D}(\varepsilon)$ be the ellipsoid $x^2 + y^2 + z^2/\varepsilon^2 = 1$, $\mathscr{D}(0) = \Delta$ is the circle $x^2 + y^2 = 1$, $h = 2\sqrt{1 - x^2 - y^2}$. Then (10) takes the form

$$\nabla^2 v + \lambda v - \frac{x\dfrac{\partial v}{\partial x} + y\dfrac{\partial v}{\partial y}}{1 - x^2 - y^2} = 0 \text{ in } \Delta.$$

Since $h = 0$ on $\partial\Delta$ one requires that v be bounded near $\partial\Delta$.

III. ADDITIONAL REMARKS

1. If $h(p)$ vanishes at the boundary $\partial\Delta$ one has a selfadjoint but not uniformly elliptic operator $\operatorname{div}(h(p)\nabla v) = h\Delta v + (\nabla h, \nabla v)$. Let us prove that the spectrum of the ratio $(\int_\Delta h|\nabla v|^2 \, dp)/(\int_\Delta h|v|^2 \, dp)$ is still discrete regardless of the rate at which $h(p)$ goes to zero when p approaches $\partial\Delta$. It is sufficient to prove that the embedding operator i: $H_1 \to H_0$ is compact, where $H_1 = H_1(\Delta)(H_0)$ is the Hilbert space of functions defined on Δ with the inner product $(u, v)_1 = \int_\Delta (h\nabla u\overline{\nabla v} + hu\bar{v})\, dp$ $((u, v)_0 = \int_\Delta hu\bar{v}\, dp)$. Let $\Delta' \subset \Delta$ be any strictly inner subdomain of Δ, $\inf_{p\to\Delta'} h(p) > 0$. It is well known that the embedding: $H_1(\Delta) \to H_0(\Delta')$ is compact. One proves that i: $H_1(\Delta) \to H_0(\Delta)$ is compact if the estimate:

$$\int_\delta h|v|^2 \, dp < \varepsilon \int_\Delta h\{|v'|^2 + |v|^2\}\, dp,$$

holds for any $\varepsilon > 0$ where δ is a sufficiently thin strip near $\partial\Delta$, e.g. $\delta = \{p\colon \rho(p, \partial\Delta)\} < \eta$, $\eta > 0$ is a sufficiently small number, and $\rho(p, \partial\Delta)$ is the distance between p and $\partial\Delta$. But the above estimate is obvious since $\int_\Delta h|v|^2 \, dp < \infty$. Note that the degeneracy of h at $\partial\Delta$ could lead to a continuous spectrum of the ratio $\int_\Delta h|\nabla v|^2 \, dp \cdot (\int_\Delta |v|^2 \, dp)^{-1}$ if h goes to zero fast enough as $p \to \partial\Delta$.

2. Let us show that

$$\lambda_j(\varepsilon) \to \lambda_j, \quad \varepsilon \to 0 \tag{16}$$

where $\lambda_j(\varepsilon)(\lambda_j)$ are the consecutive minima of the ratio (4) ((5)). From the argument after formula (4) one derives that

$$J(\varepsilon) = J + o(1) \quad \text{as} \quad \varepsilon \to 0, \tag{17}$$

where J is defined in (5). Therefore the equality

$$\lambda_j(\varepsilon) - \lambda_j = o(1) \quad \text{as} \quad \varepsilon \to 0 \tag{18}$$

holds because $o(1)$ is uniform in u running through an appropriate set dense in the unit ball in the Sobolev space H^1. Indeed, in this case

$$\min J - \max o(1) \leqslant \min J(\varepsilon) \leqslant \min J + \max o(1),$$

and $o(1)$ is uniform in $u \in M$, where M is a unit ball in H^q, H^q is the

Sobolev space $W_2^q(\mathscr{D}_1)$, $\mathscr{D}_1 = \mathscr{D}(1)$, and q is so chosen that the embedding i: $H^q(\mathscr{D}_1) \to C^2(\mathscr{D}_1)$ is bounded. For example, this is the case if $q > 2 + d/2$, where d is the dimension of the space. Since H^q is dense in H^1, $\min_{u \in H^1} J = \inf_{u \in H^q} J$. If the boundary $\partial\mathscr{D}$ is smooth enough, one knows that the $\inf_{u \in H^q} J$ is attained and therefore inf can be substituted by min. To see that $o(1)$ is uniform in $u \in M$ one can estimate $o(1)$ as follows:

$$o(1) = \int_a^b \mathrm{d}x \left| u^2(x, \xi) - u^2(x, 0) \right| \leqslant$$

$$\leqslant c \max |u| \cdot \max \mathrm{k} \nabla u| \cdot \varepsilon \max \kappa h(x)| \leqslant c_1 \varepsilon$$

since $\max |u| + \max |\nabla u| \leqslant \text{const}$ if $|u|_{H^q} \leqslant 1$. Similarly one estimates the terms with the derivatives $\nabla u(x, \xi) - \nabla u(x, 0)$.

3. It is clear from our argument that the eigenvalues of the problem

$$\nabla^2 u + \lambda u = 0 \text{ in } \mathscr{D}(\varepsilon), \frac{\partial u}{\partial N} + hu = 0 \text{ on } \Gamma(\varepsilon) \tag{19}$$

go to infinity as $\varepsilon \to 0$ provided for instance that h is a continuous function on Γ, $h \not\equiv 0$. Indeed, the ratio corresponding to problem (19) is of the form

$$\frac{\int_{\mathscr{D}(\varepsilon)} |\nabla u|^2 \,\mathrm{d}x\,\mathrm{d}y + \int_{\Gamma(\varepsilon)} h|u|^2 \,\mathrm{d}s}{\int_{\mathscr{D}(\varepsilon)} |u|^2 \,\mathrm{d}x\,\mathrm{d}y}. \tag{20}$$

As $\varepsilon \to 0$, the denominator in (20) is of order ε while the term $\int_{\Gamma(\varepsilon)} h|u|^2 \,\mathrm{d}s$ is of order 0(1). Thus, the ratio (20) is unbounded as $\varepsilon \to 0$.

Remark. The results of this appendix allow one to show easily the well-known fact that the eigenvalues of the Neumann Laplacian do not increase monotonically when the domain decreases, i.e., in contrast with the Dirichlet Laplacian $-\Delta_D$ for which $D \supset D'$ implies $\lambda_n(D) \leqslant \lambda_n(D')$, where $\lambda_n(D)$ are the eigenvalues of $-\Delta_D$, for the Neumann Laplacian the above implication does not hold. Indeed, if $\mu_n(D)$ are the eigenvalues of the Neumann Laplacian in D, and if

$$D \supset D' \text{ implies that } \mu_n(D) \leqslant \mu_n(D'), \tag{*}$$

then the limit of the spectrum of the Neumann Laplacian for the sector in Example 3, Section 1, should be the set $\{n^2\pi^2\}$ as in Example 1 in this section. Since in fact this limit is different, we conclude that (*) is false. Let us show that (*) implies that the limit of the spectrum in Example 3 is the set $\{n^2\pi^2\}$. Take a rectangular R inscribed into the sector S with height η and base $1 - \delta$, where $\eta > 0$ and $\delta > 0$ are small numbers. If (*) holds then $\mu_n(R) \geqslant \mu_n(S)$. Pass to the limit $\eta \to 0$ and $\varepsilon \to 0$, where $\varepsilon > 0$ is the same as in Example 3, Section 1. The result is $\pi^2 n^2/(1 - \delta)^2 \geqslant x_{0n}^2$ where x_{0n} are the roots of $J_0'(x) = 0$. The same argument with the circumscribed rectangular R with base 1 and height η shows that $x_{0n}^2 \leqslant \pi^2 n^2$. Since $\delta > 0$ is arbitrarily small we conclude that $x_{0n}^2 = \pi^2 n^2$, which is absurd. This contradiction shows that (*) is false. The author did not see any published proof of the fact that (*) is false.

Appendix 6. Recovering of the Surface from the Knowledge of Its Principal Curvatures

1. Let Γ be a closed compact oriented surface in $\mathbb{R}^3$ and $\psi: \Gamma \to S^2$ be such a map that the vector $\psi(s)$ when placed at the point s is the positive unit normal to Γ. Here $\psi(s)$ is the vector with the initial point at the origin (the center of the unit ball with the boundary S^2) and the end point at the $\psi(s)$. Let us assume that ψ is a diffeomorphism of Γ onto some part $\tilde{S}^2$ of S^2 and let $\Phi: \tilde{S}^2 \to \Gamma$ be the inverse mapping ψ^{-1}. Let $(e_1(s), e_2(s), e_3(s))$ be a moving orthonormal frame on Γ, $e_3(s) = \psi(s)$, $e_1(s)$ and $e_2(s)$ be the eigenvector of $-\mathrm{d}e_3(s)$ with eigenvalues $K_1(s)$ and $K_2(s)$ respectively, where $K_j(s)$, $j = 1, 2$ are the principal curvatures of Γ at the point s. Let us define differential forms ω_1, ω_2 by the equations $\omega_i(e_j) = \delta_{ij}$, $1 \leqslant i, j \leqslant 2$. Let $\mathrm{d}e_i = \omega_{ij} e_j$. Here and below one should sum over the repeated indices. It is known Hsiung [1] that $\mathrm{d}\omega_{ij} = \omega_{im} \wedge \omega_{mj}$, $\mathrm{d}\omega_1 = -\omega_2 \wedge \omega_{12}$, $\mathrm{d}\omega_2 = \omega_1 \wedge \omega_{12}$, where $\wedge$ denotes the wedge product of the differential forms, $\omega_{13} = K_1\omega_1$, $\omega_{23} = K_2\omega_2$, and the fundamental forms on Γ are $I = \omega_1^2 + \omega_2^2$, $\mathrm{II} = K_1\omega_1^2 + K_2\omega_2^2$, $\mathrm{III} = K_1^2\omega_1^2 + K_2^2\omega_2^2$. Let us pull back to S^2 the basic differential forms using the mapping $\Phi = \psi^{-1}$. The pulled back forms ω_i^*, ω_{ij}^* satisfy the equations

$$\begin{gathered}\mathrm{d}\omega_{ij}^* = \omega_{im}^* \wedge \omega_{mj}^*,\ \mathrm{d}\omega_1^* = -\omega_2^* \wedge \omega_{12}^*,\ \mathrm{d}\omega_2^* = \omega_1^* \wedge \omega_{12}^*, \\ \omega_{ij}^* = -\omega_{ji}^*,\ \omega_{13}^* = K_1\omega_1^*,\ \omega_{23}^* = K_2\omega_2^*. \end{gathered} \tag{1}$$

Here $K_i = K_i(\theta, \phi) = K_i(\Phi(\theta, \phi))$. These equations imply that

$$\mathrm{d}K_1 \wedge \omega_1^* = \sigma_1(K_1 - K_2)\omega_1^* \wedge \omega_2^* \tag{2a}$$

$$\mathrm{d}K_2 \wedge \omega_2^* = \sigma_2(K_1 - K_2)\omega_1^* \wedge \omega_2^* \tag{2b}$$

where σ_1 and σ_2 are defined by the equation

$$\omega_{12}^* = \sigma_1\omega_1^* + \sigma_2\omega_2^* \tag{3}$$

Assume that $K_1(\theta, \phi)$ and $K_2(\theta, \phi)$ are given. If one finds the forms ω_i^*, then determines vectors e_i^* such that

$$\omega_i^*(e_j^*) = \delta_{ij}, \quad 1 \leqslant i, \quad j \leqslant 2, \tag{4}$$

and finally uses the equation

$$\mathrm{d}\Phi = -\frac{\omega_1^* e_1^*}{K_1} - \frac{\omega_2^* e_2^*}{K_2} \tag{5}$$

then, assuming that K_1 and K_2 are the principal curvatures of a surface, one concludes that $\mathrm{d}\Phi$ is a closed differential. Integrating this differential one recovers the surface. In what follows the equations for the forms ω_i^*, $1 \leqslant i \leqslant 2$, are derived and analyzed, and a constructive method for finding the surface is given under the assumption that K_1 and K_2 are the principal curvatures of a surface. Let

$$\omega_1^* = a_1\,\mathrm{d}\phi + a_2\,\mathrm{d}\theta, \quad \omega_2^* = b_1\,\mathrm{d}\phi + b_2\,\mathrm{d}\theta. \tag{6}$$

The e_j^* are the eigenvectors of the mapping $\mathrm{d}\Phi$ with the eigenvalues $-1/K_j$. One can check, using the orthonormality of the vectors e_j^*/K_j with respect to III* metric, or the fact that $\mathrm{III}^* = \sin^2\theta(\mathrm{d}\phi)^2 + (\mathrm{d}\theta)^2$ (see Section 2) that

$$b_1 = \mp a_2 \frac{K_1}{K_2} \sin\theta, \quad b_2 = \pm a_1 \frac{K_1}{K_2} \frac{1}{\sin\theta}, \tag{7}$$

$$\frac{K_1^2 a_1^2}{\sin^2\theta} + K_1^2 a_2^2 = 1. \tag{8}$$

One has

$$\begin{aligned} &\omega_{13}^* = K_1\omega_1^*, \quad \omega_{23}^* = K_2\omega_2^*, \\ &\omega_{12}^* = \sigma_1\omega_1^* + \sigma_2\omega_2^* \equiv \alpha\,\mathrm{d}\phi + \beta\,\mathrm{d}\theta, \end{aligned} \tag{9}$$

where α and β are to be found. Furthermore

$$\begin{aligned} \mathrm{d}\omega_{12}^* &= \omega_{13}^* \wedge \omega_{32}^* = -K\omega_1^* \wedge \omega_2^* = -K(a_1 b_2 - a_2 b_1)\,\mathrm{d}\phi \wedge \mathrm{d}\theta \\ &= -K\frac{\sin\theta}{K}\,\mathrm{d}\phi \wedge \mathrm{d}\theta = -\sin\theta\,\mathrm{d}\phi \wedge \mathrm{d}\theta, \end{aligned} \tag{10}$$

where $K = K_1 K_2$. If one uses the definition of ω_i^* in terms of a_i and b_i then one finds

$$a_1\sigma_1 - \frac{K_1}{K_2}\sin\theta\, a_2\sigma_2 = \alpha,$$

$$a_2\sigma_1 + \frac{K_1}{K_2\sin\theta}\, a_1\sigma_2 = \beta.$$

Solving this system for σ_1 and $-\sigma_2$ one obtains:

$$\begin{pmatrix} \sigma_1 \\ -\sigma_2 \end{pmatrix} = T \begin{pmatrix} a_1 \\ a_2 \end{pmatrix}, \quad T \equiv \begin{pmatrix} \dfrac{K_1^2 \alpha}{\sin^2 \theta} & K_1^2 \beta \\ -\dfrac{K\beta}{\sin \theta} & \dfrac{K\alpha}{\sin \theta} \end{pmatrix}. \tag{11}$$

Here the equation

$$\det \begin{vmatrix} a_1 & -\dfrac{K_1}{K_2} \sin \theta \; a_2 \\ a_2 & \dfrac{K_1}{K_2 \sin \theta} \; a_1 \end{vmatrix} = \frac{\sin \theta}{K}$$

was used. Using the Equations (2) one obtains

$$Q \begin{pmatrix} a_1 \\ a_2 \end{pmatrix} = \frac{K_2 - K_1}{K} \sin \theta \begin{pmatrix} \sigma_1 \\ -\sigma_2 \end{pmatrix} \tag{12}$$

where

$$Q = \begin{pmatrix} -\dfrac{\partial K_1}{\partial \theta} & \dfrac{\partial K_1}{\partial \phi} \\ \dfrac{K_1}{K_2 \sin \theta} \dfrac{\partial K_2}{\partial \phi} & \dfrac{K_1 \sin \theta}{K_2} \dfrac{\partial K_2}{\partial \theta} \end{pmatrix}. \tag{13}$$

Therefore

$$\left(Q - \frac{K_2 - K_1}{K} (\sin \theta) T \right) \begin{pmatrix} a_1 \\ a_2 \end{pmatrix} = 0. \tag{14}$$

This equation can be written as

$$\begin{pmatrix} A & B \\ C & D \end{pmatrix} \begin{pmatrix} a_1 \\ a_2 \end{pmatrix} = 0, \quad \begin{aligned} A &= -\frac{\partial K_1}{\partial \theta} - \frac{K_2 - K_1}{K} \frac{K_1^2 \alpha}{\sin \theta} \\ B &= \frac{\partial K_1}{\partial \phi} - \frac{K_2 - K_1}{K} \sin \theta \; K_1^2 \beta \end{aligned} \tag{15}$$

where A and D are linear in α, B and C are linear in β. Since a_1 and a_2 are not zeros simultaneously, one has

$$AD - BC = 0. \tag{16}$$

If $K_2 - K_1 \neq 0$ this is a quadratic equation in α, β. Equations (15) can be written as

$$Aa_1 + Ba_2 = 0, \quad Ca_1 + Da_2 = 0. \tag{17}$$

Therefore

$$A^2a_1^2 = B^2a_2^2, \quad K_1^2 \sin^2\theta\, A^2a_1^2 = B^2 \sin^2\theta \left(1 - \frac{K_1^2a_1^2}{\sin^2\theta}\right). \tag{18}$$

Assume that $B \neq 0$. Then

$$a_1 = \pm \frac{\sin\theta}{K_1} \frac{1}{[1 + (A^2/B^2)\sin^2\theta]^{1/2}}, \quad a_2 = -\frac{A}{B} a_1. \tag{19}$$

Therefore a_1, a_2 (and b_1, b_2) are determined by α, β.

From (10) it follows that

$$\omega_{12}^* = -\cos\theta\, \mathrm{d}\phi + \mathrm{d}g, \tag{20}$$

where $g(\theta, \phi)$ is a scalar function. Therefore

$$\alpha = \frac{\partial g}{\partial \phi} - \cos\theta, \beta = \frac{\partial g}{\partial \theta}, \tag{21}$$

so that α and β are determined by g. If one solves the quadratic equation (16) for α and write $\alpha = H(\theta, \phi, \beta)$, then (19) can be written as

$$a_j = F_j\left(\theta, \phi, \frac{\partial g}{\partial \theta}\right), \quad j = 1, 2. \tag{22}$$

The equation $\mathrm{d}\omega_1^* = -\omega_2^* \wedge \omega_{12}^*$ can be written as

$$\frac{\partial a_2}{\partial \phi} - \frac{\partial a_1}{\partial \theta} = \frac{K_1}{K_2} \sin\theta \left(a_2\beta + a_1\alpha \frac{1}{\sin^2\theta}\right) = G\left(\theta, \phi, \frac{\partial g}{\partial \theta}\right). \tag{23}$$

From (22) and (23) one obtains

$$\frac{\partial F_2}{\partial \theta} + \frac{\partial F_2}{\partial \theta}\frac{\partial}{\partial \theta}\left(\frac{\partial g}{\partial \phi}\right) - \frac{\partial F_1}{\partial \theta} - \frac{\partial F_1}{\partial \beta}\frac{\partial^2 g}{\partial \theta^2} = G\left(\theta, \phi, \frac{\partial g}{\partial \theta}\right). \tag{24}$$

From (21) it follows that

$$\frac{\partial}{\partial \theta}\frac{\partial g}{\partial \theta} = -\sin\theta + \frac{\partial H}{\partial \theta}\left(\theta, \phi, \frac{\partial g}{\partial \theta}\right) + \frac{\partial H}{\partial \beta}\cdot\frac{\partial^2 g}{\partial \theta^2}. \tag{25}$$

From (24) and (25) one obtains

$$\left(\frac{\partial F_2}{\partial \beta}\frac{\partial H}{\partial \beta} - \frac{\partial F_1}{\partial \beta}\right)\frac{\partial^2 g}{\partial \theta^2} + \frac{\partial F_2}{\partial \phi} - \frac{\partial F_1}{\partial \theta} + \left(\frac{\partial H}{\partial \theta} - \sin\theta\right). \tag{26}$$

$$\frac{\partial F_2}{\partial \beta} = G.$$

This equation can be written as

$$\frac{\partial^2 g}{\partial \theta^2} = G_1\left(\theta, \phi, \frac{\partial g}{\partial \theta}\right). \tag{27}$$

Similarly, the equation $d\omega_2^* = \omega_1^* \wedge \omega_{12}^*$ leads to

$$\frac{\partial^2 g}{\partial \theta^2} = G_2\left(\theta, \phi, \frac{\partial g}{\partial \theta}\right), \tag{28}$$

where G_2 is explicitly determined. From (27) and (28) one obtains the equation for $\partial g/\partial\theta$

$$G_1\left(\theta, \phi, \frac{\partial g}{\partial \theta}\right) = G_2\left(\theta, \phi, \frac{\partial g}{\partial \theta}\right). \tag{29}$$

Assuming that this equation is solvable one finds $\partial g/\partial\theta = \beta$, then $\alpha = H(\theta, \phi, \beta)$, then A and B from (15), then a_1 and a_2 from (19), then b_1 and b_2 from (7), then ω_1^* and ω_2^* from (6), then e_1^* and e_2^* from (4), then $d\Phi$ from (5). Integrating (5) one finds the mapping Φ which determines the parametric equation of the surface Γ (parameters are θ, ϕ). The above construction is valid under the assumption that K_1, K_2 are the principal curvatures of a surface.

It was assumed above that $K_1 \neq K_2$. If $K_1 = K_2$ then it is easily seen that $K_1 = K_2 = \text{const}$. Indeed (12) shows, that in this case

$$Q\begin{pmatrix} a_1 \\ a_2 \end{pmatrix} = 0.$$

Since a_1 and a_2 are not both zero one concludes that $0 = \det Q = -(\partial K_1/\partial\theta)^2 \sin\theta - (\partial K_1/\partial\phi)^2\, 1/\sin\theta$. Thus $\partial K_1/\partial\theta = \partial K_1/\partial\phi = 0$, $K_1 = \text{const}$. If $K_1 = K_2 = \text{const}$ then the surface is spherical, its radius is K_1^{-1}.

In conclusion of this section let us formulate some open problems:

(1) Characterization problem: what are the necessary and sufficient conditions on $K_1(\theta, \phi)$, $K_2(\theta, \phi)$ in order that these functions be principal curvatures of a closed surface of class c^2?

The following problem is closely connected with characterization problem:

(2) If K_1 and K_2 are the principal curvatures of a surface then $d\Phi$ (see (5)) is a closed differential. Is the converse true? What is the relationship between the condition that $d\Phi$ is closed an the solvability conditions for Equations (16) and (29)?

2. DERIVATION OF FORMULAS (7), (8)

The identity $\text{III}^* = \sin^2\theta(d\phi)^2 + (d\theta)^2 \equiv K_1^2(a_1\,d\phi + a_2\,d\theta)^2 + K_2^2(b_1\,d\phi + b_2\,d\theta)^2$ is the starting point. One obtains the system

$$K_1^2 a_1^2 + K_2^2 b_1^2 = \sin^2\theta$$
$$K_1^2 a_1 a_2 + K_2^2 b_1 b_2 = 0$$
$$K_1^2 a_2^2 + K_2^2 b_2^2 = 1.$$

Let $K_1/K_2 = \lambda$, $K_2 \neq 0$, $b_i/\lambda = c_i$. Then the system takes the form

$$a_1^2 + c_1^2 = \sin^2\theta,\ a_1 a_2 + c_1 c_2 = 0,\ a_2^2 + c_2^2 = 1.$$

One has

$$\frac{a_1}{c_2} = -\frac{c_1}{c_2}, \quad c_1^2\left(1 + \frac{c_2^2}{a_2^2}\right) = \sin^2\theta, \; \frac{c_1^2}{a_2^2} = \sin^2\theta,$$

$$c_1 = \mp a_2 \sin\theta, c_2 = \pm\frac{a_1}{\sin\theta}.$$

Thus,

$$b_1 = \mp\frac{K_1 \sin\theta}{K_2} a_2, \quad b_2 = \pm\frac{K_1}{K_2 \sin\theta} a_1, \quad \frac{K_1^2 a_1^2}{\sin^2\theta} + K_1^2 a_2^2 = 1.$$

The choice of the sign in (31) is $(-+)$. One could chose $(+-)$ without any changes in the argument.

Remark. The unit sphere is the only compact closed surface such that III* = I*, that is the pull-backs of the third and first fundamental forms are equal on this surface. Indeed, III* = I* iff $K_1^2\omega_1^{*2} + K_2^2\omega_2^{*2} = \omega_1^{*2} + \omega_2^{*2}$. The last equation implies that $K_1^2 = 1$, $K_2^2 = 1$. This means that $K_1 = 1$, $K_2 = 1$, i.e. the surface is the unit sphere. The case $K_1K_2 = -1$ does not correspond to any compact closed smooth surface as is clear from the geometrical reasons.

Appendix 7. Calculating Resonances (Natural Frequencies) and Extracting Them from Transient Fields

Mathematical formulation and analysis of numerical methods for calculating the natural frequencies (resonances) are given. Stability of these methods towards roundoff errors and small perturbations of the obstacles is established. Some formulas for the variations of the natural frequencies due to small perturbations of the surface of the obstacle are given. A simple new method for extraction of resonances from transient fields is given.

I. INTRODUCTION

Let $\mathscr{D}$ be a finite obstacle with a smooth surface Γ, and let Ω be the exterior domain. The obstacle (scatterer) is three dimensional. The smoothness of Γ is of the type that ensures the applicability of Green's formulas. Roughly speaking, cusps-type singular points of the surface are not admissible, but edges (as in a cube) or conical points are admissible. We will discuss for simplicity scalar wave scattering, but the results and arguments are valid for electromagnetic wave scattering. The Green's function for a reflecting obstacle satisfies the equations

$$(-\nabla - k^2)G(x, y, k) = \delta(x - y) \text{ in } \Omega, \quad k > 0, x = (x_1, x_2, x_3), \tag{1.1}$$

$$G = 0, \text{ if } x \in \Gamma \tag{1.2}$$

$$r\left(\frac{\partial G}{\partial r} - ikG\right) \to 0 \text{ as } r = |x| \to \infty. \tag{1.3}$$

Here y is the position vector of the source and ∇^2 is the Laplacian. The function G is uniquely determined by the conditions (1.1)–(1.3) and can be continued analytically on the whole complex plane of k as a meromorphic function of k. Its poles lie in the half-plane $\operatorname{Im} k < 0$ and are called resonances, natural frequencies, or complex poles. The meromorphic nature of G as a function of k and the (closely connected with it) behavior of solutions of the time-dependent wave equation as $t \to +\infty$ were studied in the series of papers starting with Ref. 1. In Ref. 2 there is a bibliography of the subject. In Ref. 3 one can find a collection of papers and an extensive bibliography of the singularity and eigenmode expansion methods. In Refs. 4 and 5 there are reviews of the subject for engineers. The connection of the complex poles asymptotic with the behavior of solutions to the time-dependent wave equation is the foundation of the singularity expansion method (SEM). If

$$\begin{aligned} &\nabla^2 u = u_{tt} \text{ in } \Omega,\ t > 0,\ u = 0 \text{ on } \Gamma, \\ &u(x, 0) = 0,\ u_t(x, 0) = f(x), \end{aligned} \tag{1.4}$$

then the function v, defined as

$$v(x, k) = \int_0^\infty \exp(ikt) u(x, t)\, \mathrm{d}t, \tag{1.5}$$

satisfies the equations

$$\begin{aligned} &(\nabla^2 + k^2) v = -f,\ v = 0 \text{ on } \Gamma, \\ &r\left(\frac{\partial v}{\partial r} - ikv\right) \to 0,\ r \to \infty, \end{aligned} \tag{1.6}$$

$$v = \int_\Omega G(x, y, k) f\, \mathrm{d}y, \tag{1.7}$$

$$u = (2\pi)^{-1} \int_{-\infty}^{\infty} \exp(-ikt) v(x, k)\, \mathrm{d}k. \tag{1.8}$$

Assume that f is a smooth function which vanishes outside of a bounded domain (compactly supported). In the engineering literature (e.g., in Ref. 4) the complex variable $s = -ik$ is often used. In the physical and mathematical literature k is usually the complex variable. The half-plane $\operatorname{Im} k < 0$ (which we use in this paper) corresponds to the half-plane Re $s < 0$ on the s plane. If one knows that[2]

$$v \text{ is meromorphic (and analytic if Im } k \geqslant 0), \tag{1.9}$$

$$|v| \leqslant C(b)(1+|k|)^{-a},\ a > \tfrac{1}{2},\ b = \text{Im } k,\ |\text{Re } k| \to \infty, \tag{1.10}$$

$$|\text{Im } k_j| < |\text{Im } k_{j+1}| \to \infty \text{ as } j \to \infty, \tag{1.11}$$

then one can move the contour of integration in (1.8) in the k plane down and obtain the SEM expansion[2] (cf. (III.1.15))

$$u(x,t) = \sum_{j=1}^{N} c_j(x,t)e^{-ik_j t} + O(e^{-|\text{Im } k_{N+1}|t}),\ t \to +\infty. \tag{1.12}$$

Here k_j are the complex poles of $v(x, k)$, $c_j(x, t) = i \text{ Re } s_{k=k_j}\{e^{-ikt}v(x, k)\}$ and N is the number of the poles in the strip $0 > \text{Im } k \geqslant \text{Im } k_N$. Usually it is assumed by engineers[4] that the poles k_j are simple, in which case $c_j(x, t) = c_j(x)$. If $m_j + 1$ is the multiplicity of the pole k_j, then $c_j(x, t) = O(t^{m_j})$, and one can write (1.12) as

$$u(x,t) = \sum_{j=1}^{N} \sum_{m=0}^{m_j} c_{jm}(x)t^m e^{-ik_j t} + O(e^{-|\text{Im } k_{N+1}|t}) \text{ as } t \to \infty. \tag{1.13}$$

Expansions (1.12) and (1.13) were called in Ref. 2 asymptotic SEM expansions. These expansions are proved under the assumption that $\mathscr{D}$ is strictly convex. The expansion

$$u = \sum_{j=1}^{\infty} c_j(x,t)e^{-ik_j t}, \tag{1.14}$$

which one can see in the literature, is not proved and is not valid in general. Formula (1.13) is not valid, in general, for nonconvex obstacles. For example, in Ref. 6 it is claimed that if the obstacle consists of two strictly convex bodies, then $G(x, y, k)$ has countably many complex poles on the line Im $k = c_0$. These poles asymptotically are equidistant and the distance between the poles depends on the distance between the bodies and the curvatures and principal directions of the surfaces Γ_1 and Γ_2 of the bodies $\mathscr{D}_1$ and $\mathscr{D}_2$ at the two closest points $a_1 \in \Gamma_1$ and $a_2 \in \Gamma_2$. (That is, $|a_1 - a_2| = \min_{s_1 \in \Gamma_1, s_2 \in \Gamma_2} |s_1 - s_2|$.) This result shows that SEM expansion (1.13) is not valid for two convex bodies. It suggests also that (1.13) is not valid for a single body with nonconvex boundary which can hold a trapping mode (i.e., a standing wave in the geometrical optics approximation). This, however, is not proved yet.

In principle, one can tell the difference between convex obstacles and nonconvex obstacles, capable of holding a trapping mode, by the behavior of complex poles k_j for large j: for convex osbtacles (1.11) holds and $|\mathrm{Im}\, k_j| \to +\infty$, while in the other case there exist infinitely many poles on a line $\mathrm{Im}\, k_j = c_0 < 0$.

The significance of the complex poles is manyfold. We mention only two areas important in applications. First, one can tabulate the complex poles and use them for target identification. Practically, it is expected that different scatterers produce different sets of complex poles. Although this is not proved, there are some supporting arguments (see Ref. 2, pp. 585 and 586). Second, systems theory uses representations of impulse responses as sums of exponentials. The problem is to find these exponentials from transient fields.

It is a long-standing open problem to prove that there exist infinitely many complex not purely imaginary poles of G for any reflecting obstacle. So far it was proved that there exist infinitely many purely imaginary poles (this is a result from Ref. 7, a simple proof one can find in Ref. 2).

The objectives of this paper include (1) formulation of the mathematical methods for numerically calculating the complex poles, (2) analysis of convergence and stability of these methods, and (3) formulation of a simple technique for extracting resonances (natural frequencies) from transient fields.[8]

An extensive bibliography on the third question can be found in Ref. 9. The techniques used the literature and reviewed in Ref. 9 are based mostly on Prony's method. Some other methods were also used.[9,10] Here we present a very simple numerical technique which seems to be new and does not require solving nonlinear or even linear equations. The question most difficult in this problem, that of noisy data, is discussed. The paper is organized as follows: Section II deals with questions (1) and (2). Section III deals with question (3).

Appendix A contains some formulas for perturbations of complex poles under perturbations of the surface of the scatterer. Numeration of formulas is separate in each of the sections. The material in Sec. III appeared in Ref. 8.

II. METHODS OF CALCULATING NATURAL FREQUENCIES

A. *Basic Equations*

From the Green's formula one obtains

$$G(x, y, k) = g(x, y, k) - \int_{\Gamma} g(x, s')h(s', y)\, ds', \tag{2.1}$$

$$g = e^{ik|x-y|}/(4\pi|x-y|), \tag{2.2}$$

$$h = \frac{\partial G}{\partial N_s}, \tag{2.3}$$

where N_s is the outer normal to Γ at the point s, and the dependence on k is suppressed in some of the function for brevity. Let $x = s \in \Gamma$ in (2.1). Then

$$\int_{\Gamma} gh\, ds' = g. \tag{2.4}$$

If k_j is a pole of G then it is a pole of h, so that $h = \psi/(k - k_j)^m$. Multiply (2.4) by $(k - k_j)^m$ and let $k = k_j$ to obtain

$$Q\psi \equiv \int_{\Gamma} g(s, s', k_j)\psi\, ds' = 0. \tag{2.5}$$

Therefore the complex poles are the points k_j at which Eq. (2.5) has a nontrivial solution.

Let us differentiate (1) in the direction N_s and then take $x \to s \in \Gamma$ to obtain

$$[I + A(k)]h \equiv h + Ah = 2\frac{\partial g}{\partial N_s}, \tag{2.6}$$

$$Ah = \int_{\Gamma} 2\frac{\partial g}{\partial N_s} h\, ds'. \tag{2.7}$$

This gives the second way to characterize the complex poles: They are the points at which the equation

$$B\psi \equiv [I + A(k)]\psi = 0 \tag{2.8}$$

has a nontrivial solution.

B. *Projection Methods for Calculating the Poles*

First, consider Equation (2.8). Take a complete, in $L^2(\Gamma) = H$, set of linearly independent functions $\{\phi_j\}$. The linear span of the first n functions is a linear subspace H_n, $H_n \subset H_{n+1}$. Since the system $\{\phi_j\}$ is complete in H one concludes that the system of subspaces H_n is limit dense in H, that is, $\text{dist}(\psi, H_n) \to 0$ as $n \to \infty$ for any $\psi \in H$. This property is crucial for the analysis below. Here dist is the distance between the element ψ and the subspace H_n. Let $\psi_n = \sum_{j=1}^{n} c_j^{(n)} \phi_j$. Consider the projection method for solving (2.8):

$$\begin{aligned} &(B\psi_n, \phi_m) = 0, \qquad 1 \leqslant m \leqslant n, \\ &\sum_{j=1}^{n} (B\phi_j, \phi_m) c_j = 0, \quad 1 \leqslant m \leqslant n. \end{aligned} \tag{2.9}$$

The necessary and sufficient condition for (2.9) to have a nontrivial solution is

$$\begin{aligned} &\det b_n(k) = 0, \\ &b_n(k) \equiv [b_{jm}(k)]_{j, m=1,\ldots,n}, \quad b_{jm} \equiv (B\phi_j, \phi_m). \end{aligned} \tag{2.10}$$

The parentheses denote the inner product in $L^2(\Gamma)$, $(u, v) = \int_\Gamma u\bar{v}\, ds$. The elements $b_{jm}(k)$ are entire functions of k since the operator A in (2,7) is an entire analytic operator function of k. Therefore (1) it is not obvious that Equation (2.10) has zeros [e.g., $\exp(k)$ does not have zeros] and (2) if (2.10) has zeros $k_j^{(n)}$, $j = 1, 2, \ldots$, then one should prove convergence of this method, that is, one should prove that

$$\lim_{n \to \infty} k_j^{(n)} = k_j, \tag{2.11}$$

where k_j are the complex poles of G, and that all of the complex poles can be obtained in this way. This will be done later.

Consider Equation (2.5). In the same way as in the previous section one can derive the equation $\det Q_n(k) = 0$,

$$Q_n(k) \equiv [Q_{jm}(k)]_{j, m=1,\ldots,n}, \; Q_{jm} = (Q\phi_j, \phi_m). \tag{2.12}$$

This equation is of the same structure as (2.10), and the same questions (1) and (2) should be investigated for Equation (2.12). The difference between operators Q and B is that Q is compact while B is of the

Fredholm type, so that Equation (2.5) is of the first kind, while (2.8) is of the second kind. The elements of Q_{jm} are easier to compute than b_{jm}.

C. *Variational Methods for Calculating the Poles*

Consider the problem

$$|Qf| = \min, \ |f| = 1, \tag{2.13}$$

where $|f|$ is the $L^2(\Gamma)$ norm, $|f|_p$ is the Sobolev space $W^{2,p}(\Gamma) = H^p$ norm, $|f| = |f|_0$, $|f|_p^2 = \int_\Gamma \{|u|^2 + |Du|^2 + \ldots + |D^p u|^2\}\, ds$, and D denotes the first-order derivative on Γ. For $p < 0$ the space $W^{2,p}$ is defined as a dual to $W^{2,|p|}$.

Take $f_n = \sum_{j=1}^n c_j^{(n)} \phi_j$, substitute in (2.13), and obtain the problem

$$\sum_{j=1}^{n} q_{mj}^{(n)} c_j^{(n)} = \lambda c_m^{(n)}, \ 1 \leqslant m \leqslant n, \ q_{mj}^{(n)} \equiv (Q\phi_j, Q\phi_m), \tag{2.14}$$

where λ is an eigenvalue of the matrix $q_{mj}^{(n)}$. This matrix is an entire function of k. Its minimal eigenvalue $\lambda_1^{(n)}(k)$ is the minimum of the function $|Qf|$ under the constraint (13). The points $k_j^{(n)}$, which are zeros of $\lambda_1^n(k)$,

$$\lambda_1^{(n)}(k_j^{(n)}) = 0, \tag{2.15}$$

converge to the complex poles of $G(x, y, k)$ and all of the complex poles can be obtained as limits of $k_j^{(n)}$ as $n \to \infty$,

$$k_j = \lim_{n \to \infty} k_j^{(n)}. \tag{2.16}$$

A similar idea was used in Ref. 11. Convergence of the methods given in Section IIB and a study of their stability are given in the next subsection.

D. *Convergence and Stability of the Methods for Calculating the Poles*

The basic ideas and methods of the analysis and proofs are taken from Ref. 12 (see also Refs. 2, 11, and 13). The basic results consist of a proof of convergence and stability of the methods given in Section II B towards roundoff errors and perturbations of the data.

(1) We start with the method given in the first part of Section II B. Let us assume that there exists a countable discrete set $\mathscr{P}$ of points k_j at which Equation (2.8) has a nontrivial solution. In Section II A we

proved that any complex pole of G belongs to $\mathscr{P}$. Let us show that any point $k_0 \in \mathscr{P}$ is a complex pole of G. Let ψ be a nontrivial solution to (8). Define the simple layer potential $v = \int_\Gamma g\psi \, ds'$. From the known formula (see Ref. 12, p. 240) $\partial v/\partial N_i = (A\psi + \psi)/2$ [in which $\partial/\partial N_i$ denotes the limit value of the normal derivative on Γ from the interior and A is given in (2.7)] and from Equation (2.8) it follows that $\partial v/\partial N_i = 0$. We know that $(\nabla^2 + k_0^2)v = 0$ in $\mathscr{D}$. Since k_0^2 is a complex number and the spectrum of the interior Neumann Laplacian consists of positive numbers only, we conclude that $v = 0$ in $\mathscr{D}$. Therefore $V = 0$ on Γ. If G does not have a pole at $k = k_0$ then the problem $(\nabla^2 + k_0^2)v = 0$ in Ω, $v = 0$ on Γ, $v(x, k_0)$ is the limit value of a function $v(x, k)$ analytic in k in a neighborhood of k_0 and belonging to $L^2(\Omega)$ when $\text{Im } k > 0$, has only the trivial solution. Thus, $v = 0$ in Ω if k_0 is not a pole in G. Therefore $\psi = \partial v/\partial N_i - \partial v/\partial N_e = 0$, where $\partial v/\partial N_e$ is the limit value of the normal derivative on Γ from the exterior domain. This contradicts the assumption that $\psi \not\equiv 0$. Therefore k_0 is a pole of G.

Let us prove now that for sufficiently large n, (1) Equation (2.10) has solutions, (2) Equation (2.11) holds, (3) all the complex poles can be obtained as limits (2.11), (4) complex poles k_j are stable towards small perturbations of the data; the notion of small perturbation will be specified.

Equation (2.9) can be written as an operator equation $P_n B P_n \psi = 0$, or

$$(I + P_n A)\psi_n = 0, \tag{2.17}$$

where P_n is the orthoprojection onto H_n, $\psi_n = P_n\psi \in H_n$, I is the identity. Since $A(k)$ is compact in H for any k and $P_n \to I$ as $n \to \infty$, where the arrow denotes strong convergence, one has $\|B - P_n B\| \to 0$ as $n \to \infty$. Therefore the operator $I + P_n A = I + A(k) - P^{(n)}A(k)$ is invertible for sufficiently large n in a neighborhood of any point k_0 at which $I + A(k_0)$ is invertible. Here $P^{(n)} \equiv I - P_n$, $P^{(n)} \to 0$ as $n \to \infty$. This argument shows that if k_0 is not a complex pole then there are no roots $k_j^{(n)}$ of Equation (2.10) in a neighborhood of k_0. It remains to be proved that if k_0 is a pole of G, then for sufficiently large n there exists a root $k_j^{(n)}$ of Equation (2.10) which lies in the circle C_δ: $|k - k_0| \leqslant \delta$, where $\delta > 0$ is an arbitrary small number. Suppose that for some $\delta > 0$ and all n there are no roots $k_j^{(n)}$ of Equation (2.10) in the circle C_δ. Then the operator

$I + P_n A(k)$ is invertible in C_δ, the operator $(I + P_n A(k))^{-1}$ is analytic in k in C_δ, and therefore $\|(I + P_n A(k))^{-1}\| \leqslant c$, where c is a constant which does not depend on n. On the other hand, $(I + A(k))^{-1} = (I + P_n A(k) + P^{(n)} A(k))^{-1} = (I + P_n A(k))^{-1}(I + P^{(n)} A(k) (I + P_n A(k))^{-1})^{-1}$. Since $\|P^{(n)} A(k)\| \to 0$ and $\|(I + P_n A)^{-1}\| \leqslant c$ one concludes that $(I + A(k))^{-1}$ is a bounded operator in C_δ. This is a contradiction since k_0 is a pole of the operator $(I + A(k))^{-1}$. The contradiction proves that for any $\delta > 0$ and sufficiently large n there is a root $k_j^{(n)}$ of Equation (2.10) in the circle C_δ.

(2) The above argument settles also the question about stability of the poles towards small perturbations of the data and roundoff errors. Indeed, small perturbations of the data and roundoff errors are equivalent to small perturbations of the matrix $b_{jm}(k)$.

Let us assume that a small perturbation of the matrix $b_{jm}(k)$ is caused by a small perturbation of the operator $B = I + A$. Let us denote $\tilde{B} = I + \tilde{A} = I + A + T$ as the perturbed operator. In this formulation the perturbed matrix $\tilde{b}_{jm}^{(n)}$ is the matrix of the operator $P_n \tilde{B} P_n$. The perturbation T can describe both the perturbation of Γ and the roundoff errors in the computing matrix $b_{jm}^{(n)}$. Our aims is to prove that in any finite domain on the k plane the poles $\tilde{k}_j^{(n)}$ of the perturbed operator $(I + P_n \tilde{B}(k) P_n)^{-1}$ differ from the poles $k_j^{(n)}$ of the unperturbed operator $(I + P_n B(k) P_n)^{-1}$ a little: $|k_j^{(n)} - k_j^{(n)}| \leqslant \varepsilon(n, \|T\|)$, $\varepsilon \to 0$ if $\|T\| \to 0$ and $n \to \infty$. Since we have already established the convergence property (2.11), it is sufficient to prove that

$$|\tilde{k}_j - k_j| \to 0, \text{ if } \|T\| \to 0, \; T = \tilde{B} - B. \tag{2.18}$$

Let k_j be a pole of $(I + A(k))^{-1}$ and there are no other poles of this operator in the circle C_δ. One has $(I + A(k) + T)^{-1} = (I + A(k))^{-1}(I + T(I + A(k))^{-1})^{-1}$. Suppose k_j is a pole of multiplicity m. Then

$$\|(I + A(k))^{-1}\| \leqslant c/|k - k_j|^m, \; k \in C_\delta, \; c = \text{const}.$$

Thus

$$\|T(I + A(k))^{-1}\| \leqslant c\|T\| \, |k - k_j|^{-m}.$$

If $c\|T\|\delta^m < 1$, then the perturbed pole $\tilde{k}_j$, corresponding to the unperturbed pole k_j, lies inside the circle C_δ, that is, $|\tilde{k}_j - k_j| < \delta$. In other words $|\tilde{k}_j - k_j| = 0(\|T\|^{1/m})$ where m is the multiplicity of the pole k_j and $O(\|T\|^{1/m})$ means $\leqslant$ const $\|T\|^{1/m}$.

The smallness of the perturbation of the surface is described in terms of the smallness of the norm $\|T\|$. One can give a relationship between the equation of the perturbed surface and the norm of T. This is cumbersome and is done in Appendix A.

Let us study the method based on Equation (2.12). The results will be the same: (1) Equation (2.12) has roots $k_j^{(n)}$ for sufficiently large n, (2) Equation (2.11) holds, (3) all the complex poles can be obtained as limits (2.11) and (4) small perturbations of the data lead to small perturbations of the complex poles uniformly on any bounded domain on the complex k plane.

Analysis of Equations (2.5) and (2.12) is more complicated than that of Equations (2.8) and (2.10) because (2.5) is an equation of the first kind. The basic tool in our analysis is the factorization formula

$$Q(k) = Q_0(I + V),\ Q_0 \equiv Q(0),\ V \equiv Q_0^{-1}(Q(k) - Q_0). \tag{2.19}$$

Here $Q_0 f = \int_\Gamma f \, ds'/(4\pi|s - s'|)$ is a self-adjoint positive definite operator on $H = L^2(\Gamma)$. This operator is an isomorphism between $H = H^0$ and H^1, while the operator V is compact in any space H^p (see Refs. 2 and 12 for details). Therefore the bilinear form $(Q_0 u, v)$ defines an inner product equivalent to the inner product $(u, v)_{-1/2}$ in $H^{-1/2}$. The matrix

$$Q_{jm} = (Q_0(I + V)\phi_j, \phi_m) = ((I + V)\phi_j, \phi_m)_{-1/2}.$$

Our previous arguments in Section II D are fully applicable to this matrix because (1) V is compact in $H^{-1/2}$ and depends analytically on k and (2) if a system $\{\phi_j\}$ is complete in H^0 then it is complete in H^p for any $p < 0$. Compactness of V was already mentioned. To explain the second statement assume that $f_p \in H^p$, $p < 0$. It is well known that $H^p \subset H^q$ if $p > q$, H^p is dense in H^q (that is, for any $\varepsilon > 0$ and any $f \in H^q$ there exists an $f_\varepsilon \in H^p$ such that $|f_\varepsilon - f|_q < \varepsilon$, where $|\cdot|_q$ is the norm in H^q), and $|f|_p \leqslant |f|_q$ if $p < q$. Let $f \in H^p$, $p < 0$, and $\varepsilon > 0$ is fixed. Find $f_\varepsilon \in H^0$ such that $|f - f_\varepsilon|_p < \varepsilon/2$. Use completeness of the system $\{\phi_j\}$ in H^0 to find $h_\varepsilon = \sum_{j=1}^{n(\varepsilon)} c_j(\varepsilon)\phi_j$ such that $|h_\varepsilon - f_\varepsilon|_0 < \varepsilon/2$. Then

$$|f - h_\varepsilon|_p \leqslant |f - f_\varepsilon|_p + |f_\varepsilon - h_\varepsilon|_p \leqslant$$
$$\leqslant \varepsilon/2 + |f_\varepsilon - h_\varepsilon|_0 < \varepsilon, \text{ if } p < 0.$$

Therefore the system $\{\phi_j\}$ is complete in $H^{-1/2}$ and the matrix Q_{jm} is a matrix of the operator $I + V(k)$ in $H^{-1/2}$, where $V(k)$ is compact in $H^{-1/2}$ and analytic in k. The rest of the argument is the same as in Sections. II D (1) and II D (2) and the conclusions are formulated in the beginning of this section.

(4) In this section we study the variational method given in Section II C. If $\lambda_1^{(n)}(k_j^{(n)}) = 0$, then Equation (2.14) corresponds to the projection method for the equation $Q^*Qf = 0$. The factorization in Section II D (3) is sufficient for the arguments of that section to hold for the operator Q^*Q [the reason is that $Q^*Q = (I + V^*)Q_0^2(I + V) = Q_0^2(I + V_1)$ where V_1 is compact]. Here we used the self-adjointness of Q_0. Compactness of V_1 follows from simple properties of pseudo-differential operators: ord $Q_0 = -1$, ord $V = -3$, $V_1 \equiv Q_0^{-2}V^*Q_0^2 + V + Q_0^{-2}V^*Q_0^2V$, ord $V_1 \leqslant$ ord $V < 0$. Here ord Q is the order of the pseudodifferential operator Q. One can find properties of pseudo-differential operators, e.g., in Ref. 14. Thus, one concludes that the results (1)–(4) in Section II D (3) hold for the variational method described in Section II C.

III. EXTRACTING NATURAL FREQUENCIES FROM TRANSIENT FIELDS

A. *Preliminaries*

Consider the problem

$$u_{tt} = \nabla^2 u,\ t \geqslant 0,\ x \in \Omega \subset \mathbb{R}^3,\ u = 0 \text{ on } \Gamma,$$
$$u(x, 0) = 0,\ u_t(x, 0) = f(x), \tag{3.1}$$

where $\mathscr{D} = R^3 \setminus \Omega$ is a bounded connected domain with a smooth strictly convex boundary, $f \in C_0^\infty$. In Ref. 2 the basic results on the asymptotic behavior of u as $t \to +\infty$ are described. In particular, the following asymptotic SEM (singularity expansion method) formula holds:

$$u = \sum_{j=1}^{N} \sum_{m=0}^{m_j} c_{jm}(x) t^m e^{-ik_jt} + O(e^{-|\mathrm{Im}\, k_{N+1}|t}),\ t \to +\infty, \tag{3.2}$$

where $c_{jm}(x)$ do not depend on t, $k_j = a_j - ib_j$, $b_j > 0$, are complex poles of the resolvent kernel G of the Dirichlet Laplacian in Ω

$$(\nabla^2 + k^2)G(x, y, k) = -\delta(x - y) \text{ in } \Omega,\ G = 0 \text{ on } \Gamma,$$

$$|x| \left(\frac{\partial G}{\partial |x|} - ikG \right) \to 0 \text{ as } |x| \to \infty,$$

and $m_j + 1$ is the multiplicity of the pole k_j, $\operatorname{Im} k_j < 0$, $|\operatorname{Im} k_j| \to \infty$ as $j \to \infty$. The poles k_j are called resonances or natural frequencies. The signal (3.2) is the transient field that can be observed experimentally. Knowledge of the resonances k_j may serve for target identification: the scatterers of various shapes produce various sets of resonances, and this is one of the reasons to be interested in resonances.

The other reason is that in systems theory one often models impulse responses as a sum of the type given in formula (3.2). The important problem of system identification can be formulated as follows: From the observation of the transient field (3.2) find the numbers k_j and m_j. There is an extensive literature on the subject. Many researchers contributed to the field (Prony, Bruns, Dale, Lagrange, Kühnen, and quite a few modern researchers). A large bibliography can be found in Refs. 9 and 15. Only the case $m_j = 0$ (simple poles) was treated in the literature as far as I know.

The purpose of the following subsection is (1) to give a simple numerical procedure for computing the number m_j and k_j, $1 \leqslant j \leqslant N$ for any fixed N from the exact transient data, (2) to discuss this problem for noisy data, and (3) to briefly review the classical methods (Prony, Bruns, Lagrange, and Dale).

B. *A Simple Method for Extracting Resonances from the Transient Field*

Assume first that the scatterer is a strictly convex reflecting body so that (3.2) holds. By $u(n)$ let us denote the sequence $u(x, nh)$, where $h > 0$ is a fixed number. It follows from (3.2) that

$$u(n) = c_{1m_1} h^{m_1} n^{m_1} e^{-ia_1 nh} e^{-b_1 nh}(1 + O(1/n)) \quad \text{as } n \to \infty, \text{ provided that } b_1 < b_2 < \dots. \tag{3.3}$$

From (3.3) one obtains

$$\frac{u(n+1)}{u(n)} = e^{-ia_1 h} e^{-b_1 h}(1 + O(1/n)) \text{ as } n \to \infty. \tag{3.4}$$

Thus,

$$b_1 = \frac{1}{h} \ln \left| \frac{u(n)}{u(n+1)} \right| + O\left(\frac{1}{n}\right) \text{ as } n \to \infty, \tag{3.5}$$

$$b_1 + ia_1 = \frac{1}{h} \ln \frac{u(n)}{u(n+1)} + O\left(\frac{1}{n}\right) \text{ as } n \to \infty. \tag{3.6}$$

Suppose that k_j, m_j, and c_{jm}, $1 \leqslant m \leqslant m_j$, $1 \leqslant j \leqslant N$, are computed. In what follows, a method for computing c_{jm} is given. Let u_N denote the sum in formula (3.2), $u - u_N \equiv \omega_N$. Then, as above,

$$b_{N+1} = \frac{1}{h} \ln \left| \frac{\omega_N(n)}{\omega_N(n+1)} \right| = O\left(\frac{1}{n}\right) \text{ as } n \to \infty, \tag{3.7}$$

$$\begin{aligned} b_{N+1} + ia_{N+1} &= \frac{1}{h} \ln \frac{\omega_N(n)}{\omega_N(n+1)} \\ &= O\left(\frac{1}{n}\right) \text{ as } n \to \infty. \end{aligned} \tag{3.8}$$

If a_1 and b_1 are found, then m_1 can be found by the formula

$$m_1 = \frac{\ln\{u(n)e^{ia_1 nh + b_1 nh}\}}{\ln n} + O\left(\frac{1}{\ln n}\right) \text{ as } n \to \infty. \tag{3.9}$$

Similarly,

$$m_{j+1} = \frac{\ln\{\omega_j(n)e^{ia_{j+1} nh + b_{j+1} nh}\}}{\ln n} + O\left(\frac{1}{\ln n}\right) \tag{3.10}$$

as $n \to \infty$, $j = 1, 2, \ldots$.

If a_1, b_1, and m_1 are found then

$$c_{1m_1} = \frac{u(n)e^{ia_1 nh + b_1 nh}}{(nh)^{m_1}} + O\left(\frac{1}{nh}\right) \text{ as } n \to \infty. \tag{3.10$'$}$$

In the literature[15–22] the case of simple resonances ($m_j = 0$) only was discussed. In this case $O(1/n)$ in (3.11) can be substituted by $O(e^{-(b_2-b_1)nh})$ as $n \to \infty$.

In a similar way all the coefficients c_{1m} can be computed. Practically, one takes n large, neglects the remainder in formulas of the type (3.10′), and uses the main term in the right-hand side of (3.10′) as the formula for c_{1m}, etc.

If k_1, a_1, b_1, and C_{1m}, $1 \leqslant m \leqslant m_1$, are found then one works with $\omega_1 = u - u_1$ and so on. This is a method for computing the coefficients c_{jm} from the transient field.

An alternative method is the usual least squares method. If m_1, a_1, and b_1 are found then c_{1m}, $1 \leqslant m \leqslant m_1$, can be found from the requirement

$$\sum_{n=1}^{\infty} \left| u(n) - \sum_{m=0}^{m_1} c_{1m}(nh)^m e^{(-ia_1-b_1)nh} \right|^2 = \min.$$

This leads to a uniquely solvable linear system for the coefficients c_{1m}, $1 \leqslant m \leqslant m_1$. If k_1, a_1, b_1, m_1, and c_{1m}, $1 \leqslant m \leqslant m_1$, are found then one works with $u - u_1 = \omega_1$ and considers ω_1 as the transient field. The method previously described is quite simple and does not require solving linear systems.

Formulas (3.5)–(3.11) give a simple method for extracting resonances and their multiplicities from the exact transient field. The much more complicated case of noisy data is discussed below.

In systems theory u, being an impulse response of a system, often does not depend on x.

We will discuss the case of noisy data. Assume that $y(n) = u(n) + \varepsilon(n)$ is measured instead of $u(n)$. Here $\varepsilon(n)$ is noise. Let us assume that $\varepsilon(n)$ is uniformly distributed on the interval $[-\varepsilon, \varepsilon]$, $\varepsilon > 0$ is a given number. In practice the level of noise is not known exactly since the noise comes not only from the errors in measurements but also from the unknown background noise in the environment of the scatterer. But without some assumptions about the noise nothing can be derived. One has

$$\begin{aligned} y(n) &= c_{1m_1} h^{m_1} n^{m_1} e^{-ia_1nh-b_1nh}(1 + O(1/n)) + \varepsilon(n) \\ &= c_{1m_1} h^{m_1} n^{m_1} e^{-ia_1nh-b_1nh}\{1 + O(1/n)) + \\ &\quad + \varepsilon(n) c_{1m_1}^{-1} (hn)^{-m_1} e^{ia_1nh+b_1nh}\}. \end{aligned} \tag{3.11}$$

From (11) it follows that, regardless of the method used, the extraction of the complex poles k_j from noisy data is highly unstable and depends on the magnitude of $\alpha_n \equiv O(1/n) + \varepsilon(hn)^{-m_1} e^{b_1 nh} c_{1m_1}^{-1}$. If there exists n such that $\alpha_n \ll 1$ (say $\alpha_n < 0.1$), then the pole $k_1 = a_1 - ib_1$ can be computed by formulas (3.5) and (3.6) in which $y(n)$ (α_n) should take the place of $u(n)$ $(O(1/n))$. Similar considerations hold for other poles. Since $b_j > 0$ the factor $e^{b_j nh}$ is growing as $n \to \infty$. Therefore ε should be small in order that α_n be small and k_j could be computed. In this case it is not advisable to take n too large because for large n the second term in α_n becomes large. Since the bound on $O(1/n)$ is not available it is not worthwhile to compute the optimal n, but practically, n should be taken as a value for which $(1/h) \ln |y(n)/y(n+1)|$ is stationary when one computes b_1, and for which $(1/h) \ln |z_j(n)/z_j(n+1)|$ is stationary when one computes b_{j+1}.

If the constant c_{1m_1} is small then the second term in α_n is large unless ε is sufficiently small. Therefore it is difficult to compute resonance with small Laurent coefficients (coupling coefficients) in front of the singular terms $(k - k_j)^{-m}$. All these arguments are very simple but they show clearly the nature of the difficulties for which noise is responsible and the limitations of any method of resonances extraction from noisy data.

We assumed that the scatterer was convex. This assumption implies the basic result: the validity of Equation (3.2). In the paper Ref. 6 it is claimed that for the scatterer consisting of two strictly convex reflecting bodies (3.2) does not hold: there exist countably many poles k_j on some line $\operatorname{Im} k = \text{const}$. Therefore one cannot order the poles by the rule $|\operatorname{Im} k_j| < |\operatorname{Im} k_{j+1}|$ in the case of two disjoint convex reflecting bodies. If the scatterer is just one strictly convex reflecting body then it is known that $|\operatorname{Im} k_j| \to +\infty$ as $j \to \infty$ and (3.2) holds.[2]

C. *A Brief Review of the Existing Methods for the Resonances Extraction*

The most popular method is Prony's method.[8,16] One assumes that $u = u(t) = \sum_{j=1}^{N} c_j e^{s_j t}$, $s_j \equiv -ik_j$, $c_j = \text{const}$. One observes $u(t)$ experimentally and wants to compute s_j and c_j. If the data is exact (there is no noise) then Prony's method consists in the following. Let $f_n \equiv u(nh)$ where $h > 0$ is a fixed number, $e^{s_j h} \equiv z_j$. Then $f_n = \sum_{j=1}^{N} c_j z_j^n$, $n \geqslant 0$. An obvious linear algebra argument shows that det $A_{pq}^{(m)} = 0$, $m \geqslant 0$, $0 \leqslant p, q \leqslant N$, where $A_{p0}^{(m)} \equiv f_{p+m}$, $A_{pq}^{(m)} = z_q^p$, $0 \leqslant p \leqslant N, 1 \leqslant q \leqslant N$. Therefore,

$$0 = \sum_{p=0}^{N} f_{p+m} A_p, \; m \geqslant 0, \tag{*}$$

where A_p are the cofactors corresponding to the elements f_{p+m} of the matrix $A_{pq}^{(m)}$. Notice that A_p do not depend on m. Write $N + 1$ equations (*) taking $m = 0, 1, \ldots, N$, and find a nontrivial solution $(A_0, A_1, \ldots, A_N)$ to the $N + 1$ simultaneous equations (*). Consider the equation

$$\sum_{p=0}^{N} A_p z^p = 0. \tag{**}$$

From the structure of the matrix $A_{pq}^{(0)}$ it is clear that Equation (**) has solutions $z_j = e^{s_j h}$. Thus, $s_j = h^{-1} \ln z_j$. If one does not know the number N (and this is usually the case in practice) then there is a problem of choosing the right N. In Ref. 14, p. 140, there is a method (due to Kühnen) for choosing N. If the data are noisy then one faces the difficulties explained in Section III B and reported in the literature.[9] If the data are noisy then the matrix f_{p+m}, $0 \leqslant p, m \leqslant N$ is nonsingular and system (*) with $0 \leqslant m \leqslant N$ has only the trivial solution $A_p = 0$, which cannot be used since Equation (**) in this case gives no information. Therefore in practice one takes A_p, $0 \leqslant p \leqslant N$, the components of an eigenvector corresponding to the minimal eigenvalue of the matrix F^*F. Here F^* is the adjoint matrix, and F is the matrix of the noisy data, $F_{p+m} = f_{p+m} + \varepsilon_{p+m}$, where ε_{p+m} is noise. If there are several eigenvectors corresponding to the minimal eigenvalue, one has no rule to pick up any particular eigenvector. But this situation is not generic in the sense that a small perturbation of the matrix will split up the multiple eigenvalue into a number of simple ones. However, the simple eigenvalues will be close to each other and it will be difficult to find the minimal eigenvalue numerically. (Recall that an eigenvalue is called simple if there is only one linearly independent eigenvector corresponding to this eigenvalue.) One can find an extensive bibliography and a discussion of Prony's method in Ref. 9. The Bruns' method described in Ref. 15 is essentially the Prony method for real resonances.

Let us outline another method for extracting the resonances. Let $f(t) = \sum_{j=1}^{N} c_j e^{s_j t}$ and N is assumed known. Then $f^{(m)}(t) = \sum_{j=1}^{N} c_j s_j^m e^{s_j t}$. Taking $t = 0$ and $m = 0, \ldots, N, \ldots$, one obtains

$$\begin{vmatrix} f(0) & 1 & \ldots & 1 \\ f'(0) & s_1 & & s_N \\ \vdots & \vdots & & \vdots \\ f^{(N)}(0) & s_1^N & & s_N^N \end{vmatrix} = 0, \quad \begin{vmatrix} f'(0) & 1 & 1 \\ \vdots & \vdots & \vdots \\ f^{(N+1)} & s_1^N & s_N^N \end{vmatrix} = 0, \ldots . \tag{3.12}$$

Therefore

$$\sum_{p=0}^{N} f^{(m+p)}(0) A_p = 0, \; m = 0, 1, 2, \ldots . \tag{3.13}$$

Here A_p is the cofactor corresponding to the element $f^{(p)}(0)$ in the first matrix in (3.12). The argument above is very similar to that given in the previous method and $f^{(m+p)}(0)$ plays the role of f_{m+p} in (*) in that section. Taking $m = 0, 1, \ldots, N$ in (3.13) one obtains a homogeneous system of linear equations. If $A_0, \ldots, A_N$ is a solution to this system then the N roots of equation (**) for this method are equal to s_j, $1 \leqslant j \leqslant N$. This gives a method of extracting the resonances s_j from the exact transient data. It is interesting that only the data near $t = 0$ is used in this procedure, while the procedure in the previous method needs the data at large t. On the other hand, the procedure based on Equation (3.13) is very sensitive to the noise in the data because one needs to differentiate the data.

We assumed above that the number N of resonances was known. If N is not known, then one can find N as the smallest number for which $\det f_{p+m} = 0, 0 \leqslant p, m \leqslant N$.

The simple algorithm in Section III B *requires that* $b_1 < b_2 < b_3 \ldots$. In practice the poles $a_j - ib_j$ occur in pairs $\pm a_j - ib_j$ and the measured transient field is a real-valued function. Assuming that the poles are simple, i.e., $m_j = 0$, one has

$$u(t) = \sum_{j=1}^{N} c_j \exp(-b_j t) \cos(a_j t + \phi_j) + O(e^{-b_{N+1} t})$$

as $t \to +\infty$. Therefore for large t one obtains $u(t) = c_1 \exp(-b_1 t)$

$\cos(a_1 t + \phi_1) + O(e^{-b_2 t})$, $t \to +\infty$. If the values $u_n \equiv u(nh)$ are measured, then the values c_1, b_1, a_1, and ϕ_1, $0 \leqslant \phi_1 < 2\pi$, can be numerically obtained from the requirement

$$F(c_1, b_1, a_1, \phi_1) \equiv \frac{1}{m} \sum_{n=n_0}^{n_0+m} |c_1 \exp(-b_1 hn) \times \cos(a_1 hn + \phi_1) - u_n|^2 = \min. \tag{3.14}$$

Here n_0 is a large number such that $\exp(-b_1 hn_0) \gg \exp(-b_2 hn_0)$, $m > 4$ is a fixed number, and the function $F(c_1, b_1, a_1, \phi_1)$ is to be minimized numerically. If this minimization problem is solved one can consider $\omega_1 \equiv u(t) - u_1(t)$, where $u_1(t) \equiv c_1 \exp(-b_1 t) \cos(a_1 t + \phi_1)$, and apply the same procedure for finding c_2, b_2, a_2, ϕ_2. Each step requires minimization of a function of four variables only. The basic new idea in our method is to use the asymptotic behavior as $t \to +\infty$ of the transient field. One should have in mind that the basic asymptotic SEM expansion (3.2) is proved only if the scatterer is convex[1] (or, more generally, star shaped). It does not hold, for example, when the scatterer consists of two convex obstacles. In this case there exists infinitely many poles k_j, such that $|k_j - ic_0 - \pi d^{-1} j| \leqslant c(1 + |j|)^{-1/2}$ for all large j, $j = \pm j_0, \pm(j_0 + 1), \ldots$. Here d is the distance between the two obstacles and c_0 depends on d, on the principal curvatures, and principal directions of the surfaces Γ_1 and Γ_2 of the two obstacles at the points $s_1 \in \Gamma_1$ and $s_2 \in \Gamma_2$, such that $|s_1 - s_2| = d$. This result was given recently by Ikawa.[6]

IV. BIBLIOGRAPHICAL REMARKS

First we mention some of the old papers. Of these only Prony's paper[16] is often mentioned by modern authors. There is a translation of this paper in Ref. 9. Bruns[17] used practically the same idea as Prony. His work is discussed in Ref. 14. There are several authors, astronomers mostly, which were interested in detection of hidden periodicities.[17–22] Although only the case $m_j = 0$, $b_j = 0$ was discussed in these papers, the basic questions (extracting the resonances from the transient field, determining the number N of resonances, etc.) were actually identical with the questions discussed in a very recent review[9] of the state of the art in this field. References[17–22] are not cited by modern western authors in the field. A review of these papers can be found in Ref. 15.

There exists a very extensive modern literature on the extracting of resonances. One can find a large bibliography and a review of the basic results in Ref. 9. We did not discuss here some of the methods mentioned in Ref. 9.

There are many reasons for being interested in the extracting of resonances. We mention only two major theories: singularity expansion method (see Refs. 2, 12, and 23 for the mathematical results) and systems identification (see the bibliography in Ref. 9).

APPENDIX A: PERTURBATION OF RESONANCES

1. *Abstract Scheme*

First let us present an abstract scheme. Assume that a compact operator function $A(k)$ on a Hilbert space H is analytic in k in a domain on the complex plane k, and $A(k_0)$ has an eigenvalue -1. Then $(I + A(k))^{-1}$ has a pole at $k = k_0$. Suppose that $A(k, \varepsilon)$ is a compact operator function such that $A(k, 0) = A(k)$, which is analytic in k and ε, $|k - k_0| < \delta_0$, $|\varepsilon| < \delta_1$. Assume that k_0 is an isolated pole of $I = A(k)$. [This is the case if $I + A(k)$ is invertible for at least one k in the disk $|k - k_0| < \delta_0$.] Then, in a neighborhood of k_0, there exist a finite number m_0 of points $k_j(\varepsilon) = k_0 + \sum_{m=1}^{\infty} a_{mh}\varepsilon^{m/p}$, $1 \leqslant j \leqslant j_0$, such that the operator $(I + A(k, \varepsilon))^{-1}$ has poles $k(\varepsilon)$. Here j_0 is the multiplicity of the pole k_0, and the meaning of the integer p will be explained in the proof which is based on the idea in Ref. 2, p. 582.

Let $\phi_1, \ldots, \phi_n$ be an orthonormal basis of $N(I + A(k_0))$, where $N(A)$ is the null space of an operator A. Let $\psi_1, \ldots, \psi_n$ be an orthonormal basis of $N(I + A^*(k_0))$, where the star denotes the adjoint operator. Let $Th \equiv \sum_{j=1}^{n} (h, \phi_j)\psi_j$. The operator $I + A(k_0) + T$ is invertible in H. Indeed, $(I + A(k_0) + T)h = 0$ implies that $(Th, \psi_j) = 0$, $1 \leqslant j \leqslant n$. This leads to $(h, \phi_j) = 0$, $1 \leqslant j \leqslant n$, i.e., $Th = 0$ and $(I + A(k_0))h = 0$. Thus $h \in N(I + A(k_0))$ and $h \perp N(I + A(k_0))$. Therefore, $h = 0$, and by Fredholm's alternative $(I + A(k_0) + T)^{-1} = \Gamma$ is bounded. Consider $(I + A(k, \varepsilon))^{-1} = (I + A(k_0) + T + A(k, \varepsilon) - A(k_0) - T)^{-1} = (I + a(k, \varepsilon))^{-1}$. Then $\Gamma(k, \varepsilon)$, where $\Gamma(k, \varepsilon) = (I + A(k_0) + T + A(k, \varepsilon) - A(k_0))^{-1}$, is analytic in k and ε in a neighborhood Δ of $(k_0, 0)$, and $a(k, \varepsilon) = -\Gamma(k, \varepsilon)T$ is a finite-dimensional operator analytic in k and ε in Δ, $ah = -\sum_{j=1}^{n}(h, \phi_j)\Gamma(k, \varepsilon)\psi_j$. Since $\Gamma(k, \varepsilon)$ is an isomorphism of H onto H for $k, \varepsilon \in \Delta$, the elements $\psi_j(k, \varepsilon) = -\Gamma(k, \varepsilon)\psi_j$ are linearly independent and analytic in $k, \varepsilon \in \Delta$. Therefore the operator $(I + a(k, \varepsilon))^{-1}$ can be constructed explicitly. If $(I + a)h = f$, then $h + \sum_{j=1}^{n} h_j\psi_j(k, \varepsilon) = f$, $h_j \equiv (h, \phi_j)$. Multiply by ϕ_m to obtain $h_m + \sum_{j=1}^{n} c_{mj}h_j = f_m$, where $f_m = (f, \phi_m)$, $c_{mj} = (\psi_j(k, \varepsilon), \phi_m)$, c_{mj} are analytic in $k, \varepsilon \in \Delta$. Thus, $h_m = d_m(k, \varepsilon)/d(k, \varepsilon)$, where d_m and $d = \det(\delta_{mj} + c_{mj})$ are analytic in Δ. One has $(I + a(k, \varepsilon))^{-1}f = f - \frac{1}{d}\sum_{j=1}^{n} d_j(k, \varepsilon)\psi_j(k, \varepsilon)$. From this formula it is clear that the poles of $(I + a(k, \varepsilon))^{-1}$ can occur only at the zeros of $d(k, \varepsilon)$. Thus the equation for the perturbed poles is

$$\begin{aligned} d(k, \varepsilon) &= \det[\delta_{jm} - (\Gamma(k, \varepsilon)\psi_j, \phi_m)] = 0, \\ \Gamma(k, \varepsilon) &= (I + T + A(k, \varepsilon))^{-1}. \end{aligned} \tag{A1}$$

For $\varepsilon = 0$ the function $d(k) = d(k, 0)$ has, by assumption, a zero of multiplicity j_0. By the Weierstrass' preparation theorem (see, e.g., Ref. 2, p. 583) one has $d(k, \varepsilon) = [k^{j_0} + \sum_{j=1}^{j_0-1} c_j(\varepsilon)k^j]g(k, \varepsilon)$, where $g(0, 0) \neq 0$, $c_j(0) = 0$, and c_j and $g(k, \varepsilon)$ are holomorphic functions. Therefore Equation (A1) is equivalent to

$$k^{j_0} + \sum_{j=1}^{j_0-1} c_j(\varepsilon)k^j = 0. \tag{A2}$$

This equation has j_0 roots. These roots can be divided into several groups so that the p roots $(k_1(\varepsilon), \ldots, k_p(\varepsilon))$ in the νth group can be expanded in a Puiseux series $k_j(\varepsilon) = k_0 + \sum_{m=1}^{\infty} a_{mj}\varepsilon^{m/p\nu}$, $\sum_\nu p_\nu = j_0$. The number of groups and the integers p_ν can be computed by the algorithm known as Newton's diagram method (see, e.g., Ref. 24). Let us summarize our arguments: a method for computing the poles $k_j(\varepsilon)$ of the perturbed operator $(I + A(k, \varepsilon))^{-1}$ is given. The method is valid under the following assumptions: $A(k, \varepsilon)$ is a compact analytic (in k and ε) linear operator function on a Hilbert space, the operator $(I + A(k, 0))^{-1}$ has an isolated pole at $k = k_0$. To use the method computationally one needs to compute (or to know) the bases in the subspaces $N(I + A(k_0))$ and $N(I + A^*(k_0))$. This is a linear algebra problem.

2. *Reduction of a Concrete Perturbation Problem to the Abstract One*

Suppose that the surface Γ of the obstacle is perturbed. Let $x_j = x_j(u, v)$, $1 \leqslant j \leqslant 3$ be a parametric equation of Γ, and $z_j = x_j(u, v) + \varepsilon y_j(u, v)$ be the equation of the perturbed surface Γ_ε, where ε is a small parameter. Assume that the functions x_j and y_j, $1 \leqslant j \leqslant 3$, $u, v \in S \equiv \{u, v\colon 0 \leqslant u, v \leqslant 1\}$ are smooth. Consider Equation (2.8). Assume that k_0 is a pole of the operator $(I + A(k))^{-1}$. Suppose that the bases of the subspaces $N(I + A(k_0))$ and $N(I + A^*(k_0))$ are computed. Consider the problem corresponding to the perturbed surface Γ_ε. The operator $A(k, \varepsilon)$, associated with this problem, is of the form

$$\int_{\Gamma_\varepsilon} 2 \frac{\partial g}{\partial N_s} h \, \mathrm{d}s = \int_S A(u, v, u', v', k, \varepsilon) h \, \mathrm{d}u' \, \mathrm{d}v'.$$

Here $A(u, v, u', v', k, \varepsilon)$ is the kernel of the integral operator in the variables u, v, u', v'. If $\varepsilon = 0$ then $\Gamma_\varepsilon = \Gamma$ and we assume that the sets $\{\phi_j\}$ $(\{\psi_j\})$, $1 \leqslant j \leqslant n$ of all linearly independent solutions of the equation $\phi + \int_S A(u, v, u', v', k_0, 0)\phi \, \mathrm{d}u' \, \mathrm{d}v' = 0$ $(\psi + \int_S A(u', v', u, v, k_0, 0)\psi \, \mathrm{d}u' \, \mathrm{d}v' = 0)$ are known. Then the abstract scheme is applicable.

Since small perturbations of the kernel cause small perturbations of the poles, one can approximate the kernel A by a degenerate kernel and consider the corresponding matrix problem.

As an example consider a simple case when the matrix is 2×2. Let

$$\begin{pmatrix} 1 & e^{\pi k} \\ e^{\pi k} & 1 \end{pmatrix} \begin{pmatrix} h_1 \\ h_2 \end{pmatrix} = \begin{pmatrix} f_1 \\ f_2 \end{pmatrix}.$$

Then the inverse matrix is

$$\begin{vmatrix} 1 & -e^{\pi k} \\ -e^{\pi k} & 1 \end{vmatrix} \frac{1}{1 - e^{2\pi k}}.$$

It has simple poles $k_m = im$, $m = 0, \pm 1, \ldots$. Consider, for example, pole $k_0 = 0$. The set $\{\phi_j\}$ corresponding to this pole is the set of linearly independent solutions to the equation

$$\begin{cases} c_1 + c_2 = 0 \\ c_1 + c_2 = 0 \end{cases}.$$

Thus, there exists one linearly independent solution

$$\phi_1 = \begin{pmatrix} -1 \\ 1 \end{pmatrix}.$$

In our example the matrix

$$\begin{pmatrix} 1 & e^{\pi k} \\ e^{\pi k} & 1 \end{pmatrix}$$

is self-adjoint for $k = 0$. Thus

$$\psi_1 = \phi_1 = \begin{pmatrix} -1 \\ 1 \end{pmatrix}.$$

Consider the perturbed matrix

$$\begin{pmatrix} 1 + \varepsilon a_{11} & e^{\pi k} + \varepsilon a_{12} \\ e^{\pi k} + \varepsilon a_{21} & 1 + \varepsilon a_{22} \end{pmatrix}.$$

The poles of this matrix are the roots of the equation $(1 + \varepsilon a_{11})(1 + \varepsilon a_{22}) - (e^{\pi k} + \varepsilon a_{12})(e^{\pi k} + \varepsilon a_{21}) = 0$. We are interested in the root $k(\varepsilon)$ such that $k(\varepsilon) \to 0$ as $\varepsilon \to 0$. One has $1 - e^{2\pi k} + \varepsilon[a_{11} + a_{22} - e^{\pi k}(a_{12} + a_{21})] + \varepsilon^2(a_{11}a_{22} - a_{12}a_{21}) = 0$. Let $e^{2\pi k} = z$, $a_{11} + a_{22} = a$, $a_{12} + a_{21} = b$, $a_{11}a_{22} - a_{12}a_{21} = c$. Then

$$z^2 + z\varepsilon b - 1 - \varepsilon a - \varepsilon^2 c = 0,$$

$$z(\varepsilon) = -\varepsilon b/2 + \sqrt{(\varepsilon^2 b^2/4) + 1 + \varepsilon a + \varepsilon^2 c} \approx$$

$$\approx 1 + (\varepsilon/2)(a - b).$$

The plus sign in front of the radical is chosen because $z = 1$ if $k = 0$. Thus, in this example the perturbed pole $k_0 = 0$ can be computed for small ε as $k(\varepsilon) = \pi^{-1} \ln z(\varepsilon) = (a - b)\varepsilon/2\pi$. Depending on the values of a and b the perturbed pole can move in any direction. If $a \neq b$ then $k(\varepsilon) = O(\varepsilon)$. If $a = b$ then $k(\varepsilon) = O(\varepsilon^2)$. The bifurcation theory and the Newton diagram method solve the following problem. Given an equation $F(k, \varepsilon) = 0$, find its solutions $k(\varepsilon)$ such that $k(\varepsilon) \to 0$ as $\varepsilon \to 0$. It is assumed that $F(0, 0) = 0$. If $F_k(0, 0) \neq 0$ then the solution is well known and is given by the standard implicit function theorem. If $F_k(0, 0) = 0$ then the solution is more complicated. Methods and algorithms for solving this problem can be found in Ref. 24.

APPENDIX B: EXTRACTION OF RESONANCES FROM THE TRANSIENT FIELD

The case of simple poles symmetrical with respect to the imaginary axis is as follows.

Let the poles be simple and occur in pairs $\pm a_j - ib_j$, $0 < b_1 < b_2 < \dots$. Then the transient field is $u = \sum_{j=1}^{N} u_j + O(\exp(-b_{N+1}t))$, as $t \to +\infty$, $u_j = c_j \exp(-b_j t) \cos(a_j t + \phi_j)$, $c_j > 0$, $a_j > 0$, $b_j > 0$, $0 \leqslant \phi_j < 2\pi$. We sketch a method for finding c_j, b_j, a_j, and ϕ_j, which is simpler than the one given by formula (3.14).

Step 1: One has $u(t) = u_1 + O(\exp(-b_2 t))$, so $u(t) \approx u_1(t)$ as $t \gg 1$. Thus

$$\frac{1}{mh} \ln \frac{u[(n+m)h]}{u(nh)} \approx \approx -b_1 + \frac{1}{mh} \ln \frac{\cos(a_1 nh + a_1 mh + \phi_1)}{\cos(a_1 nh + \phi_1)} \cdot h > 0. \tag{B1}$$

Therefore for $m \gg 1$ one can find b_1 from (B1). Some caution is needed: if $\cos(a_1 nh + a_1 mh + \phi_1) \approx 0$ then the ln term is large. But this can happen rarely. One can compute the left side of (B1) for a number of consecutive m and neglect large values of this quantity.

Step 2: If b_1 is found then a_1 can be found by the formula $a_1 \approx (-\ddot{v}/v)^{1/2}$, $t \gg 1$, where $v(t) \equiv u(t)\exp(b_1 t)$, and $\ddot{v} = \mathrm{d}^2 v/\mathrm{d}t^2$.

Step 3: If a_1 and b_1 are found then $c_1 \approx (\dot{v}^2/a_1^2 + v^2)^{1/2}$, $t \gg 1$.

Step 4: If a_1, b_1, and c_1 are found, then ϕ_1 can be found from the equations

$$c_1^{-1} v(2\pi n/a_1) = \cos\phi_1, \ -c_1^{-1} v(2\pi n/a_1 + \pi/2a_1) = \sin\phi_1, \ n \gg 1.$$

Since $0 \leqslant \phi_1 < 2\pi$, these equations determine ϕ_1 uniquely. If b_1, a_1, c_1, and ϕ_1 are found one can use $u - u_1$ for finding u_2, etc. The basic idea is the same as in Section III. The difficulties are similar to the ones discussed after formula (3.11).

APPENDIX C: EXISTENCE OF INFINITELY MANY PURELY IMAGINARY RESONANCES

If $-ib$, $b > 0$, is a resonance, i.e. a complex pole of Green's function of the Dirichlet Laplacian in Ω, then the equation

$$q(b)f = \int_\Gamma \frac{\exp(b|s-t|)}{|s-t|} f(t)\,\mathrm{d}t = 0 \tag{1}$$

has a nontrivial solution and vice versa. In order to prove that there are infinitely many purely imaginary resonances, it is sufficient to prove that Equation (1) has nontrivial solutions for infinitely many values of b, $b = b_n$, $b_n \to +\infty$. The operator $q(b)$ in (1) is compact in $H = L^2(\Gamma)$, selfadjoint for real-valued b, and analytic in b. Thus, its eigenvalues $\lambda_n(b)$ are analytic in b about the real axis. If $b \leqslant 0$ then $q(b)$ is positive definite in H, and $\lambda_n(b) > 0$. If we show that there are $N(b) \to +\infty$, as $b \to +\infty$, negative eigenvalues of $q(b)$, then there are $N(b)$ points b_n at which $\lambda_n(b_n) = 0$, i.e. $q(b_n)f = 0$, $f \not\equiv 0$.

The number of negative eigenvalues of $q(b)$ is

$$N(b) = \sup_M \dim M,\ M = \{f: (q(b)f, f) < 0\} \tag{2}$$

(see Lemma 1 in Appendix 2). Let M be a linear span of the functions $h_j(x) \in H$, $h_j(x) = 1$ if $x \in \Delta_j$, $h_j(x) = -1$ if $x \in \Delta_{-j}$, $h_j(x) = 0$ otherwise. Here Δ_j is a small neighborhood of the points $s_{\pm j} \in \Gamma$, $j = 1, 2, \ldots N$, $\Delta_j \cap \Delta_i = \emptyset$ if $j \neq i$. The points $s_{\pm j}$ are chosen so that

$$\max_{i \neq -j} |s_i - s_j| < \min_{1 \leqslant j \leqslant N} |s_j - s_{-j}|. \tag{3}$$

Let a be the area of Δ_j and $a^{1/2}$ be the linear dimension of Δ_j. One has (the bar denotes complex conjugate)

$$\left(q(b)\left(\sum_{j=1}^N c_j h_j\right), \sum_{m=1}^N c_m h_m\right) = \sum_{j,m=1}^N q_{jm}(b)\bar{c}_m c_j,\ q_{jm} = \bar{q}_{mj} \tag{4}$$

where

$$q_{jm}(b) = \int_\Gamma \int_\Gamma \frac{\exp(b|s-t|)}{|s-t|} h_j(t)h_m(s)\,\mathrm{d}t\,\mathrm{d}s$$

$$= \left(\frac{\exp(b|s_j - s_m|)}{|s_j - s_m|} - \frac{\exp(b|s_j - s_{-m}|)}{|s_j - s_{-m}|} - \right.$$

$$\left. - \frac{\exp(b|s_{-j} - s_m|)}{|s_{-j} - s_m|} + \frac{\exp(b|s_{-j} - s_{-m}|)}{|s_{-j} - s_{-m}|} \right) \times \tag{5}$$

$$\times\, a^2(1 + o(1)),\ a \to 0, j \neq m;$$

$$q_{jj}(b) = -2\,\frac{\exp(b|s_j - s_{-j}|)}{|s_j - s_{-j}|}\, a^2(1 + o(1)) +$$

$$+\, 0(a^{3/2}\exp(ba^{1/2})), \quad a \to 0.$$

It follows from (3)—(5) that

$$q_{jj}(b) < 0, \quad \text{and} \quad |q_{jj}| > \sum_{i \neq j,\, i=1}^{N} |q_{ji}(b)|, \quad \text{as} \quad b \to +\infty. \tag{6}$$

Inequalities (6) imply that the matrix $q_{jm}(b)$, $1 \leqslant j, m \leqslant N$, is negative definite if b is sufficiently large positive. This proves

THEOREM. *There exists infinitely many purely imaginary poles $k_n = -ib_n$, $0 < b_n \to +\infty$, of $G(x, y, k)$, the resolvent kernel of the Dirichlet Laplacian in Ω.*

One can prove similarly that the resolvent kernel of the Schrödinger operator in $L^2(R^3)$ with a real-valued compactly supported potential $V(x) \in L^2(R^3)$, $V(x) \geqslant 0$ has infinitely many poles $k_n = -ib_n$, $0 < b_n \to +\infty$ [25].

APPENDIX D: A BOUND ON THE LOCATION OF THE RESONANCES

Let $V(x) \in L^2(R^3)$, $V(x) = 0$ if $|x| \geqslant R$ where $R > 0$ is an arbitrary large fixed number. Then the resolvent kernel $G(x, y, k)$ of the

Schrödinger operator $-\nabla^2 + V(x)$, i.e. the kernel of $(-\nabla^2 + V(x) - k^2)^{-1}$, is meromorphic in k on the whole complex k-plane. If $V = \overline{V}$ then the poles of $G(x, y, k)$ are symmetric with respect to the imaginary axis, and if, in addition, $V \geqslant -c$, where $c =$ const, then there exist infinitely many poles $k_n = -ib_n$, $b_n > 0$, $b_n \to +\infty$. In the half-plane $\operatorname{Im} k > 0$ there may be only a finite number of poles $k_n = iE_n^{1/2}$, $E_n > 0$ is the energy of the n-th bound state. The point $k = 0$ may or may not be a pole of G.

The aim of this appendix is to prove

THEOREM. *If* $\nabla V \in L^1(R^3)$, $V = 0$ *for* $|x| \geqslant R$ *then* $G(x, y, k)$ *has no poles in the region*

$$\Delta(a, b, d) = \{-a \ln|k| - b < \operatorname{Im} k < 0, |\operatorname{Re} k| > 2\} \cup \cup \{-d < \operatorname{Im} k < 0, |\operatorname{Re} k| \leqslant 2\}$$

where a, b, d are positive constants.

Proof. There are no poles in the region where the equation

$$u = -\int_{|y| \leqslant R} \frac{\exp(ik|x - y|)}{4\pi|x - y|} V(y) u \, dy \equiv T(k)u$$

has only the trivial solution. Thus, there are no poles in the region where

$$\|T^2(k)\| \leqslant \max_{|y| \leqslant R, |x| \leqslant R} \left| \int_{|z| \leqslant R} \frac{\exp\{ik(|x - z| + |z - y|)\}}{16\pi^2 |x - z| \, |z - y|} \cdot V(z) \, dz \right| \int_{|y| \leqslant R} |V(y)| \, dy < 1. \tag{1}$$

Here the operator norm is taken in $C(B_R)$, $B_R = \{x: |x| \leqslant R\}$.

If we prove that

$$|A(x, y, k)| \equiv \left| \int_{|z| \leqslant R} \frac{\exp\{ik(|x - z| + |z - y|)\}}{|x - z| \, |z - y|} V(z) \, dz \right| \to 0 \tag{2}$$

when

$$|k| \to \infty, \; k \in \Delta(a, b, d),$$

then (1) holds in a region $\Delta(a, b, d)$ with a sufficiently small d, b and a.

To prove (2) one uses the coordinates s, t, ψ defined by the formulas

$$z_1 = \ell st + \frac{x_1 + y_1}{2}, \quad z_2 = \ell \sqrt{(s^2 - 1)(1 - t^2)} \cos\psi + \frac{x_2 + y_2}{2},$$

$$z_3 = \ell \sqrt{(s^2 - 1)(1 - t^2)} \sin\psi + \frac{x_3 + y_3}{2}$$

where $\ell = |x - y|/2$ and $|x - z| + |z - y| = 2\ell s$, $|x - z| - |z - y| = 2\ell t$, $J = \ell^3(s^2 - t^2)$ is the Jacobian of the transformation $(z_1, z_2, z_3) \to (s, t, \psi)$. In the new coordinates one has

$$A(x, y, k) = \ell \int_0^{2\pi} d\psi \int_{-1}^{1} dt \int_1^{\infty} \exp(2ik\ell s) q(s, t, \psi)\, ds$$

where $q(s, t, \psi)$ is $V(z)$ in the new coordinates. Define

$$p(s) = \int_0^{2\pi} d\psi \int_{-1}^{1} dt\, q(s, t, \psi).$$

Then

$$A(x, y, k) = \ell \int_1^{\infty} \exp(2ik\ell s) p(s)\, ds. \tag{3}$$

The function $p(s)$ is compactly supported and differentiable since $V(z)$ is. Thus

$$A(x, y, k) = -\frac{\exp(2ik\ell)}{2ik} p(1) - \frac{1}{2ik} \int_1^{\infty} \exp(2ik\ell s) p'(s)\, ds. \tag{4}$$

Let $k = k_1 - ik_2$, $k_2 > 0$. Then

$$|A(x, y, k)| \leqslant c \frac{\exp(2k_2\ell) + \exp(c\ell k_2)}{(k_1^2 + k_2^2)^{1/2}} \tag{5}$$

where $c > 0$ denotes various constants which do not depend on k, x, and y. It follows from (5) that, uniformly in x and y running through any fixed compact set, the estimate holds for $|k_1| \to \infty$:

$$|A(x, y, k)| \to 0 \quad \text{if} \quad k_2 < c_1 \ln|k_1|, |k_1| \to \infty,$$

and

$$c_1 < \max\left(\frac{1}{2\ell}, \frac{1}{c\ell}\right) \tag{6}$$

From (6) and (1) the conclusion of the theorem follows for sufficiently small a, b and d.

This argument is taken from Ramm [6h]. □

NOTES

[1] A. G. Ramm, *Usp. Mat. Nauk.* **19**, 192 (1964).
[2] A. G. Ramm, *J. Math. Anal. Appl.* **86**, 562 (1982).
[3] A. G. Ramm, *Electromagnetics* **1**, 385 (1981).
[4] C. Baum, *Proc. IEEE* **64**, 1598 (1976).
[5] C. Dolph and R. Scott, 'Recent developments in the use of complex singularities in electromagnetic theory and elastic wave propagation', in *Electromagnetic scattering*, edited by P. Uslenghi (Academic, New York, 1978).
[6] M. Ikawa, *J. Math. Kyoto Univ.* **23**, 127, 795 (1983).
[7] P. Lax and R. Phillips, *Commun. Pure Appl. Math.* **22**, 737 (1969).
[8] A. G. Ramm, *IEEE Trans.* **AP-33**, 223 (1985) © 1985 IEEE.
[9] J. Auton and M. Van Blaricum, AFWL Math. Notes **79** (1981).
[10] T. Senior and J. Pond, 'Pole extraction in the frequency domain', Interaction Note 411, AFWL, Dec. 1981.
[11] A. G. Ramm, *J. Math. Phys.* **21**, 2502 (1980); **23**, 1112 (1982).
[12] A. G. Ramm, *Theory and Applications of Some New Classes of Integral Equations* (Springer, New York, 1980).
[13] A. G. Ramm, *J. Math. Phys.* **23**, 1123, 2408 (1982); **24**, 2619 (1983) (the last paper is joint with G. Kristensson and S. Ström).
[14] H. Kumano-go, *pseudodifferential Operators* (MIT, Cambridge, MA, 1981).
[15] M. Serebrennikov and A. Pervozvanskii, *Detection of Hidden Periodicities* (Nauka, Moscow, 1965) (in Russian).
[16] R. Prony, *J. Ec. Polytech.* **1**, 24 (1795).
[17] H. Bruns, *Astron. Nachr.* **188** (1911).
[18] J. Dale, *Mon. Not. R. Astron. Soc.* **74**, 628 (1914).
[19] H. Plumer, *Mon. Not. R. Astron. Soc.* **76**, 60 (1913).
[20] F. Hopfner, *Wien. Sitzungsber.* **119**, 351 (1910).
[21] S. Oppenheim, *Wien, Sitzungsber.* **118** (1909).

[22] F. Kühnen, *Astron. Nachr.* **182** (1909).
[23] A. G. Ramm, *IEEE Trans.* **AP-28**, 897 (1980).
[24] M. Vainberg and V. Trenogin, *Bifurcation of Solutions of Nonlinear Equations* (Noordhoff, Leiden, 1974).
[25] A. G. Ramm, *Phys. Lett.* **101A**, 187 (1984).

Research Problems

Here we formulate several open problems. A different list of problems is in Ramm [9].

(1) (a) Does the two-dimensional scattering data $f(n_0, \nu, k_0)$, $n_0 \in S^2$ is fixed, $k_0 > 0$ is fixed, $\nu \in \tilde{S}^2$, determine uniquely the surface of a bounded obstacle and the boundary condition of one of the types $u = 0$; $u_N = 0$; or $u_N + \eta u = 0$ on Γ, $\operatorname{Im} \eta \geqslant 0$?

(b) Similar question for obstacles with infinite boundaries.

(2) (a) Is it true that the scattering amplitude $f(n, \nu, k)$ for a bounded obstacle D has infinitely many complex (not purely imaginary) poles?

Comment. It is not proved that f has even finite number of complex (not purely imaginary) poles. The poles $k_j(D) = k_j = \pm a_j - ib_j$, $b_j > 0$, $a_j > 0$ are symmetric with respect to the imaginary axis of the k-plane (see (I.4.3)).

It is proved that f has infinitely many purely imaginary poles Lax, Phillips [2], Ramm [61]; see also Petkov [2] and references therein.

(b) It is known that for a convex smooth bounded obstacle D there is a strip $-b < \operatorname{Im} k < 0$ which is free from the poles of f. Let $M(R, r)$ be the set of all convex obstacle $D \subset B_R = \{x\colon |x| \leqslant R\}$ with Gaussian curvatures bounded from below by $r > 0$.

Problem: estimate $\inf_{D \subset M(R, r)} b(D) \equiv a(R, r)$ where $b(D) = |\operatorname{Im} k_1(D)|$, and $k_1(D)$ is a pole of f with the minimal absolute value of its imaginary part, i.e. $|\operatorname{Im} k_1(D)| \leqslant |\operatorname{Im} k_j(D)|$.

Can one drop the restriction on the Gaussian curvature? That is, if $M(R)$ is the set of all convex obstacles $D \subset B_R$, then is $\inf_{D \subset B_R} b(D) \equiv a(R) > 0$? What is the behavior of $a(R, r)$ as $R \to +\infty$?

(c) To what extent the set of the poles of f and the multiplicities of these poles determine the shape of the obstacle D, assuming the Dirichlet boundary condition?

(d) Is it true that for a convex obstacle and the Dirichlet boundary condition all the complex (not purely imaginary) poles of f are simple?

Comment. Some remark about question (c) is in Ramm [61]. Similar questions are open and of interest in the scattering by a compactly supported potential (see Ramm [8e]).

(3) The following is a problem with a history of half a century. See Simon, Ursell [1] for a review and references. The problem is basic in water-wave theory.

Let D be a finite body with a smooth boundary Γ, $D \subset R^3_- = \{x: x_3 < 0\}$, $\Omega = R^3_- \setminus D$, N is the unit normal on Γ pointing into Ω. Consider the problem

$$\nabla^2 u = 0 \text{ in } \Omega, \quad \frac{\partial u}{\partial N} = 0 \text{ on } \Gamma,$$

$$\frac{\partial u}{\partial x_3} = ku \quad \text{at } x_3 = 0, \quad k = \text{const} > 0$$

$$|u| + |\nabla u| \to 0, |x| \to \infty,$$

where k is a given positive constant.

Does this problem have a nontrivial solution?

Comment. There are some geometries for which the answer is no (see Simon, Ursell [1] for references).

(4) Prove inequality (0.16) without using the results on the exponential decay of the solutions to the time-dependent problem.

(5) Derive formulas for the S-matrix for the elastic wave scattering by small bodies of arbitrary shapes. For some known results see Gubernatis [1], Olsson [1], Rose [1] and references therein.

(6) (a) If D is an arbitrary compact perturbation of the half cylinder, then is it true that the Dirichlet Laplacian in D does not have positive eigenvalues on its continuous spectrum?

(b) Let C be an infinite cylinder and $D \subset C$ be a finite obstacle. Does the Dirichlet Laplacian have positive eigenvalues on its continuous spectrum?

(7) For a given obstacle D and incident field $u_0 = \exp(ikn \cdot x)$ find the region of convergence of the expansion $u_0 + \sum c_n \psi_n$ of the solution to the scattering problem in Ω in terms of the geometry of D.

Comment. This problem is closely related to the Rayleigh hypothesis. Some partial results can be found in the literature (Barantsev *et al.* [1], Millar [1], Berg, Fokkema [1], De Santo [1] and references therein). If Γ is bounded then the problem was studied for ellipse only and the arguments depend strongly on the fact that ellipse is an analytic curve. If Γ is unbounded the prroblem was studied in R^2 for the sinusoidal Γ and for a special gating (see the cited papers).

(8) Let $q(x) \in L^2(B_R)$, $q(x) = 0$ if $|x| \geqslant R$, $x \in R^3$. Does the scattering amplitude $f(n, \nu, k_0)$ given for all $n, \nu \in S^2$ and for a fixed $k_0 > 0$, determine the potential $q(x)$ uniquely?

Does $f(n, \nu_0, k)$ given for all $n \in S^2$, fixed ν_0, and all $k > 0$ determine $q(x)$ uniquely?

Comment. In both cases it does in the Born approximation.

Bibliographical Notes

Chapter I. Wave scattering by obstacles is very important in applications and from the mathematical point of view. Therefore the literature of the subject is enormous. Integral equation methods were used by Kupradze [1] in 1934 and in his later works, Vekua [1] in 1943, by Werner [1], Leis [3], Müller [1], Ramm [1], and many others. The properties of the potentials of single and double layers are well known for domains with Liapunov boundaries. Many of these properties were established by A. M. Liapunov (see Günter [1]). For nonsmooth boundaries potential theory in R^2 was developed by J. Radon [1] in 1920, and in R^n by Burago, Maz'ya and Sapozhnikova in 1962—1966 (Burago, Maz'ya [1]), by Kral (Kral [1]), and by Fabes *et al.* [1] (see also Kresin, Maz'ya [1]). Existence and uniqueness of the solutions to exterior boundary value problems were established by Kupradze [1] in 1934, and Vekua [1] in 1943. Kupradze used the classical approach, but modified the integral equations for the exceptional values of k for which the adjoint homogeneous integral equation has nontrivial solutions. He obtains the solution of the exterior boundary value problems at the exceptional values of k as a sum of the single and double layer potentials (see also Weyl [1]). His analysis is complicated. Brakhage, Werner [1], Leis [3] and Panich [1] used representation (I.1.15) for the solution of the exterior Dirichlet problem. One can look for a solution of the exterior Neumann problem in this form but the resulting equation is complicated because the normal derivative of the double layer potential is an unbounded operator. The radiation conditions was introduced in Sommerfeld [1], in 1912. Until 1943 this condition was used in the form $r(v_r - ikv) = o(1)$, $v = 0(r^{-1})$, $r \to \infty$. In 1934 Vekua (see Vekua [1]) proved that the second equation, $v = 0(r^{-1})$, is redundant, and that both equations follow from the radiation condition of the form $\int_{S_r} |v_r - ikv|^2 \, ds \to 0$, $r \to \infty$.

Werner [1] used representation (I.1.20) and proved existence of the solution to the exterior boundary value problems. For computational purposes the two-dimensional boundary integral equations are used mostly. A review of the boundary integral equations is Kleinman, Roach [1]. Numerical methods based on the boundary integral equations were suggested by Greenspan and Werner [1], Burton, Miller [1], Kussmaul [1], Terai [1] and others.

The short proof of the existence of solutions to the exterior boundary value problems in Lemma I.1.3 is given by the author. The other short proof, in I.1.3, is taken from Ramm [1]. The existence proof in I.1.6 was discussed by Jones and Ursell (Ursell [1]). The existence proof in I.1.7 was suggested in Krein [1]. Lemma I.2.1 was proved in Kupradze [1], Freudenthal [1] and Vekua [1]. The second proof is given in Ramm [3]. The result of Lemma I.2.4 is given in Lubič [1]. Our argument is taken from Leis [2]. The result in I.2.6 is taken from Ramm [8d]. The argument in I.2.8 belongs to R. Phillips. The argument in I.2.9 is standard. Lemma I.2.11 is taken from Ramm [2d]. In I.2.11 we followed Ramm [1, 6, d]. The limiting absorption principle was first established by the argument used in the proof of Theorem I.2.3 in Eidus [3]. Similar arguments were used afterwards by many authors (e.g. Saito [1]) in the study of the general second order elliptic operators. This argument has been well known for a long time in the Fredholm–Riesz theory. The operator $(-\nabla^2 - k^2 - i0)^{-1}$, where $-\nabla^2 = -\Delta_D$, is unbounded in $L^2(\Omega)$, but is bounded from $L^2(\Omega, p(x))$ into $L^2(\Omega, p^{-1}(x))$ with a suitable weight $p(x)$, e.g. $p(x) = (1 + |x|)^a$, $a > 1$.

The solution to the problem $(-\nabla^2 - k^2)v = f$ in Ω, v satisfies conditions (0.2) and (0.5), where $f \in L_0^2(\Omega)$, can be obtained as follows: $v = \lim_{n\to\infty} v_n$, where v_n solves the problem $[-\nabla^2 - k^2 + q_n(x)]v_n = 0$ in R^3, v_n satisfies (0.5), $q_n(x) = \begin{cases} 0 \text{ in } \Omega \\ n \text{ in } D, n > 0 \end{cases}$. The limit $\lim_{n\to\infty} v_n = \begin{cases} 0 \text{ in } D \\ v \text{ in } \Omega \end{cases}$, is attained in the norm $L^2(R^3, P^{-1}(x))$. This result was proved first in Ramm [1g] and later in Lax, Phillips [2]. In Ramm [1h] a generalization to the case of interface problems is given. Wilcox [1] used the limiting absorption principle for a proof of the existence of solutions to the scattering problem with the homogeneous Neumann boundary condition. This approach allows one to assume less about the smoothness of the boundary than the integral equation approach. In particular, in Wilcox [1] a certain condition on Γ called the local compactness property of Γ, is used. This condition is sufficient for the proof of the existence result in L^2_{loc} (see Appendix 1, Section 7). Necessary and sufficient codnitions on the boundary for the existence of solutions to elliptic equations of the second (and higher) order for various boundary conditions, including the Neumann condition, in L^p and $W^{p,1}$ classes were given in Maz'ya [1] and Maz'ya, Shaposhnikova [1] (see also Schulze, Wildenhain [1] and Appendix 1). The arguments in Wilcox [1] depend crucially on the assumption that the boundary condition is homogeneous. If it is not, then some additional smoothness is required from Γ so that the embedding theorems, in particular the trace theorems, be valid. The presentation in I.3 follows that in Ramm [1a—d]. The first proof along these lines was given in Povzner [1]. A different proof can be found in Ikebe [1] (see also Shenk [1], Shenk, Thoe [1], Wilcox [1]). The properties of the scattering amplitude given in Theorem I.4.1 are well known (Newton [1], Lax, Phillips [1]). The properties (I.4.5) and (I.4.6) are called "optical theorem" (see Newton [5] for historical remarks). The proof of the meromorphic character of the scattering amplitude in k-variable was first given in Ramm [2a, b, f]. Lemma I.4.2, is taken from Ramm [12]. A weaker lemma was used in the proof of the meromorphicity in Ramm [2a, f]. The meromorphic nature of the scattering amplitude for the Schrödinger operator in R^3 was proved independently in Steinberg [1] by means of a lemma similar to that in Ramm [2f], p. 462. (see also Ribaric, Vidav [1], Tamarkin [1], Gohberg—Krein [1], Theorem I.5.1, Dunford Schwartz [1], Reed, Simon [1, b], p. 107). Lemma I.4.3 is the classical Schmidt lemma (Vainberg—Trenogin [1]). It was used in Ramm [2a, f] for the proof of the meromorphic dependence of the scattering amplitude on k (see also Ramm [6l]). Another method of proving the meromorphicity of the scattering amplitude is given in Lax, Phillips [1]. The argument in I.4.3 is taken from Ramm [8k, a]. The material in I.5 is standard (see e.g. Kuroda [1]). Theorem I.6.1 is proved in Alber [1]. We follow his argument. Inequality (I.6.30), a maximum principle for solutions to Helmholtz's equation, is proved in Ramm [2d]. The results and proofs in I.7 are taken from Ramm [2m]. Boundary value problems for pseudodifferential elliptic equations are studied in Eskin [1].

There is an extensive literature on the high frequency asymptotics of solutions to Helmholtz's equation, on justification of the geometrical optics (Alber [1], [2], Alber, Ramm [1], Babich, Buldyrev [1], Buslaev [1], Fock [1], Filippov [1], Guillemin, Sternberg [1], Hönl et al. [1], Majda [1], Majda, Taylor [1], Morawetz, Ludwig [1], Morawetz, Ralston, Strauss [1], Petkov [1], Ramm [8b, 1], Taylor [1], Ursell [2] et al.). We do not discuss wave scattering by moving bodies (see Cooper, Strauss [1] and references therein).

Chapter II. The inverse diffraction problem is of great practical and theoretical interest and therefore the literature of the subject is extensive. By the diffraction problem we mean the scattering problem for a bounded obstacle. The basic questions to consider are: (1) uniqueness of the solution to the inverse problem and (2) actual recosntruction of the obstacle from the data. In the uniqueness problem the basic result is due to M. Schiffer. His argument is used in the proof of Lemma II.1.1. The scattering data in this lemma is three-dimensional and the object to recover, the surface in R^3, is two-dimensional. In practice the data is often given at a fixed frequency. Theorem II.1.1., which handles this case, is due to the author. In Colton—Sleeman [1] a uniqueness theorem is formulated which says that an acoustically soft connected obstacle, which is situated in a known ball B_R of known radius R, is uniquely determined by the scattering data $f(n_j, \nu, k_0)$ where $k_0 > 0$ is fixed, $\nu \in S^2$ and n_j are distinct unit vectors, $j = 1, 2, \ldots, m$, and $m = m(R)$ is the number of eigenvalues of the Dirichlet Laplacian in D which do not exceed k_0^2. The argument in this paper is not quite satisfactory since the authors assume that the other obstacle with the same scattering data lies also in B_R, which is actually an additional assumption. The initial assumption that D lies in the known ball is also not satisfactory. There is an error in the proof of Theorem 2.2 in Colton—Sleeman [1], and in formulas (3.1) and (3.6). In Ramm [8e] an attempt is made to prove the uniqueness theorem with the two-dimensional data $f(n_0, \nu, k_0)$, $n_0 \in S^2$ is fixed, $\nu \in \tilde{S}^2$, $k_0 > 0$ is fixed. The argument in this paper is given in 2.1.3. The result is not sufficient for the uniqueness theorem, since the mapping $f(n_0, \nu, k)$ constructed in 2.1.3 depends on D. Therefore, it is an open problem if the above data determine the obstacle uniquely or not. Engineers assume implicitly that the above two-dimensional data determine the obstacle uniquely and tried to suggest some numerical procedures for actual reconstruction of the obstacle (see e.g. Imbriale, Mittra [1]). Some recent reviews on the inverse scattering by obstacles are Colton [1] and Sleeman [1]. Practically the high frequency data are used for inversion. The first inversion of the high frequency back-scattering data was based on the Kirchhoff approximation. This procedure allows one to compute the Fourier transform of the characteristic function of D. The three-dimensional Fourier inversion gives the characteristic function of D (see Bojarski [1], Lewis [1]). Another idea, practically more effective, was given in Ramm [8b] (see also Ramm [3] and [8j]). In this paper the support function of the obstacle is calculated from the scattering data $f(n, \nu, k_0)$, where k_0 is fixed and large enough so that the Kirchhoff approximation is applicable, and $n, \nu \in S^2$ run through such subsets of S^2 that the vector $\ell = (n - \nu)|n - \nu|^{-1}$ runs through all of S^2. This solution allows one to estimate the error of the reconstruction procedure even in the case of noisy data. The results and the arguments in II.2.3 are taken from Ramm [8b, j, k]. Lemma II.3.3 is from Firey [1]. In all of the above papers it is assumed that the obstacle is convex. In Majda [1] and Majda, Taylor [1] some of the formulas from Ramm [8b] were rediscovered and proved. In Taylor [1] the Kirchhoff approximation is justified as an asymptotic high frequency approximation for the solutions of diffraction problems (see also Melrose [2]). It was known long ago (Fock [1]) that the scattering data $f(n, -n, k_0)$, k_0 large fixed, determines the Gaussian curvature of the convex obstacle. The classical Minkowski result says that the smooth convex compact surface is uniquely determined by its Gaussian curvature given as a function of a unit normal. For existence results see Nirenberg [1], where sufficient conditions for a given function of a unit normal to be the Gaussian curvature of a convex smooth compact surface are formulated. The solution to Minkowski's problem is not constructive: the problem of finding a convex compact smooth surface from its Gaussian curvature is reduced to a nonlinear elliptic boundary value problem. Lemma II.2.2. can be found, e.g., in Blaschke [1]. The stable differentiation procedure is II.2.3 is taken from Ramm [11a, d] (see Appendix 4). There are very few results on the high frequency asymptotics of the solutions to the exterior problems for non-convex obstacles (see Alber [2], Petkov [1]). The inverse scattering problem for a class of non-convex obstacles was solved in Alber, Ramm [1]. We do not discuss the inverse scattering problem for potential scattering (see Chadan—Sabatier [1], Newton [1]—[3], Marchenko [1], Ljance [1] and references therein.

Chapter III. Behavior of the solutions to the time-dependent problems for large t was discussed in many papers, often in connection with the limiting amplitude principle and exponential decay of solutions to the wave equation in exterior domains (Vainberg [1], [2], Eidus [1], Ramm [2a—l], Melrose [1], Morawetz [1], [2], Morawetz, Ludwig [1], Morawetz, Ralston, Strauss [1], Lax, Phillips [1] *et al*). The results and arguments in 2.1—2.3 are taken from Ramm [2], in particular [2d, k]. The presentation in 2.4 is influenced by the work Eidus [1]. There are some minor inaccuracies in this paper, which resulted in different constants in the asymptotics of $I(t, a)$ (see formula (1.31) in Eidus [1] and (III.2.82)). The argument and results in III.3 are essentially from

Ramm [1a, d], [2d]. In Asakura [1] there is an example of the solution to the exterior Robin problem for the wave equation which does not decay as $t \to \infty$. The arguments in III.4.1. are closely connected with these in III.2.4. Similar arguments can be found in Sanchez-Palencia [1]. The singularity expansion method (SEM) associates the analytic properties of the resolvent kernel of a Laplacian on the complex k-plane with the behavior of the solutions to the corresponding time-dependent problem as $t \to \infty$. The resolvent kernel of the Dirichlet Laplacian in R^3 is meromorphic on the k-plane, and in R^2 it is meromorphic on the k-plane with the cut along the negative imaginary semiaxis (or on the Riemannian surface of $\ln k$). This leads to different behavior of the solutions to the exterior time-dependent problems in R^2 and R^3. Indeed, in R^3 for star-shaped (or, more generally, non-trapping) obstacles the solution to problem (0.11) decays exponentially as $t \to \infty$. In R^2 the solution decays not faster that $0(t^{-1})$, $t \to \infty$, in general (Ramm [2a]). In Ramm [2l] the diffusion of waves in R^2 was explained by the presence of the logarithmic branch point in the analytic continuation of the resolvent kernel in k-plane as well as the order $0(t^{-1})$ of the decay.

It is quite difficult to prove exponential decay of the solution to (0.11). In Morawetz [1], Lax, Phillips, Morawetz [1], Morawetz, Ralston, Strauss [1], Melrose [1] the exponential decay was established for the Dirichlet boundary condition for $L = -\nabla^2$. In Melrose [1] it is stated that the same holds for the Neumann boundary condition. It is easier to prove such results when $L = -\nabla^2 + q(x)$, $q(x) \in C_0(R^3)$ (or $L_0^2(R^3)$) and the wave equation is considered in all of R^3. In Ramm [6h] it was proved that the poles of the analytic continuation of the resolvent kernel of the Schrödinger operator L lie under a logarithmic curve on the k-plane, $k_2 \leqslant -a \ln|k| - b$ where a and b are some positive constants, $k = k_1 + ik_2$, and there are no poles in some strip $-b_0 < k_2 < 0$, $b_0 > 0$. This bound on the location of the poles is sharp: for some spherically symmetric potentials the bound is attained. In Ramm [2g] the exponential decay was proved for the Dirichlet Schrödinger operator in the exterior of a star-shaped obstacle. The argument in this paper depends on the result in Lax, Phillips, Morawetz [1].

The eigenmode expansion method (EEM) consists in solving equation (III.5.4) using the root function expansion. For the self adjoint operator B this method goes back to Picard (1891) (see e.g. Goursat [1]). If B is nonselfadjoint then the problem is very difficult in general:

(1) B might be quasinilpotent (e.g. Volterra operator) and then it has no eigenvalues, (2) the closure of the linear span of the eigenvectors (or root vectors) of B may not coincide with the whole space H, (3) if it does, the root system of B, i.e. the set of all linearly independent root vectors of B, may not form a basis of H. There is a considerable interest in applications to EEM (see Voitovic, Kacenelenbaum, Sivov [1], Baum [1]). At this time there is an extensive bibliography of the engineering work on EEM and SEM, hundreds of references can be found in the special issue of the journal "Electromagnetics" 1, N4, (1981). Engineers (see the above references) believed that the complex poles of the analytic continuation of the resolvent kernel of $L = -\nabla^2$ are all simple, and that EEM is applicable to Equation (III.5.2) in the form (III.5.6). Mathematical analysis of the problem was initiated in Ramm [6a, b, c]. This work was stimulated by the questions asked by the author's colleague B. Kacenelenbaum, professor of electrical engineering in Moscow, who initiated the modern applications of the EEM method in electrical engineering in 1968—69. In the USA C. Baum initiated in (1971—72) applications of the SEM and EEM primarily for the target identification problems. Dolph [1] was one of the early papers which promoted interest in applications of the theory of nonselfadjoint operators to electrical engineering and physics (see also Dolph, Scott [1] and Dolph, Cho [1]). Completeness of the root system of the operator (III.5.3) was proved first in Ramm [6c]. Theorems III.5.1 and III.5.2 and their prof are taken from this paper. Theorem III.5.3 and its proof are based on the works Marcus [1], Kacnelson [1], Agranovich [1] (see Appendix in Voitovich *et al.* [1]).

Chapter IV. The T-matrix approach is one of the most popular numerical methods for solving the scattering problems. Varadan, Varadan [1] is a good source of references. It is a collection of papers presented at the international symposium on the T-matrix approach held in 1979. In the USA the T-matrix approach is often associated with the papers Waterman [1], [2]. However Barantsev [1] used this method earlier (see also Barantsev *et al.* [1]). Although the T-matrix approach was used at least from 1962 there was no mathematical analysis of its convergence, validity and dependence of the rate of convergence on the geometry of the obstacle. The method was used practically in the acoustic, electromagnetic, and elastic wave scattering problems without theoretical justification. Many speakers at the 1979 symposium called for such a justification of the T-matrix approach and its analysis.

Theoretical analysis of the *T*-matrix approach was given first in Ramm [7]. It was refined and illustrated by numerical results in Kristensson, Ramm, Ström [1]. The results and arguments in Chapter IV are taken from these papers. Some abstract theorems on convergence of projection methods for solving linear equations in Banach spaces are taken from Gohberg—Feldman [1], Krasnoselski *et al.* [1], Kantorovich, Akilov [1], and Mikhlin [1] in a modified form. The first part of the example in Remark IV.2.4 is taken from Ize *et al.* [1], the second is new. The notion of the Riesz basis and some related facts from linear functional analysis one can find in Gohberg, Krein [1] and Young [1]. We did not discuss the finite difference and finite elements methods for solving the exterior boundary value problems (see Goldstein [1] and references). We mention also paper Aziz [1], Alber [4], Limič [1], and papers in parts 3 and 4 of Varadan—Varadan [1].

One of the interesting problems in the *T*-matrix approach is that of the validity of the Rayleigh hypothesis. A review of this problem is Millar [1] where many references can be found (see also Barantsev *et al.* [1] and Berg, Fokkema [1]).

Chapter V. Wave scattering by small bodies is of great interest in many applications. There is enormous literature of the subject. The applications include: light scattering in muddy water, in cosmic dust, in colloidal solutions, radiowave scattering by rain and hail, radiation from small holes, particle distribution in smoke fog, etc. The pioneering work was done by Rayleigh in 1871 and he had been working on this problem until his death (Rayleigh [1]). Rayleigh understood that the field scattered by a small body of an arbitrary shape is basically a dipole radiation. J. J. Thomson (in 1898) realized that if the smal body is a perfect conductor than the magnetic dipole radiation is of the order of magnitude of the electric dipole radiation. If $ka \sim 0.1$, where k is the wave number and a is the diameter of the body, then the multiple radiations are negligible compared with the dipole radiations. Therefore, physics of the phenomenon has been well understood for a long time (nearly a century). The problem is to compute the dipole moments in terms of the geometry of the body. The body is assumed homogeneous. The other problem is to study wave scattering by many small bodies. This problem is subdivided into two: scattering by r bodies with 1) $r \sim 10$, and 2) $r \sim 10^{23}$. The second case is the case of wave scattering in a medium consisting of many small bodies. The mathematical problem of finding the expansion of the scattering amplitude about $k = 0$ is of interest. The results and the arguments in V.1—3 are taken from Ramm [10] and Ramm [4]. Only a small part of the author's results on wave scattering by small bodies of arbitrary shapes is presented here. Many related problems and results from Ramm [4] and Ramm [10] are not discussed (two-sided estimates of capacitances and polarizability tensors; stable solution of integral equations at an eigenvalue: such equations arise in electro- and magnetostatics, elastostatics, hydrodynamics of incompressible ideal fluid; numerical examples; estimates of the capacitancy of a perfect conductor placed in an inhomogeneous medium, the inverse radiation problem for small bodies, radiation from small holes, skin-effect for thin wires, etc.) V.4 is the paper Ramm [10v]. In MacCamy [1] the low frequency behavior of the solutions to the exterior Dirichlet and Neumann problems was studied for the obstacle whose boundary consists of one connected component. In Werner [4] the case of multiconnected boundaries is treated by a different method. Werner's method is based on some suitable Fredholm's integral equations (see also Muravei [1] and Favorin [1]; the last paper is the only one in which the Robin boundary condition was considered). Kress (see Colton—Kress [1], p. 23) proposed in 1979 a method to study the dependence of the solution to the equation $[I - A(k)]u(k) = f(k)$ on k under the assumptions: $\|A(k) - A(0)\| \to 0$, $\|f(k) - f(0)\| \to 0$, $k \to 0$, $A(k)$ is a compact operator function, $N(I - A(k)) = \{0\}$ if $k = 0$, dim $N(I - A(0)) \geqslant 1$. Kress obtained from his abstract some results for the Helmholtr equation R^3 which were obtained earlier by different methods. In Werner [5], the low frequency behavior of the solutions to Maxwell's equations is studied. The method in V.4 is new and the results obtained include previously known results on this topic.

In the monograph Marchenko, Khruslov [1] and in Khruslov [1] some necessary and sufficient conditions are given for the solutions of the Dirichlet and Neumann boundary vlaue problems in the exterior of r obstacles to have a limit when $r \to \infty$ and the obstacles become smaller in a suitable way. There are many possible applications of the results of this type. A related area of research is the homogeneization theory for partial differential equations (see Sanchez—Palencia [1] and references therein). In Albeverio *et al.* [1], [2], the scattering theory is discussed for the Schrödinger operators with point interactions. The inverse radiation problem mentioned on p. 9 is solved in Ramm [4] p. 110.

We mention papers Jones [2], Keller *et al.* [1], Kleinman [1], Senior [1] which deal with low frequency scattering.

Chapter VI. In this chapter we are interested in finding coefficients of some elliptic equations from the knowledge of the solutions to these equations on certain manifolds. Such type of problems are of prime interest in explorational geophysics. Various aspects of these problems are discussed in Anikonov [1] and Lavrientiev *et al.* [1]. These inverse problems are often ill-posed. Various aspects of the theory of ill-posed problems are discussed in Fedotov [1], Ivanov *et al.* [1], Tanana [1], Tikhonov, Arsenin [1], and in a number of other books and papers (there are more than 300 references in the cited books). Newton [6] is a review of some inverse problems of geophysics which can be reduced to the inverse scattering problem for the one-dimensional Schrödinger equation. Weglein [1] is a review of the inverse scattering theory methods in geophysical exploration. Chapter VI is the paper Ramm [13].

The presentation in VI.1 is based partly on the works Ramm [81] and Martin, Ramm [1]. The properties of the Clebsch–Gordan coefficients are well known (see e.g., Landau, Lifschitz [1] or Lubarskii [1]). Expansions of the type (VI.1.108) were used in Norton, Linzer, [1] for inversion of the data in imaging problems. Section 1.11 is based on the papers Ramm [13b, c]. The inversion procedure in section 1.11 differs from that in Norton, Linzer [1]. Some unclear point in Norton, Linzer [1] is mentioned in Ramm [13c]. VI.2 is the paper by Ramm, Weglein [1]. VI.3 is the work Ramm [8h]. Formulas and error estimates for approximation by entire functions whose Fourier transform has support in a given ball are taken from Ramm [3], where references to his earlier papers on this topic and applications to apodization theory (synthesis theory in optics) are given. VI.4 is the paper Ramm [8m]. In Coen, Yu [1], Oristaglio [1] and Minerbo [1] one can find many references on the induction logging problem.

Chapter VII. Physical intuition dictates the statement of the scattering problem by a compact obstacle. If the boundary of the obstacle is infinite then it is not clear what to take as the "plane wave" u_0 in D. The spectrum of the Dirichlet Laplacian L is not necessarily continuous as in the case of the Laplacian in the exterior of a compact obstacle. Therefore, first one should study the spectral properties of L in $L^2(D)$.

The material in VII.2 is well known (see e.g., Glazman [1]) except Lemma 1′ which is new. Lemmas VII.2.2 and VII.2.4 are taken from Rellich [2] and Rellich [1] respectively. Lemmas VII.2.5 was used in Ramm [1d]. In VII.3 the results from Rellich [2] are given (see also Jones [1]). Our presentation differs from the presentation in Rellich [2].

The arguments and results in VII.4 are taken from Ramm [1a, d]. The arguments follow closely the arguments in Kato [1]. The basic new idea in Ramm [1a, d] was to study the operator Sturm – Liouville equation (VII.4.3) not in the fixed Hilbert space $L^2(S^2)$ as in Kato [1], but in a family of the Hilbert spaces depending on the parameter r. This idea was generalized in Mishnaevskii [1], [2], who developed a very powerful theory and obtained the strongest results on the spectral analysis of the general second-order elliptic differential operators, not necessarily selfadjoint, in some domains with infinite boundaries. His results contain the results in Eidus [1], Jäger [1], Tayoshi [1], Ramm [1d], Eastham, Kalf [1], Vinnik, Eidus [1], Ikebe, Uchiyama [1], Constantin [1] and other papers. The arguments in Mishnaeskii [1], [2] are quite complicated.

The results in VII.5 are new. The results in VII.6 are taken from Ramm [1d]. We do not discuss wave scattering by periodic structures and random surfaces (see Alber [3], Wilcox [2], Bass, Fuks [1], DeSanto [1], [2], and references therein). Existence of the wave operators for the scattering problems in some domains with infinite boundaries was discussed in Ramm [1k], Saenz [1], Tayoshi [2], and Combes, Weder [1].

Appendices. The material in Appendix 1 is taken from Kantorovich, Akilov [1], Günter [1], Fabes *et al.* [1]. More material on embedding theorems can be found in Triebel [1]. Lemma 8 is proved, e.g., in Ramm [2k]. The argument on the top of p. 326 was used in Wilcox [1]. The results in Appendix 2 are standard. Lemmas 1—5 are proved in Glazman [1]. The results of section 2 in Appendix 3 are from Ramm [5c, d, e] and [3].

Appendix 4 is the paper Ramm [11d]. Appendix 5 is the paper Ramm [6n]. Appendix 6 is part IV in Ramm [8g]. The results of this appendix are from Miller [1]. Appendix 7 is the paper Ramm [6m] and can be read as a separate paper.

Bibliography

Agranovich, M. [1] 'Summability of series in root vectors of nonselfadjoint elliptic operators', *Funct. Anal. and Appl.*, **10**, 65—74, (1976).

Akhiezer, N. [1] *Theory of approximation*, Ungar, New York, 1956. [2] *The classical moment problem and some related questions in analysis*, Oliver & Boyd, Edinburgh, 1965.

Alber, H. [1] 'Konvergenz der geometrischen Optik Approximation and Konstruktion des ersten gliedes einer Parametrix fur die Wellengleichung bei Tangentialstrahlen höherer Ordnung', Thesis, Univ. of Bonn, (1982), *J. f. die reine und angew. Math.*, **345**, 39—62, (1982). [2] 'Justification of geometrical optics for non-convex obstacles', *J. Math. Anal. Appl.*, **80**, 372—386, (1981). [3] 'A quasi-periodic boundary value problem for the Laplacian and the continuation of its resolvent', *Proc. Roy. Soc. Edinburgh*, **82a**, 251—272, (1979). [4] 'Numerical computation of waves at high frequency by iterated WKB method', *Math. Meth. in the Appl. Sci.*, **6**, 520—526, (1984).

Alber, H. and Ramm, A. G. [1] 'Scattering amplitude and algorithm for solving the inverse scattering problem for a class of non-convex obstacles', Univ. of Bonn, Inst. Angew. Math, preprint (1984); *Phys. Lett.*, 108A, 238—240, (1985).

Albeverio, S. *et al.* [1] 'Low-energy parameters in nonrelativistic scattering theory', *Ann. of Phys.*, **148**, 308—326, (1983). [2] 'Schrödinger operators with point interactions and short range expansions', *Physica* **124A**, 11—28, (1984).

Anikonov, Ju. [1] 'Methods for studying multidimensional inverse problems for differential equations', Nauka, Novosibirsk, (1978).

Asakura, F. [1] 'On the Green's function for $\Delta + \lambda$ with the boundary condition of the third kind in the exterior of a bounded obstacle', *J. Math. Kyoto Univ.*, **18—3**, 615—625, (1978).

Aziz, A., Dorr, M., and Kellog, B. [1] 'A new approximation method for the Helmholtz equation in an exterior domain', *SIAM J. numer. anal.*, **19**, 899—908, (1982).

Babich, V. and Buldyrev, V. [1] 'Asymptotic methods in the short-wave diffraction problems', Nauka, Moscow, (1972). (in Russian)

Barantsev, R. [1] 'The method of separation of variables in the problem of scattering by a body of arbitrary shape', Sov. Phys. Doklady, 7, N11, 974—975, (1963); (Russ. original Doklady, 147, 569—570, (1962).

Barantsev, R. *et al.* [1] 'Concerning the Rayleigh hypothesis in the problem of scattering from finite bodies of arbitrary shapes', Vestnik Leningradskogo Univ., *Mathem., Mech. Astron*, **7**, 56—62, (1971); *Ibid.* **7**, 71—77, (1968).

Bass, F. and Fuks, I. [1] 'Wave scattering from statistically rough surfaces', Pergamon Press, New York, 1979.

Baum, C. [1] 'The singularity expansion method, in Transient electromagnetic fields', Chapter 3, Felsen L. (ed.), Springer-Verlag, New York, 1976.

Bellman, R., Kalaba, R., and Lockett, J. [1] *Numerical inversion of the Laplacian transform*, Elsevier, New York, (1966).

Berezanskii, Ju. [1] *Expansions in eigenfunctions of selfadjoint operators*, AMS, Providence, (1968).

Berg, P. van der and Fokkema, J. [1] 'The Rayleigh hypothesis in the theory of diffraction by a cylindrical obstacle', Trans. *IEEE* **AP-27**, 577—583, (1979).

Beylkin, G. [1] 'The fundamental identity for iterated spherical means and the inversion formula for diffraction tomography and inverse scattering', *Math. Phys.* **2M**, 1399—1400, (1983).

Blaschke, W. [1] *Vorlesungen über differential Geometry*, Springer-Verlag, Berlin, (1924).

Bojarski, N. [1] 'A survey of the physical optics inverse scattering identity', *IEEE* **AP-30**, 980—988, (1982).

Borovikov, V. and Kinber, B. [1] *Geometric theory of diffraction*, Svjaz, Moscow, (1978) (in Russian).

Brakhage, h. and Werner, P. [1] 'Über das Dirichletsche Aussenraum problem für die Helmholtzsche Schwingungsgleuchung', *Arch. Math.*, **16**, 325—329, (1965).

Burago, Maz'ya, V. [1] *Potential theory and function theory for irregular regions*, Consult. Bureau, New York, (1969).

Burton, A. and Miller, G. [1] 'The application of integral equation methods to the numerical solution of some exterior boundary-value problems', *Proc. Roy. Soc. London* **A323**, 201—220, (1971).

Buslaev, V. [1] 'On the asymptotic behavior of the spectral characteristics of exterior problems for the Schrödinger operator', *Math. USSR*, **9**, 139—223, (1975).

Chadan, K. and Sabatier. [1] *Inverse problems in quantum scattering*, Springer-Verlag, New York, 1977.

Choquet-Bruhat, Y., DeWitt-Morette, C., and Dillard-Bleik, U. [1] *Analysis, manifolds and physics*, North Holland, Amsterdam, (1982).

Coen, S. and Yu, M. [1] 'The inverse problem of the direct current conductivity profile of a layered earth', *Geophysics*, **46**, 1702—1713, (1981).

Colton, D. and Kress, R. [1] *Integral equation methods in scattering theory*, John Wiley, New York, 1983.

Colton, D. [1] 'The inverse scattering problem for time-harmonic acoustic waves', *SIAM Review*, **26**, 323—350, (1984).

Colton, D. and Sleeman, B. [1] 'Uniqueness theorems for the inverse problem of acoustic scattering', *IMA J. of Appl. Math.*, **31**, 253—259, (1983).

Combes, J. and Weder, R. [1] 'Criterion for existence and completeness of wave operators and applications to scattering by unbounded obstacles', *Comm. PDE*, **6**, 1129—1263, (1981).

Constantin, P. [1] 'Scattering for Schrödinger operators in a class of domains with noncompact boundaries', *J. Func. Anal.*, **44**, 87—119, (1981).

Cooper, J. and Strauss, W. [1] 'Representation of the scattering operator for moving obstacles', *Ind. Univ., Math. Jour.*, **28**, 643—671, (1979).

Courant, R. and Hilbert, D. [1] *Methods of Mathematical Physics*, vol. 2, Interscience, New York, 1962.

Devaney, A. [1] 'Inversion formula for inverse scattering within the Born approximation', *Opt. Lett.* **7**, 111—112, (1982).

De Santo, J. [1] 'Scattering from a periodic corrugated structure: thin comb', *J. Math. Phys.* **12**, 1913—1923, (1971); **13**, 336—341, (1972). [2] 'Scattering from a perfectly reflecting periodic surface: an exact theory', *Radio Sci.*, **16**, 1315—1326, (1981).

Dolph, C. [1] 'Recent developments in some nonselfadjoint problems of mathematical physics', *Bull. Am. Math. Soc.*, **67**, 1–69, (1961).

Dolph, C. and Scott, R. [1] 'Recent developments in the use of complex singularities in electromagnetic theory and elastic wave propagation', in *Electromagnetic scattering*, Uslenghi, P. (ed.), Acad. Press, New York, pp. 503–570, (1978).

Dolph, C. and Cho, K. [1] 'On the relationship between the singularity expansion method and the mathematical theory of scattering', *IEEE* **AP-28**, 888–896, (1980).

Dolph, C., McLeod, J., and Thoe, D. [1] 'The analytic continuation of the resolvent kernel and scattering operator associated with the Schrödinger operator', *J. Math. Anal. Appl.*, **16**, 311–332, (1966).

Dunford, N. and Schwartz, J. [1] *Linear operators, I*, Interscience, New York, (1958).

Eastham, M. and Kalf, H. [1] *Schrödinger-type operators with continuous spectra*, Pitman, Boston, (1982).

Eidus, D. [1] 'The limiting amplitude principle', *Russ. Math. Survey*, **24**, 51–64, (1969). [2] 'Completeness properties of scattering problem solutions', *Comm. in PDE*, **7**, 55–75, (1982). [3] 'The principle of limiting absorption', *Amer. Math. Sco. Transl.* **47**, 157–191, (19965).

Erdelyi, A. *et al.* [1] *Tables of integral transforms*, McGraw-Hill, New York, (1954).

Eskin, G. [1] *Boundary value problems for elliptic pseudodifferential equations*, Am. Math. Soc. Providence, RI, (1981).

Ewing, W., Jardetzky, W., and Press, F. [1] *Elastic waves in layered media*, McGraw-Hill, New York, (1957).

Fabes, E., Jodeit, M., and Riviere, N. [1] 'Potential techniques for boundary value problems on C^1 domains', *Acta. Math.* **141**, 165–186, (1978).

Favorin, V. [1] 'Behavior as $k \to 0$ of the Green's function of the exterior Robin problem for the two-dimensional Helmholtz's equation', *Dif. Uravneniya*, **15**, 709–716, (1979). e.t. *Dif. equations* **15**, 497–502, (1979).

Fedotov, A. [1] *Linear ill-posed problems with random errors in data*, Nauka, Novosibirsk, 1982.

Filippov, A. [1] 'Justification of the high frequency asymptotics in the three dimensional diffraction problem', *Siber. Math.J.* **10**, 1406–1421, (1969).

Firey, W. [1] 'The determination of convex bodies from their mean radius curvature function', *Mathematika*, **14**, 1–13, (1967).

Fock, V. [1] *Electromagnetic diffracation and propagation problems*, Pergamon Press, New York, (1965)

Freudenthal, H. [1] 'Über ein Beugung problem aus der electromagnetischen Lichttheorie', *Comp. Math.*, **6**, 221–227, (1938).

Glazman, I. [1] *Direct methods of qualitative spectral analysis of singular differential equations*, Israel Sci. Transl., Davey, New York, (1965).

Gohberg, I. and Feldman, I. [1] *Convolution equations and projection methods for their solutions*, Transl. of Math. Monographs, Vol. 41, Amer. Math. Soc., Providence, RI, (1974).

Gohberg, I. and Krein, M., [1] *Introduction to the theory of linear nonselfadjoint operators*, Transl. of Math. Monographs Vol. 18, Amer, Math. Soc., Providence, RI, (1969).

Goldstein, C. [1] 'The finite element method with nonuniform mesh sizes for unbounded domains', *Math. of Computation*, **36**, 387–404, (1981); *Numer. Math.*,

38, 61—82, (1981). [2] 'Eigenfunction expansions associated with the Laplacian for certain domains with infinite boundaries', *Trans. Amer. Math. Soc.,* **143**, 283—301, (1969).

Goursat, E. [1] *Course d'analyse mathematique*, Vol. 3, Dunod, Paris, (1927).

Gradshteyn, I. and Ryzhik, I. [1] *Tables of integrals series and products*, Acad. Press, New York, (1980).

Greenspan, D. and Werner, P. [1] 'A numerical method for the exterior Dirichlet problem for the reduced wave equation', *Arch. Rat. Mech. Anal.,* **23**, 288—316, (1966).

Gubernatis, J. [1] 'Long wave approximations for the scattering of elastic waves from flaws with applications to ellipsoidal voids and inclusions', *J. Appl. Phys.,* **50**, 4046—4058, (1979).

Guillemin, V. and Sternberg, S. [1] *Geometric asymptotics*, Amer. Math. Soc., Providence, (1977).

Günter, N. [1] *Potential theory and its applications to basic problems of mathematical physics*, Ungar, New York, (1967).

Hardy, G. [1] *Divergent series*, Oxford Univ. Press, New York, (1956).

Hönl, G., Maue, A., and Westpfahl, K. [1] *Theorie der Beugung*, Springer-Verlag, Berlin (1961), Handbuch der Physik 35/1, pp. 281—573.

Hsiung, C. [1] *A first course in differential geometry*, Wiley, New York, (1981).

Ikawa, M. [1] 'On the poles of the scattering matrix for two strictly convex obstacles', *J. Math. Kyoto Univ.,* **23**, 127—194, (1983). [2] 'On the distribution of the poles of the scattering matrix for two strictly convex obstacles', *Hokkaido Math. Jour.,* **12**, 343—359, (1983).

Ikebe, T. [1] 'Eigenfunction expansions associated with the Schrödinger operator and their applications to scattering theory', *Arch. Rat. Mech. Anal.,* **5**, 1—35, (1960).

Ikebe, T. and Uchiyama, J. [1] 'on the asymptotic behavior of eigenfunctions of second order elliptic operators', *J. Math. Kyoto Univ.,* **11**, 425—448, (1971).

Imbriale, W. and Mittra, R. [1] 'The two-dimensional inverse scattering problem', *IEEE* **Ap-18**, 633—642, (1970).

Ivanov, V. K. *et al.* [1] *Theory of linear ill-posed problems and applications*, Nauka, Moscow, (1978).

Ize, J., England, R., and Sabina, F. [1] *Theoretical and numerical study of diffraction of waves by inhomogeneous obstacles*, Univ., Nac. Aut, Mexico, Tech. Rep. 291, (1981).

Jäger, W. [1] 'Ein gewöhnlicher differential Operator zweiter Ordnung für Funktionen mit Werten in Hilbertraum', *Math. Z.,* **113**, 68—78, (1970).

John, F. [1] *Plane waves and spherical menas applied to partial differential equations*, Interscience, New York, (1955). [2] *Partial differential equations*, Springer-Verlag, New York, (1982).

Jones, D. [1] 'The eigenvalues of the Laplacian when the boundary conditions are given on semi-infinite domains', *Proc. Cambr. Phil. Soc.* **40**, 668—683, (1953). [2] 'Low frequency electromagnetic radiation', *J. Inst. Math. Appl.,* **23**, 421—447, (1979).

Kacnelson, V. [1] 'Conditions under which systems of eigenvectors of some classes of operators form a basis', *Funct. Anal. and Appl.,* **1**, 122—132, (1967).

Kantorovich, L. and Akilov, G. [1] *Functional analysis*, Pergamon Press, New York, (1980).

Kato, T. [1] 'Growth properties of solutions of the reduced wave equation with a variable coefficient', *Comm. Pure Appl. Math.*, **12**, 403—425, (1959).

Keller, J., Kleinman, R., and Senior, T. [1] 'Dipole moments in Rayleigh scattering', *J. Inst. Math. Appl.*, **9**, 14—22, (1972).

Khruslov, E. [1] 'Asymptotic behavior of solutions of the Neumann problem', *Math. Sborn.*, **106**, 604—621, (1978).

Kleinman, R. [1] 'Low frequency solution of electromagnetic scattering problems', in *Electromagnetic wave theory, II*, Brown, J. (ed.), pp. 891—905, Pergaman Press, New York, (1967).

Kleinman, R. and Roach G. [1] 'Boundary integral equations for the three-dimensional Helmholtz equation', *SIAM Review*, **16**, 214—236, (1974).

Kral, J. [1] *Integral operators in potential theory*, Springer-Verlag, New York, (1980).

Krasnoselskii, M. *et al.* [1] *Approximate solutions of operator equations*, Wolters-Noordhoff, Groningen, (1972).

Krein, S. [1] *Linear equations in Banach spaces*, Birkhauser, Boston, 1982.

Kresin, G., Maz'ya, V. [1] 'On the essential norm of an operator of double layer potential type in the space C_m', *Sov. Math. Dokl.*, **20**, 459—462, (1979).

Kristensson, G., Ramm, A. G., and Ström, S. [1] 'Convergence of the T-matrix approach in scattering theory, II', *J. Math. Phys.*, **24**, 2619—2631, (1983).

Krylov, V. and Skoblya, N. [1] *Handbook of numerical inversion of Laplace transform*, Israel Program for Sci. Transl., Jerusalem, (1969).

Kupradze, V. [1] *Randwertaufgaben der Jehwingungs Theorie und Integralgleichungen*, VEB, Berlin, (1956). [2] *Potential methods in the theory of elasticity*, Israel Sci. Transl., Jerusalem, 1965. [3] 'On Sommerfeld's radiation principle', *Doklady Acad. Nauk. SSSR*, **2**, 1—7, (1934).

Kuroda, S. [1] *An introduction to scattering theory*, Lecture notes N51, Aarhus Univ., (1978).

Kussmaul, R. [1] 'Ein numerische Verfahren fur Lösung des Neumannischen Aussenproblems für die Helmholtzsche Schwingungsgleichung', *Computing*, **4**, 246—273, (1969).

Ladyzhenskaja, O. [1] 'On the limiting amplitude principle', *Uspechi Math. Nauk*, **12**, 161—164, (1957).

Landau, L. and Lifschitz, E. [1] *Quantum mechanics.* Nonrelativistic theory, Pergamon Press, New York, 1959.

Lavrientiev, M. *et al.* [1] *Multidimensional inverse problems for differential equations*, Springer-Verlag, New York, (1970), Lecture notes in Math. N. 167. [2] *Ill-posed problems of mathematical physics and analysis*, Nauka, Moscow, (1980) (in Russian).

Lax, P. and Phillips, R. [1] *Scattering theory*, Acad. Press, New York, 1967. [2] 'Decaying modes for the wave equation in the exterior of an obstacle', *Comm. Pure Appl. Math*, **22**, 737—787, (1969). [3] 'A logarithmic bound on the location of the poles of the scattering matrix', *Arch. Rat. Mech. Anal.*, **40**, 268—280, (1971).

Lax, P., Morawetz, C., and Phillips, R. [1] 'Exponential decay of solutions of the wave equation in the exterior of a star-shaped obstacle', *Comm. Pure Appl. Math.*, **16**, 477—486, (1963).

Leis, R. [1] *Vorlesungen über partielle Differentialgleichungen Zweiter Ordnung*, Bibliographisches Institut, Mannheim, (1967). [2] 'Zur Monotonie der Eigenwerte selbstadjungierter elliptischer Differentialoperatoren', *Math. Zeitschr.*, **96**, 26–32, (1967). [3] 'Zur Dirichletschen Randwertaufgabe des Aussenraums der Schwingungsgleichung', *Math. Zeitschr.*, **90**, 205–211, (1965).

Lewis, R. [1] 'Physical optics inverse diffraction', *IEEE* **AP-17**, 308–314, (1969); correction **AP-18**, 194, (1970).

Limič, N. [1] 'The exterior Neumann problem for the Helmholtz equation', *Glaznik Math.* **16**, 51–64, 245–260, (1981).

Ljance, V. [1] 'Analog of the inverse scattering problem for a nonselfadjoint operator', *Math. Sborn.*, **72**, 537–557, (1967).

Lubarskii, G. [1] *Group theory and its applications to physics*, Fizmatgiz, Moscow, 1957.

Lubič, Ju. [1] 'On fundamental solutions of linear elliptic equations', *Math. Sborn*, **39**, N1, 23–36, (1957).

MacCamy, R. [1] 'Low frequency acoustic oscillations', *Quart. Appl. Math.* **23**, 247–256, (1965).

Majda, A. [1] 'High-frequency asymptotics of the scattering matrix and the inverse problem of acoustical scattering', *Comm. Pure Appl. Math.*, **29**, 261–291, (1976).

Majda, A. and Taylor, U. [1] 'Inverse scattering problems for transparent obstacles and hyperbolic systems', *Comm. PDE*, **2**, 395–438, (1977).

Marchenko, V. [1] 'Stability of the inverse scattering problem', *Math. Sborn.*, **77**, 139–162, (1968); **78**, 475–484, (1969).

Marchenko, V. and Khruslov, E. [1] *Boundary value problems in domains with granular boundary*, Naukova Dumka, Kiev, (1974) (in Russian).

Marcus, A. [1] 'Expansion in root vectors of weakly perturbed selfadjoint operators', *Sov. Math. Doklady*, **142**, 538–541, (1962).

Martin, P. and Ramm, A. G. [1] 'Inverse scattering for geophysical problems III. On the velocity-inversion problems of acousics', *Proc. Roy. Soc. London*, **A399**, 153–166, (1985).

Maz'ya, V. [1] 'On Neumann's problem in irregular domains', *Sib. Math. Jour.*, **9**, 1322–1350, (1968).

Maz'ya, V. and Shaposhnikova, [1] 'On conditions on the boundary in L_p theory of elliptic problems', *Sov. Math. Doklady*, **251**, 1055–1059, (1980).

Melrose, R. [1] 'Singularities and energy decay in acoustical scattering', *Duke Math. Journ.*, **46**, 43–59, (1979). [2] 'Forward scattering by a convex obstacle', *Comm. Pure Appl. Math.*, **23**, 461–499, (1980).

Mikhlin, S. [1] *Numerical performance of variational methods*, Walters-Noordhoff, Groningen, (1971).

Millar, R. [1] 'The Rayleigh hypothesis and a related least-squares solution to scattering problems for periodic surfaces and other scatterers', *Radio Sci.*, **8**, 785–796, (1973).

Miller, F. [1] *The construction of a surface from its principal curvatures*, preprint, Kansas St. Univ., 1983.

Minerbo, G. [1] *Radial inversion of TE measurements in a borehole*, preprint, (1983).

Mishnaevskii, P. [1] 'On the spectral theory for the Sturm-Liouville equation with operator coefficient', *Math. USSR, Izvestija*, **10**, 145–180, (1976). [2] 'A study of elliptic second order operators in domains with infinite boundaries by means of the operator Sturm-Liouville equation, I, II', *Dif. Uravnenija,* **19**, 445—457, 645—653, (1983).

Molchanov, A. [1] 'Conditions for discreteness of spectra for differential operators of second order', *Proc. Moscow, Math. Soc.* **2**, 169—200, (1953).

Morawetz, C. [1] 'Exponential decay of solutions of the wave equation', *Comm. Pure Appl. Math.,* **19**, 439—444, (19966). [2] 'The limiting amplitude principle', *Comm. Pure Appl. Math.,* **15**, 349—361, (1962).

Morawetz, C. and Ludwig, D. [1] 'An inequality for the reduced wave operator and the justification of geommetrical optics', *Comm. Pure Appl. Math.,* **21**, 187—203, (1968).

Morawetz, C. Ralston, J., and Strauss, W. [1] 'Decay of solutions of the wave equation outside non-trapping obstacles', *Comm. Pure Appl. Math.,* **30**, 447—508, (1977).

Müller, C. [1] *Foundations of the mathematical theory of electromagnetic waves,* Berlin, Springer-Verlag, (1969). [2] 'Radiation patterns and radiation fields', *J. Rat. Mech. Anal.,* **4**, 235—246, (1955).

Muravei, L. [1] 'Analytic continuation with respect to a parameter of the Green's functions of exterior boundary value problems for the two-dimensional Helmholtz equation, III', *Math. USSR, Sbornik,* **34**, 55—98, (1978).

Mushelishvili, N. [1] *Singular integral equations,* Noordhoff, Groningen, (1972).

Newton, R. [1] *Scattering of waves and particles,* McGraw-Hill, New York, (1966). [2] 'Inverse scattering, I, II, III, IV', *J. Math. Phys.,* **21**, 493—505, (1980); **21**, 1698—1715, (1980); **22**, 2191—2200, (1981); **23**, 594—604, (1982). [3] 'On a generalized Hilbert problem', *J. Math. Phys.,* **23**, 2257—2265, (1982). [4] 'Marchenko and Gelfand-Levitan methods in the inverse scattering problem in one and three dimensions', pp. 1—74, in *Conference on inverse scattering: theory and applications,* Bednar J. *et al.* (eds.), SIAM, Philadelphia, (1983). [5] 'Optical theorm and beyond', *Am. J. of Phys.,* **44**, 639—642, (1976). [6] 'Inversion of reflection of data for layered media: a review of exact methods', *Geophys. J. R. Ast. Soc.,* **65**, 191—215, (1981).

Nirenberg, L. [1] 'The Weyl and Minkowski problems in differential geometry in the large', *Comm. Pure Appl. Math.,* **6**, 337—394, (1953).

Noble, B. [1] 'Integral equation perturbation methods in low frequency diffraction', pp. 323—360, in *Electromagnetic waves,* R. Langer (ed.), Univ. of Wisconsin Press, Madison, (1962).

Norton, S. and Linzer, M. [1] 'Ultrasonic reflectivity imaging in three dimensions: exact inverse scattering solutions for plane, cylindrical, and spherical apertures', *IEEE Translon Biomedical Engineering,* **BME-28**, 202—220 (1981).

Nussenzveig, H. [1] 'The poles of the S-matrix of a rectangular potential well or barrier', *Nucl. Phys.* **11**, 499—521, (1959).

Olsson, P. [1] 'Elastostatics as a limit of elastodynamics — a matrix formulation', *Appl. Sci. Res.,* **41**, 125—151, (1984).

Oristaglio, M. [1] *Born inversion of the induction log,* preprint (1983).

Panich, O. [1] 'On the question of the solvability of thc exterior boundary value problems for the wave equation and Maxwell's equation', *Russ. Math. Surveys,* **20**, 221—226, (1965).

Petkov, V. [1] 'High frequency asympotitics of the scattering amplitude for non-convex bodies', *Comm. PDE*, **6**, 293—329, (1980). [2] 'Distribution of poles of the S-matrix for trapping obstacles', *C. R. Acad. Sci., Paris, I. Math.*, **295**, 443—445, (1982).

Povzner, A. [1] 'On the expansions of arbitrary functions in terms of eigenfunctions of the operator $-\Delta + c(x)$', *Math. Sborn.*, **32**, 109—156, (1953); *Amer. Math. Transl. Series* **2**, 60, 1—49, (1967); *Soviet Math., Doklady*, **104**, 360—363, (1955).

Radon, J. [1] *Über die Randwertaufgaben beim logarithmischen Potential*, Sitzungsber, Akad. Wiss. Math. Naturw. Kl., 128, 5.B.IIa, (1920).

Ralston, J. [1] 'Propagation of singularities and the scattering matrix', in Garnir, H. (ed.), *Singularities in boundary value problems*, D. Reidel Publ., Co., Dordrecht, Holland, (1981).

Ramm, A. G. [1] (a) 'Spectral properties of the Schrödinger operator in some domains with infinite boundaries', *Soviet Math. Doklady*, **152**, (1963), 282—285. 27 # 3930. (b) 'Investigation of the scattering problem in some domains with infinite boundaries I, II', *Vestnik*, 7 (1963), 45—66; 19 (1963), 67—76. 27 # 483, 23 # 374. (c) 'Absence of the discrete positive spectrum of the Dirichlet Laplacian in some infinite domains', *Vestnik Leningrad. Univ.* **13**, (1964), 153—156; N1, (1966), 176. 30 # 1295. (d) 'Spectral properties of the Schrödinger operator in some infinite domains', *Math. Sbor.*, **66**, (1965), 321—343. 30 # 3297, 34 # 7994. (e) 'Statement of the diffraction problem in domains with infinite boundaries', *Proc. 3 all-union wave diffraction symp.*, Nauka, Moscow, (1964), 28—31. (f) 'Green's function study for differential equation of the second order in domains with infinite boundaries', *Diff. eq.*, **5**, (1969), 1509—1516. 40 # 6078; e.t. 1111—1116. (g) 'On a method of solving the Dirichlet problem in some infinite domains', *Ibid*, **5**, (1965), 124—127. 32 # 7993. (h) 'Exterior boundary value problems as limits of interface problems', *J. Math. Anal. Appl.*, **84**, (1981), 256—263. (i) 'On some integral operators', *Diff. eq.*, **6**, (1970), 1430—1451. E.t. 1096—1106. 43 # 8339.

[2] (a) 'On the analytic continuation of the solution of the Schrödinger equation in the spectral parameter and the behavior of the solution to the nonstationary problem as $t \to \infty$', *Uspechi Mat. Nauk*, **19**, (1964), 192—194, (b) 'Analytical continuation of the resolvent kernel of the Schrödinger operator in the spectral parameter and limit amplitude principle in some infinite domains', *Doklady Ac. Sci. Azerb. SSR*, **21**, (1965), 3—7. (c) 'Necessary and sufficient conditions for validity of limit amolitude principle', *Doklady* **163**, (1965), 584—586. 33 # 7673. (d) 'Behavior of the solution to a nonstationary problem as $t \to \infty$', *Mathematics, Izvestija vuzov*, **1**, (1966), 124—138. 33 # 7674. (e) 'On limit amplitude principle', *Diff. eq.*, **4**, (1968), 714—720. 37 # 1759. (f) 'Some theorems on analytic continuation of the Schrödinger operator resolvent kernel in the spectral parameter', *Izv. Ac. Nauk. Armjan. SSR, Mathematics*, **3**, (1968), 443—464. 42 # 5563. (g) 'Exponential decay of solutions of the hyperbolic equation', *Ibid.*, **6**, (1970), 2099—2100. 44 # 631. E.t. 1598—1599. (h) 'Necessary and sufficient conditions for the validity of the limit amplitude principle', *Diff. eq.*, **7**, (1971), 366—367. 45 # 5523; E.t. 284—285. (i) 'Necessary and sufficient conditions for the validity of the limit amplitude principle', *Mathematics, Izv. vuzov*, **5**, 1978, 96—102. (j) 'Mathematical foundations of the singularity and eigenmode expansion methods', *J. Math. Anal. Appl.*, **86**, 1982, 562—591. (k) 'Representations of solutions to Helmholtz's equation', *J. Math. Phys.*, **25**, N4, 807—809, (1984). (l) 'On wave diffusion', *Mathematics, Izvestija vuzov*, **2**, (1965), 136—138. 32 # 1451.

[3] *Theory and applications of some new classes of integral equations*, Springer-Verlag, N.Y. 1980.

[4] *Iterative methods of calculating static fields and wave scattering by small bodies*, Springer-Verlag, N.Y., 1982.

[5] (a) 'Eigenfunction expansion for nonselfadjoint Schrödinger operator', *Doklady*, **191**, (1970), 50—53. 42 # 703. (b) Eigenfunction expansions for exterior boundary problem', *Ibid.*, 7, (1971), 737—742. 44 # 2094; E.t. 565—569. (c) 'Spectral properties of some nonselfadjoint operators', *Bull. Am. Math. Soc.*, **5**, N3, (1981), 313—315. (d) 'Spectral properties of some nonselfadjoint operators and some applications', in *Spectral theory of differential operators*, Math. Studies, North Holland, Amsterdam, 1981, I. Knowles and R. Lewis (eds.), 349—354. (e) 'Perturbations preserving asympototics of spectrum with a remainder', *Proc. A.M.S.*, **85**, N2, (1982), 209—212. (f) 'On the basis property for the root vectors of some nonselfadjoint operators', *Jour. Math. Anal. Appl.* **80**, (1981), 57—66.

[6] (a) 'On exterior diffraction problems', *Ibid.*, 7, (1972), 1362—1365. 51 # 4864; E.t. 1064—1067. (b) 'Calculation of the quasistationary states in quantum mechanics', *Doklady*, **204**, (1972), 1071—1074. 56 # 14326. (c) 'Eigenfunction expansion corresponding to the discrete spectrum', *Radiotech. i Electron.*, **18**, (1973), 496—501. 50 # 1641 E.t. 364—369. (d) 'Nonselfadjoint operators in diffraction and scattering', *Math. Methods in Appl. Sci.*, **2**, (1980), 327—346. (e) 'Perturbation of resonances', *J. Math. Anal. Appl.*, **88**, (1982), 1—7. (f) 'Asymptotics of resonant states (with P. A. Mishnaevsky)', *J. Math. Anal. Appl.*, **87**, (1982), 323—331. (g) 'A variational principle for resonances', *J. Math. Phys.* **21**, (1980), 2052—53; 'Variational principles for resonances II', *J. Math. Phys.*, **23**, N6, (1982), 1112—1114. (h) 'Domain where the resonances are absent in the three dimensional scattering problem', *Doklady*, **166**, (1966), 1319—1322. 34 # 3902. (i) 'Theoretical and practical aspects of singularity and eigenmode expansion methods', (Tech. Report, Math. Res. Center, Univ. Wisconsin-Madison, (1979). IEEE A-P, 28, N6, (1980), 987—901. (j) 'On the singularity and eigenmode expansion methods', *Electromagnetics*, **1**, N4, (1981), 385—394. (k) 'Variational principles for eigenvalues of compact nonselfadjoint operators, II', *J. Math. Anal. Appl.*, **91**, (1983), 30—38. (l) Mathematical foundations of the singularity and eigenmode expansion methods', *J. Math. Anal. Appl.*, **86**, (1982), 562—591. (m) 'Calculation of resonances and their extraction from transient fields', *J. Math. Phys.*, **26**, N5, (1985), 1012—1020, (n) 'Limit of the spectra of the interior Neumann problems when a solid domain shrinks to a plane one', *J. Math. Anal. Appl.*, **108**, 107—112, (1985).

[7] (a) 'Convergence of the T-matrix approach to scattering theory', *J. Math. Phys.*, **23**, N6, (1982), 1123—1125. (b) 'Justification of the T-matrix approach, Proc. of the internat', *IEEE symp. on Antennas and Propagation*, Albuquerque 1982, Vol. 1, pp. 13—14, (1982). (c) 'Convergence of the T-matrix approach in the potential scattering', *J. Math. Phys.* **23**, (1982), 2408—2409. (d) 'Convergence of the T-matrix scheme', *Proc. of the IEEE intern. symp. on antennas and propagation*, May, 1983, Houston.

[8] (a) 'Reconstruction of the domain shape from the scattering amplitude', *Radiotech. i Electron.* **11**, (1965), 2068—2070. (b) 'Reconstruction of the shape of a reflecting body from the scattering amplitude', *Radiofisika*, **13**, (1970), 727—732. (c) 'On a property of the set of radiation patterns', *J. Math. Anal. Appl.*, **98**, (1984), 92—98.

(d) 'Scattering by a penetrable body', *J. Math. Phys.,* **25**, N3, (1984), 469—471. (e) 'Existence of infinitely many purely imaginary resonances in the problem of potential scattering', *Phys. Lett.,* **101A**, (1984), 187—188. (f) 'Description of the degree of non-uniqueness in inverse source problem', *J. Math. Phys.,* **25**, N6, (1984), 1791—1793. (g) 'Inverse diffraction problem', in Proc. NATO-ARW, *Inverse methods in electromagnetic imaging,* Boerner, W. (ed.), Reidel, Dordrecht, (1985), pp. 231—249. (h) 'An inversion formula in scattering theory', *Phys. Letters.,* **99A**, (1983), 201—204. (i) 'Reconstruction of the shape of a domain from the scattering amplitude', [1e], 143—144. (j) 'On inverse diffraction problem', *J. Math. Anal. Appl.,* **103**, (1984), 139—147. (k) 'Remarks about inverse diffraction problem', *J. Math. Phys.,* **25**, N9, (1984), 2672—2674. (l) 'Inverse scattering for geophysical problems', *Phys. Letters.,* **99A**, (1983), 258—260. (m) 'Inverse scattering for geophysical problems IV'. Inversion of the induction logging measurements. *Geophys. Exploration,* (1986) (to appear)

[9] 'Topics in scattering and spectral theory', *Notices AMS,* **29**, 1982, 327—329.

[10] (a) 'Iterative solution of the integral equation in potential theory', *Doklady,* **186**, (1969), 62—65. 41 # 9462. (b) 'Calculation of the scattering amplitude for the wave scattering from small bodies of an arbitrary shape', *Radiofisika,* **12**, (1969), 1185—1197. 43 # 7131. (c) 'Approximate formulas for tensor polarizability and capacitance of bodies of arbitrary shape and its applications', *Doklady,* **195**, (1970), 1303—1306. 55 # 1947; E.t. 15 (1971), 1108—1111. (d) 'Calculation of the initial field from the scattering amplitude', *Ibid.,* **16**, (1971), 554—556. (e) 'Approximate formulas for polarizability tensor and capacitances for bodies of an arbitrary shape', *Radiofisika,* **14**, (1971), 613—620. 47 # 1386. (f) 'Iterative methods for solving some heat transfer problems', *Eng. Phys. Journ.* **20**, (1971), 936—937. (g) 'Electromagnetic wave scattering by small bodies of arbitrary shapes', *Proc. of 5 all-union symp. on wave diffraction,* Trudy. Math. Inst. Steklova, Leningrad, 1971, 176—186. (h) 'Calculation of the magnetization of thin films', *Microelectronics* **6**, (1971), 65—68. (with Frolov). (i) 'Calculation of the scattering amplitude of electromagnetic waves by small bodies of an arbitrary form II', *Radiofisika,* **14**, (1971), 1458—1460. (j) 'Electromagnetic wave scattering by small bodies of an arbitrary shape and related topics', *Proc. Intern. symp. URSI,* Moscow, 1971, 536—540. (k) 'Calculation of thermal fields by means of iterative processes', *Proc. 4 all-union conference on heat and mass transfer. Minsk,* (1972), 133—137. (l) 'Calculation of the capacitance of a parallelepiped', *Electricity,* **5**, (1972), 90—91, (with Golubkova, Usoskin). (m) 'On the skin-effect theory', *Journ. of Tech. Phys.,* **42**, (1972), 1316—1317. (n) 'Calculation of the capacitance of a conductor placed in anisotropic inhomogeneous dielectri', *Radiofisika, Izvesti a vuzov,* **15**, (1972), 1268—1270, 47 # 2284. (o) 'Light scattering matrix for small particles of an arbitrary shape', *Opt. and Spectroscopy,* **37**, (1974), 125—129. (p) 'Scalar scattering by the set of small bodies of arbitrary shapes'. *Radiofisika,* **17**, (1974), 1062—1068. (q) 'New methods of calculation of the static and quasistatic electromagnetic waves', *Proc. 5 Intern. sympos. Radioelectronics-74 Sofia,* **3**, (1974), 1—8 (report 12). (r) 'Influence of the shape of cosmic particles on the scattering amplitude', *Proc. all-union sympos.* Interaction of cosmic dust with the atmosphere. Ashabad, 1974, 11—12. (s) 'Estimates of some functionals in quasistatic electrodynamics', *Ukrain. Phys. Journ.,* **5**, (1975), 534—543. 56 # 14165. (t) *Electromagnetic wave scattering by small*

bodies. In Varadan, Varadan [1] 537—546. (u) 'Electromagneic wave scattering by small bodies', *Nonlinear Vibr. Probl.,* **20**, (1981), 109—120. (v) *Behavior of solutions to exterior boundary value problems at low frequencies*, preprint (1984); *J. Math. Anal. Appl.* (to appear) (w) 'Wave scattering by small bodies', *Reports on Math. Phys.,* **21**, 69—77, (1985).

[11] (a) 'On numerical differentiation', *Mathem., Izvestija vuzov,* **11**, (1968), 131—135. 40 # 5130. (b) 'Simplified optimal differentiators', *Radiotech. i Electron.* **17**, (1972), 1325—1328; E.t. 1034—1037. (c) 'Stable solutions of some ill-posed problems', (Tech. report of Naval Res. Lab., 1980) *Math. Meth. in Appl. Sci.,* **3**, 1981, 336—363. (d) 'Estmates of the derivatives of random functions', *J. Math. Anal. Appl.,* **102**, (1984), 244—250. (e) 'Estimates of the derivatives of random functions II'. (with T. Miller). *J. Math. Anal. Appl.,* **110**, (1985), 429—435.

[12] *Singularities of the inverses of Fredholm's operators.* (to appear in Proc. Roy. Soc. Edinburgh).

[13] (a) 'Some inverse scattering problems of geophysics', *Inverse Problems,* **1**, 133—172, (1985); (b) *ibid.,* **1**, L35—37, (1985); (c) *Phys. Lett.,* **113 A**, 172—176, (1985).

Ramm, A. G., Weglein, A. [1] 'Inverse scattering for geophysical problems, II', *J. Math. Phys.,* **25**, 3231—3234, (1984).

Rayleigh, J. [1] *Scientific papers*, Cambridge Univ. Press, Cambridge, (1922). [2] *The theory of sound*, Dover, New York, (1945).

Reed, M. and Simon, B. [1] (a) *Scattering theory*, Acad. Press, New York, (1979). (b) *Analysis of operators*, Acad. Press, New York, (1978).

Rellich, F. [1] 'Über das asympototische Verhalten der Lösungen von $\Delta u + \lambda u = 0$ in unendlichen Gebieten', *Jahresber. Deutsch. Math. Verein.,* **53**, 57—65, (1943). [2] *Das Eigenwetproblem von* $\Delta u + \lambda u = 0$ *in Halbröhren*, Studies and essays presented to R. Courant, New York, (1948).

Ribaric, M. and Vidav, I. [1] 'Analytic properties of the inverse $A^{-1}(z)$ of an analytic operator valued function $(A(z))$', *Arch. Rat. Mech. Anal.,* **32**, 298—310, (1969).

Rose, J. [1] 'Inverse scattering at long wavelength', pp. 131—135, in *Review of progress in quantitative nondestructive evaluation*, Vol. 1, Thompson (ed.), Plenum Publ., New York, (1982).

Saenz, A. [1] 'Scattering by periodic surfaces', *J. Math. Phys.,* **22**, 2872—2884, (1981); **23**, 1115—1122, (1982).

Saito, Y. [1] 'The principle of limiting absorption for second-order differential operators with operator valued coefficients', *Publ. Res. Inst. Math. Sci.,* **7**, 581—619, (1971, 72). [2] 'Some properties of the scattering amplitude and the inverse scattering problem', *Osaka J. Math.,* **19**, 527—547, (1982). [3] 'Extended limiting absorption method and analyticity of the S-matrix', *J. f. die reine und angew. Math.,* **343**, 1—22, (1983).

Sanchez-Palencia, E. [1] *Nonhomogeneous media and vibration theory, Lecture notes in Phys.,* 127, Springer-Verlag, New York, (1980).

Senior, T. [1] 'Low frequency scattering by a dielectric body', *Radio Sci.,* **11**, 477—482, (1976).

Shenk, N. [1] 'Eigenfunction expansions and scattering theory for the wave equation in an exterior region', *Arch. Rat. Mech., Anal.,* **21**, 120—150, (1966).

Shenk, N. and Thoe, D. [1] 'Eigenfunction expansions and scattering theory for perturbations of $-\Delta$', *Rocky Mountain J. of Math.*, **1**, 89—125, (1971).

Shulze, B. and Wildenhain, G. [1] *Methoden der Potenzialtheorie für elliptische Differentialgleichungen beliebiger Ordnung*, Akad. Verlag, Berlin (1977).

Simon, M. and Ursell, F. [1] 'Uniqueness in linearized two-dimensional water-wave problems', *J. Fluid Mech.*, **148**, 137—154, (1984).

Sleeman, B. [1] 'The inverse problem of acoustic scattering', *IMA Jour. Appl. Math.*, **29**, 113—142, (1982).

Sommerfeld, A. [1] 'Die Greensche Function der Schwingungsgleichung', *Jahresber. der deutsche Math. Verein*, **21**, 309—353, (1912).

Stechkin, S. and Subbotin, Ju. [1] *Splines in numerical analysis*, Nauka, Moscow, (1976) (in Russian).

Steinberg, S. [1] 'Meromorphic families of compact operators', *Arch. Rat. Mech. Anal.*, **31**, 372—379, (1968).

Stevenson, A. [1] 'Solution of electromagnetic problems as power series in the ratio (dimension of scatterer)/(wave-length)', *J. Appl. Phys.*, **24**, 1134—1152, (1953).

Tamarkin, J. [1] 'On Fredholm's integral equations whose kernels are analytic in a parameter', *Math. Ann.*, **28**, 127—152, (1927).

Tanana, V. [1] *Methods of solving operator equations*, Nauka, Moscow, 1981.

Taylor, M. [1] *Pseudodifferential operators*, Princeton Univ. Press, Princeton, 1981.

Tayoshi, T. [1] 'The asymptotic behavior of the solutions of Helmholtz's equation in a domain with unbounded boundary', *Publ. RIMS, Kyoto Univ.*, **8**, 375—391, (1972, 73). [2] 'Wave operators in the scattering problem related to non-finite obstacles', *Ibid.*, **16**, 627—634, (1980).

Terai, T. [1] 'On calculation of the sound speed around three dimensional objects by integral equation method', *Journ. of Sound and Vibr.*, **69**, 71—100, (1980).

Tikhonov, A. and Arsenin, V. [1] *Solutions of ill-posed problems*, Winston, Washington, (1977).

Titchmarsh, E. [1] *Theory of Fourier integrals*, Oxford Univ. Press, Oxford, 1948.

Triebel, H. [1] *Interpolation theory, function spaces, differential operators*, VEB Verlag, Berlin, (1978).

Ursell, F. [1] 'On the exterior problems of acoustics', *Proc. Cambridge Phil. Soc.*, **74**, 117—125, (1973); **84**, 545—548, (1978). [2] 'On the rigorous foundation of short-wave asymptotics', *Ibid.*, **62**, 227—244, (1966).

Vainberg, B. [1] 'Principles of radiation, limiting amplitude and limiting absorption in the general theory of partial differential equations', *Russian Math. Surveys*, **21**, 115—194, (1966). [2] 'Short wave asymptotics of solutions to the stationary problems and asymptotics $t \to \infty$ of solutions to time-dependent problems', *Russ. Math. Surveys*, **30**, 3—55, (1975).

Vainberg, M. and Trenogin, V. [1] *Bifurcation of solutions of nonlinear equations*, Noordhoff Int., Leiden, (1974).

Varadan, V. and Varadan, V. (ed.). [1] *Acoustic, electromagnetic and elastic wave scattering-Focus on the T-matrix approach*, Pergamon, New York, (1980).

Vekua, I. [1] *New methods for solving elliptic equations*, North-Holland, Amsterdam, (1968), Appendix; 'Metaharmonic functions', *Trudy Tbilis. Math. Inst.*, **12**, 105—174, (1943). [2] 'On completeness of a system of metaharmonic functions', *Dokl. Akad. Nauk. USSR*, **90**, 715—718, (1953); 'On completeness of a system of harmonic polynomials in space', *Ibid.*, **90**, 495—498, (1953).

Vinnik, A. and Eidus, D. [1] 'On radiation conditions for domains with infinite boundaries', *Soviet. Math. Doklady,* **15**, 12—15, (1974).

Voitovich, N., Kacenelenbaum, B., and Sivov, A. [1] *Generalized method of igenoscillations in diffraction theory,* Nauka, Moscow, (1977).

Waterman, P. [1] 'Matrix formulation of electromagnetic scattering', *Proc. IEEE,* **53**, 805—812, (1965). [2] 'New formulation of acoustic scattering', *J. Acoust. Soc. Amer.,* **45**, 1417—1429, (1969).

Weglein, A. [1] 'The inverse scattering concept and its seismic application', in *Developments in geophysical exploration methods,* Vol. **6**, Appl. Sci. Publ., London, (1985).

Werner, P. [1] 'Zur mathematischen Theorie akustischer Wellenfelder', *Arch. Rat. Mech. Anal,* **6**, 231—260, (1960). [2] 'On the behavior of stationary electromagnetic wave fields for small frequencies', *J. Math. Anal. Appl.,* **15**, 447—496, (1966). [3] Über das Verhalten electromagnetischer Felder für kleine Frequenzen in mehrfach zusammenhängenden Gebieten', *J. f. reine u. angew. Math.,* **1**, 278/279, 365—397, (1975), **2**, 280, 98—121, (1976). [4] *Low frequency asymptotics for the reduced wave equation in two-dimensional exterior spaces, Math. Meth. in the Appl. Sci.,* **7**, (1985).

Weyl, H. [1] 'Kapazität von Strahlungsfeldern', *Math. Z.,* **55**, 187—198, (1952).

Wiener, N. [1] *Extrapolation interpolation and smoothing of stationary time series,* Wiley, New York, (1949).

Wilcox, C. [1] *Scattering theory for the d'Alembert equation in exterior domains, Lecture notes in Mathematics,* N492, Springer-Verlag, New York, 1975. [2] *Scattering theory for diffraction gratings,* Springer-Verlag, New York, 1984.

Young, R. [1] *An introduction to nonharmonic Fourier series,* Acad. Press, New York, (1980).

Explanation of some abbreviations:
41 # 9462 is the reference to the review in Mathematical reviews

e.t., E.t.	— English translation
Doklady	— Doklady Acad. Nauk USSR
Vestnik	— Vestnik Leningradskogo Universit.
Math. Sbor.	— Mathematicheskii Sbornik.
Diff. Eq.	— Differential Equations.

List of Symbols

D —	a domain,
$\Gamma = \partial D$ —	its boundary
Ω —	the exterior domain, $D \cup \Omega = R^3$
R^n —	the Euclidean n-dimensional space, $R^n_+ = \{x: x \in R^n, x_n \geqslant 0\}$, $R^1_+ = R_+$, $R^n = R^n \backslash R^n_+$, S^{n-1} — the unit sphere in R^n
S^2 —	the unit sphere in R^3, S^2 — a solid angle, i.e., an open set in S^2
$n(\nu)$ —	unit vectors in the direction of propagation of the scattered (incident) wave
$B_r =$	$\{x: \lvert x\rvert \leqslant r\}$, $D_r = D \cap B_r$, $D'_r = D \cap \Omega_r$,
$\Omega_r =$	$\{x: \lvert x\rvert \geqslant r\}$, $\Omega'_r = \{x: \lvert x\rvert \leqslant r, x \in \Omega\}$, $\Gamma_r = \Gamma \cap B_r$, $\Gamma'_r = \Gamma \cap \Omega_r$
$S_r =$	$\{x: \lvert x\rvert = r\}$ $P_r = \{x: x \in D, x_3 = r\}$, $S_r(D) = S_r \cap D$
$\tilde{\Gamma}_R =$	$\{x: x \in \Gamma, x_3 \geqslant R\}$
$D\mu$, $S\sigma$, $Q\sigma$, $T\rho$,	Section I.1
g, g_ε, g_M	Section I.1
A, A_ε	Section I.1
A'	Section I.1
$\bar{A}$	Section I.1
ψ_{nm}	Section I.1
j_n, $h_n^{(j)}$	Section I.1
$\partial/\partial N_\pm$	Section I.1, N is the exterior normal, N points into Ω.

If A: $B_1 \to B_2$ is a linear operator from a Banach space B_1 into a Banach space, B_2, im A is closed, dim ker $A < \infty$ and dim coker $A < \infty$ then $i(A) = \text{ind } A = \dim\ker A - \dim\operatorname{coker} A$.

ker A =	$N(A) = \{f: Af = 0\}$,
im A =	$R(A) = \{f: f = Au\}$,
$C^m(D)$ $(C_0^m(D))$ —	the space of m times continuously differentiable (compactly supported) functions in D
$H^s = W^{2,s}(D)$ —	the Sobolev spaces
$\tilde{H}^s =$	$H^s(\Gamma)$
$\mathring{H}^s(D)$ —	the closure of $C_0^\infty(D)$ in the norm $H^s(D)$, $\mathring{H}^{2s}(D) = H^{2s}(D) \cap \mathring{H}^s(D)$
$H_0^2(D)$ —	the set of $H^s(D)$ functions with compact support in D
$L^p(D)$ —	the Lebesgue spaces
$\lVert u\rVert_{L^p(D)} =$	$(\int_D \lvert u\rvert^p \, dx)^{1/p}$
$W^{m,p}(D)$ —	the Sobolev spaces
Im (Re) —	imaginary (real) part
$\gamma(r) =$	$r^{-1}\exp(ikr)$
ℓ^p —	the space of sequence $f = \{f_n\}$ with the norm
$\lVert f\rVert_{\ell^p} =$	$(\sum_{n=1}^\infty \lvert f_n\rvert^p)^{1/p}$
iff =	if and only if
$\forall$ —	for all
$\exists$ —	there exists

$\equiv$ —	identically equal, equal by definition
$\Rightarrow$ —	implies
$\to$ —	converges
$\rightharpoonup$ —	weakly converges
l.i.m. —	limit in mean (in L^2) p. 27
$\delta_{ij} =$	$\begin{cases} 0 & i \neq j \\ 1 & i = j \end{cases}$
$\alpha_{ij}(\gamma), \beta_{ij}$ —	polarizability tensors
SEM, EEM —	singularity expansion method, eigenmode expansion method
cap —	capacity
σ_∞ —	the set of all linear compact operators
σ_p —	p. 387
W_R—	the set of entire functions of exponential type $\leqslant R$, p. 261.
a.e.	almost every.

Index

Mathematics and Its Applications

Mathematics and Its Applications is devoted to interrelations such as:

- a central concept which plays an important role in several different specialized areas of mathematics and/or science;
- new applications of the results and ideas from one area of scientific endeavour into another;
- the influence which the results, problems and concepts of one field of enquiry have, and have had, on the development of another.

Present-day (applied) mathematics has some noteworthy features compared with even a few decennia ago:

- Rapidly increasing mathematization of a large number of areas such as, e.g., physical chemistry, electrical engineering, and geology.
- The enormous power of mathematical modeling which, when combined with powerful computers, enables us to carry out experiments which otherwise "in reality" would be too dangerous, too expensive, take too long, and be simply impossible.
- The era of increasing specialization seems to be over. The time has come to start using all the powerful tools that have been forged.

All in all it looks like the next decennia will witness an unequalled flowering in the mathematical sciences.

Based on these facts and this philosophy, the *Mathematics and Its Applications* book series is being developed quite vigorously. Given its success we plan to continue along the same lines often by means of specially commissioned volumes covering new areas of interaction likely to generate new kinds of problems, or covering a more established topic together with all its interrelations with others.

MAIN SERIES

W. KUYK. **Complementarity in Mathematics.** A First Introduction to the Foundations of Mathematics and Its History.

P. H. SELLERS. **Combinatorial Complexes.** A Mathematical Theory of Algorithms.

J. CHAILLOU. **Hyperbolic Differential Polynomials and Their Singular Perturbations.**

S. FUČÍK. **Solvability of Nonlinear Equations and Boundary Value Problems.**

W. L. MIRANKER. **Numerical Methods for Stiff Equations and Singular Perturbation Problems.**

P. M. COHN. **Universal Algebra.**

V. I. ISTRĂŢESCU. **Fixed Point Theory.** An Introduction.

N. E. HURT. **Geometric Quantization in Action.** Applications of Harmonic Analysis in Quantum Statistical Mechanics and Quantum Field Theory.

P. M. ALBERTI and A. UHLMANN. **Stochasticity and Partial Order.** Doubly Stochastic Maps and Unitary Mixing.

L. LANGOUCHE, D. ROEKAERTS, and E. TIRAPEGUI. **Functional Integration and Semiclassical Expansions.**

C. P. BRUTER, A. ARAGNOL, and A. LICHNEROWICZ (eds). **Bifurcation Theory, Mechanics and Physics.** Mathematical Developments and Applications.

J. ACZÉL (ed.). **Functional Equations: History, Applications and Theory.**

A. V. ARKHANGEL'SKII and V. I. PONOMAREV. **Fundamentals of General Topology.** Problems and Exercises.

P. E. T. JØRGENSEN and R. T. MOORE. **Operator Commutation Relations.**

R. BELLMAN and G. ADOMIAN. **Partial Differential Equations.** New Methods for Their Treatment and Solution.

N. K. BOSE (ed.). **Multidimensional Systems Theory.** Progress, Directions and Open Problems in Multidimensional Systems.

R. BELLMAN and R. VASUDEVAN. **Wave Propagation – an Invariant Imbedding Approach.**

R. A. ASKEY, T. H. KOORNWINDER, and W. SCHEMPP (eds.). **Special Functions: Group Theoretical Aspects and Applications**

L. ARNOLD and P. KOTELENEZ (eds). **Stochastic Space-Time Models and Limit Theorems.**

Y. BEN-HAIM. **The Assay of Spatially Random Material.**

Y. CHERRUAULT. **Mathematical Modelling in Biomedicine: Optimal Control of Biomedical Systems.**

R. E. BELLMAN and R. S. ROTH. **Methods in Approximation.** Techniques for Mathematical Modelling.

V. KOMKOV. **Variational Principles of Continuum Mechanics with Engineering Applications.** Volume 1: Critical Points Theory.

C. CUVELIER, A. SEGAL, and A. A. VAN STEENHOVEN. **Finite-Element Methods and Navier-Stokes Equations.**

J. D. LOUCK and N. METROPOLIS. **Symbolic Dynamics of Trapezoidal Maps.**

A. G. RAMM. **Scattering by Obstacles.**

EAST EUROPEAN SERIES

F.-H. VASILESCU. **Analytic Functional Calculus and Spectral Decompositions.**

W. KECS. **The Convolution Product and Some Applications.**

K. REKTORYS. **The Method of Discretization in Time** and Partial Differential Equations.

J. MŎCKOŘ. **Groups of Divisibility.**

L. IXARU. **Numerical Methods for Differential Equations and Applications.**

H. WALTHER. **Ten Applications of Graph Theory.**

A. WAWRZYŃCZYK. **Group Representations and Special Functions.**

D. S. MITRINOVIĆ and J. KEČKIC. **The Cauchy Method of Residues.** Theory and Applications.

V. BARBU and T. PRECUPANU. **Convexity and Optimization in Banach Spaces.**

I. BUCUR. **Selected Topics in Algebra** and its Interrelations with Logic, Number Theory and Algebraic Geometry.

I. M. STANCU-MINASIAN. **Stochastic Programming with Multiple Objective Functions.**

J. SZÉP and F. FORGÓ. **Introduction to the Theory of Games.**

L. BERAN. **Orthomodular Lattices.** Algebraic Approach.

A. MARCINIAK. **Numerical Solutions of the N-Body Problem.**

S. ROLEWICZ. **Metric Linear Spaces.**

A. PAZMÁN. **Foundations of Optimum Experimental Design.**

K. WAGNER and G. WECHSUNG. **Computational Complexity.**

SOVIET SERIES

A. V. SKOROHOD. **Random Linear Operators.**

B. S. RAZUMIKHIN. **Physical Models and Equilibrium Methods in Programming and Economics.**

N. H. IBRAGIMOV. **Transformation Groups Applied to Mathematical Physics.**

A. S. DAVYDOV. **Solitons in Molecular Systems.**